$62.50

Avoiding Liability in Premises Security

A CASEBOOK

The Honorable
R. Keegan Federal, Jr.
EDITOR-IN-CHIEF

Jennifer L. Fogleman
MANAGING EDITOR

Strafford Publications, Inc.
ATLANTA

Publisher: Richard M. Ossoff
Editor-in-Chief: R. Keegan Federal, Jr.
Managing Editor: Jennifer Fogleman Vaughan

This publication is designed to provide accurate and authoritative information in regard to the subject matter covered. It is distributed with the understanding that the Publisher and Editors are not engaged in rendering legal, accounting or other professional service. If legal advice or other expert assistance is required, the services of a competent professional person should be obtained. (From a Declaration of Principles jointly adopted by a Committee of the American Bar Association and a Committee of Publishers.)

Strafford Publications, Inc.
1375 Peachtree Street, N.E.
Suite 235
Atlanta, Georgia 30367
(404) 881-1141

ISBN: 0-9616858-3-2

Printed in the United States of America

Table of Contents

How to Use this Book

This book was prepared to help the executive with security and loss prevention responsibility to comprehend the scope and trends of premises liability litigation, in order that he or she might better prepare and manage his own security programs. Compiled from the archives of *Private Security Case Law Reporter* (*PSCLR*), the cases have been selected for their diversity, the breadth of issues they address, and the varying perspectives they offer.

The cases are organized by chapters which describe the setting of the case. A citation above the case directs the reader to the issue of *PSCLR* in which the case originally appeared. The cases are listed within each chapter in the order they appeared in *PSCLR*.

A Table of Cases and a Subject Matter Index have been included to help guide you to particular cases.

The Table of Cases is in two sections: (a) an alphabetical listing by the first party listed in the court's opinion, e.g., *Stalzer v. European American Bank* would be listed alphabetically "Stalzer v. European American Bank" and (b) an alphabetical listing of the first opposing party listed in the court's opinion, e.g., the case above would be listed alphabetically "European American Bank, Stalzer v." These sections are useful in finding cases involving a specific company.

The Subject Matter Index is a comprehensive listing of topics and subtopics covered by each case. Each case in the book may appear under several different topics. In searching for a case, we suggest you think in terms of (a) setting or industry, e.g., Retail, Shopping Center, (b) the legal issue involved, e.g., Robbery, Sexual Assault, and (c) the "actors" involved, e.g., Employees, Security Guards. The page numbers listed are the pages on which each case begins, rather than the page on which the topic actually appears.

Although the book has been organized in chapters describing setting (e.g., Hotels/Motels, Restaurants, Schools), the reader should not restrict his review to that chapter which directly describes his own operation. The situation in a particular case may have occurred in a setting different than your own, but the issues and legal duties discussed may be applicable to your operations as well.

The law affecting private security operations is evolving rapidly, and most of it is being "made," not in the legislature, but in the courts. Staying abreast of these developments is key to avoiding liability, and we hope this casebook contributes toward that objective.

Table of Cases

Table of Cases

A. by first party listed

Table of Cases

B. by opposing party

Apartments

PSCLR—NOV-DEC 82 (10)

Property Owner Not Liable For Vandalism To Car In Parking Lot, Even With Knowledge Of Prior Criminal Acts In Lot:

Jewel Jubert v. Shalom Realty et al., *Civ. A. No. 15221, Appellate Department, Superior Court, Los Angeles County, decided July 27, 1982, reported at 185 Cal. Rptr. 641.*

When tenant Jewel Jubert's car was destroyed by a fire started by vandals in his apartment building's parking lot, he sued the building's owners and managers to recover for his loss, claiming steps should have been taken to protect tenants' cars. Jubert's lawsuit alleged the landlord had been put on notice to secure the lot after receiving reports of unauthorized persons in the lot and of previous acts of vandalism in the lot. Before trial, the suit was summarily dismissed by the trial court and judgment was entered in favor of the apartment owners:

> The liability of a landlord for criminal acts of third persons ... is limited, and no such liability existed in this case where defendants did not undertake to provide security parking, made no representations regarding the providing of such security, and where the criminal acts did not result from a dangerous condition created by the landlord or from one which he was obligated to remedy.

On appeal by the tenant, the court pointed out that a landlord could be held liable to tenants for damage caused by a hazardous condition which the landlord knew of and was obligated to correct. In this case, however, the lot had neither been represented to tenants as a protected lot, nor did a known hazardous condition or defect exist which endangered tenants or their property. Stating that the landlord's only duty was "to take reasonable precautions to prevent the reoccurrence of criminal acts which are known to be likely to reoccur," the court determined that the landlord had fulfilled its obligation to its tenants.

In affirming judgment for the apartment landlord, the court said:

> To hold otherwise would require landlords who provide parking areas for tenants to install bars and/or electronic gates, or to provide full-time security personnel to patrol the garage area, once he is placed on notice that vandalism has taken place in that area. Such a requirement clearly exceeds the duty to take 'reasonable' precautions to prevent the reoccurrence of criminal acts.

COMMENTARY:

The Jubert case should not be read as authority for landlords to fail or refuse to provide security measures for the protection of tenants. However, it does stand for the proposition that merely because a landlord has actual notice of prior acts of vandalism on his property, this fact standing alone will not suffice to permit a recovery by an injured tenant.

It should also be noted that Jubert here could not argue that the landlord was negligent for failing to warn tenants of the possibility of vandalism in the parking lot because Jubert himself also had prior knowledge of such prior acts of vandalism.

PSCLR – MAY 83 (5)

Landlord's Failure to Repair Burglar Alarm in Accordance With Lease Agreement Is No Basis for Tenant's Suit Arising From Criminal Assault:

Brenda K. Carrigan f/k/a Brenda K. Nunn v. New World Enterprises, Ltd., a corporation; and Layco, Inc., a corporation, *No. 82-221, in the Appellate Court of Illinois, Third District, decided February 17, 1983, reported at 446 N.E.2d 265.*

During the evening of July 25, 1980, Brenda Nunn stayed out late with some girl friends in Peoria, Illinois, celebrating the birthday of one of them. She arrived home at about 3:00 AM, and upon opening her apartment door she heard a tapping noise which she thought was caused by her live-in boyfriend, Terry Carrigan. She called out, but received no response and walked down the hallway to her bedroom where she was accosted by an armed intruder.

At gunpoint, she was forced into a car trunk and taken to an unknown place where she was twice raped. She later escaped and notified the Peoria Police.

Investigation revealed a removed screen and pry marks on a first floor window of Nunn's apartment which indicated that the assailant had gained entry to the apartment through the window. Also, the burglar alarm in the victim's apartment was not operating at the time she was abducted; otherwise, it would have emitted an audible sound. The evidence was somewhat contradictory as to how long before the abduction the landlord learned that the alarm was not working.

The victim Nunn brought her lawsuit against the apartment owner and operator on the theory that the latter breached their duty to repair the burglar alarm, thereby permitting the rapist to enter her apartment through a window, abduct her and rape her. In support of her contention that the defendants had such a duty, she introduced evidence that the defendants, in marketing the apartment to her, had emphasized the presence of a burglar alarm in the apartment, and that a sales brochure also emphasized that each apartment was equipped with an individually controlled electronic burglar alarm and fire and smoke detector.

After trial by jury, the plaintiff was awarded $125,000 in compensatory damages, plus $50,000 in punitive damages. The landlord and its agent appealed.

The appellate court stated:

> The general rule in Illinois is that a landlord does not owe a tenant a duty to protect him from criminal acts. ...

The installation of burglar alarms cannot be construed as absolute protection against crime. Even if the alarm had been operable, it can only be speculated that the plaintiff would have made it operable on the night the crime occurred and again one must engage in speculation that when the rapist activated the alarm he would have fled.

The court noted that limited duties have been imposed on landlords in previous Illinois cases in certain situations. In one such case, a criminal gained entry into an apartment three times in the same way – the court held that the first burglary put the landlord on notice to secure that entry. In another case, a landlord was held to be negligent for <u>increasing</u> danger to persons on his property by providing guard protection from 9:00 AM to 1:00 AM It was proved that the landlord subsequently discovered that criminal activity substantially increased after 1:00 AM but did not take steps to alleviate the problem.

The court felt in this case, however, that there was no evidence that the inoperative alarm was the "proximate," or direct, cause of Brenda's injuries. The judgment of $175,000 against the apartment owners was therefore reversed.

PSCLR–OCT 83 (1)

Landlord Must Provide Adequate Security To Protect Tenants From Foreseeable Criminal Actions Of Others:

Samuel Feld and Peggy Feld v. John W. Merriam and Thomas Wynne, Inc., Co-Venturers, t/a Cedarbrook Joint Venture and Globe Security Systems, Inc., *in the Superior Court of Pennsylvania, decided May 13, 1983, reported at 461 A.2d 225.*

Cedarbrook, an apartment complex containing 1,000 residential units and located on a 36-acre tract in Cheltenham, was owned by defendant John Merriam. The property could be entered by vehicle through two different entrances, each of which was manned by security guards from 3 PM to 9 AM daily. In June 1975, Samuel and Peggy Feld lived in Building One of the Cedarbrook Complex. They also leased an automobile parking space in a garage under the building. Entry to the garage could be gained through two entrances, neither of which had doors. Lighting in the guard was apparently not good.

Cedarbrook contracted with Globe Security Services to provide security for the apartment complex. (Globe was found not negligent in the case, and no issue concerning Globe's liability was raised in this appeal.) The security services included placing doormen at the entrances to the buildings, and at the two main entrances to the complex. These doormen were not trained as security guards. Globe also provided one guard who patrolled on foot and one other guard who patrolled by car. The garage area was not patrolled by guards, nor was there an attendant present.

The crime rate in the Cheltenham area rose from 1970 to 1975. In fact, during the three months immediately preceding this incident, Globe received reports of 21 separate incidents of criminal activity in the apartment complex. The tenant's association complained to the complex management and requested additional security services.

Additional services requested by the association included better lighting in the garage, and the installation of automatic doors in the garage which could be opened only by the use of a special key or card. These improvements would have cost Cedarbrook approximately $20,000. The tenants also suggested that additional, and trained guards be provided. (The evidence showed that Globe could have provided more trained guards if the apartment complex had agreed to pay a higher hourly wage.) The apartment management declined to implement any of the suggested security improvements.

At approximately 9 PM on June 27, 1975, the Felds parked their car in the garage and began walking to the pedestrian exit. Suddenly, they were surrounded by three men with guns, who robbed them and forced them to return to their car. Two of the assailants drove them away from the complex, with the third following in another car.

The assailants indicated that they were going to lock Mr. Feld in the trunk. Mrs. Feld, worried that he would be unable to breath because he suffered from a lung disease, told the men that she would cooperate fully with them if her husband was not harmed. Thereupon, she was raped and sodomized several times by the three men, before she and her husband were released.

Eventually, the three men were all caught and convicted for their crimes.

The Felds brought suit against Cedarbrook, alleging that it had a duty to provide adequate security for its tenants and that Cedarbrook's breach of its duty caused their injuries. After an eight-day jury trial, Mrs. Feld was awarded $2,000,000 in compensatory damages, Mr. Feld was awarded $1,000,000 in compensatory damages, and each was awarded an additional $1,500,000 in punitive damages, for a total jury award of $6,000,000. Cedarbrook appealed.

Cedarbrook first argued that there was insufficient evidence to establish its negligence in providing security at the apartment complex. The appellate court noted that the issue of whether a landlord owes tenants a duty to provide security (and if so, the extent of that duty) had never been resolved in the state of Pennsylvania. The appellate court, however, had no difficulty in concluding:

> While we do not consider a landlord to be an insurer of his tenants, we hold that in all areas of the leasehold, particularly in the areas under his control, the landlord is under a duty to provide adequate security to protect his tenants from the foreseeable criminal actions of third persons.

In order to prove that a landlord breached a standard of care, the plaintiff-tenant must present evidence showing that the landlord (1) had knowledge of criminal activity which presented a danger to its tenants, (2) had the opportunity to take measures to protect the tenants against the risk, and (3) failed or refused to do so, and such failure caused the injuries suffered by the tenant.

Sufficient evidence was presented at trial, according to the appellate court, to prove that Cedarbrook management was aware of criminal activity at the complex. The jury could therefore conclude that the apartment management had sufficient notice that criminal activities were likely to occur which could endanger the tenants.

The appellate court also reviewed evidence comparing Cedarbrook with other, similar apartment complexes. Two experts had testified that the security provided by Cedarbrook did not conform to the accepted standards in buildings of the same type at that time.

The defendants also argued that the amount of damages awarded was excessive. On this issue, plaintiffs testified that they had believed during the episode that they would be killed, and were extremely distressed even after the incident. Mrs. Feld was receiving psychotherapy treatment to assist her in dealing with the emotional trauma she had experienced. A psychiatrist testified that she had undergone a major personality change and was deeply depressed. She had discontinued her many social and community service activities because she was afraid to leave her home. It was his opinion that she would never fully recover from the emotional trauma.

Evidence further indicated that the Felds' marriage and family had been seriously harmed – Mr. Feld stated that they would continue to live together out of compassion rather than love. Mrs. Feld no longer enjoyed the company of her children and grandchildren.

Mr. Feld suffered other after-effects: he testified that he could not sleep and constantly felt nervous and afraid. Because he was not able to discuss the details of the episode with anyone, not even his wife, he had not sought psychological help. Family members testified that he seemed unhappy and constantly "barked" at others.

The court concluded that the evidence amply supported the jury's conclusion that the injuries suffered by the Felds were severe. The court also concluded that punitive damages were authorized by the evidence and by Pennsylvania law, which provided that punitive damages be awarded for "outrageous conduct." Outrageous conduct was defined as "acts done with a bad motive or reckless indifference to the interests of others."

However, the appellate court did decide to reduce the $1,500,000 punitive damages award to Samuel Feld to $750,000. With that modification, the award of $5,250,000 was affirmed.

PSCLR—NOV-DEC 83 (3)

Foreseeability Is the Test in Determining the Extent of Landlord's Duty to Protect Tenants from Criminal Acts of Third Persons:

Georgia Shepard v. Drucker & Falk, a partnership; Fairway Co., a limited partnership; David Falk, individually and as a general partner in Fairway Co., a limited partnership, and as a partner in Drucker & Falk, a partnership, Bernard Kayden, individually and as a general partner in Fairway Co., a limited partnership, E.E. Falk and Erwin B. Drucker, individually and as partners in Drucker & Falk, a partnership, *No. 3210SC992, in the Court of Appeals of North Carolina, decided September 6, 1983, reported at 306 S.E.2d 199.*

The plaintiff Georgia Shepard was a tenant in an apartment complex owned and operated by the defendants. She was sexually assaulted in the parking lot and in her apartment at Fairway Apartments in Raleigh, North Carolina, and this action was to recover for personal injuries incurred in the assaults. She claimed that the defendants negligently failed to warn her of dangerous conditions in the parking lot, negligently failed to provide adequate lighting in the parking lot, and negligently failed to provide adequate security for the tenants.

The incident occurred on the fifth of May, 1979. Shepard got out of her car in the parking lot of the Fairmont Apartments at approximately 1:45 AM. She was immediately seized by a man with a gun. She was then raped in her car and again later in her apartment. The assailant was subsequently convicted of rape and kidnapping.

The evidence presented by the plaintiff included testimony from five employees of the local sheriff's department who described various crimes which they had investigated at the Fairway Apartments, including larceny, breaking and entering, and sex offenses. The police officers testified that they had notified the apartment management each time. She also presented a number of witnesses who described the parking lot lighting as inadequate and a former employee of the complex had admitted having received a number of complaints from tenants concerning the inadequate lighting.

The plaintiff also presented expert testimony from Dr. William Bopp who testified as an expert in the fields of crime prevention and security criminology. He testified that adequate lighting has a deterrent effect on intruders and that additional lighting was needed to provide minimum security.

The defendant apartment owners and operators presented former managers of the property who testified that they remembered some property crimes having occurred but no assaults. Several of the defense witnesses testified that lighting was adequate and comparable to other apartment complexes in the area. Three witnesses testified that it was not customary in the Raleigh and North Carolina area for apartment complexes to hire security guards. Furthermore, Bruce Marshburn, the Deputy Director of the Governor's Crime Commission, gave his opinion that the assault in this case was not predictable and that there was no relationship between crimes which might be deterred by lighting, such as vandalism of cars, and the crime of rape.

The jury found in favor of the defendants, and the plaintiff Shepard appealed.

The Court of Appeals reviewed the principles of law applicable to the landlord's duty to his tenants:

> —A landlord is under a duty to its tenant to inspect and to repair the common areas of the premises.
>
> —A tenant is normally seen as an invitee and the liability of a landlord for physical harm to its tenant depends on if it knows of the danger.
>
> —[The landlord] is ordinarily under no duty to exercise any care until he knows or has reason to know that the acts of the third person are occurring or are about to occur ... .

> —If the place or character of [the landlord's] business, or his past experience, is such that he should reasonably anticipate careless or criminal conduct on the part of third persons, either generally or at some particular time, he may be under a duty to take precautions against it ... .
>
> —... foreseeability is the test in determining the extent of a landowner's duty to safeguard his business invitees from the criminal acts of third persons.

The appellate court specifically cited the case of *Urbano v. Days Inn* (reported at page 6 of the November/December 1982 **PSCLR**) as precedent for the legal principle that a landowner has a duty to protect its customers/invitees from the criminal acts of third persons on its premises, specifically in a motel parking lot.

After reviewing the evidence which was presented by both sides, as summarized above, the court found that the jury was authorized to find that the particular assaults suffered by the plaintiff Shepard were not foreseeable, and that since the jury had so found, the court declined to set aside its verdict and affirmed the trial court's judgment in favor of the defendants.

COMMENTARY:

One of the interesting, if not surprising, evidentiary rulings made by the trial court and affirmed by the appellate court was to the effect that it was not erroneous to permit testimony comparing the lighting facilities at the Fairway Apartments with the lighting facilities at other apartment complexes. The court found that this was a proper effort by the landlord to show the common practice among apartment complexes in general which would bear on the issue as to whether reasonable care was exercised by the defendants in this case. At the same time, it does not furnish a test which is conclusive or controlling, and a jury may find that negligence existed notwithstanding that the security methods employed were customary in the business.

A similar rationale was applied by the court to the landlord's testimony concerning the use of security guards by other apartment complexes. Such evidence tends to establish the industry's standard of care for similar apartment complexes, and was proper for the jury to consider, but was not conclusive or binding on the jury.

The lesson here is, of course, that it is advisable to keep abreast of security practices which are employed by others engaged in your industry in general, and especially by similar industries in your particular locality.

PSCLR—MAY 84 (6)

Apartment Landlord Must Protect Visitors In Parking Lot From Criminal Assaults If Assaults Are Foreseeable:

Celina Aisner v. Lafayette Towers, a limited partnership and Daniel E. Levin and Robert E. Johnson, jointly and severally, *Docket No. 66146, in the Court of Appeals of Michigan, decided October 11, 1983, released for publication December 15, 1983, reported at 341 N.W.2d 852.*

Celina Aisner was robbed and sexually assaulted in the Lafayette Towers apartment building parking lot. Aisner was not a tenant at Lafayette Towers, but was on her way to visit a friend who was a tenant at the apartment complex. She filed suit against the apartment owners on the grounds that they failed in their obligation to keep the parking lot reasonably safe for visitors, specifically by failing to provide sufficient lighting and security guards. She also claimed that the apartment owners had knowledge of the high amount of criminal activity in the apartment building area, and that they knew of two other attacks on women occurring at the apartment building within 5 days before her assault. The trial judge dismissed her action, finding that the apartment owners did not owe any duty to her because the apartment building was not in fact located in a high crime area and the danger to the plaintiff was not therefore foreseeable. The trial judge was reversed on appeal.

There is no question that a landlord owes a duty to his tenants to protect them from unreasonable dangers resulting from the foreseeable criminal misconduct of other parties on the landlord's premises. That is to say the landlord's duty to protect his tenant arises when and if the risk of harm is reasonably foreseeable by a prudent landlord, and the perceived risk to the tenant is unreasonable. Evidence that only 3 felonies had occurred previous to the assault on the plaintiff in the Lafayette Towers parking lot, that no rapes had ever occurred there, and that the plaintiff's "expert witness" stated that the apartment building grounds could not be considered a "high crime area" is insufficient in and of itself to prove that criminal activity was not foreseeable to the apartment owners. The trial judge, according to a unanimous appellate court, should not have dismissed the case on summary judgment but rather should have submitted to a jury the question of foreseeability of the danger and the unreasonableness of the risk. The case therefore was remanded for trial.

PSCLR – MAY 84 (9)

Landlord/Tenant Ruled Not "Special" So As To Impose Liability On Landlord For Criminal Attack On Tenant:

Erlene Morgan v. Dalton Management Company and David Hunter, *No. 82-2878, in the Appellate Court of Illinois, First District, Fourth Division, decided September 8, 1983, reported at 454 N.E.2d 57.*

The issue presented by the appeal in this case required the appellate court to decide whether a landlord owes a duty to a tenant so as to prevent other tenants from injuring the tenant in the common areas of an apartment building.

Erlene Morgan sued her landlord, Dalton Management Company, and her attacker, David Hunter, for personal injuries and permanent disfigurement caused by Hunter. She and Hunter were both tenants in an apartment building owned and managed by Dalton Management Company. A Tony McClerty lived with Hunter, but had not signed a lease with Dalton Management Company. In September 1979, McClerty forced open the locked door of the plaintiff and she reported the incident to the apartment manager and to the police. Hunter threatened the plaintiff Morgan on two occasions, and she reported these incidents to a maintenance engineer employed by Dalton.

On October 31, 1979, Hunter threw acid on Morgan while on an elevator in the apartment building, which caused Morgan to be permanently scarred. This incident was the basis of her suit against Dalton Management Company, claiming it owed a legal duty to her to prevent the attack by Hunter.

The general rule in Illinois is that a "landlord generally has no duty to safeguard its tenants from the criminal acts of third persons." The appellate court further ruled that Dalton had no duty to protect the plaintiff from the specific injuries received, since such injuries were not <u>reasonably</u> foreseeable, and since negligent maintenance of the building did not contribute to the injuries. A mere possibility that Hunter would harm the plaintiff was not sufficient to make the subsequent attack "reasonably foreseeable by a reasonably prudent person" and lastly, the court reiterated the Illinois Supreme Court's previously-stated position that the landlord/tenant relationship is not such a "special relationship" so as to make a landlord responsible and liable for the criminal acts of a third party. Accordingly, the trial judge's dismissal of the plaintiff's cause of action against the defendant Dalton was affirmed.

PSCLR – JUN-JUL 84 (6)

Landlord Faces Negligence Action For Criminal Assault Of Tenant Where Assault Was Foreseeable:

Scott Penner v. Joe Falk, et al., *Civ. 69318, in the California Court of Appeal, Second District, Division 4, decided March 28, 1984, 200 Cal. Rptr. 661.*

The plaintiff in this case, Scott Penner, brought this damages suit against the owners and operators of his apartment building after he was assaulted and robbed by two unknown assailants in the building. Penner sought compensatory and punitive damages based on the defendant owner's negligence, and compensatory damages based on the defendant's breach of warranty of habitability. The suit was dismissed before trial and he appealed.

The California Court of Appeal reviewed the suit and noted that the defendants were aware of certain conditions of the building which gave imposed a duty on them to protect the plaintiff. These conditions included tenants' reports of criminal incidents in the building, tenants' complaints concerning the premises accessibility to non-tenants, and the defendant's knowledge that "acts of trespass, robbery, burglary, physical assault, battery and rape had been committed on the premises and in the apartments." The court concluded that this evidence supported Penner's lawsuit on the negligence charge and should be presented to a jury for determination.

In addition, the court held that if Penner could prove that "the defendant was aware of the probable dangerous consequences of his conduct, and that he willfully and deliberately failed to avoid those consequences," then punitive damages would be justified.

However, the court did not find a basis for an action based on a breach of warranty of habitability. There was no written lease; an implied warranty of habitability of rented property does not require that a landlord provide a perfect living environment, but rather "bare living requirements." If the building's security had been at tighter when Penner first entered into the landlord/tenant relationship than it was when he was attacked, then he could have alleged "an implied obligation on the landlord [to continue] to provide those protective measures which were within his reasonable capacity." Since this was not the case, there was no basis for an implied warranty on the part of the landlord to provide security. The warranty of habitability allegation was properly dismissed by the trial judge, according to the appeals court.

The judgment dismissing the negligence action was therefore reversed and the action will proceed to trial on that issue.

PSCLR–SEP 84 (8)

$1,000,000 Judgment Against Landlord For Tenant's Eye Injuries Due To Firecracker Explosion In Doorway Peephole Is Reversed; Peephole Manufacturer Dismissed Without Liability:

Carletha Bridges v. Riverbay Corporation and Auth Electric Company, Inc., *in the New York Supreme Court, Appellate Division, First Department, decided June 28, 1984, reported at 477 N.Y.S.2d 157.*

On the Fourth of July holiday, the doorbell at Carletha Bridge's apartment rang. She asked who was there and no one responded, though she heard someone running away from her door. When she looked through the "peephole" eyeviewer in her door, a firecracker, which had been placed in the eyeviewer, exploded. Her left eye was permanently blinded. She sued the owner of the apartment complex, Riverbay Corporation, and the manufacturer of the eyeviewer, Auth Electric Company. Bridges was granted a judgment in the amount of $1,000,000 against Riverbay Corporation, and the suit against Auth was dismissed.

It was Bridges' position that Riverbay had failed to take reasonable security precautions to protect her as a tenant and further that the defendant had actually encouraged the use of firecrackers on its premises without obtaining a permit, in violation of New York City law. The New York Supreme Court rejected this argument, stating that the law to which Bridges referred was concerned with <u>public</u> displays of firecrackers. Riverbay had not conducted a public display of fireworks on its property but had permitted individuals who wished to celebrate the holiday with fireworks to do so in the parks of the apartment complex property, an area called "Greenways." The court stated that Bridges had failed to prove a direct, causal connection between her injuries and Riverbay having allowed individuals to use fireworks in a specific location which was a block away from Bridges' apartment building. The cause of her injuries was an "intervening criminal act of the unknown person," and there was no evidence that the criminal act was reasonably foreseeable. The court therefore reversed the judgment against Riverbay Corporation.

PSCLR–OCT 84 (11)

Apartment Complex Liable To Tenant Sexually Assaulted In The Presence Of Her Son, Resulting From Negligent Hiring Of Employee And Failing To Provide Adequate Security:

Keyes v. Dalewood Associates, *in the United States District Court for the Southern District of New York, No. 81-CIV-162(CSH), decided January 18, 1984.*

The jury trial of this case resulted in a verdict of $600,000 for the 31-year-old woman and $900,000 for her ten-year-old son who suffered severe psychological trauma subsequent to a sexual assault on the woman which took place in her apartment and in the presence of her son. The assailant was the apartment complex maintenance man who had been hired and was supervised by the apartment complex's superintendent and management company. The jury accepted plaintiffs' contention and evidence and found that the defendants were negligent in failing to investigate the background of the maintenance man prior to hiring him and in failing to have adequate security at the apartment complex. The evidence showed that all prospective employees are interviewed and must complete an application form to establish their background, but that the maintenance man was hired without any screening. Although the defendants denied having any knowledge of the employee's prior convictions for felonious sex crimes, nor any knowledge concerning his psychiatric problems or alcoholism, the employee himself testified that his superintendent had known him for years and was probably aware of his criminal history. There was also evidence concerning prior crimes which had taken place in the apartment complex, and the jury determined that the assault on the plaintiff was therefore foreseeable.

COMMENTARY:

We would normally wait for the appellate review and an appellate decision to be rendered in this case before publishing our report on the case. However, we have learned from the plaintiff's attorney that the appeal has been dropped and the case has been settled for the full amount of the $1,500,000 verdict.

PSCLR – NOV-DEC 84 (7)

Apartment Building Owner/Operator Is Not Liable For Assault In Vacant Apartment Where Both Rapist And Victim Were Trespassers:

Gale Nixon, Individually and as next friend of Rhonda Marie Vanderburg v. Mr. Property Management Company, Inc., and Brett Davis, No. 05-83-00957, *in the Court of Appeals of Texas, Dallas, decided July 13, 1984, rehearing denied August 8, 1984, reported at 675 S.W.2d 585.*

This is an unsuccessful appeal of a summary judgment granted the defendants in a lawsuit brought by the mother of a ten-year-old girl who was taken to the defendants' apartment building and raped by an unknown attacker. The child, Rhonda Marie Vanderburg, was kidnapped while on the grounds of her resident apartment building in Dallas, Texas, by an unknown man who grabbed her and threatened to kill her if she screamed. He then took her across the street to the Chalmette Apartment Building where he raped her three times in an empty apartment. Rhonda was later found as she was crawling out the window of the apartment by a police officer who had been called by her mother when the child was discovered missing.

Rhonda's mother brought suit against the owners and managers of the apartment building, alleging that they were liable to the child for failing to exercise reasonable care to prevent the attack in the apartment building. The suit claimed that the defendants' duty to Vanderburg did not arise because of the traditional reasons: the child was not an invitee, a licensee, or a trespasser. Instead, the suit proposed that a new rule be created to impose a duty on any property owner in a "high crime" district toward any persons who were on that property unwillingly. In addition, the suit included a complaint that the rape was proximately caused by the defendants' violation of a Dallas ordinance which required property owners to "keep doors and windows of a vacant structure or a vacant portion of a structure securely closed to prevent unauthorized entry," because the door to the apartment in which the child was raped was off its hinges.

The Court of Appeal of Texas disagreed with the contention that the young girl was <u>not</u> a trespasser, stating that "[t]respassing does not depend upon Rhonda's volition but upon the knowledge and consent of Chalmette." Chalmette had neither knowledge of nor granted consent to the child's presence on its property and therefore it had no duty other than to not willfully or wantonly, or with gross negligence the minor trespasser. The court concluded:

> We decline to expand that duty of a property owner to one who, by whosever volition, is on the owner's property without the owner's knowledge or consent because the property owner can never know when such a duty might come into existence or the extent and degree of measures reasonably necessary to discharge that duty.

The court further refused to impose a duty on the defendants to protect the child merely because the apartment building was located in a "high crime area" since the assault was not foreseeable.

The complaint that the defendants' alleged failure to comply with a Dallas ordinance proximately caused the rape was also rejected by the court since the kidnapping and rape was not a reasonably foreseeable consequence of the failure to keep an apartment door on its hinges.

The appellate court affirmed the judgment granted to the defendants.

COMMENTARY:

American common law has evolved distinctions as to the duties owed by a landowner to persons who enter his land. These are based to some degree on the knowledge and consent of the landowner to the person's presence. It is helpful in reading the cases to know that these distinctions are generally divided into three categories: trespasser, licensee, and invitee.

A trespasser is a person who enters land in the possession of another without a privilege to do so. The landowner owes the least duty to trespassers since they have no right to demand a safe place to trespass. A landowner is not usually held liable for injuries to trespassers, although there are some exceptions to the general rule.

A licensee is a person who has a privilege to enter the land for his own purposes, often with express or implied consent. A social guest is an example of a licensee. To this person, the landowner owes a slightly higher degree of care in that he has an obligation to exercise reasonable care for the protection of a licensee from the landowner's known and ongoing operations.

An invitee enters the premises upon business of the landowner, with express or implied consent. The landowner, having consented to the invitee's presence, has an affirmative duty to protect against known dangers and dangers which could be discovered with reasonable care. A customer or delivery person is an example of an invitee. A landowner owes the highest duty of care to the invitee.

PSCLR–FEB 85 (3)

Landlord Liable To Tenants For Failure To Protect Or Warn Where Minor Tenant's Aberrant Behavior Presented A Known Danger To Other Minor Tenants:

E.W. Lambert, E.W. Lambert, Jr., and Robert C. Seaver v. Brenda Doe, as next friend of Steven Doe, a minor, *No. AS-152, in the District Court of Appeal of Florida, First District, decided June 26, 1984, rehearing denied August 1, 1984, reported at 453 So. 2d 844.*

The owners of Shoreline Apartments in Gulf Breeze, Florida, were sued by two of their tenants, the parents of a nine-year old male resident who had been sexually assaulted by another minor male resident. The couple asserted that the apartment owners were liable for their son's injuries because the landlord had knowledge that such attacks had taken place, but did nothing to prevent such attacks or to warn tenants of the possibility. The court awarded compensatory and punitive damages to the parents of the minor victim, and the apartment owners appealed. The decision and the compensatory damages award was affirmed, but the punitive damages award was reversed.

The plaintiff's son, Steven Doe, was sexually assaulted by Sean Roe (both names are fictitious, chosen by the court to protect the identities of the children involved) on two occasions at the Shoreline Apartments, once in the Roes' apartment although the attacker's parents were in another room of the apartment and again in the Does' apartment with the victim's parents in another room of the apartment. He was also molested by Sean on the property of a local school. Steven delayed reporting the incidents to his parents because he was afraid of Sean, who was much larger by some 10 inches in height and 75 pounds in weight. Sean also beat Steven, threatened to falsely accuse Steven of vandalism, to shoot him, and to harm his parents.

Steven's mother had noticed that Steven, normally an outgoing child, had begun to withdraw into himself and spend most of his time in the apartment. After he told her that he had been assaulted by Sean, Mrs. Doe went to the managers of Shoreline Apartments. The Roes were informed that they would have to vacate their apartment within 3 days, and they did so.

It was shown that the apartment managers had had a great deal of trouble with Sean in the past. In fact, a manager had asked the Roes to leave when their lease expired, seven months before Steven was molested, but the Roes were allowed to stay after "Mrs. Roe cried" and pleaded with the manager. Substantial evidence was admitted which showed that Sean was indeed a dangerous child, whose bizarre behavior demonstrated what appeared to be a "severe personality disorder." Residents and employees of the Shoreline Apartments testified about other incidents in which Sean molested several other minor male tenants, tortured and killed small animals, threw eggs at other tenants, vandalized property or threatened others with a gun.

Although the apartment managers admitted that they had received numerous reports about Sean for almost a year before Steven was attacked, they hesitated taking action against the boy or his parents because they felt there wasn't enough evidence and they might be exposed to a lawsuit for slander if they warned other tenants. However, a deputy sheriff who lived in the apartments was hired as a part-time security guard. He was asked to investigate the reports that Sean had sexually molested younger boys in the complex. He could not find sufficient evidence to support such charges, and filed such a report with the Sheriff's office; however, he testified that most of the children living in the complex "were hesitant to talk to him because they were afraid of Sean and Judy Wilburn (the apartment manager) was aware of this."

Before the Does moved into the Shoreline apartment complex, Mrs. Doe specifically inquired of the management whether the complex was a "nice, safe place for her family." She was assured that it was safe, that it had 24-hour security and that other tenants had not caused any trouble. The only warning concerning Sean was Wilburn's comment to Mrs. Doe that he was a "bully."

Officer Ralph Barker was allowed to testify concerning his conversations with Sean. After receiving a report of the attacks on Steven, Officer Barker went to the Shoreline complex, took Sean into custody, and took a statement from the boy. Sean told the officer of other incidents in which he forced young boys and girls to participate in sexual activity.

Reviewing the evidence, the Florida appellate court stated that "A landlord who, as in this case, recognizes and assumes the duty to protect his tenants from foreseeable criminal conduct, is required to take reasonable precautions to prevent injury from such conduct." The court pointed out that the Shoreline management had been given sufficient information on Sean's behavior for it to be reasonably foreseeable that the boy would continue to victimize other young children in the complex. Although Steven's attacks took place inside apartments, this would not absolve the landlord from the duty of warning Steven's parents that Sean presented a threat to the boy. The landlord's defense that it didn't have enough evidence to evict the family is severely weakened by the fact that the Roes' lease had expired several months before the attacks on Steven, and that the Roes had been allowed to stay in the apartment on a month-to-month basis even though they could have been asked to leave without reason.

The court affirmed judgment to the plaintiffs, ruling that the defendant apartment complex owners were indeed negligent in failing to prevent Sean's attack on Steven. However, the punitive damages award was overturned because it was not shown, as was necessary for such an award, that the apartment owners, managers, or employees had "engaged in willful, wanton or outrageous conduct while acting within the scope of their employment ... [nor were they] guilty of some fault which foreseeably contributed to the injury."

COMMENTARY:

Every landlord should insure that their resident managers are made aware of the significance of this decision. Action must be taken when a tenant is known to be guilty of abberant behavior. Also, complaints against such tenants must be accurately recorded and cataloged.

PSCLR – FEB 85 (12)

Landlord's Negligence In Failing To Repair A Stuck Door Was Not Foreseeable Cause Of Tenant's Injuries When Shot After Being Unable To Open The Door:

Maria V. Santiago v. New; York City Housing Authority, *in the Court of Appeals of New York, decided September 29, 1984, reported at 480 N.Y.S.2d 321.*

The defendant, the New York City Housing Authority, had failed to repair a jammed exterior door on the apartment in which the plaintiff Maria Santiago was a tenant. When she attempted to enter the building, and could not open the door, she was shot by an assailant and received injuries. She sued the landlord on the ground that their failure to repair the door was a proximate cause of her injuries, alleging that she would not have been injured by the attacker if she had been able to enter her building. The New York Court of Appeals disagreed, ruling that the intervening act of the assailant was unusual and unforeseeable and thus served to "break the causal connection" between the landlord's negligence and the tenant's injuries. "To hold the shooting incident a foreseeable consequence of the defendant's negligence would 'stretch the concept of foreseeability beyond acceptable limits'." Case dismissed.

PSCLR—MAY 85 (6)

Previous Criminal Incident In Apartment Complex May Make Burglary/Assault On Tenant Foreseeable And Impose Liability On Landlord:

Carol Maricle v. Florence Mick, *in the Supreme Court of New York, Appellate Division, Third Department, decided February 21, 1985, reported at 485 N.Y.S.2d 622.*

The question presented by this case is whether a landlord who leaves the rear door to a dwelling open and unlocked is liable to a tenant who is injured by burglars?

Carol Maricle was a tenant in a downstairs apartment rented from Florence Mick, her landlord, who resided upstairs. There were separate front doors to each unit, and a common hallway in the rear of the building which contain separate doors to each unit. The rear door of the building was not kept locked nor kept closed by the defendant. Also, the defendant landlord had a dog which was kept on a long chain and which used the back door to go in and out of the rear hallway at will.

The plaintiff Maricle contended that her landlord, who worked in the evenings, asked her to watch the landlord's apartment while she was away and to generally keep an eye on things. One evening at approximately 10 PM, the plaintiff Maricle heard glass breaking in the defendant's apartment and went up the hallway stairs, encountering two unidentified males leaving the defendant's apartment. Maricle was forcefully pushed down the stairs by the two men, sustaining serious permanent injuries. This lawsuit ensued, but the trial judge dismissed the case after reviewing the pleadings. Maricle appealed.

The appellate court concluded that the case should have been presented to a jury. Since the evidence showed that the rear door was knowingly left open by the landlord, that the landlord had asked the plaintiff tenant to watch over the landlord's apartment while she was away, and that there had been at least one previous theft which caused the tenant to warn the landlord of potential intruders. In addition, the landlord had been robbed at work and the thief may have learned of her address in this way and attempted the burglary at her residence. A jury could therefore conclude that the assault on the tenant was foreseeable, and that the risk of an assault outweighed the simple precaution of locking the rear door. In other words, the jury could infer from the factual situation presented that it was foreseeable that the tenant Carol Maricle would be injured by a burglar gaining entrance through the unlocked door. Since these and various inferences could be reached by a jury, it was improper for a trial judge to summarily dismiss the case and the case therefore was reversed and remanded to the trial judge for presentation to a jury.

PSCLR – JUN-JUL 85 (10)

Landlord Not Required To Protect Tenants Against Criminal Assault On Leased Premises Absent Specific Agreement:

Samuel Feld and Peggy Feld v. John W. Merriam and Thomas Wynne, Inc., Co-Venturers, t/a Cedarbrook Joint Venture, and Globe Security Systems, Inc.; Appeal of John W. Merriam and Thomas Wynne, Inc., Co-Venturers, t/a Cedarbrook Joint Venture and Cross Appeal of Samuel Feld and Peggy Feld, *in the Supreme Court of Pennsylvania, decided December 18, 1984, reported at 485 A.2d 742.*

This decision reverses an earlier appellate decision in which the court affirmed a jury verdict to tenants who were kidnapped from their apartment building's garage and assaulted by the kidnappers. The tenant's action had alleged that their landlord had owed them a duty of protection, had breached that duty, and were therefore liable for their injuries; the jury awarded a judgment of $6,000,000 in compensatory and punitive damages to the tenant/plaintiffs against the apartment building owners. The previous appellate court had only reduced the punitive damages awarded to the husband bringing the final judgment to $5,250,000. (See the October 1983 **PSCLR** at page one for a summary of the facts and the earlier appellate court's decision.) This decision results from apartment complex owner's appeal of the verdict, and Mr. Feld's appeal of the reduction of his punitive damages award to the Supreme Court of Pennsylvania. The case must now be retried in accordance with its decision.

Peggy and Samuel Feld resided in a large apartment community owned by Cedarbrook Joint Venture, and which consisted of 150 acres and 1,000 apartments in three high-rise buildings. They paid an extra rental fee to use the garage which was adjacent to the apartment buildings. One evening, after parking their car, they were kidnapped at gun point by three men who forced the elderly couple into a car and then drove off the complex premises. During this time, Mrs. Feld was assaulted and the two were released, separately, on deserted streets. The Felds sued the apartment complex owners, as stated above.

Although there was no precedent in Pennsylvania, the previous appellate court ruled that a landlord is under a duty to provide adequate security to protect his tenants from the foreseeable criminal actions of third persons. The court created a three part test to establish a landlord's liability, which stated that evidence must be presented by a plaintiff which shows (1) that the landlord knew or should have known of criminal activity on the premises which presented a danger to the tenants, (2) that the landlord had the opportunity to take action to protect the tenants against the danger, and (3) that the landlord's failure or refusal to take such action was the cause of the tenant's injuries. The previous trial court and appellate court determined that there was sufficient evidence to show that the Cedarbrook apartment owners and management was liable to the Felds for the kidnapping/assault.

Cedarbrook contended that the trial court (and appellate court in affirming) had erred in instructing the jury that the landlord had a duty to provide adequate security to protect the tenants from foreseeable criminal activity in all areas of the leased property. The Supreme Court of Pennsylvania agreed that such an instruction was erroneous. The courts which previously considered this case had treated criminal activity on leased property as a "defective condition" of the property which must be cured by the landlord to protect his tenants. The Supreme Court, however, ruled that such an extension of the landlord's duty made him an insurer of the tenant's safety, "a burden which could never be completely met given the unfortunate realities of modern society." As the court stated, "A program of security is not the usual and normal precaution that a reasonable home owner would employ to protect his property." Therefore, the landlord had no "general duty" to protect the tenants from criminal intrusion, unless he had taken on the duty voluntarily or by specific agreement. "The duty is one of reasonable care under the circumstances." Therefore, the court held that the case must be returned to the trial court for reconsideration to determine whether or not Cedarbrook's voluntary security measures were negligently exercised and whether that negligence was the proximate cause of the Felds' injuries.

The court reviewed the damages awards and determined that any award of punitive damages to the plaintiffs was improper: "While the record indicates that the security systems might have been inadequate under the circumstances, there was no evidence of an evil motive or a reckless indifference to the safety of the tenants." The award of compensatory damages was also set aside because the jury may have been prejudiced by improperly admitted evidence concerning Cedarbrook's financial net worth of $40 million.

PSCLR–OCT 85 (2)

Florida Holds Plaintiff May Recover For Sexual Assault Without Proof Of Prior Similar Incidents On Premises:

Polly Suzanne Paterson v. Kent C. Deeb, Transamerica Insurance Co., W. Fenton Langston, and Hartford Accident & Indemnity Co., *No. AY-8, in the District Court of Appeal of Florida, First District, decided June 12, 1985, rehearing denied July 22, 1985, rehearing denied August 14, 1985, reported at 472 So. 2d 1210.*

Polly Paterson rented a one-bedroom apartment owned by Kent Deeb and W. Fenton Langston in Tallahassee, Florida. During the second year she leased the apartment, she was beaten and raped by an unknown assailant who was hiding in her bathroom. She brought suit seeking compensatory and punitive damages from the owners, alleging that the attack was a foreseeable result of their breach of common law and statutory duties owed to her. The suit was dismissed, but on appeal the dismissal was reversed and the case remanded for trial.

Paterson's apartment was located in a small, older building and consisted of the entire top floor; two other apartments were located downstairs. The building had front and back doors which led into a common downstairs hallway. A stairway led up to a hallway which split Paterson's apartment: her living and bedroom quarters were on one side of the hall and her bathroom was on the other side.

After Paterson had signed the contract to renew her lease for another year, she told the leasing agents that she was concerned about the building's lack of security. She specifically mentioned that the front door lock was broken, that there was no lock on the back door, and that the vacant apartment downstairs had open windows and doors allowing trespassers to use the apartment and to enter the building. She also complained that the windows had no screens and could not be secured because they were broken. The leasing agents spoke with the owners and reported to Paterson that they felt repairs and security improvements were a waste of money because the building was to be demolished. She continued to reside in the apartment until several months later when she was attacked by the assailant who was hiding in her unlocked bathroom.

In her lawsuit, Paterson alleged that the owners knew or should have known that the neighborhood in which the apartment building was located had experienced a significant number of break-ins and assaults–higher than the average for other areas of Tallahassee. She alleged that the defendants violated a statutory duty to maintain the doors, locks, keys, and common areas of the apartment building in a reasonably safe condition, pursuant to the Florida Residential Landlord and Tenant Act. She alleged that the owners were negligent in failing to make the repairs and security improvements to the doors, windows, and common areas or to provide adequate locks to the front and back doors, and her bathroom door, in keeping with her complaints and Florida statutory law.

The defendants accurately pointed out that the plaintiff failed to allege that the criminal assault on her was a "reasonably foreseeable occurrence" at the apartment building, or that the defendants had any legal duty to protect her against the assault, or that the defendants' alleged negligence caused the plaintiff's injuries, or that the defendants were liable for punitive damages. In view of these omissions, the trial judge ruled that the plaintiff's claim was insufficient in that she failed to allege the requisite foreseeability; that is, she failed to allege that the defendant property owners had actual or constructive knowledge of prior similar acts committed upon person on the premises.

The general rule is and has been that a property owner has no legal duty to protect a tenant from a third party criminal attack over which the property owner has no control, unless the attack is reasonably foreseeable. To prove that such an attack was foreseeable, the tenant must allege and prove that the property owner had either actual or constructive knowledge of prior similar acts, the reason being that the property owner should not be required to prevent or protect against attacks which he has no reason to anticipate. However, the court noted that tenants may have certain rights to recover damages for criminal attacks based on theories other than foreseeability, including instances where the landlord undertook to provide security protection by contract, made representations concerning security to the tenant, or abolished past security practices which had provided tenants with protection from criminal assaults. Accordingly, the Florida appellate court concluded that the foreseeability theory of the general rule governing land owner and invitee liability was not the exclusive basis upon which a landlord's liability to the tenant could be grounded.

The court also reviewed the legal impact of the Florida Residential Landlord and Tenant Act on a tenant's right to recover damages for a criminal attack from a landlord by bringing a tort action, and noted that it was a first impression review. The court stated:

> [The Act] plainly imposes a continuing obligation on the landlord to provide the leased premises 'with locks and keys' in the leased areas and the common areas, and to maintain the common areas of the premises in a 'safe condition.' Without debating whether these provisions may be construed as requiring a landlord to provide security protection in the form of guards, scanning video cameras or other security measures, it cannot be doubted that the statute at least contemplates that the physical structure will be furnished with locks and maintained in a reasonably safe condition to protect the tenant from unwanted intrusion of undesirable persons. Certainly the principal reason, if not the only reason, for furnishing locks and keys is to prevent a human being from making an unauthorized entry into the premises. Similarly, means of preventing unauthorized access through doors and windows may be essential to maintaining the safe condition of common areas.

The intervening criminal act by the assailant was clearly the cause of Paterson's injuries. However, if it was a foreseeable consequence of the defendant's negligence, then the defendants may be held liable for her injuries. Although the plaintiff could not prove that "prior, similar" acts had occurred at the apartment building, the court ruled that the type harm suffered could be determined by a jury to be a foreseeable result of the landlord's failure to provide adequate security locks which is the purpose of the statutory requirement. Therefore, the appellate court ruled that the issue of the landlord's liability should have been presented to a jury.

The Florida court then noted with approval the opinion of the California Supreme Court in *Issacs v. Huntington Memorial Hospital* (reported in the May 1985 **PSCLR**, case 3, page 3). The California Supreme Court held in that case, an action against a landowner for criminal acts by third persons on the landowner's property, that a plaintiff could establish foreseeability on the part of the landowner without necessarily proving that prior similar incidents had occurred on the premises. The California court, and in turn the Florida court, rejected the recent line of court decisions which had established the rule:

> In the absence of prior similar incidents, an owner of land is not bound to

> anticipate the criminal activities of third persons, particularly where the wrongdoer was a complete stranger to the landowner and the victim and where the criminal activity leading to the injury came about precipitously.

The court also ruled that the trial court erred in dismissing the plaintiff's claim for punitive damages in addition to compensatory damages. The court stated:

> ...plaintiff out of fear for her safety, made requests of defendants to provide locks or other means of securing the premises to prevent unwanted trespassers from entering and harming her. ...defendants refused to comply with her request because they planned to demolish the building and did not want to 'waste' their money.... Defendants were under a statutory mandate to act in good faith in performing every duty under the Act. Defendant's breach of duty to provide locks and security and to keep the common areas in a safe condition is, for the reasons stated, tantamount to a willful tort.

The judgment against the plaintiff is therefore reversed and is remanded for trial on the issues.

COMMENTARY:

As the California and Florida courts noted, the rule requiring the plaintiff to prove the existence of "prior similar incidents" in order to prove that the attack on the plaintiff was or should have been foreseeable by the defendant landowner, was seriously defective in three particulars:

The imposition of such a rule results in unfair consequences: the first victim always loses, while later victims are permitted to recover for their injuries, and furthermore, the rule discourages property owners from protecting premises which they know to be dangerous but on which no attack has yet occurred. The owner should not be permitted to get "one free assault" before being held liable for criminal attacks which take place on his property.

The rule leads to arbitrary and differing results, because there is no uniform requirement as to how "similar" the prior incidents must be in order to satisfy the rule. (We have seen in previous cases reported in **PSCLR** *different judges ruling different ways on how close in time the prior incidents must be in order to be "similar" or how near in location they must be.)*

The mere fact standing alone that a certain type of incident has not happened before does not necessarily mean that such an incident is one which might not reasonably have been foreseen. And it follows that the absence of prior incidents does not justify relieving a landowner from responsibility for the foreseeable consequences of his acts.

Proof of prior similar incidents can, no doubt, be helpful in determining foreseeability, but California and Florida in the decisions above have ruled that such proof is no longer necessary. The jury, under all the facts and circumstances of a given situation, including whether there were or were not any prior similar incidents, can make its own determination as to whether the assault on the plaintiff was or was not foreseeable.

PSCLR – NOV-DEC 85 (4)

Guest Of Tenant Criminally Assaulted At Tenant's Apartment Fails To Prove Causal Link Between Assault And Landlord's Failure To Provide Adequate Security:

Rullman v. Helen Fisher, v. Stacy Greene, *No. CO-85-423, in the Court of Appeals of Minnesota, decided July 23, 1985, reported at 371 N.W.2d 588.*

While a guest at a friend's New Year's Eve party at a Minneapolis apartment, Renae Rullman was criminally assaulted and seriously injured by another guest. She sued her host and her host's landlord to recover damages for her injuries. The trial judge granted the landlord a summary judgment, and Rullman appealed.

Stacy Greene and two roommates rented the second floor of a three-story building owned by Helen Fisher. The first floor was leased to a commercial sauna and massage parlor. The two top floors, formerly for commercial use, were studios and loft apartments. Greene was among the first residential tenants who moved into the building and were taking on the job of converting the spaces from its former use. These tenants received a $2,000 grant from the "Artspace Reuse Project" for renovation costs, and received a reduction in rent to compensate for their labor. Soon after the renovation began, Greene invited a number of her friends, including Rullman, to her loft for an impromptu New Year's Eve party. Because the building was not equipped with an intercom or buzzer system, guests entered through the building's front door which was not locked. The stairway door to the lofts, another to the sauna, and the building's front door were all located in the front vestibule. The stairway door was kept locked during the early hours of the party and guests had to knock to summon someone down the stairs to let them in. Because the door did not lock automatically, Greene accompanied departing guests to ensure that the door remained locked. She testified that she was worried about security but, as the evening wore on and people were coming and going, she stopped checking the door.

Rullman and a group of friends arrived at the party just after midnight. She met, and began dancing with, a man who she introduced to Greene as "Leon." Although Greene did not know Leon, and had not invited him to the party, she was not concerned since he was not behaving improperly. She stated if anyone at the party had acted improperly, she would ask them to leave, if necessary with the assistance of some of her male friends. At some point, Rullman went to the back of the loft to use the restroom, away from the party crowd. Leon followed her and pushed her into the bathroom, where he attacked her, broke her jaw, and raped her. The attack was interrupted by another guest; Leon escaped and was never identified.

Rullman contended that Fisher was negligent for failing to provide adequate security in the form of a dead-bolt lock on the apartment and a buzzer system for the building. Generally, an unforeseeable criminal act by a third person breaks the "chain of causation." However, the appellate court noted that the Minnesota Supreme Court has sanctioned a landlord's liability for criminal actions of third persons in certain situations.

Rullman argued that the landlord could have and should have foreseen her assault

> because of the commercial and warehouse character of the neighborhood and because of past criminal activity at the first-floor sauna.

Addressing this issue, the court ruled that Rullman would have to link foreseeability of the attack with causation to prove the landlord liable; it is here that her argument proved inadequate. Although the stairway door was allowed to remain unlocked partially because there was not a buzzer, the court disagreed that there was any link between the lack of a buzzer and the attack itself:

> Whether the assailant knocked and was allowed into the loft, walked in with a group, or merely wandered in through the unlocked door, it is undisputed that he was at the party for some time, mingled with guests and the host, and, although recognized as a stranger, was not questioned or asked to leave.

Since the evidence did not show the "requisite causal connection" between a breach of any duty by the landlord and the assault on Rullman, the summary judgment granted Fisher was affirmed.

PSCLR – NOV-DEC 85 (6)

Where Landlord's Duty Of Care To Protect Tenant From Criminal Attack Must Be Established Through Foreseeability Of Crime Based On Prior Similar Incidents, How Much Similarity Must Exist?

Bayshore Company et al. v. Pruitt, *No. 70218, in the Court of Appeals of Georgia, decided July 16, 1985, rehearing denied July 31, 1985, certiorari denied September 26, 1985, reported at 334 S.E.2d 213.*

Ms. Pruitt, a tenant at an apartment building owned by the Bayshore Company, was the victim of an assault in her apartment. Her attacker crawled up to her second story apartment by stepping onto an air conditioning unit and then scaling up the gutter drain. From her balcony he was able to pry open a sliding glass door and enter the apartment. He was later apprehended and pled guilty to the related criminal charges. Pruitt brought an action against the owner and manager of the apartment building; she alleged that they were negligent in failing to secure the sliding glass door to her apartment. Bayshore made a motion for summary judgment, claiming that it could not be held liable as a matter of law because intervening criminal act of the assailant had caused her injuries. In opposing this motion, Pruitt offered evidence of 17 prior criminal incidents at the apartment complex in the three years preceding her attack. The Bayshore Company argued that none of these incidents should be admitted on consideration of the summary judgment, because they were not sufficiently similar to the subject incident. The trial court denied both the motion to bar from consideration all 17 allegedly similar crimes and the motion for summary judgment. Bayshore appealed to the Court of Appeals of Georgia. The majority of judges affirmed the trial court's orders. However, two justices wrote a concurring opinion and one justice wrote a strong dissent.

Under Georgia law an invitee may recover from a property owner for injuries received due to a dangerous condition on the property, if the invitee can prove that the owner/occupier/manager knew or had reason to know that a dangerous condition existed on the property and had a reasonable opportunity to prevent injury to persons on the property. Knowledge of a dangerous condition is established through evidence of prior similar accidents:

> All that is required is that the prior accident be sufficient to attract the owner's attention to the dangerous condition which resulted in the litigated accident. [from *McCoy v. Gay*, reported in the August 1983 **PSCLR**, case 2, page 2]

The majority opinion noted that Pruitt alleged that a dangerous condition, lack of a security device to prevent criminal entry through her apartment's sliding glass door, led to her attack. In addition to the 17 prior incidents of crime at the apartment complex, she offered evidence that the attacker had first attempted to enter her neighbor's apartment through the sliding door, but could not because that apartment's sliding door was secured with a bar, known as a "charley bar."

The court reviewed the trial judge's order denying Bayshore's motion to strike all of the 17 incidents, and stated "we need not engage in an esoteric discussion of exactly how much similarity must exist." Rather, the court found that "at least some of the 17 prior incidents ... apparently involved forced entry through a sliding glass door. In our opinion, this is clearly sufficient similarity... ." The appellate court affirmed that <u>all</u> 17 of the incidents <u>should not be excluded</u> from consideration, but not that all 17 will necessarily be admitted into evidence at trial. Rather the court determined that the specific incidents among the 17 admitted into the trial evidence should be determined at the trial court level.

The concurring opinion filed by two of the Court of Appeals judges expressed agreement with the majority's affirmation of the trial court's orders. The concurring opinion agreed with the decision, but disagreed that the admission of individual incidents should be left to the trial court. The concurring judges would have the court rule that all 17 prior crimes were admissible, not just ones selected by the trial court, to show that the Bayshore Company had knowledge of a dangerous situation, i.e., unauthorized entries into the apartments over a long period of time:

> That some of these unauthorized entries occurred on the first floor rather than on the second floor ... or that some entries were made through windows or through the front door rather than through the sliding glass door are circumstances affecting the weight of this evidence, not its admissibility.

The lone dissenting judge stated that he was baffled at the majority's conclusion, particularly in the statement that the court "need not engage in an esoteric discussion of exactly how much similarity must exist between a prior and subsequent incident in order ... to qualify as admissible evidence." This judge strongly expressed his opinion that the majority opinion provided "absolutely no guidance or clarification for the litigants and the court below." He believed, as did the concurring judges, that the majority opinion should have specifically chosen which of the 17 prior criminal incidents were admissible. Unlike the concurring judges, the dissenting judge selected only two of the 17 prior criminal incidents, both of which involved break-ins through the sliding glass door (that they occurred on the second floor was irrelevant in his opinion).

The majority opinion of course rules, and the trial judge will make the decision as to which of the 17 prior instances of crime will be admitted. It will be interesting to see if that decision ends up a focal point in a future appeal.

PSCLR – JAN 86 (3)

U.S. District Court Rules That Arkansas Supreme Court Would Recognize Landlord's Duty To Employ Reasonable Security To Protect Tenants From Foreseeable Criminal Attack:

Cheryl Lynn Jackson v. Warner Holdings, Ltd., a Canadian corporation, and Ruth Singer, *Civ. No. 85-2102, in the United States District Court, for the Western District of Arkansas, Fort Smith Division, decided September 5, 1985, reported at 617 F. Supp. 646.*

Cheryl Lynn Jackson, a tenant in the Yorkshire Apartments in Fort Smith, Arkansas, was the victim of an illegal entry into her apartment and repeated sexual assault by Charles Wilson. She sued the apartment building owner, Warner Holdings, Ltd., a Canadian corporation, to recover damages alleging that its negligence was the proximate cause of her injuries. The case was filed in the U.S. District Court in Arkansas, which must apply the law of Arkansas to determine the outcome. In response to the defendant's motion to dismiss, the court issued this opinion.

Among Jackson's allegations of the defendants' negligence were charges that they had failed to reasonably and adequately secure the apartment from unauthorized entry, "to provide adequate security devices such as deadbolt or tamperproof locks," to install curtains or to otherwise insure the privacy of the tenants within their apartments, or to warn Jackson that she should install security devices for her own protection. In determining whether or not to dismiss the suit on summary judgment, the District Court noted that the Arkansas courts have not yet dealt with the question of whether a landlord is under a duty to employ reasonable security measures to protect tenants from criminal attacks by third persons. However, the District Court found cases on analogous issues which provide guidance in its decision. The progression of the law concerning "extracontractual duties" owed to tenants by their landlords is reflected through three cases decided by the Arkansas Supreme Court between 1932 and 1983, according to the U.S. District Court. The earliest decision demonstrated the court's unequivocal refusal to require a landlord "to light common passageways for the benefit of tenants" in the absence of a statute or agreement. But in 1983, the court ruled an employment agency liable for referring a client to a prospective employer who abducted and raped her. In that case, the Arkansas Supreme Court stated:

> Specifically, where a victim of a rape can show that a defendant who owed some duty to the victim, breached that duty, and could foresee that the breach might result in injury to the victim, and that breach was significant factor in a rape, then a jury question of negligence exists. In other words, such cases should not be automatically dismissed .... [In a California case], a woman was raped in her apartment and she sued the landlord and rental agent for failing to take reasonable steps to protect her. The trial court ... dismissed the action. On appeal, the case was reversed holding that the plaintiff had a cause of action.

The court relied on the case of *Butler v. Acme Markets, Inc.* [reported in the November-December 1983 **PSCLR**, case nine], in which a woman was assaulted in a supermarket parking lot, sued the store, received a jury verdict, but the trial judge granted the defendant's motion for a judgment notwithstanding the verdict. On appeal, the judge's action was reversed, the court finding that liability for criminal attacks may be imposed where the harm is foreseeable and a reasonable person would have taken precautions against it.

The U.S. District Court interpreted this discussion to

> indicate that the Arkansas Supreme Court, if presented with the issue at bar, would recognize a duty owed by a landlord to his or her tenants to

> employ reasonable security measures to avoid foreseeable criminal attacks by third persons.

Therefore, the U.S. District Court refused to grant the defendant's motion to dismiss and the case will proceed to trial.

COMMENTARY:

During a speaking engagement at the National Crime Prevention Institute at the University of Louisville in Kentucky, we were asked by a participant to explain the value of "precedent" cases, and how we decide which cases to include in **PSCLR**.

The above case gives a good example of this: the federal judge who was called upon to decide the above case of Jackson v. Warner Holdings, Ltd., had the responsibility of determining the law of the state of Arkansas, in order to reach a proper decision on the defendant's motion to dismiss. In doing so, he was not bound by the decisions from courts of any state other than Arkansas. At the same time, he noted that the Arkansas appellate had never decided the particular issue presented by the Jackson situation.

Accordingly, he looked at the holdings in cases involving similar fact situations which had been decided in other states. One of the cases he found was Butler v. Acme Markets, Inc., decided by the New Jersey Supreme Court. **PSCLR** *had reported the Butler case, as noted above, because we felt that the holding, while not binding on any courts except those in New Jersey, nevertheless indicated a new trend in the law and would probably be adopted by courts in other states, as the district court in Arkansas did in this case.*

It therefore is readily apparent that by reading persuasive opinions by courts in jurisdictions other than one's own and noting developing trends, the reader can not only stay abreast of recent developments in the law, but also can take certain preventive measure to protect himself and his company, or clients, against possible future liability.

PSCLR – JAN 86 (5)

Property Owner And Manager Held Not Liable For Shooting Of Tenant By Co-Tenant:

Kavoos Sinai and Maliheh Sinai v. Polinger Co. and Park Southern Co., *No. 82-1139, in the District of Columbia Court of Appeals, decided August 30, 1985, reported at 498 A.2d 520.*

Dr. Kavoos Sinai suffered permanent injuries after being shot five times by a tenant at the Park Southern Apartments in Washington, DC where Sinai leased space. Sinai and his wife sued the tenant-assailant, Willard Johnson; the building management company, Polinger Co.; and the building owner, Park Southern Co. A default judgment of $1.5 million was entered against Johnson, but the jury found in favor of Polinger Co. and Park Southern Co. The judgment for those defendants was upheld by the District of Columbia Court of Appeal on Sinai's appeal.

The shooting occurred shortly after Dr. Sinai's first encounter with Johnson. As Johnson was entering the parking area (and Sinai was leaving the parking area to enter the building) he accosted Sinai, cursed at him, and slapped him. Sinai immediately went to the management office and reported the incident, while Johnson proceeded to the parking area. Employees in the management office told Sinai that they would call the police.

Sinai returned to the parking area, and saw Johnson getting into a parked car, so he moved in front of the car, and began writing down the car's license plate number. Johnson got out of the car and fired a gun at Sinai, and the two men began to struggle. Sinai attempted to hold down the arm in which Johnson held the gun, but was unable to do so. As Sinai released Johnson and attempted to flee, Johnson fired the gun at him twice, hitting him once. After Sinai fell to the ground, Johnson walked over and shot him three more times, inflicting serious permanent injuries.

The Sinais contended that the defendant owners/managers were liable for the shooting because they knew that Johnson had threatened the safety of other tenants, but did nothing to protect or warn the tenants. The Sinais offered evidence that Johnson had frequently complained the management that unnamed tenants were "watching him, harassing him, and ...had tried to run him over in the parking lot." There was also evidence that Johnson had fired shots from his apartment balcony into the parking lot, and had threatened another tenant with a gun. The police had been called after Johnson threatened the other tenant, and the manager had warned that tenant to avoid situations with Johnson.

The defendants argued at trial that the shooting of Sinai by Johnson was not foreseeable. Moreover, Sinai had voluntarily assumed the risk of injury by returning the parking lot and confronting Johnson after the first assault, instead of waiting in the management office for the police to arrive.

The jury found no negligence on the defendants part, and the Sinais argued on appeal that the jury incorrectly came to this decision because the court had given instructions on "assumption of risk." The appellate court agreed that there was evidence that Sinai had "assumed the risk" of an altercation with Johnson, but he had not assumed the risk of being shot five times. However, the appellate court ruled that the jury could have found Sinai negligent in returning to the parking lot and in grabbing Johnson who was wielding a gun, and in finding that this negligence contributed to his injuries.

The plaintiffs also argued that the trial judge had erred in instructing the jury on the landlord's duty of care to protect tenants from foreseeable crime. Sinai contended that Polinger's property manager was required to provide a higher standard of care to tenants than would be required of a typical landlord because he was a professionally certified property manager. The court disagreed, and held that the landlord's duty was "reasonable care in all the circumstances," and the jury had been properly instructed to determine whether the landlord had been negligence in failing to evict Johnson.

COMMENTARY:

This case includes an interesting discussion by the court of the legal effect of the fact that the property manager employed by the building was qualified as a "Certified Property Manager" ("CPM"). This designation is analagous to the "CPP" certification used by those private security professionals who have earned the "Certified Protection Professional" designation by the American Society for Industrial Security.

As the court noted, an individual must have three to five years of property management experience, must complete certain courses, and must pass a peer review examination in order to become a CPM. The individuals who complete these requirements can then be certified by the Institute of Real Estate Management. They are not, however, licensed by federal or state authorities, and there is no requirement that property managers be so certified.

The plaintiffs in <u>Sinai</u> argued that the defendants should be held to a <u>professional</u> negligence standard, that of a reasonable certified property manager as opposed to an uncertified property manager. Plaintiffs cited certain legal authority which stated that if the individual has more than the minimum qualities and skills, then that individual should be required to exercise the superior qualities and skills which he has in a manner reasonable under the circumstances of the situation. The standard of negligence then would be that of a reasonable person with such superior attributes.

There is no question that such a higher standard is imposed in professional malpractice cases against doctors, lawyers, architects, and the like. In those cases, the duty to exercise reasonable care requires that those with special training and experience adhere to a standard of conduct commensurate with such attributes.

The court in the <u>Sinai</u> case did not answer the question as to whether the evidence presented established the existence of a "profession" of CPM-trained property managers, nor did it answer the question of whether such persons should be held to the "superior attributes" of a CPM-trained *person, because the negligence complained of in Sinai took place prior to the time the manager obtained his CPM certification.*

However, the court concluded that the law applied by the trial judge in Sinai was correct, that no "professional negligence" law was applicable, and that there is only one standard of care which is that of the reasonable person's duty to exercise ordinary care under all of the circumstances.

The increasing use of CPM and CPP designations may well cause additional litigation on this issue, and we will await another case for the issue to be decided definitively.

PSCLR – SEP 86 (9)

Housing Authority's Duty To Protect Its Tenants Does Not Include Evicting Violent Tenants Or Providing Private Police Protection Of Its Premises:

Mrs. Mary Smith, et al., v. Mrs. Maggie Howard, et al., *No. 85-CA-0354, in the Court of Appeal of Louisiana, First Circuit, decided May 28, 1986, reported at 489 So. 2d 1037.*

The plaintiff in this case was the estate of a man killed at the Oklahoma Street Housing Project in Baton Rouge. A tenant of the Housing Authority Project, the deceased had been shot by a fellow tenant, Maggie Howard, who had mistakenly believed that he was attempting to break into her apartment.

The deceased's estate brought a wrongful death action against Mrs. Howard and the East Baton Rouge Parish Housing Authority; the court granted dismissal of the suit against the Authority. The plaintiff appealed.

The suit argued that the Authority had caused the death by failing in its duty:

> (1) to evict Mrs. Howard after her neighbors reported to the Housing Authority that she was a threat to their safety and to the safety of visitors; (2) to maintain a proper screening program so as to avoid renting to tenants with a history of violent propensities; (3) to maintain policies requiring tenants to state whether they have any dangerous weapons; (4) to have a program for following up reports of violent conduct by tenants against other tenants or visitors; (5) to insure the safety and security of guests; and (6) to warn tenants and guests on the premises of the dangers posed by Mrs. Howard.

The appellate court stated that, even if the plaintiff could prove all of the above-listed allegations, the Housing Authority would not owe a duty to evict persons suspected of violent behavior or to maintain its own private security force on the premises.

The court stated,

> It is well settled that there is no duty to control the actions of a third person and thereby prevent him from causing harm to another unless some special relationship exists to give rise to such a duty.

For example, there is a "special relationship" between the innkeeper and his guests imposing a corresponding special duty of protection on the innkeeper, whereas the landowner does not have a special relationship with people living on his property and therefore has no special duty.

Since the court found no legal basis for imposing the "special duty of protection" on the landlord to protect tenants, the court affirmed the judgment of the trial court dismissing the action against the Housing Authority.

PSCLR – FEB 87 (2)

Virginia Law Does Not Impose A Duty On Landlords To Protect Tenants From An Intentional Criminal Act Committed By An Unknown Third Party:

Laura Jean Deem v. Charles E. Smith Management, Inc., *No. 85-1996, in the United States Court of Appeals, Fourth Circuit, on appeal from the United States District Court for the Eastern District of Virginia, decided September 8, 1986, reported at 799 F.2d 944.*

Laura Deem, resident manager of the Windsor Towers apartment complex in Arlington, Virginia, was assaulted in the rear parking lot of the complex at 9 p.m. one evening. She sued her landlord and employer, Smith Management, claiming that the company's negligence had partially caused her injuries. She alleged that Smith had not corrected unsafe conditions of the parking lot, particularly the inadequate lighting, even after prior criminal incidents had occurred. She argued that Smith had breached its duty to her as a tenant.

The U.S. District Court entered judgment for the defendant and Deem appealed.

The District Court had determined that Deem was assaulted as she was returning from a personal errand, therefore she was acting as a tenant and not as an employee when she was assaulted. Further, the landlord does not have a duty to protect tenants from the criminal acts of unknown third parties, under Virginia's laws. Even if Virginia law recognized such a duty by a landlord, Deem did not provide sufficient evidence to support the conclusion that the Windsor Towers back parking lot was known to be unsafe.

The U.S. Court of Appeals affirmed the District Court's findings and conclusions.

The Court of Appeals noted that the findings in this case were governed by Virginia common law and, most particularly, by the case of *Gulf Reston, Inc. v. Rogers* (207 S.E.2d 841). In that case, the Virginia Supreme Court was dealing with a situation similar to Deems, and the court had clearly held that the defendant landlord was under no duty to protect the plaintiff tenant from an intentional criminal act committed by an unknown third party.

Deem argued that the above rule had been superseded by the Virginia Legislature in the Virginia Residential Landlord and Tenant Act which obligates the landlord to "keep all common areas ... of the premises in a clean and safe condition" and furthermore, "to comply with requirements of applicable building and housing codes materially affecting health and safety." She claimed that Smith Management failed to meet these obligations to her as a tenant by failing to provide adequate lighting in the parking areas of the Windsor Towers. The Court of Appeals disagreed with Deem's application of these statutory requirements, which are actually geared to ensure the physical safety of the rented premises, i.e., "the protection of the tenant from injuries caused by failures of the building – collapsing stairs, faulty walls, and dangerous windows." Since the Virginia Legislature has not specifically addressed the issue of a landlord's duty to provide tenants with protection from crime, the prior decision is controlling. Therefore, judgment for the defendant was affirmed.

PSCLR – MAR 87 (6)

Where Landlord Has Security Program Promising Additional Safety Factors, Tenant Has Cause Of Action For Breach Resulting In Rape Due To Inoperative Door Lock:

Lisa Reider v. Frank A. Martin, John Brzyski and J.T.B. Associates, a partnership, *in the Superior Court of Pennsylvania, filed January 5, 1987, reported 519 A.2d 507.*

Lisa Reider was a student at Lehigh University, and she rented an apartment in Bethlehem, Pennsylvania, which was owned by the defendants, who also owned two additional apartments at the same location. She was raped, beaten, and robbed in the stairway outside her apartment by Melvin Dean Smith, who had followed her into the building through an unsecured front door. Reider filed a lawsuit against the apartment owners claiming that they had failed to install an operating door lock on the apartment building front door, even though they had been informed on several prior occasions, by Reider and other tenants in the building, that the lock on the front door of the building was not working, and even though they had requested that the lock be repaired or replaced. The building owners had in fact notified the tenants, including the plaintiff, that they would replace or repair the lock as requested. However, it was not done prior to the assault and rape of Reider.

After Reider presented her evidence on the issue of the defendants' liability, showing that the apartment owners had been notified of the the broken lock, and showing that the owners had assured the tenants that the lock would be repaired, and that they had failed to do so, the trial judge dismissed the case based upon the trial judge's interpretation of the Pennsylvania Supreme Court's ruling in *Feld v. Merriam* (reported in the October 1983 **PSCLR**, case one, page one; and the June-July 1985 issue, case nine, page ten). In the *Feld* case, the Supreme Court had ruled that a landlord had no general duty to protect his tenants against the criminal acts of third parties, unless the landlord has agreed or voluntarily offered to protect the tenants, in an effort to attract or keep tenants, in which event the landlord may incur the duty to protect them against such third party criminal acts. Reider appealed the dismissal of her case, alleging that the trial judge had improperly interpreted the Supreme Court's ruling in the *Feld v. Merriam* case. In reviewing her arguments on appeal, the court therefore reviewed the case of the *Feld* case (which are summarized as noted above in **PSCLR**; which case resulted in a judgment for the plaintiff tenants in the amount of $5,250,000). The court first noted that the issue presented in the *Feld* case had never before been decided by the Pennsylvania courts.

The trial judge in the instant case based his decision to dismiss Reider's lawsuit on the following language which he found in the *Feld* decision:

> absent therefore an agreement wherein the landlord offers or voluntarily proffers a program, we find no general duty of a landlord to protect tenants against criminal intrusion. However, a landlord may, as indicated, incur a duty voluntarily or by specific agreement if to attract or keep tenants he provides a program of security. A program of security is not the usual and normal precautions that a reasonable home owner would employ to protect his property. It is, as in the case before us, an extra precaution, such as personnel specifically charged to patrol and protect the premises.

The trial judge in the Reider case concluded that there was therefore no duty on the part of the apartment owners since there was no "program of security" which had been promised to the tenants by the owners. He felt that the installation of the front door lock is something that a reasonable home owner would do himself to protect his own property, and is not an "extra precaution." He felt that the apartment owners here had merely promised to repair the front door, rather than to provide a "program of security."

The appellate court, however, disagreed. The appellate court said that while the decision in the *Feld v. Merriam* had indicated that an example of an "extra precaution" would be "personnel specifically charged to patrol and protect the premises," that language did not mean that all security systems provided by an owner must be of a complicated nature in order to constitute a "program of security." The appellate court acknowledged that, at first blush, providing a front door lock does not seem to amount to a program of security, but went on to find that a jury could conclude as a substantive matter that such a precaution would fall within the definition of a security program.

The trial judge's order was reversed and the case was remanded to him for further proceedings.

COMMENTARY:

We have here an excellent example of the manner in which courts, by subsequent interpretations of precedent setting cases, can and do establish new standards (in effect, "new laws") for the conduct of society. Prior to the decision of Feld v. Merriam, there was no duty on the part of landlords to protect their tenants from criminal assaults by third persons. The court in that decision justified imposing such an obligation on landlords, however, in the situation where the landlord had promised a program of security which would provide an additional factor of safety to the tenants.

And, now in the Reider case here, the court has said that will allow a jury to determine whether the provision or promising of the repair or replacement of a front door lock amounts to a program of security which would provide an additional factor of safety. As has been seen above, both of these cases rest on some affirmative promise having been made by the landlord to the tenants, for the purpose of attracting or keeping tenants. Of course, we can now anticipate that the Pennsylvania courts will next be confronted with the situation where no affirmative promise as such has been made by a landlord, but a criminal act by third parties has occurred because the landlord failed to provide a program of security which was reasonable under all the facts and circumstances of a given situation, and where the failure to provide a reasonable program resulted in foreseeable consequences.

In other words, it can be anticipated that the court will have to decide whether to impose liability on landlords for foreseeable consequences of failing to provide security even though the landlord has not made any expressed promise to do so.

PSCLR – MAY 87 (4)

Landlord Has Duty To Install And Maintain Front Door Locks, And May Be Held Liable To Tenants For Foreseeable Criminal Conduct If He Fails To Provide Reasonably Adequate Locks:

David Brichacek v. Eugene Hiskey and Hiskey Apartments Co., *No. 85-1502, in the Supreme Court of Iowa, decided February 18, 1987, reported at 401 N.W.2d 44.*

David Brichacek of Des Moines, Iowa, brought suit against his landlord, Hiskey Apartments, after he suffered losses due to a robbery of his apartment. He alleged that the landlord was liable because it had been negligent in the installation and maintenance of the locks, and that such negligence breached the landlord's implied warranty of habitability. Following a trial, the court dismissed the suit. Brichacek appealed to the Supreme Court of Iowa.

The Iowa higher court reviewed the evidence, and noted that a landlord is required by Des Moines housing codes to provide a front door lock, but there was no evidence concerning specifications as to the type and installation of the lock. In addition, there was no provision concerning locks or other security devices in Brichacek's lease with Hiskey Apartments. The defendant landlord testified that he had owned the building for fifteen years, and that each apartment door was secured by a standard door handle lock, which locks had passed city inspections. Although many of the tenants (including Brichacek's predecessor) had additionally installed dead bolt locks on the apartment doors, Hiskey had not received any complaints about any door locks from Brichacek or the city inspectors.

The burglary of Brichacek's apartment occurred while he was out of town. The wooden door to his apartment was chipped in the area of the dead bolt, giving the appearance that a sharp instrument was used to pry the lock out of the door jamb. Then by kicking the door in, the burglars broke the knob lock and entered the apartment. The plaintiff alleged that the locks were overcome easily because they were not adequate to safeguard the premises. He presented evidence that the dead bolt recess in the door jamb was not deep enough to accommodate the dead bolt, making the apartment more susceptible to a break-in. Moreover, a locksmith testified that a different type of door handle lock would have provided more protection against break-in. But, he also conceded that another door handle lock would have also given way to the force used to break through Brichacek's door.

The appellate court recognized that part of a landlord's warranty of habitability with his tenants is to provide a workable front door lock to protect a tenant against harm from criminal intrusion. Although a landlord is not an insurer against every conceivable criminal intrusion, the adequacy of the locks installed will be judged on whether the landlord acted reasonably and carefully given the circumstances.

In the present case, the plaintiff had always passed the city housing inspection, and had not received any complaints about the adequacy of the apartment locks from the city housing inspector. There was no evidence to indicate the common areas of the building were particularly accessible to outsiders, or that there had been other criminal incidents in the building or in the area. These facts do not establish that the Hiskey Apartments landlord was negligent or in breach of the warranty of habitability.

The appellate court affirmed the judgment of the trial court in dismissing the case, but did note that the trial court was incorrect in one of its conclusions. The trial court had held that even if the defendant had breached a duty to the plaintiff, it could not be liable because the plaintiff's harm was caused by an intervening third party criminal act. The appellate court stated instead:

> If the landlord had violated his duty to provide a lock on the door, he might have been held liable if he could or should have realized that a break-in could occur. Thus, a break-in is not necessarily a superseding cause that would cut off liability.

PSCLR – MAY 87 (5)

Landlord Is Under No Duty To Protect Tenant From Harm By Co-Tenant Absent Evidence That Landlord Should Have Foreseen Co-Tenant's Criminal Act:

Joyce Williams v. Robert Gorman, and Community Realty Management, Irwin Yeagle, and the Mansions Apartments, *in the Superior Court of New Jersey, Appellate Division, decided December 11, 1986, reported at 520 A.2d 761.*

Joyce Williams had a problem – her upstairs neighbor in the Mansion Apartments' building where she lived. From the time the neighbor, Robert Gorman, moved in the building in August 1978, Williams observed "continual drug traffic" and members of the "Pagan Motorcycle Gang" coming in and out of Gorman's apartment. Williams was also disturbed by violent fights between Gorman and his wife. She did not complain to the landlord until February 1979, after a fight occurred in Gorman's apartment between members of the motorcycle gang. During the fight upstairs, some furniture was apparently thrown, causing a chandelier in Williams' apartment to fall from the ceiling onto a glass table top. Williams also called the police, who told her that they had received other complaints. Although the police came to the building, no one was arrested at that time. Williams claims that after this incident, an employee of Community Realty Management promised that the Gormans would be evicted.

The Gormans were not evicted, and in September of 1980, a gunshot blast from above blew a hole in Williams' ceiling. She was not directly hit by the shot, but was injured by the blast. Her injuries included a lacerated eye and damage to her nose which ultimately required surgery. After she reported this incident to the police, Gorman was arrested and then convicted for discharging a firearm. He was served a prison sentence from November 1980 until late 1982.

Williams also reported this incident to the landlord, and again received assurance that they would take action against Gorman. However, the landlord only sent him a "lease Violation Report" in response.

The third and final incident took place in September 1982, after Gorman was released from prison. He and his wife had a violent fight upstairs which shook some ceramic decorations off the wall in the bedroom of Williams' daughter. The decoration hit her sleeping child in the head, and the injuries required emergency hospital care. Williams called the police, but before they arrived, Gorman tried to break into Williams' apartment and threatened to kill her.

In December 1982, Williams hired an attorney who demanded that Community Realty take action against Gorman. The landlord notified Mrs. Gorman on January 14, 1983, that her husband was barred from the premises. Community Realty also terminated the lease, and evicted the Gormans.

Williams then brought suit against Community Realty, among others, alleging that the landlord was liable for her injuries, her property losses, and lost wages. She did not include medical damages, or any damages resulting from her child's head injuries.

Williams alleged that the landlord and its managers were negligent because they failed to take appropriate action to protect her by evicting Gorman. She also alleged that the defendants breached an implied warranty of habitability and quiet enjoyment, and that the defendants willfully failed to provide adequate security.

After a trial on her allegations of breach of implied warranties, the trial judge granted judgment to Williams and awarded $133, which amount represented an abatement of rent from January 1, 1981, to October 31, 1983. The trial court granted summary judgment to the defendants on the first and third counts.

Williams appealed the summary judgments. The appellate court noted that the issues presented by the case were:

> Whether, as a matter of tort law, (1) Gorman's conduct before the February 1979 incident should have prompted defendants to evict him, and (2) Gorman's cumulative conduct before the September 1980 incident rendered negligent defendant's failure to evict him before that date.

The appellate court noted that there are only limited qualifications to the general principle that the landlord/tenant relationship does not impose a duty on the landlord to protect tenants against the criminal acts of a third party. The "crucial factor" in determining a landlord's liability for third-party criminal injury to a tenant lies in whether the harm was foreseeable to the landlord. Furthermore, the implied warranty of habitability requires a landlord "furnish reasonable safeguards to protect tenants from foreseeable criminal activity on the premises" in the common areas.

Applying these standards to the present case, the appellate court found that prior to the February 1979 incident, the landlord did not have any reason to know of Gorman's "destructive tendencies." Because this first incident was not foreseeable, the landlord was not under a duty to evict the Gormans prior to that time.

When the second incident occurred, some 18 months later, the landlord did not have any notice that Gorman would or could deliberately assault a fellow tenant with a shotgun. The prior incident was not criminal in nature and did not warn the landlord that the second, criminal incident could occur. Therefore, the failure of the landlords to evict the Gormans before the shotgun incident did not "unreasonably enhance the risk of harm to plaintiff."

Finally, the appellate court noted that the landlord's implied warranty of habitability required the landlord to maintain proper and adequate security in the common areas of the apartment building, and not in the interior of each apartment. To recover damages other than rent abatement from the landlord for the criminal conduct of a fellow tenant or any third party, the plaintiff would first have to establish the the conduct was foreseeable by the landlord and Williams failed to do this in the present case.

The appellate court therefore affirmed the judgment of the trial court in granting summary judgment to the defendants on the plaintiff's allegations of negligence and failure to provide adequate security.

PSCLR—MAY 87 (6)

Landlord Who Had Knowledge Of Criminal Activities In Complex And Of Defective Door Lock In Apartment Is Liable To Tenant Raped By Intruder Gaining Access Thereby:

Rhonda Lynn Lay v. Alvin Dworman; Daltex Capital Corporation, a Texas corporation; Willow Creek Condominiums, Inc., an Oklahoma corporation; Willow Creek I Neighborhood Association, Inc., an Oklahoma corporation; Willow Creek II Neighborhood Association, Inc., an Oklahoma corporation, *No. 60770, in the Supreme Court of Oklahoma, decided December 16, 1986, as corrected on denial of rehearing February 2, 1987, reported at 732 P.2d 455.*

Rhonda Lynn Lay was sexually assaulted by an unknown intruder while she was in her apartment. She filed suit against the owners of the complex where she lived, claiming that they were negligent and that they had breached their warranty of habitability in the premises. The trial judge dismissed the case and the plaintiff appealed to the Oklahoma Court of Appeals, which affirmed the dismissal. The plaintiff then asked the Supreme Court of Oklahoma to review the trial court's and appellate court's decisions.

The state supreme court granted Lay's request to review, and after doing so, set aside the lower appellate court's opinion and reversed a part of the trial court's decision, thereby reinstating her negligence claim against the apartment complex owners.

In her complaint Lay alleged that she had sublet an apartment in the defendants' complex and that the defendants were aware of criminal activities on the premises, including other incidents of rape in the apartment complex. She also alleged that she had complained of a defective lock on the sliding glass doors to the apartments, but that the defendants negligently failed to repair the locks, and thereby failed to protect her by providing a secure residence.

She further alleged that the result of the defendants' negligence was that an unknown intruder gained access into her residence and raped her. She sought damages for her personal injuries plus punitive damages on the basis that the defendants were guilty of gross negligence.

The court first examined the legal issue of the extent of a landlord's duty to protect a tenant from third party criminal misconduct. The plaintiff argued that the court should adopt the expanded view of the landlord's duty as set forth in the D.C. case of *Kline v. 1500 Massachusetts Avenue Apartment Corp.*, 439 F.2d 477 (1970). In that case, the United States Court of Appeals for the District of Columbia acknowledged that the general rule regarding landlord/tenant relationships had traditionally been that an individual had no duty to protect another from the criminal misconduct of a third party. That court recognized that the reasons that the rule was applied to landlords was that courts were reluctant to interfere with traditional concepts of landlord/tenant relations; an intentional criminal act by a third person would be an intervening cause of harm; the difficulty in determining foreseeability; the fact that the landlord would have to meet vague standards; the imposition of the duty to intervene to protect crime would have economic consequences to landlord; and the public policy of our society has allocated the duty of protecting individuals to the government rather than to private industry and real estate owners.

However, the D.C. court in the *Kline* decision went on to recognize that the application of the general rule, that a private individual has no duty to protect another from a third party criminal attack, is problematic when applied to modern day urban conditions. The court noted that while the landlord should not be held to be an insurer of his tenants' safety, neither should he be allowed to take the position of a mere bystander. Furthermore, the landlord has reason to believe that criminal conduct may take place again on the property because he has in fact been put on notice of repeated criminal conduct on the premises and knows that the misconduct has taken place on that area of the premises which is exclusively within his control (that is, the common areas). If then he has the power and opportunity to reduce the risk of or to prevent the criminal conduct from occurring again, it is not then unfair to place the legal burden upon the landlord to take reasonable measures to reduce the foreseeable risks to his tenants.

The District of Columbia appellate court therefore concluded in the *Kline* case that the law does place a duty on landlords to protect tenants from foreseeable criminal misconduct on the premises.

After reviewing the *Kline* decision, the Oklahoma Supreme Court noted that that case had dealt with an assault on a tenant in one of the common areas of an apartment building, and not inside one of the apartment units. The Oklahoma Supreme Court felt that there was no need to expand the potential liability of the landlords in the present situation to the magnitude of that imposed by the *Kline* case. The court further noted that Oklahoma law already placed upon the landlord a duty to use ordinary care to maintain the common areas of leased premises in a safe condition. Accordingly, the court applied Oklahoma law to the present law suit and found that the duty owed by the defendants was the duty to use

> reasonable care to maintain the common areas of the premises in such a manner to insure that the likelihood of criminal activity is not unreasonably enhanced by the condition of those common premises.

The court then said that where the property leased to the tenant is unsafe due to defective or ineffective locks or materials, then the landlord also has a duty to provide modifications and/or repairs upon notification of such defects by the tenants. The court ascribed the existence of this duty to the contract existing between the landlord and tenant, and from the implication and understanding that the landlord had to provide services under the contract diligently.

The element of foreseeability can be established either by showing a history of criminal misconduct on the property in question or merely by showing that the nature of the defect itself such as to make the subsequent criminal misconduct foreseeable as a matter of fact.

There is of course a third area of liability which can sometimes be established in landlord/tenant relationships and that is in the situation governed by the law which provides:

> One who undertakes, gratuitously or for consideration, to render services to another which he should recognize as necessary for the protection of the other's person or things, is subject to liability to the other for physical harm resulting from his failure to exercise reasonable care to perform his undertaking if his failure to exercise such care increases the risk of such harm, or if the harm is suffered because of the other's reliance upon the undertaking.

In other words, under the general principles of applicable law there is no duty to undertake the protection of another. However, if a person does undertake to protect another, whether or not for payment, then he must exercise due care in doing so. He will be held liable if his efforts actually increase the risk of harm to the person he is attempting to protect or if the person is actually injured because he relied upon the offered protection.

Based upon this analysis of the law, the Oklahoma Supreme Court declined to follow the rationale set forth by the D.C. court in the *Kline* decision, and declined to expand the landlord's duties to that extent. The court did, however, note that its own prior decisions were in conflict with its present analysis of the applicable law and to that extent disapproved of its previous holdings in those earlier cases.

Accordingly, the court ruled that the Oklahoma appellate court and trial court had erred in dismissing the plaintiff's complaint based on the negligence theory.

The court then reviewed the plaintiff's effort to recover for her injuries under a theory of breach of warranty. To sustain a recovery under this theory, she claimed in her lawsuit that the defendants had warranted to the public and to the plaintiff in particular that the apartment complex in question did have security and was secure, that she herself had relied on these representations when she sublet the apartment unit, that the defendants had breached the warranties by failing to provide a safe and secure unit for her to live in, by failing to provide sufficient security personnel in the area, and by failing to provide sufficient lighting in the area, all of which Lay alleged constituted a breach of warranty by the defendants, resulting in damages and injury to herself.

The court, however, rejected this approach, noting that to accept the plaintiff's theory would in essence establish the landlord as an insurer of the tenants' safety. The court also rejected the notion that a statement by the landlord concerning the relative security of a complex can, in and of itself, be the basis for the landlord's liability for subsequent criminal assaults on the premises, particularly without a showing of a causal relationship between the alleged breach of warranty and the injury. Here, Lay was unable to prove how the existence of a breach of warranty, if any, had resulted in her injuries. She merely asked that the conclusion be drawn that the breach was responsible for her injuries.

Accordingly, the case was returned to the trial court and the plaintiff will be permitted to pursue her cause of action based upon a negligence theory, but not upon a breach of warranty theory.

COMMENTARY:

Another case referred to by the Oklahoma Supreme Court in reaching the above decision was an older case out of California which also recognized the inherent problems of dealing with these issues:

> *In this day of an inordinate volume of criminal activity, there are a myriad of 'security devices' available to the public, including the hiring of armed guards. No one really knows why people commit crime, hence no one really knows what is 'adequate' deterrence in any given situation. While bright lights may*

> *deter some, they will not deter all. Some persons cannot be deterred by anything short of impenetrable walls and armed guards.*

It would be intolerable and grossly unfair to permit a lay jury, after the fact, to determine in any case that security measures were 'inadequate,' especially in light of the fact that the decision would always be rendered in a case where the security had in fact proved to be inadequate.

PSCLR – JUN-JUL 87 (8)

Tenant Injured In Apartment Fire May Recover From Landlord For Negligence In Storing Combustible Waste On Premises:

LaSheryl Williams v. Jerry Johns, *No. 87070, in the Court of Appeals of Michigan, decided January 6, 1987, released for publication April 16, 1987, reported at 403 N.W.2d 516.*

The defendant Jerry Johns owned an apartment building in Detroit, Michigan, which caught fire late one evening. The fire forced LaSheryl Williams, one of his tenants, to jump from a third floor window to escape the flames and she incurred serious personal injuries as a result of the fall. She initiated this lawsuit against Johns, claiming that he was responsible for her injuries due to his negligence in creating an unsafe and hazardous condition for apartment tenants by storing combustible waste material on the premises, thereby causing the fire to spread rapidly throughout the apartment building. She also alleged that the defendant violated two city ordinance regarding the storage of combustible waste material inside buildings.

The two arson investigators who investigated the fire in question both reached the conclusion that the fire had begun in trash stored at the bottom of the stairwell near the rear exit to the building. The fire progressed from the interior of the stairwell itself along the halls of the building. The stairwell acted as a chimney and vented the fire up to the other floors, according to the investigators. When the fire reached the top level of the third floor, it progressed down the hallways and one tenant described it as being "very intense." The fire itself began, according to the arson investigators, when an unknown person deliberately set afire the waste material at the bottom of the stairwell.

Johns moved for a dismissal of the case, arguing that even if everything the plaintiff Williams alleged was true, there was no negligence on the part of Johns which could be considered the cause of the plaintiff's injuries because the fire had been purposely set by an unknown person.

In response, however, Williams' position was that the apartment owner's act of permitting a fuel supply (the combustible waste material) to be stored in the apartment building stairwell in violation of two city ordinances constituted negligence on his part, regardless of the origin of the fire itself. She also tendered the testimony of another expert arson investigator who testified that the origin of the fire could have been either accidental or intentional.

The trial judge agreed with the apartment building owner and granted his motion for summary judgment and dismissed the case. Williams appealed.

On appeal, the Michigan Court of Appeals first noted the ordinances of the city of Detroit:

> Combustible waste or refuse shall be properly stored or disposed of at the end of each working day ... and whenever necessary to prevent unsafe conditions.

The fire prevention code defines a fire hazard as:

> Any situation, process, material or condition which on the basis of applicable data, may cause a fire or explosion or provide a ready fuel supply to augment the spread or intensity of the fire or explosion and which poses a threat to life or property of others.

The appellate court noted that a landlord "owes a duty to his tenants to protect them from unreasonable risks of harm resulting from foreseeable activities occurring within the common areas of the landlord's premises" and determined that the relevant question was whether the landlord's storage of combustible material in the stairwell constituted negligence on his part. The court also concluded that such was a question of fact to be decided by a jury and not by a judge alone on a motion for summary judgment, as the trial court had here done. The appellate court specifically noted that the trial judge had committed error by concluding that since the fire had been purposely set, the plaintiff would be unable to prove that the landlord's negligent conduct, if any, had caused the injuries to the tenant. The appellate court in this regard noted that it is not necessary for a plaintiff to prove that the defendant's negligence was the only causal factor resulting in harm to the plaintiff; it is sufficient to establish liability if the plaintiff can prove that the defendant's negligence was a substantial factor in causing the harm incurred. The fact that there was an intervening cause, such as the arsonist using the waste material to set the fire, does not constitute an absolute bar to the landlord's liability if the intervening act is foreseeable even though the act itself may be negligent or even criminal. And, whether any intervening negligent or criminal act constitutes a preceding cause, which would relieve the defendant from liability, is also a question of fact to be decided by the jury and not by the judge alone on motion for summary judgment. The plaintiff's evidence in the subject case could be determined sufficient by a jury to prove that the intervening act of the unknown third party could have been reasonably foreseen by the landlord. Further, the landlord knew or should have known that someone could have either accidentally or intentionally started a fire in the waste material if it were not disposed of properly and in accordance with the city ordinances. Then, as a consequence, the fire could spread rapidly and easily to the upper floors and the jury could determine that the landlord's negligence was the proximate cause of the plaintiff's injuries in her jump to escape the flames. Therefore, the jury should have the opportunity to resolve these issues and the appellate court returned the case to the lower court for trial.

COMMENTARY:

If your security operations include the protection and patrolling of buildings of almost any type or description, including apartments, hospitals, schools, office buildings, offices, and the like, another area of major concern must be not only the protection of such property from criminal activity, but also the protection of same from negligent activity which could result in loss. Furthermore, the professional security manager must be absolutely up to date on all local ordinances and amendments thereto, concerning the elimination of fire hazards and the disposal of hazardous waste materials.

The above case is clear authority for the fact that violation of such ordinances will be considered by the courts in determining whether the owners of property have or have not carried out their duties to the tenants or invitees on the premises to protect them from foreseeable dangers. Do your security officer training programs include instruction regarding fire prevention and the importance of hazardous waste disposal? Are they also trained to report suspicions regarding the improper storage of combustible material? If not, they should be, and perhaps some revisions and additional training to your security officers' program is warranted.

PSCLR—OCT 87 (1)

To Sustain Criminal Conviction For Negligent Homicide, State Must Prove Defendant Grossly Deviated From Standard of Conduct He Was Required To Meet, Thereby Causing Death:

State of Connecticut v. Gordon L. White, *No. 12749, in the Supreme Court of Connecticut, decided July 14, 1987, reported at 528 A.2d 811.*

This case involves the criminal prosecution of the defendant Gordon L. White who was charged with negligent homicide for causing the deaths of three individuals, and with failing to provide a smoke detector in a residential building which he owned and which was occupied by the deceased tenants.

As reported in the appellate court's opinion, the jury could have reasonably determined the following to be the facts from the evidence: White owned a residential building in New Britain, Connecticut, occupied by three families. In the early morning hours of Christmas day, 1982, the occupant of a first floor apartment awakened to the smell of smoke and immediately woke his family and got them outside. He called the fire department, who arrived on the scene "almost immediately" and began fighting the blaze. The firemen found the bodies of Mary Ann Jones and her two young children, Lindsay and Brandy on the second floor. It was later determined that the three of them had asphyxiated from smoke inhalation. A subsequent investigation determined that the fire apparently had been started by an electrical overload on a second floor wall outlet where a heater had been plugged in. It was estimated that the overload began smoldering for about three hours before the fire started. The fire itself was slow burning, and as a result, the second floor dwelling was covered with soot and filled with thick smoke.

There were no smoke detectors installed anywhere in the building.

There was also evidence presented at the trial about the cost, installation, effectiveness, and operation of a smoke detector. There was also evidence that, if smoke detectors had been installed in their building, the victims would have been alerted to the fire in enough time to escape. Furthermore, the state presented evidence that White owned another apartment building in East Hartford, Connecticut, and that he had been warned by the East Hartford fire marshal (only six months before the fatal fire) to install smoke detectors in the East Hartford building to comply with the fire code. White did in fact install detectors in the East Hartford apartment building but did not install them in the New Britain apartment building.

After the defendant's motion for a judgment of acquittal was denied by the trial judge, he was in fact convicted on all six counts (three counts of negligent homicide and three counts of failing to provide smoke detectors) and sentenced to one year in jail, suspended, with three years probation plus a fine of $4500. The defendant appealed his conviction and sentence.

The appellate court rejected White's argument on appeal that the fire safety code and the statute authorizing the establishment of the code were unconstitutionally vague and found that White had been "fairly warned that he had a duty to install smoke detectors in his New Britain building." However, the appellate court did agree with White's argument that the fire marshal's fire safety code regulations exceeded the scope of authority conferred by the enabling statute authorizing the fire marshal to promulgate a fire safety code.

In reaching this conclusion the appellate court examined the state fire safety code provisions and applicable statutes, including the legislative history behind them.

(Although the legislature has the exclusive power to make law, it may pass a law for a certain purpose and then expressly authorize a government agency to promulgate rules and regulations for the operation of the law. The law passed by the legislature must state legislative policy, or establish standards with which the government agency must comply. If the legislature fails to set forth the limits of the power given to the agency, then its delegation of authority is null and void. Furthermore, the agency to which the power is granted must act according to and within the limitations of the authority granted to it.)

The court first reviewed the 1947 Connecticut law, which provided:

> 'The state fire marshal shall establish a fire safety code and at any time may amend the same. The regulations in said code shall provide for reasonable safety from fire, smoke, and panic therefrom, in all buildings except private dwellings occupied by one or two families, and upon all premises except those used for manufacturing.' The purpose of the statute was to give the fire marshal the ability to enact reasonable minimum requirements for safety in new and existing buildings.

This act remained unchanged for 30 years, but in 1976, the legislature amended it by requiring that the regulations issued by the fire marshal:

> shall include provision for smoke detection systems in residential buildings designed to be occupied by two or more families for which a building permit is issued on or after October 1, 1976.

Furthermore, the fire marshal then amended the fire safety code to require that smoke detectors be installed in

> each guest room, suite, or sleeping area of hotels, motels, lodging or rooming houses, and dormitories and in each dwelling unit within apartment houses and one- and two-family dwellings ....

The appellate court concluded that the legislative intent behind the 1976 act was not entirely clear. Since the <u>statute</u> directed that the fire code should provide regulations for the installation of smoke detectors in buildings for two or more families, "for which a building permit has been issued on or after October 1, 1976," the fire safety <u>code</u> exceeded the scope of its statutory authority by attempting to require the installation of smoke detectors in <u>all</u> buildings designed for two or more families. Hence, the court concluded that the defendant White could not be legally convicted of violating the fire safety code requiring the installation of smoke detectors without proof that a building permit had been issued for the New Britain building on or after October 1, 1976. Since the state failed to present any such evidence at the trial, the appellate court ruled that the defendant's convictions for these violations would have to be reversed and set aside.

The appellate court also reversed White's convictions for the three counts of criminally negligent homicide. In order to be found guilty of criminally negligent homicide, the defendant must have caused death while failing to perceive a substantial risk that the death will occur; the state would have to prove three elements:

> (a) that there was a duty or obligation on the defendant required by law, to conform to a certain standard of behavior;
>
> (b) that the defendant on trial had grossly deviated from such standard of behavior;
>
> (c) his deviation was the direct cause of the resulting death.

Since the appellate court had determined that White was apparently not obligated under the fire safety code to install smoke detectors, the court had to conclude that he could not be convicted of criminally negligent homicide for having failed to do so. Accordingly, the appellate court reversed the defendant's convictions on all six counts.

COMMENTARY:

The above case, of course, is not a civil suit, claiming damages for negligently causing the deaths of the victims, but rather is a criminal prosecution by the state for negligent homicide of the victims. In a civil suit, the plaintiffs would be seeking monetary damages, whereas in this criminal prosecution the state was seeking (and originally obtained) a criminal conviction and criminal sentence involving jail, probation, and a fine.

If this had been the evidence presented at the trial of a civil action for the monetary damages for negligently causing the deaths of the victims, what do you think the result would have been? To answer this question, remember that to win a civil negligence lawsuit, a plaintiff must prove:

> *(1) some duty or obligation to conform to a particular standard to conduct toward another;*
>
> *(2) a breach of that duty of care;*
>
> *(3) that such breach was the direct and proximate cause of the resulting injury or loss; and*
>
> *(4) that there was in fact a resulting injury or loss.*

Even if, as the appellate court found here, there was no statutory duty on the part of the defendant landlord to install the smoke detectors, did such a duty on his part arise from some other basis? The duty or obligation to conform to a certain standard of conduct may arise

> *either from a contract, a statute, or from circumstances under which a reasonable person knew or should have known that harm of the general nature as that suffered was likely to result from his act or his failure to act.*

It is also worth noting that the appellate court in the above decision stated:

> *Had the original statute [from 1947] been in effect during this case, there would be little doubt that the challenged fire safety code regulations would be valid.*

In other words if the legislature had not amended the act in 1976 to provide for a fire code provision for smoke detection systems in residential buildings occupied by two or more families, "for which a building permit is issued on or after October 1, 1976," and if the fire marshal had simply included in the fire safety code the same regulation requiring the installation of smoke detectors pursuant to the original enabling statute of 1947, then the defendant's convictions would have been affirmed.

For the most part, and as a general rule, it would be best not to rely on such technical interpretations of the law in order to avoid civil, and perhaps criminal, liability. The prudent course for a security director is to ensure that all fire safety code regulations are complied with (or that written exemptions are obtained where appropriate and necessary). Otherwise, as can be seen from the foregoing case, the security director may well be risking not only civil liability for monetary damages for himself and his employers/clients, but also the possibility of criminal prosecution and conviction for violation of the fire code itself as well as for negligent homicide and/or destruction of property for the resulting loss.

PSCLR – NOV-DEC 87 (5)

Where Local Ordinance Requiring Dead-bolt Locks Conflicts With State Statute Concerning Same, Landlord's Violation Of Ordinance Will Not Support Claim Against Him By Tenant Raped By Intruder:

Klingbeil Management Group Company, et al. v. Barbara L. Vito, *Record No. 840404, in the Supreme Court of Virginia, decided June 12, 1987, reported at 357 S.E.2d 200.*

The issue presented to the appellate court for decision in this case was whether or not a landlord of an apartment building could be held liable to a tenant for injuries she received at the hands of an intruder, and whether the landlord's liability can be based upon his failure to comply with a local dead-bolt ordinance requiring the installation of dead-bolt locks.

The factual situation giving rise to this lawsuit is as follows: The tenant Barbara L. Vito was raped by an unknown intruder inside her apartment in Arlington, Virginia. She sued the owners and manager of the apartment complex, hereinafter referred to as the landlord. She sued for compensatory and punitive damages for the injuries she suffered, based on her allegations of negligent and wanton misconduct on the part of the landlord for its failure to provide adequate security, including "proper security procedures, proper exterior lighting, and proper locks on the doors to the buildings."

By the time the case reached trial, certain issues had been eliminated. The case was presented to the jury for it to decide whether the landlord had been negligent for failing to install a dead-bolt lock on the front to Vito's apartment. The jury returned a verdict in her favor in the amount of $150,000 compensatory damages, and the trial judge granted judgment in her favor based upon this verdict. The landlord appealed.

The record of evidence in the case showed that Vito lived on a ground floor and on the date in question, she returned home around midnight. The front door to her apartment did not have a dead-bolt lock, but had a standard "button" lock. When this lock was opened with a key from the outside, the button on the inside of the door remained in the locked position and opening the door did not move the button to the unlocked position. After Vito used her key to unlock her front door, and had entered her apartment and closed the door behind her (thereby leaving the door locked). She had been inside for some time when she heard a noise and returned to her front door at which time she noticed that the button on the inside of the door was in the "out," or unlocked, position. She was then grabbed from behind by an unseen intruder while she was still inside her apartment, and was assaulted. The intruder was not apprehended.

The record did not show with any degree of certainty the manner by which the intruder had obtained access to Vito's apartment. Subsequent investigation did not disclose any evidence of a forced entry.

There was, however, some evidence showing that the front door lock could be easily opened by slipping a plastic card, such a credit card, between the door and the jamb, and the evidence was sufficient for the jury to conclude that the intruder had entered the apartment by way of the locked front door.

On appeal, the Virginia Supreme Court first ruled that the trial judge had committed error by allowing the jury to impose liability on the landlord for injuries caused by the criminal act of the intruder. The appellate court held that a landlord has no duty to protect a tenant from criminal acts by third parties. A person would be held to owe a duty to control the conduct of a third person in order to prevent harm to another only if there was some "special relationship" between the plaintiff and the defendant such as between a common carrier and its passengers, a business owner and its invitees and an innkeeper and its guests. The Virginia Supreme Court had previously ruled (in 1974) that there was no such "special relationship" between a landlord and his tenant, and the court in the present case decided to follow that precedent ruling.

The second issue presented on appeal by the landlord was whether liability could be based upon the landlord's failure to comply with the local dead-bolt ordinance. In 1972, the "Housing Standards Ordinance of Arlington County, Virginia" was enacted, requiring, among other things, that no dwelling unit in multiple dwelling building could be leased unless the exterior doors were equipped with a "vertical bolt deadlock or a bolt with not less than a 5/8 of an inch minimum throw." At the trial, the judge had instructed the jury that they could find the landlord liable to the plaintiff if the landlord had violated this requirement set out in the local ordinance.

The appellate court found that this instruction to the jury was erroneous because, as argued by the landlord, a state statute enacted in 1974 preempted the local ordinance. That act provided that a local government could pass an ordinance requiring a landlord "who rents five or more dwelling units in any one building" to install, upon the written request of the tenant, dead-bolt locks. An examination of the record revealed that Vito may have orally asked the landlord to install a dead-bolt lock on her apartment door, but she had made no written request to that effect. In addition, the lease which she had signed specifically referred to the state statute and provided that the lease was subject to the provisions of that Act. And, since the same subject matter was addressed in both the ordinance and the state statute, as a matter of law the statute's provisions must prevail.

Therefore, the appellate court ruled that the trial judge should have held that the landlord had no liability to the tenant under the facts and circumstances of this case. The judgment in favor of the tenant was reversed and final judgment was entered in favor of the landlord.

PSCLR–FEB 88 (6)

Landlord Liable For Failing To Warn Tenant Of Third Parties' Violation Of Tenant's Privacy, Where Similar Invasions Had Been Reported To Landlord:

New Summit Associates Limited Partnership et al. v. Sylvia Nistle, *No. 466, Sept. Term, 1987, in the Court of Special Appeals of Maryland, decided December 7, 1987, reported at 533 A.2d 1350.*

This is an action by the plaintiff/tenant Sylvia Nistle for both punitive and compensatory damages, against her former landlord, the defendant New Summit Associates Limited Partnership, and its agent, the defendant Dreyfuss Brothers, Inc., Nistle claimed in her lawsuit that her former landlord New Summit had allowed an invasion of her privacy to occur during the course of the renovation of the apartment immediately adjacent to her apartment. At the trial of the case, she recovered a verdict against the defendants in the amount of $20,000 compensatory damages, plus $90,000 punitive damages against New Summit, plus $60,000 punitive damages against Dreyfuss Brothers. The defendants appealed.

The record of the evidence presented at the trial revealed that Nistle was a 25-year old divorced woman when she rented an apartment at the New Summit apartment complex in Rockville, Maryland. New Summit Associates Limited Partnership owned the apartments, and hired Dreyfuss Brothers, Inc., to manage them. During the year Nistle moved into her apartment, several apartments in the complex were being renovated as each apartment was vacated by its tenants. Artery Organization, Inc., a corporation affiliated with New Summit, was the general contractor for the rehabilitation project.

A number of acts of vandalism and thefts took place in the various vacant apartments while they were undergoing renovation; in spite of this, the locks on the apartment doors were frequently removed while the work was going on, and the apartments were not secured while they were being renovated.

The bathroom in Nistle's apartment shared a common wall with the bathroom in the adjacent residence, and bathroom mirrors in each apartment covered an open space in the common wall. Renovation of the residence immediately adjacent to Nistle's apartment began shortly after she moved in, and the workers removed the wall mirror from the adjacent apartment's common wall in the bathroom, exposing the rear side of Nistle's bathroom mirror.

Nistle subsequently found two marks which appeared to be scratches on her mirror, but when she attempted to clean the mirror, she was not able to remove the marks. She reported them to the Dreyfuss Brothers office. She did not know at the time that the back of her mirror was exposed by virtue of the renovation in the adjacent apartment.

A few weeks later a neighbor reported to Nistle that he had been inside the vacant apartment being renovated next to hers, and he found that the interior of Nistle's bathroom could be viewed through two circular holes which had been scratched in the back of her bathroom mirror. Nistle went into the adjacent apartment, which was unlocked, and saw for herself that the interior of her bathroom was visible through the scratches. This realization caused her great anxiety, and she reported the situation to the police later that same day. The police investigated but took no further action in the matter. The next day, Nistle notified the resident manager, Karen Geier-Smith, asking her to inquire into an invasion of the privacy of her bathroom, although she did not specifically mention the mirror scratches. The resident manager examined the adjacent apartment, discovered the scratches on the mirror, and apologized to Nistle, explaining that the management was aware of similar situations which had occurred in other apartments during the course of restoration. She also told Nistle that the management was afraid that warning the workers not to scratch the mirrors would only exacerbate the problem; instead, the Dreyfuss Brothers company had assigned an individual to paint the rear of any mirrors which he found to have been scratched.

At the time Nistle made her complaint of the scratches on her mirror, the Dreyfuss Brothers company were aware that at least two other bathroom mirrors in the apartment complex had been likewise scratched. These other incidents had been discussed by personnel from New Summit, Dreyfuss, and Artery.

The record further showed that Nistle immediately vacated her apartment, and moved into her parents' residence, experiencing nausea, diarrhea, and insomnia for several weeks after she found that someone may have watched her through the scratches. She subsequently had to undergo psychiatric counselling.

The defendants/appellants argued first on their appeal that they should not be held liable for any invasion of Nistle's privacy because (1) Nistle failed to prove that her privacy was in fact invaded, and (2) any invasion of her privacy was the act of unknown third parties. In response to this argument, the appellate court found that Nistle was not required to prove that any particular person did in fact observe her while she used her bathroom facilities, and the court concluded that her privacy had undoubtedly been invaded by the unknown person who had scratched the peepholes in the rear of her bathroom mirror. However, the court ruled that the defendants could not be held liable for any invasion of Nistle's privacy because the law requires that, for liability to attach, there must be an intentional intrusion by the defendant upon the plaintiff's privacy, and Nistle had not proved that her privacy had been invaded by any agent, servant, or employee of either of the defendants. Accordingly, without evidence of intentional intrusion by an agent of the defendants, there can be no legal basis upon which the defendants can be held liable for invading the plaintiff's privacy.

With respect to the plaintiff's claim that the defendants were negligent in failing to warn her, the defendants claimed that there was insufficient evidence to support this claim because the law imposes no special duty on a landlord to protect its tenants against criminal acts of third parties. But, the court noted, the law does impose a duty on a landlord to warn a tenant of latent defects in the rented property. The court went on to conclude that the peepholes scratched in the mirror were a latent defect, and this defect in the rented property was an unreasonable risk to the plaintiff, which defect and risk the defendants should have known was not discernible by the tenant. The court also found that it was entirely foreseeable that Nistle would suffer severe emotional shock when she found the peepholes. Previous reports of similar incidents had put the defendants on notice of the defect, and they owed a duty to Nistle to disclose the defective condition of her apartment to her once they became aware of it. Since they failed to do so, they are liable for her resulting injuries. The court therefore concluded that Nistle's claim of negligence on the part of the defendants was properly submitted to the jury for its consideration.

The court then examined the defendant's/appellant's argument that the punitive damages should be set aside because the plaintiff had not shown that the defendants had acted with actual malice toward her. And, here, the record of evidence does not show that they acted with any evil motive in an effort to willfully or deliberately harm Nistle. While they failed to advise her of the latent defect in her apartment, this failure on their part does not support the conclusion that the defendants were acting out of evil or rancorous motive, or out of any deliberate intention to harm the tenant.

PSCLR–MAR 88 (4)

Negligence Is Not Proven By Landlord's Violation Of City Ordinance Which Requires Certification Of Alarm Systems And Prohibits Use Of Uncertified Equipment, But Fails To Specify Procedures For Installation, Inspection, Or Defects To Be Avoided:

Dr. Saundra M. Hunter v. Joanne B. Whitney, wife of/and David L. Whitney and Alarm Device Manufacturing Company (ADEMCO), *No. CA-7261, in the Court of Appeal of Louisiana, Fourth Circuit, decided December 15, 1987, reported at 517 So. 2d 1249.*

Dr. Saundra M. Hunter leased her residence on Esplanade Avenue in New Orleans from David and Joanne Whitney. The Whitneys had renovated the house prior to leasing it to Dr. Hunter, and David Whitney had personally installed an alarm system in the house, protecting all doors and windows to the house.

A couple of weeks prior to the assault, Dr. Hunter found that she was unable to "arm" the alarm system, and she sent a note to the Whitneys along with her rent around the first of November 1982. On November 10, David Whitney inspected the alarm system, and discovered that the wiring was broken where it had been gnawed through. He told Dr. Hunter that he could not fix the system because he did not have the necessary equipment with him, but that he would return in a few days to do so. Dr. Hunter reportedly replied, "No sweat."

On that same night, an armed assailant entered her home and criminally assaulted and raped Dr. Hunter. Police were not able to definitely establish the point of entry, but they did find that the kitchen door at the rear of the house was unlocked and may have been the point of entry.

Dr. Hunter brought suit against the Whitneys, seeking to obtain money damages for the injuries and mental suffering she had incurred as a result of the assault. The suit was tried before a jury, whose verdict was in favor of the Whitneys, finding them free of any negligence and not at fault in causing the damages. Dr. Hunter then filed this appeal.

As grounds for her appeal, Dr. Hunter argued that the trial judge had erred by refusing to admit into evidence a New Orleans ordinance, which in the relevant part required the certification of alarm systems, and prohibited the use of uncertified alarm equipment. The ordinance also provided that the owners of then-existing alarm systems would have to obtain a certification of their systems by a licensed person pursuant to the ordinance. The plaintiff contended that the judge's refusal to admit the ordinance into evidence prevented her from establishing that the Whitneys were guilty of negligence by being in violation of the ordinance.

The appellate court disagreed with the plaintiff's argument, stating that while the ordinance does provide certain criteria for the certification of alarm systems, it nevertheless failed to provide any procedures for the installation of alarm systems, or the inspection of such systems, or to detail what defects, if any, an inspection of the system should detect or avoid. The ordinance also failed to provide any standards for determining whether an alarm system had been improperly installed: "In short, the ordinance is so vague and incomplete that proving a violation of its provisions would have little if any probative value on the issue of negligence."

Therefore, the appellate court ruled that the trial judge was correct in his decision to refuse to admit the ordinance for consideration by the trial jury. Therefore, the trial judge's judgment in favor of the defendants, based on the jury verdict, was affirmed.

COMMENTARY:

Another issue dealt with by the appellate court in the above case was the admission of certain testimony by the plaintiff's expert witness who was qualified in the field of alarm systems and the installation of such systems.

The expert, Jerry Martinez, had inspected the alarm system at the rented residence in preparation for his trial testimony. The appellate court noted that the purpose of his inspection and testimony was to try to show that the landlord had failed to use the proper type of wire when he installed the alarm system.

During his testimony, the expert stated that the industry as a whole had recommended the use of 24-gauge wire, instead of 22-gauge wire as was used by the landlord. However, the defendants' objection was sustained because the expert had not testified concerning his knowledge of such standards in the industry, nor had he given the factual basis for his expert opinion concerning the gauge wire which the alarm industry recommends.

This emphasizes—for those readers who, from time to time, act as consultants to lawyers, and as expert witnesses—the need to remember that merely having an opinion concerning a subject is not sufficient under the law for that opinion to carry evidentiary weight. One must have a factual basis which support opinions, and must be prepared to state those facts when called up to testify.

PSCLR—MAR 88 (5)

Absent Evidence Of Prior, Actual Child Molestation, Landlord Has No Duty To Prevent Child Residents From Being Molested:

O.H., a minor who sues by and through his guardian and next friend, T.E.B. v. Ballard Realty Company and Bayou Bend, Ltd., *86-166, in the Supreme Court of Alabama, decided August 21, 1987, rehearing denied November 20, 1987, reported at 516 So. 2d 519.*

The minor plaintiff in this case (called "O.H." in the opinion) was eleven years old at the time of the alleged incident. His lawsuit contained allegations that O.H. was assaulted and sexually molested near the Bayou Bend apartments, and he further claimed that the assault was performed by a former resident of the Bayou Bend apartments. The suit claimed that the former resident had led O.H., who is mildly mentally retarded, from the swimming pool area of the apartment complex to a nearby wooded area where the alleged assault took place.

At the time of the assault, O.H. was a resident of the apartment complex, living with his half-sister's father. The suit was brought against Ballard Realty, a management company with the responsibility of operating the Bayou Bend apartments at the time in question, and also against the owners of the apartment complex itself. Various claims were included in the complaint, including negligence, wantonness, and fraud in the maintenance of security by the defendants for the residents. After reviewing the allegations and supporting affidavits from all parties, the trial court dismissed the claim and the plaintiffs appealed.

On appeal, the supreme court examined the issue as to whether, under the facts presented, there was any duty on the part of the defendants to protect the minor O.H. from the alleged sexual attack on the defendants' premises. In examining the applicable law, the court quoted from the previous case of *Ortell v. Spencer Companies* (reported in the February 1986 **PSCLR**, case nine, page six) where the court had ruled as follows:

> This court has recognized that a duty may be imposed on a store owner to take reasonable precautions to protect invitees from criminal attack in the exceptional case where the store owner possessed actual or constructive knowledge that criminal activity which could endanger an invitee was a probability. *Henley v. Pizitz Realty Co.*, ... [reported in the January 1985 **PSCLR**, case nine, page nine]. Nevertheless, we have found that 'it is difficult to impose liability on one person for an intentional criminal act committed by a third person.' ... This court has not yet decided a case whose facts command such a duty, ... *Latham v. Aronov Realty Co.*, ... [reported in the August 1984 **PSCLR**, case eight, page seven]; *Berdeaux v. City National Bank of Birmingham* ... [reported in the May 1983 **PSCLR**, case three, page three].

The minor O.H. here argued that the apartment owners and operators did have knowledge of the danger which existed and also that the defendants had made certain representations as to the degree and level of security at the apartment complex.

On this latter point, the record indicated that employees of the defendant Ballard Realty Company did make certain representations about security to the father of O.H.'s half-sister (the tenant with whom O.H. was living) when he enquired as to the previous criminal activity which had taken place in the apartment complex area. In response, the Ballard Realty employees had told the tenant that there was some type of security person present at the complex, and they informed him as to the frequency of property crimes which had occurred in the area.

The plaintiff also claimed that the defendants knew of criminal activity which would endanger O.H., because there had been rumors around the apartment complex concerning the alleged assailant before the assault on O.H. The alleged assailant had been evicted from the complex for failing to pay rent and for possessing a marijuana plant.

It appeared that this third person had spent a considerable amount of time on the property, from time to time visiting friends, in and around the apartment complex, after his eviction. The evidence was not clear concerning how much information the management had concerning the rumors that the alleged assailant had followed children in and around the complex, and there was no proof that the alleged assailant had actually molested any children. The only evidence presented by the plaintiff in this regard was that there were rumors circulating among the residents that the alleged assailant had been seen following children.

The appellate court concluded that the "mere existence of rumors and generalized statements concerning security" is not sufficient to support or create a duty to protect a resident from a sexual assault by a third person. Accordingly, the trial judge's dismissal of the plaintiff's case was affirmed.

COMMENTARY:

A review of the decisions previously rendered by Alabama Supreme Court, as cited in the above summary and reported in previous issues of **PSCLR**, *shows that the court takes a particularly conservative view on premises liability issues. It is not entirely clear by any means that all courts and all jurisdictions would have taken this same view of the factual situation, and the legal obligations of the apartment complex owners and operators as did the Alabama Supreme Court in this case.*

PSCLR – MAY 88 (9)

Locksmith's Act In Installing Lock On Apartment Building For Landlord Does Not Impose Duty On Locksmith To Prevent Intruder From Entering Apartment Building And Raping Tenant's Guest:

Laura Einhorn and Kenneth Einhorn v. David Seeley, REM Discount Security Products, Inc., and 15 East 21st Street Co., *in the Supreme Court of New York, Appellate Division, First Department, decided March 8, 1988, reported at 525 N.Y.S.2d 212.*

On June 1981, an assailant accosted Laurie Einhorn on the second floor landing of the 156 E. 21st Street Apartments, and dragged her to her fiancee's apartment on the third floor, where he raped her.

She subsequently sued her fiance's landlord. She also sued the landlord's locksmith for improper installation or maintenance of the front door lock. Her complaint sought relief in tort as well as contract law.

The trial court denied the locksmith's motion for summary judgment. The Appellate Court reversed, one judge dissenting.

The majority first noted the lack of direct contractual relationship between the plaintiff and the locksmith. The plaintiff was a guest of the tenant of the landlord who had hired the locksmith, and she was criminally assaulted by a stranger to all parties. Under these circumstances, the majority felt that the relationship between the locksmith and the plaintiff was too attenuated for the plaintiff to benefit from the contract between the landlord and the locksmith.

The court cited a previous New York case, *Moch Company v. Rensselaer Water Co.*, which had denied a claim by a citizen harmed by the failure of the company to perform under its contract with the city to provide water. When the water failed, the plaintiff's property burned as, apparently, the fire department watched. The judge in *Moch* pointed out the fact that the plaintiff and the defendant lacked "privity" of contract. He then reasoned no one contracting to supply water to a city could reasonably be held to a duty to refrain from negligent behavior as to each potential fire victim.

The *Moch* court considered the benefit granted any fire victim by the city's water supplier to be incidental, rather than intended, as far as the contract between the supplier and the city was concerned. The court feared that attaching a duty towards an incidental beneficiary of contract upon the party to the contract who bestowed that benefit would create an endless and senseless chain of potential liability.

> The assumption of one relation will mean the involuntary assumption of a series of new relations, inescapably hooked together. Again we may say in the words of the Supreme Court of the United States, 'The law does not spread its protection so far'.

The *Moch* court thus denied the victim's claim against the water supplier.

The majority in the present case applied *Moch*'s reasoning and reversed the denial of the locksmith's motion for summary judgment as to the contract and third party beneficiary counts. The court held the plaintiff an incidental rather than intended beneficiary.

As for an ordinary tort duty on the part of the locksmith to the plaintiff, the court found a special relationship would be required in cases where third parties intervened in an intentionally harmful way to the plaintiff. The landlord was an example of someone in a special relationship with the plaintiff.

The court held the locksmith was a stranger to both the victim and the assailant and had no duty to the victim to prevent the assailant's rape of the victim. The court, voicing fear of crushing liability unleashed by a contrary decision, held the locksmith was not liable for the plaintiff's injuries:

> Under these circumstances, to hold a locksmith responsible for the alleged consequences of an allegedly defective lock would be to enlarge the obligations of such artisans far beyond the existing law and beyond sound public policy.

COMMENTARY:

A dissenting opinion was filed in this case which expressed another view of the locksmith's potential liability to the rape victim. The dissent argued first that privity between plaintiffs and defendants had long ago been eliminated as a requirement for plaintiff to establish a duty owed him by defendant.

The dissent then chided the majority for focusing on the lack of the special relationship and ignoring the question of whether the locksmith might have anticipated danger to the tenants of the building, if its negligence caused the lock to fail on the front door.

According to the dissent, the class of victims here, tenants in one building, was quite discrete and thus could not be compared to the limitless class of potential fire victims in New York City found in Moch.

The dissent would not have found it unreasonable for a security company, such as the locksmith here, to assume that risks would be created if its duty were performed negligently, particularly in New York City. While a company supplying water to a city would normally not be able to pinpoint the occurrence of a fire in that city so that it might ensure that all fire fighters had adequate water, a security company such as the locksmith here knew exactly which buildings and areas it had purported to secure.

The dissent, however, did not prevail here.

PSCLR—AUG 88 (7)

Landlord Who Breaches Contractual Duty Imposed By Lease to Change Apartment Lock May Be Liable To Tenant Raped By Intruder:

Debbie Blaustein v. Gilbert-Dallas Co., Inc., et al., *No. 11-87-124-CV, in the Court of Appeals of Texas, Eastland, decided April 28, 1988, rehearing denied May 26, 1988, reported at 749 S.W.2d 633.*

Debbie Blaustein was raped by an unidentified man who gained access to her apartment, rented to her by Gilbert Properties. Blaustein sued Gilbert Properties for injuries sustained due to the rape, but the trial court dismissed her case prior to trial, finding that Gilbert Properties owed no duty to Blaustein for injuries suffered from the criminal acts of a third party, in this case, a rapist. Blaustein appealed, and the appellate court reversed and remanded the case for trial on the merits.

When Blaustein leased the apartment from Gilbert Properties, she requested "Sonny" Cohen, the apartment manager, to change the locks. Cohen replied that it was not necessary to change the locks since only he and the maintenance man had a key. Blaustein lived alone and had not made any duplicate keys.

On the evening of the incident, Blaustein returned home, locked the doors, and went to bed. During the night, she was awakened by the assailant, who covered her mouth with his hand and raped her. According to the victim, the assailant told her he had a key to her apartment and all of the other apartments. Blaustein called the police after the assailant left. When the police arrived, Blaustein went to answer the door and found the door unlocked. There was no evidence of forcible entry.

The main issue at this stage of the case is whether the landlord owed a duty to Blaustein. The appellate court defined duty as the "obligation to conform to a particular standard of conduct toward another." The court found persuasive a recent article in a legal publication, making "a strong case for not imposing a duty of security on the landlord." The author of the article reasoned that, when harm to a tenant results from the criminal act of another, a duty should not be imposed on a landlord because the criminal, not the landlord, is culpable. Moreover, since the harm did not arise directly out of the landlord-tenant relationship, any duty owed by a landlord to a tenant should be based only upon special circumstances in a particular case.

In this case, however, the appellate court found that the tenant's apartment lease created an affirmative duty on the landlord. The lease required that "[U]pon payment of reasonable charge, resident may require owner to change (or re-key) a door lock." The landlord was required to use ordinary care in carrying out such a duty.

The appellate court stated the general rule in Texas that a negligent actor is not liable for the criminal conduct of a third party, unless the criminal conduct is a foreseeable result of such negligence. A result is foreseeable if "a person of ordinary intelligence and prudence should anticipate the danger to others created by his negligent act even though it is not required that he anticipate exactly how injuries might arise." (The court here cited the Texas cases of *Nixon v. Mr. Property Management Company*, Inc., reported in the November-December 1984 **PSCLR**, case seven, page five and *Castillo v. Sears, Roebuck & Co.*, April 1984 **PSCLR**, case ten, page eight.)

Applying this rule to the facts of this case, the Texas Court of Appeal disagreed with the landlord's argument that the rape was unforeseeable.

> [T]he purpose of a locked door is obviously to secure the premises, possessions, or person from intrusion by others. After a request to change the door locks and the failure to do so with an assurance that it was unnecessary, a material fact question exists as to whether or not Cohen realized or should have realized that others who were unauthorized may have had access to Blaustein's apartment and anticipated the danger created by his refusal to change the lock.

The court thus reversed the judgment of the trial court in favor of the landlord, Gilbert Properties, and remanded the case for trial.

COMMENTARY:

An easy solution to the problem posited by the above fact situation would be to eliminate that language in the lease which imposes the duty on the landlord to change the locks on tenant's apartments upon their request to do so.

However, the question then arises as to whether the courts would find a duty to exist notwithstanding the omission of such language, and regular readers of **PSCLR** *know that many landlords have been held liable to tenants assaulted by third parties even absent such a language in the lease.*

Landlords should recognize their relationship with, and consequent duty to, their tenants and the reasonable steps to protect them against foreseeable dangers, including foreseeable assaults by third parties.

Review the following cases in previous issues of **PSCLR** *which discuss landlord liability for criminal assaults on their tenants—Arkansas: Jackson v. Warner Holdings, Ltd, January 1986, case three, page two; California: Penner v. Falk, June-July 1984, case five, page four; Jubert v. Shalom Realty, November-December 1982, case ten, page seven; Florida: Paterson v. Deeb, October 1985, case two, page three; Georgia: Bayshore Company v. Pruitt, November December 1985, case six, page five; Iowa: Brichaceck v. Hiskey, May 1987, case four, page four; Minnesota: Rullman v. Fisher, November-December 1985, case four, page three; New Jersey: Williams v. Gorman, May 1987, case five, page five; New York: Maricle v. Mick, May 1985, case six, page seven; North Carolina: Shepard v. Drucker & Falk, November-December 1983, case three, page three; Oklahoma: Lay v. Dworman, May 1987, case six, page six; Pennsylvania:*

Feld v. Merriam, October 1983, case one, page one, and a later appeal at June-July 1985, case ten, page ten; Virginia: Klingbeil Management Group Co. v. Vito, November-December 1987, case five, page eight.

PSCLR – JAN 89 (8)

Landlord Has No Duty To Insure Tenant Safety; Only To Provide Reasonable Security In Common Areas Of Apartment Complex:

Carmichael v. Colonial Square Apartments, *No. 86AP-819, in the Court of Appeals for Franklin County, Ohio, decided April 14, 1987, reported at 38 Ohio App. 3d 131.*

Paul Carmichael, Sr. was awakened one evening by the sound of his doorbell. As he opened the door, an unknown intruder shoved a shotgun between the door and the door frame, forced his way into the apartment, and assaulted and robbed Carmichael. Carmichael had opened the door without looking through the peephole or securing the safety chain.

Colonial Square Apartments had leased a second-floor apartment to Carmichael. The only entrance to the apartment was from an interior common hallway. Colonial Square had equipped the apartment with a safety chain, lock, peephole, and deadbolt, in addition to providing some lighting in the hallways.

Approximately three months before the assault and robbery, Colonial Square notified Carmichael of its intent to install a security system comprised of locked outside doors connected to an alarm and buzzer system that would permit someone inside the building to release the front door for visitors. At the time of Carmichael's assault, side door locks had been installed, but not the buzzer system at the front door.

Carmichael sued Colonial Square Apartments, alleging that Colonial was negligent in failing to provide security in the common areas of the building and in failing to complete the installation of the planned security system, and that Colonial's negligence caused him to be robbed. In order to bring the matter of Colonial's negligence to trial, Carmichael had to identify a duty, or duties, owed to him by Colonial and to produce sufficient evidence to allow a jury to conclude that a specific duty was breached and that the breach of duty caused his injuries.

In similar cases addressing a landlord's duty to provide security in an apartment complex, Ohio courts have held that "while the landlord has some duty to provide security to common areas in an apartment complex, he is not an insurer of the premises against criminal activity." Thus, "the duty on the landlord is only to take some reasonable precautions to provide reasonable security."

The court noted that, in this case, Colonial did take reasonable steps to provide for Carmichael's security: Colonial was in the process of installing a security system for outside entrances to the building; had installed locks for outside doors, though it had not yet completed installation of the buzzer system; and had equipped Carmichael's apartment with two locks, a chain, and a peephole.

Carmichael urged the court to impose a mandatory duty on landlords to provide common areas with a locked security system. The court, however, did not feel that such a duty should be imposed in this case because it was not a situation where the landlord was on notice as to specific instances of criminal acts against particular tenants, and then did nothing to secure the common areas. In this case, Carmichael raised only general references to criminal acts at the apartment complex. Further, Carmichael was assaulted in his apartment and not in a common area of the complex.

The court distinguished this case from cases in which the landlord's security measures deteriorated from the time that the tenant leased the apartment. In such cases, courts have required a landlord "to maintain the same degree of security as existed in the building at the time the tenant leased the premises." In this case, however, Colonial was in the process of improving the security in the common areas of the building.

The court held that, to the extent that a landlord is required to provide reasonable security to tenants, Colonial had complied with that duty. In addition, the court held that there was insufficient evidence to prove that, even if Colonial had breached its duty to provide reasonable security, such breach could be said to have caused Carmichael's injuries.

There was no evidence to show how the unknown intruder entered the building or that a locked front door would have prevented such entry. The only evidence that the intruder entered from an outside door was mere speculation on the part of Carmichael. Further, Carmichael himself opened the door and, knowing that the security system was not yet complete, failed to use the security devices provided by the landlord.

The court dismissed the case against Colonial without a trial due to insufficient evidence to raise a question for the jury as to Colonial's failure to provide reasonably adequate security and whether such security, if provided, would have prevented the assault and robbery of Carmichael.

COMMENTARY:

Carmichael had urged the court to adopt a rule imposing a mandatory duty on landlords to provide common areas with a a locked security system. He relied on two cases: the District of Columbia case of Kline v. 1500 Massachusetts Ave. Apt. Corp., *493 F.2d 477 (C.A. D.C. 1970) and the Florida case of* Paterson v. Deeb, *reported in the October 1985* **PSCLR**, *case two, page three.*

PSCLR – JAN 89 (9)

Private Security Guard And His Employer, Under Contract To Protect Apartment Complex Only, Is Not Liable For Personal Injury To Apartment Complex Guest:

Deeds v. American Security, et al., *No. C-860483, in the Court of Appeals for Hamilton County, Ohio, decided August 5, 1987, reported at 39 Ohio App. 3d 31.*

Suzanne Deeds drove to the Olde Montgomery Apartment complex early one July morning to visit her fiance. A man armed with a knife accosted her in the parking lot, forced her to the ground and raped her. While the rape was still in progress, Robert Justice, a security guard employed by American Security, which had contracted with Olde Montgomery to provide security at the complex, was patrolling the parking lot in his car. He stopped his vehicle within five feet of Deeds and her assailant, observed what he perceived to be a consensual sexual act, and advised Deeds and her assailant that behavior of that nature would not be tolerated. He left, promising to return.

When Justice returned, Deeds freed herself from the rapist, ran to Justice, and stated that she had been raped. The assailant assured Justice that he and Deeds were acquainted and that the sexual conduct that Justice had witnessed was consensual. Justice allowed the assailant to leave the premises. He then contacted the authorities and provided a description of the assailant and the license number of his vehicle. The assailant was subsequently apprehended and convicted.

Deeds filed suit against American Security and Robert Justice, seeking damages for physical injuries and psychological trauma suffered as a result of Justice's alleged negligent execution of his duties as a security guard. The case was tried before a jury, and the jury returned a verdict in favor of American Security and Justice.

During the course of the trial, the trial judge refused to permit expert testimony from a professional private security guard. In addition, the judge refused to instruct the jury at the close of trial that a private security guard has a duty to protect the public.

Deeds appealed from the judgment entered against her on the basis of the jury's verdict, challenging the trial court's refusal to recognize a specialized duty to the general public on the part of a private security guard to protect both persons and property. Deeds contended that the trial court should have allowed the security expert to testify to the generally accepted standard of conduct for a private security guard and to express his opinion as to whether Justice, in his conduct with respect to Deeds, adhered to that standard.

A claim of negligence is predicated upon injury caused by the failure to discharge a duty owed to the injured party. To sustain an action for negligence, a plaintiff must demonstrate: (1) that the defendant had a duty, recognized by law, requiring him to conform his conduct to a certain standard for the protection of the plaintiff; (2) that the defendant failed to conform his conduct to that standard; and (3) that the defendant's conduct proximately caused the plaintiff to sustain actual loss or damage.

Expert testimony and expert opinion will be permitted on an ultimate issue to be decided by the judge or jury as trier of fact if the witness is qualified as an expert with "specialized knowledge" and if his testimony or opinion will assist or help the trier of fact to understand the evidence or decide an issue of fact.

Ohio law does not impose a specialized standard of care upon a private security guard in the execution of his duties. However, a duty to protect residents of an apartment complex and their guests from crimes against persons may arise under contract.

In this case, the agreement between Olde Montgomery and American Security was reflected in "on the job instructions" issued by American Security to its employees. This document set forth patrolling and reporting procedures and expressly limited its security purpose to the reduction of vandalism, or crimes against property. Therefore, the court held that Deeds failed to demonstrate the existence of a duty on the part of Justice in the execution of his employment duties as a security guard that would extend to her or would encompass the protection of her person on the premises. As a result, the expert testimony and opinion that Deeds sought to present at trial would not have been helpful to the jury, according to the appellate court which therefore held that the expert testimony was properly excluded.

The trial court also refused to instruct the jury on the standard of care of a "reasonably prudent security guard," but did instruct the jury on the general legal duty to exercise ordinary care to avoid injuring another. The court further charged that one who undertakes an act which he has no legal duty to perform must, in the performance of that act, exercise ordinary care and may be liable for injuries if a failure to exercise ordinary care increased or caused the risk of harm.

The court stressed that Ohio law does not recognize a different duty on the part of a private security guard to protect members of the general public from offenses against their persons. Therefore, the requested instruction was properly omitted. The appellate court, thus, upheld the trial court's judgment in favor of the security guard and his employer based upon the jury's verdict finding no breach by him of the general duty to exercise ordinary care to avoid injuring another.

COMMENTARY:

This case emphasizes the general rule that a person does not have an affirmative duty to act to protect another. Such a duty would arise in the case of a private security officer based upon a contract, which provides that the guard is employed for the purpose of protecting the safety of individuals. In this case, however, the security contract explicitly limited the security guards' duties to guarding against vandalism, or injury to property.

On the other hand, if the security guard had undertaken to protect the plaintiff in this case, and had done so negligently, the security guard would have been liable for his negligence in failing to carry out the voluntarily-assumed duty. Since this security guard did nothing to attempt to aid the plaintiff, and the applicable contract did not require him to protect her, there was no basis for finding a duty on his part to act affirmatively to aid the plaintiff.

PSCLR – FEB 89 (2)

Landlord Not Liable For Wrongful Death Of Child Who Drowned In Unguarded Swimming Pool, When Landlord Did Not Voluntarily Assume Duty To Supervise The Pool Area:

Alabama Farm Bureau Mutual Casualty Insurance Co. v. David L. Hixon and Debbie Hixon, *No. 86-1241, in the Supreme Court of Alabama, decided July 8, 1988, rehearing denied September 14, 1988, reported at 533 So. 2d 518.*

Five-year-old David Lee Hixon drowned in the swimming pool at the Village West Trailer Park in Montgomery, Alabama. The child's parents, David and Debbie Hixon, sued Alabama Farm Bureau Mutual Casualty Insurance Co., the owner of the park and pool, for the wrongful death of their son. At trial, the jury returned a $750,000 verdict for the Hixons, and the defendant appealed.

David Lee Hixon lived at the Village West Trailer Park with his parents and younger siblings. On the afternoon of David's death, his mother had stopped by the trailer park office adjacent to the pool and signed a document stating that she had read the "Pool Rules" established by the trailer park management.

Mrs. Hixon, the two younger children, David Lee, and some neighbors went to the pool. After the neighbors had left, Hixon opened the pool gate and watched her son leave the pool area to walk home. She went back to the pool for the other children and, on returning to her trailer, discovered that David Lee was not there.

After checking with neighbors, Mrs. Hixon returned to the pool to search for her son. A Eugene Thomas stated that he thought "that something he brushed in the pool 15-20 minutes earlier might have been a child," and dived into the pool, found David Lee, and pulled his body from the water. Efforts to resuscitate David Lee failed.

No lifeguard was on duty at the time David Lee's body was found. A seven-foot-high fence topped with barbed wire surrounds the pool. Cement and wire at the bottom of the fence prevent anyone from crawling under it. A six-foot-high gate topped with barbed wire provides the only access to the pool. There are four spring-loaded hinges on the gate, and the latch is 65-70 inches above the ground. The gate closes automatically and has a self-closing latch. The trailer park office adjacent to the pool closes at 4:30. David Lee drowned at approximately 6:00 p.m.

David Lee's parents alleged two theories as a basis for recovery. First, they alleged that the trailer park and pool owner voluntarily assumed the duty to supervise the pool area and that they breached this duty by improperly and inadequately supervising the pool. Second, they asserted that the trailer park and pool owner breached an alleged duty to maintain the pool by allowing the water in the pool to become so dirty that David Lee Hixon's body could not be seen on the bottom of the pool by the people in the area.

A landlord who retains control over an area of common use has a duty to make safe any condition that presents an unreasonable risk and could be discovered by the exercise of reasonable care. The Hixons asserted that the cloudy condition of the pool water presented an unreasonable risk that the defendant could have discovered by reasonable care and that the failure to remedy that condition caused the death of their son.

The appellate court disagreed. The court reasoned that the cloudiness of the pool water did not present an unreasonable risk, when David Lee had been swimming in the same pool only a short time earlier under the supervision and with the approval of his mother. Because the risk was not unreasonable, the defendant had no duty to make the condition safe and is not liable for the child's death.

Further, the appellate court reasoned that if the condition of the pool water were an unreasonable risk, then the injury resulted from a condition that was obvious and should have been observed in the exercise of reasonable care by the Hixons. The obvious nature of the risk would, therefore, relieve the trailer park and pool owner from liability for the drowning.

The appellate court also found that the Hixons' contention that there was inadequate supervision of the pool was not supported by the evidence. The court noted the protection of the pool provided by the fence and the gate and the fact that there was a sign at the pool that read: "No lifeguard. Swim at your own risk." Mrs. Hixon testified that she had read the sign and was aware of the pool rules because she had signed a copy of them on the same day. The rules required that children under twelve years old be accompanied by a parent.

The appellate court stated that there is no absolute duty to provide a lifeguard or other supervision at a swimming pool. The lack of a lifeguard is a condition that can be discovered by reasonable and customary inspection, and there is no liability for the mere failure to provide a lifeguard.

The appellate court overturned the jury verdict and directed the trial court to enter judgment in favor of the defendant.

PSCLR – FEB 89 (9)

Landlord May Be Liable For Criminal Act Of Assailant Who Gained Entry To Tenant's Apartment Via Building's Common Fire Escape And Inoperative Window Latch:

Lorri Aaron v. David Havens, *No. 70366, in the Supreme Court of Missouri, En Banc, decided October 18, 1988, reported at 758 S.W.2d 446.*

Lorri Aaron was raped and sodomized by an intruder who entered her apartment through a window adjoining the building's common fire escape. The trial court dismissed her case, holding that the landlord owed no duty to protect Aaron from criminal assault by a third party in her apartment. Aaron appealed.

Aaron occupied the apartment on the third floor of a building owned by David Havens at 5560 Cates in St. Louis, Missouri. At the time that she rented the apartment, the inside latch on the rear window was broken, which would permit an intruder to enter from the fire escape. A burglar had previously entered Aaron's apartment through the front balcony doors, also accessible to the fire escape, and the landlord was aware of this.

Aaron repeatedly reported the broken window latch to her landlord and told him that she feared for her safety. She felt that she was in particular danger because the landlord had designated her to collect rents from the other tenants, and the neighborhood was in a high-crime area.

Aaron charges that the landlord was negligent in that he failed to install a lock to the rear window of Aaron's apartment after she repeatedly requested he do so; failed to secure a lock to the door along the six-foot fence surrounding the back yard of the apartment building, thus allowing anyone entry; and failed to take precautions so that an intruder could not easily gain access to the fire escape and, therefore, the windows to Aaron's apartment.

To succeed in a negligence action, a plaintiff must allege facts demonstrating: (1) a duty owing from the defendant to the plaintiff; (2) a breach of duty; and (3) resulting damages.

The court in this case rejected the landlord's argument that he had no duty to tenants to make the premises safe against the criminal conduct of intruders. The court noted that the owner of an apartment building has a "recognized duty to use due care to make common premises safe, as against foreseeable risks." The landlord knew or should have known of the dangerous condition of the fire escape, through which a criminal intruder could obtain easy access to Aaron's apartment. The court observed, "There is no logical reason why a dangerous condition of common premises, proximately causing injury on private premises, could not be the occasion for liability."

The court stated that the abstract rule that there is no duty to protect against criminal misconduct has been substantially attenuated in recent cases. For example, it is not necessary to allege that past crimes involving entry into unauthorized places are of the same general nature as the one which gave rise to the claim. Thus, the fact that a burglar had previously entered Aaron's apartment provides evidence that the intrusion by the rapist was foreseeable.

The court restricted its analysis to Aaron's allegation that the landlord had failed to take available precautions to prevent easy access to the fire escape. Because the fire escape is part of the common premises under the landlord's control, tenants did not have the right to make repairs or modifications, thus giving rise to the landlord's duty to use due care to protect against dangerous conditions upon it.

The court felt that the Missouri statutory requirement of a fire escape did not relieve the landlord from liability for maintaining a fire escape that reaches the ground, "if the jury is persuaded that a design which will give the tenants better protection is reasonably available and should be used in the exercise of ordinary care."

The court ordered that the case be tried so that a jury might consider whether the landlord had breached his duty to use "ordinary care" to make the premises "reasonably safe." The court stated that the jury may consider economic factors in making these determinations.

COMMENTARY:

The dissent in this case noted that the "dangerous condition" of the fire escape, a common area, is dangerous in that it permits access to an entrance to the plaintiff's apartment. Thus, he argued, this case moves one step beyond a landlord's liability to make safe known unsafe conditions in common areas, and to impose liability when the design and construction of an apartment building's common area merely provides access for intruders to points of entry into the private apartments.

The dissent stressed the "corollary to the landlord's duty to correct dangerous conditions in common areas: the tenant bears responsibility for conditions in those parts of the leasehold under the tenant's control."

The dissent went on to argue that the tenant is in a better position to know about and remedy unsafe conditions in the leased premises. Thus, in the absence of a contractual obligation on the part of the landlord to effect repairs, the responsibility should fall on the tenant.

At the very least, the lease agreement should specify whether the tenant or the landlord is responsible for repairs to the private areas of the leased premises, and the agreement should also require written notification of defects to put the responsible party on notice to remedy them. Of course, the responsible party cannot ignore notice received other than in writing, such as through inspections, oral complaints, observations, or otherwise.

PSCLR—FEB 89 (11)

Landlord Is Liable For Failure To Protect Tenant From "Reasonably Foreseeable Criminal Conduct" On Tenant's Leased Premises:

L.K. v. Water's Edge Association, *No. 87-1513, in the District Court of Appeals of Florida, Third District, decided October 4, 1988, rehearing denied November 15, 1988, reported at 532 So. 2d 1097.*

L.K. was sexually assaulted in the condominium leased to her by Water's Edge Association. The intruder entered her condominium through a kitchen window.

L.K. sued Water's Edge Association for negligent failure to provide adequate security. At trial, L.K. presented evidence concerning crime in the area, assurances by Water's Edge of excellent security, the Association's disregard of recommendations by a detective agency as to when a guard should be on duty, and the inadequacy of the condominium's security.

L.K. requested the court to give standard jury instructions on duty and negligence as follows:

> 1. The court has determined and now instructs you, as a matter of law, that the circumstances at the time and place of the incident complained of were such that the Defendant, Water's Edge Association, had a duty to employ reasonable security measures to protect [L.K.] from reasonably foreseeable criminal activity....
>
> 2. The first issue for your determination on the claim of the plaintiff, [L.K.], against defendant ...is whether the defendant was negligent in failing to employ reasonable security measures and, if so, whether such negligence was a legal cause of the damages sustained by [L.K.]....

The trial judge refused to give the requested instructions and charged the jury instead that the issues for its determination were whether Water's Edge

> negligently failed to correct a dangerous condition, of which the defendant ...either knew or should have known by the use of reasonable care or negligently failed to warn the plaintiff ...of a dangerous condition concerning which the defendant ...had or should have had knowledge greater than that of the plaintiff.

The jury returned a verdict against L.K. and the trial court entered judgment pursuant to the verdict. L.K. appealed, complaining of the trial court's failure to give her requested instructions.

The appellate court held that the evidence supported L.K.'s theory of the case as reflected in her requested jury instruction and that the trial court's instruction had misled the jury. Under L.K.'s theory, Water's Edge had breached its duty to keep the common areas reasonably safe and to provide adequate security measures to protect L.K. from reasonably foreseeable criminal activity. The appellate court agreed that: "As a matter of law, Water's Edge was obliged to protect L.K. from reasonably foreseeable criminal conduct."

The appellate court stated that the instruction given by the trial court might have been appropriate when a landlord's failure to repair results in a tenant's slip and fall, "but is misleading when, as here, the landlord's negligence is based on a failure to provide adequate security."

The court, therefore, reversed the judgment and ordered the case to be retried.

COMMENTARY:

The importance of this decision is that it emphasizes the fact that the responsibility of landlords has been expanded significantly by court decisions in recent years. These decisions, and the above opinion, make it clear that a landlord now has an affirmative duty to protect tenants from reasonably foreseeable criminal conduct by third parties. The old rule—that the criminal act of another which causes loss to a tenant constitutes an intervening cause which would shield the negligent landlord from liability—is dead.

PSCLR—APR 89 (7)

Apartment Managers May Be Criminally Liable For Failing To Provide, Install, And Maintain Acceptable Window Guards In Buildings That They Control:

The People of the State of New York v. Ike Portnoy, *in the Criminal Court of the City of New York, Bronx County, Part AR-9, decided November 4, 1988, reported at 535 N.Y.S.2d 305.*

The defendants in this case, individuals who controlled or managed apartment buildings in the Bronx in New York City, were criminally charged with violating the New York City Health Code by failing to provide, install, and maintain acceptable window guards in apartments and in surrounding public hallways in the buildings that they managed or controlled. The defendants attempted to have the charges dismissed on the grounds that the statutes defining the offenses charged are unconstitutional.

The defendants were charged with violating the "Window Guard Law" of the New York City Health Code, imposing criminal penalties for failing to provide, install, and maintain acceptable window guards where a child under eleven years of age resides. The defendants asserted that the law unconstitutionally imposes criminal liability based upon unpublished New York City Department of Health ("DOH") Regulations requiring installation of specific types of DOH-approved window guards, in violation of the New York City Charter.

The "Window Guard Law" was enacted after the DOH determined that installation of window guards in apartment buildings significantly reduced the number of window fall accidents involving young children. The law provides:

> The owner, lessee, agent or other person who manages or controls a multiple dwelling shall provide, and maintain, a window guard, of a type deemed acceptable by the Department, installation to be made pursuant to specifications provided by the Department, on windows of each apartment in which a child or children ten years of age and under reside, and on the windows, if any, in the public halls of a multiple dwelling in which such children reside, except that this section shall not apply to windows giving access to fire escapes or to a window on the first floor that is a required means of egress from the dwelling unit. It shall be the duty of each such person who manages or controls a multiple dwelling to ascertain whether such child resides therein.

The court held that the law did not violate the New York City Charter by imposing criminal liability based upon unpublished DOH regulations, because the types of window guards deemed acceptable and the installation requirements issued by the DOH were merely specifications, and not regulations as urged by the defense. Further, the specifications were readily obtainable from DOH.

The defendants also argued that the "Window Guard Law" was "void for vagueness" and, therefore, invalid under the United States Constitution. To avoid invalidation for vagueness, a statute must afford a "person of ordinary intelligence fair notice that [the] contemplated conduct is proscribed" and the statute must provide "explicit standards" to avoid arbitrary or discriminatory enforcement. The defendants claim that the statute is vague because the term "required means of egress" is undefined. The Constitution, however, only requires that statutes be reasonably precise.

The court held that, while the terms "required means of egress" are not precisely defined in the statute, the words retain their plain "everyday, common sense meaning; and when construed accordingly [are] sufficiently clear as to preclude a vagueness challenge."

The court also held that the "Window Guard Law" applied "explicit standards" for proper enforcement by stating that failure to provide, install, and maintain required window guards constitute a violation of the statute.

The court also upheld the charge against the defendants for allegedly failing "to reasonably act and take necessary precautions to protect human life and health by failing to provide, install, and maintain acceptable window guards at the subject premises." The court held that the statute upon which the defendants were charged, providing that "no person shall fail to do any reasonable act or take any necessary precaution to protect human life and health," was constitutionally valid and not void for vagueness.

The court further upheld the charge against the defendants for allegedly maintaining "a nuisance by failing to provide, install, and maintain acceptable window guards" at the apartment buildings that they maintained or controlled, thereby "constituting a condition dangerous to human life and detrimental to health." A nuisance is defined under New York law as it relates to buildings or erections as follows:

> The word 'nuisance' shall be held to embrace . . . whatever is dangerous to human life or detrimental to health;
>
> Whatever building or erection, or part or a cellar thereof, is overcrowded by occupants, or is not provided with adequate ingress and egress to and from the same or the apartments thereof, or is not sufficiently supported, ventilated, sewered, drained, cleaned, or lighted in reference to its intended or actual use; and whatever renders the air or human food or drink, unwholesome. All such nuisances are hereby declared illegal.

The court found that the law gave a person of ordinary intelligence fair notice that contemplated conduct, such as failing to provide acceptable window guards where children ten years old or younger reside, constitutes a "nuisance" and would therefore be unlawful. The court further held that the law provided explicit standards for non-arbitrary or non-discriminatory enforcement.

In sum, the court found that all of the statutes under which the defendants were charged were constitutionally valid and not vague as applied in the circumstances of this case. Accordingly, the case was allowed to proceed to trial against the defendants on all charges.

COMMENTARY:

This case, once again, emphasizes the need for security directors to be fully aware of the local ordinances in each and every municipality and local jurisdiction in which they have properties for which they are responsible.

In addition, as is noted in the above case, the ordinances themselves may not provide sufficient information as to what is required to be in compliance with the law, and the diligent security director will ensure that all rules and regulations pertaining to his or her properties are also obtained, reviewed, and reviewed again on a periodic basis to ensure that the security systems under his or her control are in full compliance at all times.

PSCLR – JUN-JUL 89 (10)

Landlord Is Liable For Negligence In Failing To Provide Adequate Window Locks For Apartment Windows:

Albert Benser, d/b/a Carrier Square Apartments v. Cynthia Johnson, Individually & as Next Friend of Alysia Johnson, a Minor, *No. 05-87-00692-CV, in the Court of Appeals of Texas, Dallas, decided March 21, 1988, rehearing denied May 11, 1988, reported at 763 S.W.2d 793.*

Cynthia Johnson and her daughter Alysia Johnson lived in an apartment in a complex known as Carrier Square Apartments, owned by Albert Benser. Johnson noticed that the locks on the living room and her daughter's bedroom windows were inoperable. After she complained about the locks, the complex gave her a "screw-type" lock for both the living room and the bedroom windows. There was also a stick in one of these windows to secure the window in lieu of a lock. Johnson placed the stick in the living room window and installed the "screw-type" locks provided by the complex. Approximately one year later, an intruder drilled a small hole in the bottom of the living room window and knocked the stick out of the way. The hole drilled was not large enough for the intruder to stick his hand through and unlock the window. Because the lock provided to Johnson by the complex did not work, the intruder was nonetheless able to open the window and enter the apartment. The intruder raped Cynthia Johnson in the presence of her daughter. Johnson sued Benser, the owner of the apartment complex, for negligence in failing to provide adequate locks for her apartment windows.

The case was tried before a jury, and the jury entered a verdict in the amount of $70,000 to Cynthia Johnson, individually, and on behalf of her daughter Alysia. Benser appealed, claiming that there was insufficient evidence to support the jury's verdict.

Specifically, Benser claimed that Johnson had submitted insufficient evidence that acts attributable to him "proximately" caused Johnson's injuries, as required in a negligence action. The two elements of proximate cause are cause-in-fact and foreseeability. (The court cited *Nixon v. Mr. Property Management Company, Inc.*, reported in the November-December 1984 **PSCLR**, case seven, page five.) Cause-in-fact is satisfied if the negligent act or omission was a substantial factor in bringing about the injury, and without which no harm would have occurred. Foreseeability requires that the actor, as a person of ordinary intelligence, should have anticipated the dangers that his negligent act created for others. The criminal conduct of a third party will relieve a negligent actor from liability unless the criminal conduct is a foreseeable result of such negligence.

The record in this case reflected that Benser was aware for a long period of time that Johnson's window locks were broken and yet refused to install working locks, in violation of Texas law. Texas law requires that a landlord make a diligent effort to repair or remedy a condition if the tenant provides notice of the condition, the tenant is not delinquent in the payment of rent, and the condition materially affects the physical health or safety of the tenant. Texas law further provides that the landlord install, change, or re-key a security device after the landlord receives a request from the tenant of a dwelling. This obligation is limited to installing one window latch on each exterior window.

The court noted that Benser was aware that the complex was located in a high crime area and that there had been previous instances of criminal activity in the complex. In addition, a police officer testified that the stick in the living room window would indicate to a potential intruder that the window could not be locked and would encourage an intruder to target such a home.

A property owner may be liable for negligent actions which create an opportunity for a third person to commit an intentional tort or crime. The chief factors courts consider in determining whether criminal activity is a foreseeable result of the property owner's negligence are whether the property is located in a high crime area and whether previous criminal activity has occurred on the property. The court found both of these factors present in this case. Therefore, the court held that there was sufficient evidence in the record to support the jury's finding that the landlord's negligence was a proximate cause of Johnson's injuries. The court, thus, affirmed the verdict in favor of Johnson.

COMMENTARY:

The above case points out the need for systematic record-keeping of all tenants' requests for repairs and/or additions, particularly as they relate to security precautions. Commonly, the lease itself should specifically require that all such requests by tenants be made in writing in order to avoid any subsequent conflict as to the nature of the request.

The system established by the landlord should of course include a filing system for such written requests, and a procedure for responding to the requests, which again should be in writing. While such a system does indeed increase the volume of paperwork involved in the management of rental complexes, the professional security manager in charge of the protection of property and persons in such complexes will insist on the establishment of such a system, and will train his or her personnel to follow such a system scrupulously. Only in that way can adequate reasons for the denial of a tenant's request for safety precautions be justified, and only in that way will liability for injuries and/or assaults of tenants be avoided.

Furthermore, with regard to the court's above recognition of the fact that the subject property was in a "high-crime" area, the cautious security manager will include in his or her security plan a procedure for notifying all tenants of criminal activity on the property or in the area, and advising them to take specific precautions to protect themselves and their property.

PSCLR – SEP 89 (3)

Landlord Must Pay Punitive Damages For Failure To Repair Broken Door And Intentional Or Reckless Disregard Of Complaints By Tenant:

McWilliams v. Hayes, *No. 77298, in the Court of Appeals of Georgia, decided February 7, 1989, rehearing denied March 13, 1989, reported at 379 S.E.2d 528.*

This case arose out of the efforts of a landlord to evict a tenant. The landlord sued the tenant for eviction because she had not paid her rent. However, the tenant countersued the landlord for damages she had suffered due to his failure to maintain the sanitation and security of the property.

The tenant had informed the landlord on several occasions of problems with her apartment, including a loose fitting door and doorknob on the front of the apartment, and miscellaneous sanitation problems. Although he had been informed of the problems and the tenant had threatened to, and had, withheld her rent because of the problems, the landlord did not make the necessary repairs.

The tenant claimed specifically that her apartment had been burglarized and that the burglar had gained entry through the unsecured front door. The jury awarded the tenant $1,346 in actual damages and $5,000 in punitive damages. The landlord appealed.

On appeal, the court upheld the judgment, noting that punitive damages were appropriate because the landlord's inaction showed "a reckless disregard for or a conscious indifference to consequences." Conscious indifference is an "aggravating circumstance" for which punitive damages may be awarded. The evidence, which showed that the landlord had knowledge of the problems with the apartment and failed to take any action after repeated complaints, justified punitive damages.

The jury's award was upheld, and the landlord's appeal was denied.

PSCLR – SEP 89 (11)

Absent A "Special Relationship," Landlord Has No Duty To Protect Tenants From Crimes Committed By Third Parties:

Patricia Cornelius v. Housing Authority of New Orleans, ABC Insurance Company, Sidney Cates, and E. Dupre, *No. 88-CA-0075, in the Court of Appeal of Louisiana, Fourth Circuit, decided March 14, 1989, rehearing denied April 11, 1989, reported at 539 So. 2d 1250.*

This case arose out of the murder of a resident of a housing development owned and managed by the Housing Authority of New Orleans (HANO). On the day of the homicide, five-year-old Timothy Ramon Cornelius returned home from kindergarten at approximately 3:15 in the afternoon. With his mother's permission, Timothy went out to sit on the front porch, which bordered a children's play area. At about 3:45 PM, Timothy's mother discovered him missing from the porch. She and some neighbors searched extensively but could not locate him. At 5:30 PM, Cornelius called the police. At 8:00 the next morning, Timothy's body was discovered on a stairway of a nearby building. The coroner ruled the death a homicide, but no suspect was ever arrested for the murder.

Timothy's mother filed suit against HANO, claiming that HANO had control of the area where the murder occurred and that the area was dangerous because of poor lighting and security. She claimed that HANO knew of these unsafe conditions but failed to correct them. She alleged that HANO's failure to provide safe conditions was the cause of her son's death.

The court rejected Cornelius' claims. Under Louisiana law, a landlord does not have a duty to protect its tenants from the violent acts of third parties except where there is a "special relationship" between the landlord and tenant, or where the landlord has "assumed an implied or express obligation" to provide security. Since HANO had not assumed an obligation to provide security, it had no duty to do so even if it knew of the unsafe conditions. Under the law, where there is no duty to protect, there can be no liability for failure to protect. Cornelius' case was, therefore, dismissed.

PSCLR – SEP 89 (18)

Landlord Has No Duty To Protect Invitee Of Tenant From Dangerous Conditions Or Acts Occurring In Tenant's Apartment:

Michael Conroy, Appellant v. Solon Gershman, Inc., and Gershman Investment Corp., *Respondents, No. 55245, in the Missouri Court of Appeals, Eastern District, Division Three, decided April 4, 1989, reported at 767 S.W.2d 381.*

The defendant in this case, Solon Gershman, Inc. owned rental apartments in St. Louis, Missouri, one of which was rented to a man named Tom Rafferty. The plaintiff, Michael Conroy, was visiting with Rafferty, in Rafferty's apartment, when the police raided the apartment. During the drug raid, Rafferty pulled a gun and accidentally shot Conroy.

Conroy brought this suit against the owner of the apartment complex, claiming that the owner was negligent for failing to (1) screen the tenants before renting, (2) take action against the tenants when the owner knew that the tenants were drug dealers, and (3) take action to protect guests in the apartment complex from the drug-dealing tenants.

The court began its analysis of the case by noting that the general rule under Missouri law is that a landlord has no liability to his tenants or tenants' invitees for injuries caused by dangerous conditions. The courts, of course, have created exceptions to that rule. One such exception is when a person who is known to be violent is on the premises, has acted in such a way as to indicate danger, and the landlord has had sufficient time to act to prevent that person from causing harm. There may also be liability when there has been a pattern of crimes on the premises that would lead a reasonable person to take precautions to prevent future occurrences. As the court noted, however, such exceptions have always involved cases where the injury that the landlord is blamed for occurred in the public or common areas of the premises.

In this case, the criminal act through which Conroy was injured did not occur in a common area of the apartment complex. Instead, it took place inside one of the privately leased areas – Rafferty's apartment – of the premises. The landlord did not have a duty to protect Conroy from an injury that occurred as a result of activities inside the apartment. The court may have ruled differently, however, if the plaintiff could have proven that the landlord had "exclusive control over the premises rented by the tenant and that the landlord had the exclusive power to take preventive action but failed to do so." The plaintiff was not able to provide such proof.

The plaintiff also claimed that the landlord knew that some of the tenants in the complex were involved in dangerous and illegal activities and that the landlord failed to take action against them. According to the court, however, even if this allegation was true, there was no evidence that the landlord knew that Rafferty was involved in such activities. Since the landlord had no knowledge of Rafferty's activities, he could not have been expected to take action against Rafferty. The appellate court therefore affirmed the dismissal of Conroy's case.

Colleges and Universities

PSCLR—JUL 82 (10)

University, As Lessee Of City Property, Not Liable To Plaintiff Injured On Property For Failure To Provide Security Guards Where Terms Of Lease Required City To Provide Security:

Victor Andruskevics v. President and Fellows of Harvard College, *Appeals Court of Massachusetts, decided February 10, 1982, reported at 430 N.E.2d 1241.*

In this case, a pedestrian was injured by a hit-and-run bicyclist in the Arnold Arboretum, a property leased to Harvard College by the City of Boston for use as a public park. Since the city has a rule against bicycling in city parks, plaintiff sued Harvard College for not maintaining proper security personnel to enforce the city's rules and thereby prevent injury to pedestrians. The trial court ruled in favor of Harvard.

The lease provided it was the duty of the city to "maintain a proper and sufficient police in and about the Arnold Arboretum...for the preservation of order and good conduct and the observance of the rules [as to which the city and Harvard were to agree] hereinafter mentioned." The Appeals Court of Massachusetts affirmed that the college had no duty to provide security personnel under the terms of the lease. The duties of Harvard were to study, maintain, and review the botanical collection. The duties of the city included maintaining means of public access and providing sufficient personnel to preserve law and order.

The Appeals Court found no duty, even though the lease allowed Harvard to join the city in rule-making for the park and in face of Harvard's occasional practice of hiring security guards to protect the botanical collection for certain events.

PSCLR—AUG 82 (4)

Landowner's Duty To Protect Invitees From Third Party Criminal Assault

Pearl Dorothea Relyea, Individually and as Administratrix of the Estate of Donna J. Fink, deceased v. State of Florida et al.; George A. Mahnke and Arlene Mahnke, Individually and as Administratrix of the Estate of Marlene Mahnke, deceased v. State of Florida et al., *Docket Nos. 76-1619, 76-1620, District Court of Appeal of Florida, Fourth District, decided July 23, 1980, reported at 385 So.2d 1378.*

This case, referred to and quoted in *Anderson* (**PSCLR**, August 1982, case three, page three), is an earlier consideration in Palm Beach County of the question of a landowner's duty to protect invitees from criminal assault by a third party, but the outcome was dramatically different.

Two female students of Florida Atlantic University (FAU) were kidnapped, raped, and murdered by three unknown assailants after the pair left a class together at 7:20 PM. The class was located in an isolated building about three-quarters of a mile from the main campus.

The parents and estate administrators of the victims sued the State of Florida, the school, and its insurance company to recover for the wrongful deaths. The plaintiffs argued that the school's negligence in failing to provide security patrols, gates on roads entering FAU, sufficient lighting or other adequate protective measures at this particularly "remote" area led to the students' deaths. The trial ended in a mistrial when the jury failed to reach a decision. The court subsequently granted judgment to the defendants, and plaintiffs appealed.

Using the same analysis as the *Anderson* court, the *Relyea* court said an owner cannot be held responsible if the injury was caused by interceding, reasonably unforeseeable criminal conduct by a third party.

The attack and murder of the two FAU students was not preceded by "prior, similar criminal acts." The only reported crimes in the past 17 years at FAU consisted of petty theft, slight damage from hit-and-run car wrecks, and other minor crimes. No duty was created on the part of FAU to protect invitees from violent assault, since no such incidents had occurred or could be anticipated to occur, unlike the preceding case. Judgment for the defendants was therefore affirmed.

COMMENTARY:

The two different results in Anderson and Relyea point out the courts' general reluctance to impose liability on a property owner for a third party's criminal acts. However, that fine line can and will be drawn where the landowner knows, or reasonably should know, of ongoing criminal conduct. The landowner must take steps to remedy such problems after they have come to his attention.

Maintaining close contact with on-site personnel, insisting on regular inspections of the premises, and adequate reporting mechanisms will provide essential information for the landowner to assess the need for security changes.

PSCLR—AUG 83 (10)

College's Failure to Provide Adequate Protection of Students From Criminal Acts Renders It Liable to Student Who Is Assaulted on Campus:

Lisa Mullins v. Pine Manor College and William P. Person, *in the Supreme Judicial Court of Massachusetts, Norfolk, decided May 2, 1983, reported at 449 N.E.2d 331.*

Pine Manor College is a four-year college for women in Brookline, Massachusetts. Although located in an area with few incidents of violent crime, in December of 1977, a resident student was raped on the campus. The student, Lisa Mullins, brought a lawsuit against the school and its vice president of operations alleging that Pine Manor's negligence in securing the dormitories against criminal entry was a direct cause of her rape. The jury agreed with her and she was awarded $175,000, which was later reduced to $20,000 by the trial judge. Pine Manor College, and the co-defendant vice president of the college, appealed.

Pine Manor argued that the rape of the student was not foreseeable, because there were no prior incidents of criminal assaults on campus. Moreover, the school felt that its security precautions fulfilled its reasonable duty of care to protect the students' safety.

Evidence was offered to show that the campus was surrounded by a chain-link fence which stood six feet high, except for an area immediately around the entrance where it stood only four feet high. The dormitories were clustered in villages which could be entered only through each village's commons building or through an exterior gate; both entrances were locked between 5 PM and 7 AM. Each student had her own key which unlocked her village's commons building, her dormitory, and her individual room. After 8 PM, visitors would first have to pass a security guard at the main entrance and then would have to register with a student receptionist at the commons building they visited. The student hostess of the visitor would then be called to escort her guest. After 1 AM, visitors were not allowed to be on campus without a student escort.

In December of 1977, two guards were on duty after midnight. One guard observed the main entrance and the other guard patrolled the campus. The guard who patrolled the campus was supposed to check each of the dormitory villages every 15 to 30 minutes. The guards were not formally supervised, but Pine Manor's director of security randomly checked on their work.

On December 11, Lisa Mullins returned to her dormitory at 3 AM. She and two friends who lived in the same village entered through the exterior gate to the village, which they found to be unlocked. After visiting a friend's room next door to her own, Mullins locked her door and went to bed. She was awakened at about 4:30 AM by an intruder who threatened her and made her accompany him out of the dormitory building. They left the village through a gate which was not securely locked. They walked to the college's dining hall, entered it, walked out, and then back in at which point Mullins was raped by her kidnapper. Mullins later testified that she was led around the campus for at least 20 minutes of the 60 to 90 minutes she was being held by her attacker.

Other evidence was presented, including testimony by a security expert (the chief of campus police at Wellesley College) who testified that the room locks were inadequate and could be opened with a credit card.

The court reviewed these facts and determined that the college definitely had a duty to protect its students from criminal acts of third persons – "a duty finds its 'source in existing social values and customs'." Pine Manor College enrolled young women who were on their own for probably the first time, required that they live in college facilities for at least the first and second years, and encouraged them to continue living there throughout their college years. The school undertook to provide the students secure and safe living quarters, which the court saw as "an indispensable part of the bundle of services which colleges ... afford their students" and this was an important consideration to prospective students and their parents in choosing a college.

In addition, the court felt that criminal attacks on the campus were a foreseeable risk – in fact, students were warned by the college staff at the beginning of the year "of the dangers inherent in being housed at a women's college near a metropolitan area only a short distance from bus and train lines which lead directly to Boston."

The foreseeability of crimes taking place on the campus, the negligence of the guard in patrolling the grounds, the lack of supervision by the school to insure that the guards were performing their duties, the inadequate locking system which consisted of easily picked locks but no chains or deadbolts, the general accessibility of the campus and its buildings to outsiders, and (according to the expert testimony) the overall inadequacy of Pine Manor's security system, was ample evidence for a jury to find that Pine Manor was negligent in protecting its students, a duty which it had taken on voluntarily. Therefore, the court (with the exception of one justice) affirmed the judgment that Pine Manor's negligence was the proximate cause of Mullins' rape and affirmed the award of damages to her.

COMMENTARY:

There are several interesting aspects of this case which are worth noting:

The appellate court found that the basis for the college's duty of care to its students could be found in either of two well-established principles of law. The first principle is that the college has the duty of care because the social values and customs of the times have caused colleges around the country to undertake the duty of protecting their students. The college community itself has recognized its obligation in this regard. As testified to by the Wellesley campus police chief, the job of designing security systems for campuses is being recognized as a separate profession. The general community has a reasonable expectation, fostered in part by the colleges themselves, that reasonable prudence will be used to protect students from foreseeable danger.

A second principle upon which a college's duty towards its students can be based is the principle that a duty which has been voluntarily undertaken must then be performed with reasonable care. Colleges generally undertake to protect their students from criminal acts, as did Pine Manor

College. If a college then fails to exercise due care and thereby increases the risk of harm, or if the injury is suffered because the student relied on the college's performing its duty properly, the college will be held liable for the injury.

Also noteworthy is the fact that the court rejected the college's argument that the student must show that prior criminal acts had occurred on campus. The court ruled that prior cases which held that previous criminal acts had to be shown were cases involving the general duty of any owner or occupier of land to persons lawfully on the property. But, the court said, the Pine Manor case "concerns only the distinctive relationship between colleges and their students."

It is very clear, therefore, that colleges have a higher duty to protect their students than does the average property owner to protect his guests or customers.

Finally, it may be useful to our college and university security subscribers to know exactly what errors or omissions the college may have committed. The court said the jury could have based its finding of negligence by the college on the following deficiencies in the security system:

(1) The observation post near the main entrance was situated at such a distance from the fence that an intruder could climb the fence without being detected.

(2) The gates to the courtyard were easy to scale (a design defect) or to open.

(3) The walls around the courtyards were too low.

(4) A single key system was used which allowed the same key to open the doors to the commons building, to the dorm, and to the student's room.

(5) Only two guards were on duty at a time.

(6) There was no system to check that the guards were performing their patrols.

(7) The locks on the dormitory and room doors were easy to pick.

(8) Neither deadbolts nor chains were used on the dorm doors or the room doors.

(9) The guards failed to perform their duties.

Perhaps this list should be used as a basic checklist to review the security system on your campus.

PSCLR – MAR 84 (7)

College Cannot Be Held Liable For Assault On Student Unless Same Reasonably Foreseeable:

Sallye Brown, Substitute Administratrix of the Estate of Whelete Venita Collins v. North Carolina Wesleyan College, Inc., *No. 827SC1275, in the Court of Appeals of North Carolina, decided December 20, 1983, reported at 309 S.E.2d 701.*

Whelete Collins was raped and murdered by Kermit Smith on December 3, 1980. Collins, a student and a cheerleader at North Carolina Wesleyan College, had attended a basketball game on that date at the campus. After the game, she and two other cheerleaders were kidnapped from the campus by Smith. She was forced to drive to a rock quarry near Roanoke Rapids, North Carolina, where she was killed. Smith's convictions for the offenses have been upheld.

Collin's estate filed this lawsuit against the college alleging that the negligence of the Wesleyan Collegeofficials permitted the incident to occur. The trial court reviewed the Estate's evidence and determined that it was insufficient to allow the case to proceed to trial against the college. The Estate appealed the judgment in favor of the college, but the North Carolina Court of Appeals agreed with the trial court.

In reaching this conclusion, the court reviewed the law regarding premises liability. As a general rule, a property owner's duty to protect invitees on his property from assaults by third persons does not arise until and unless such an assault is reasonably foreseeable:

> a possessor of land who holds it open to the public for entry for his business purposes is subject to liability to members of the public while they are upon the land for such a purpose, for physical harm caused by the accidental, negligent, or intentional harmful acts of third persons or animals, and by the failure of the possessor to exercise reasonable care to (a) discover that such acts are being done or are likely to be done, or (b) give a warning that will enable the visitors to avoid the harm or otherwise to protect them against it.

The court therefore looked to the plaintiff Estate's evidence to determine whether the assault on Collins could have been reasonably foreseen by the defendant college. That evidence showed only that there had been a break-in at the college business office over 10 years previously, a vending machine vandalism five years previously, and an alleged attempted rape two years previously. Also there was some evidence of a fight between students and local youths in 1980, and some evidence that inmates from a local detention center may have attended the basketball game on December 3, 1980. The court concluded that these random incidents of criminal activity were not sufficient to warrant a jury finding that the abduction and murder of Collins was a reasonably foreseeable occurrence. Furthermore, the court noted that the evidence showed that the defendant's security staff was composed of three full-time students, two full-time nonstudents and a director of security, and that security was reasonably adequate in light of the criminal activity in the past. The court distinguished this situation from an earlier North Carolina case in which the plaintiff's evidence showed 31 reported incidents of criminal activity in the defendant's parking lot in the 12 months immediately preceding the plaintiff's assault. In that instance, it was ruled that the plaintiff did raise a genuine issue of material fact as to whether the assault was reasonably foreseeable.

COMMENTARY:

In reaching the above decision, the North Carolina appellate court also referred to the previous decision of Urbano v. Days Inn of America, reported in **PSCLR**, *November-December 1982, page six, where the trial judge's decision to dismiss the case was reversed because the plaintiff had presented sufficient evidence to show that the assault in the motel's parking lot could have been reasonably foreseen by the defendants.*

It is clear that the owner/possessor of property is not an insurer of an invitee's safety. He is usually under no duty whatever to exercise any care until he knows or should know that the dangerous acts of a third person are occurring or are about to occur on his property. However, he may be held accountable for such knowledge, based on past experience that there is a probability of misconduct on the part of third persons in general which may endanger the safety of invitees, even though he may have no reason to expect such misconduct on the part of any particular individual. If his past experience or the character of his business is such that he should reasonably anticipate misconduct on the part of others, then he may have a duty to protect against it, including the duty to provide a reasonably sufficient number of security guards or other preventive measures to afford reasonable protection to invitees. This is particularly true if past experience shows that the public police cannot be relied upon to provide adequate protection.

PSCLR – OCT 84 (12)

Security Officer Found Liable For Failing To Notify Police Of Presence Of Assailant Thereby Permitting Assault On College Student:

Phyllis Tucker v. Ellis Sandlin, d/b/a Sterling Secret Service, Inc., and Robert Rothgeb v. Charles Stewart Mott Community College, third party defendant, *Docket No. 63219, in the Court of Appeals of Michigan, decided June 22, 1983, rehearing denied August 9, 1983, released for publication August 18, 1983, reported at 337 N.W.2d 637.*

At the trial of this case, the plaintiff presented sufficient evidence to support the following findings: that Robert Rothgeb was a security guard employed by the defendant Sandlin doing business as Sterling Secret Service, Inc., assigned to Charles Stewart Mott Community College; that Rothgeb had actual notice of a life-threatening assault on the fifth floor of the student parking ramp; that Rothgeb promised the victim of that first assault that he would notify the police of the presence of the assailant on the parking ramp; and that Rothgeb failed to notify the police immediately. Furthermore, certain expert testimony showed that there was a breach of the standard of care by Rothgeb in failing to notify the police immediately. The court noted that it was reasonable for the jury to conclude that Rothgeb's omission was a direct cause of the plaintiff's injury in that the security officer's failure to notify the police permitted the assailant to remain on the parking lot ramp and to carry out his assault on the plaintiff.

The plaintiff was permitted during the trial to introduce testimony by Deputy Police Chief Crawford as to how a police officer would have responded to the situation which confronted the security officer. While the appellate court assumed for the purpose of discussion that the standards of care applicable to police officers may differ from those applicable to private security officers, the court nevertheless noted that Deputy Chief Crawford testified clearly that a security guard would not be expected to attempt an apprehension of a suspect as would a police officer, but that the security officer would only be expected to call the police for assistance.

And the appellate court found that the jury's verdict of $80,000 was reasonable for the substantial injuries and severe mental anguish incurred by the plaintiff as a result of the attack on her.

Accordingly, the verdict and judgment in favor of the plaintiff was affirmed.

COMMENTARY:

The legal principle upon which the result of the above case is based is not new to the security professional:

> *An act or omission may be negligent if the actor realizes or should realize that it involves unreasonable risk of harm to another through the conduct of...a third person which is intended to cause harm, even though such conduct is criminal.*

One who undertakes gratuitously or for consideration to render services to another which he should recognize as necessary for the protection of a third person or his things, is subject to liability to the third person if physical harm results from his failure to exercise reasonable care to perform his undertaking, if

> *(a) his failure to exercise reasonable care increases the risk of such harm, or*
>
> *(b) he has undertaken to perform a duty owed by the other to the third person, or*

(c) the harm is suffered because of reliance of the other or the third person upon the undertaking.

PSCLR – NOV-DEC 84 (6)

State Government Held To Similar Duty Of Care To Student Residents Of State-Operated University Housing As Private Landlord Owes To Tenants:

Madelyn Miller v. State of New York, *in the Court of Appeals of New York, decided June 14, 1984, reported at 478 N.Y.S.2d 829.*

A student of the Stony Brook campus of the State University of New York (SUNY), who was raped in a SUNY dormitory by an unknown assailant, sued the State of New York to recover for her injuries and suffering. She was awarded $25,000 at trial; the judgment was reversed on appeal by the New York Appellate Division, and now the Court of Appeals of New York has reversed and again remanded the case to the Appellate Division for reconsideration.

Miller was attacked by an unknown man who confronted her in the laundry room of the SUNY dormitory in which she lived. The man threatened her with a large butcher knife, blindfolded her, and forced her to accompany him outside through the entrance to the laundry room, which was unlocked, and then back into the dormitory through another unlocked entrance. He led her up the stairs into a third floor dormitory room where, under threats of mutilation or death, he raped her twice. He then took her back outside and abandoned her in the parking lot.

Before Miller's assault, there was evidence that non-residents, men in particular, loitered in the hallways of campus dormitories. The plaintiff herself had filed complaints concerning this problem in her dormitory and others had reported the presence of men in the women's restroom. Moreover, the SUNY student newspaper had published reports of crimes committed in dormitories by intruders, including armed robbery, burglaries, criminal trespass and rape. In spite of all this, all ten of the entrances to Miller's dormitory were kept unlocked at all times.

Miller's suit against the State of New York was based on two theories of liability: the State had negligently failed to provide adequate police protection on the campus of SUNY; and, alternatively, the State was liable because it had breached its duty as a landlord to protect its tenants, the on campus student residents, from "reasonably foreseeable criminal assaults by outsiders." The trial court rejected the first claim because police protection is a governmental activity, subject to sovereign immunity. However, Miller was granted judgment on the second theory of liability and she was awarded $25,000 for her emotional damages.

The award and judgment were reversed by the Appellate Division which ruled that the State was immune from liability as a landlord because this claim merely restated the inadequate police protection claim.

On review, the New York Court of Appeals reversed again, stating that the public immunity for inadequate police protection could be overcome by a showing of a "special relationship" between the State and the injured party, in which the State had assumed a specific duty to protect the party and that party relied on the State to fulfill that duty. In this case the landlord/tenant relationship established between the State and the resident of the State-operated campus housing gave rise to a specific duty of protection by the State: "It is not disputed that when the State acts in a proprietary capacity as a landlord, it is subject to the same principles of tort law as a private landlord."

The court recognized that the issue of liability was confused by the defendant's dual role, on the one hand, as a landlord by the ownership and control of the SUNY campus and, on the other, as a governmental entity by the appointment of campus security officers to maintain law and order on the campus. The court stated:

> A governmental entity's conduct may fall along a continuum of responsibility to individuals and society deriving from its governmental and proprietary functions. ... Consequently, any issue relating to the safety or security of an individual claimant must be carefully scrutinized to determine the point along the continuum that the State's alleged negligent action falls into, either a proprietary or governmental action.

Applying this test here, the court ruled that the State's duty to maintain minimal security measures in the SUNY dormitory where criminal intrusion was foreseeable, fell under the State's proprietary function as a landlord. Further, there was sufficient evidence to support the trial court's conclusion that the State's failure to keep the dormitory entrances locked was a breach of that duty and a proximate cause of Miller's rape.

The court remanded the case to the Appellate Division for consideration of the facts and issues not considered in the earlier appeal.

PSCLR–FEB 85 (7)

College Owes Duty To Protect Students From Reasonably Foreseeable Assaults On Campus; May Be Liable To Student Victim For Failing To Warn Of Known Danger:

Kathleen Peterson v. San Francisco Community College District, et al., *S.F. 24587, in the Supreme Court of California, in bank, decided September 6, 1984, reported at 685 P.2d 1193.*

Kathleen Peterson, the victim of an attempted rape while on the campus of the City College of San Francisco where she was enrolled as a student, sued the college. The case was dismissed by the trial court; the dismissal was reversed by the Supreme Court and remanded for trial.

Peterson was attacked as she was using a stairway in the school's parking lot. An unknown man leapt from behind "unreasonably thick and untrimmed foliage and trees" and grabbed her, attempting to commit rape. She was physically and emotionally injured by the attack, and suffered economic losses. Because similar attacks had occurred on that same stairway previously, the college had attempted to improve the security in the parking area and its stairways; Peterson claimed to have relied on this increased security.

Peterson's suit alleged two causes of action: First, the school district knew of the danger in the parking area and it had a duty to protect and/or warn her of the danger. And, second, the school district was liable for her injuries because the "dangerous condition" maintained on the campus, together with the third party's criminal conduct, had caused her injuries. The court considered the nature of the relationship between Peterson and the school district, the question of whether the school owed a duty to protect Peterson, and whether that alleged duty was affected by the fact that the defendant was a public entity.

Although there is no general duty to warn potential victims of (or to protect others from) third party criminal conduct, a duty may arise in situations where there is a "special relationship" between, for example, a property owner and the victim of an attack on his property which imposes a duty on the property owner to protect the victim. A duty may also be imposed on the property owner to control a third party's behavior if there is a special relation between such third party and the property owner.

In considering whether the school owed a duty to Peterson, the court recognized that Peterson, as an enrolled student who paid a fee to use the parking lot, would be considered the invitee of the college. It further noted that several factors to determine the extent of a property owner's duty to an injured invitee must be weighed, including:

> the foreseeability of harm to the plaintiff, the degree of certainty that the plaintiff suffered injury, the closeness of the connection between the defendant's conduct and the injury suffered, the moral blame attached to the defendant's conduct, the policy of preventing future harm, the extent of the burden to the defendant and consequences to the community of imposing a duty to exercise care with resulting liability for breach, and the availability, cost, and prevalence of insurance for the risk involved.

In this situation, all the elements may indeed be present: Assuming that the college was aware of previous attacks on the parking area stairway, it knew that women were likely to use that stairway, thereby becoming potential victims of assault. It is not disputed that Peterson suffered injuries. The college's failure to warn students of the prior assaults and to exercise extra caution in the area, to trim the foliage, or to take other measures to protect students, provides the "close connection" between the school's conduct and Peterson's injuries. If the preceding factors are established by the evidence, then the defendant's failure to prevent a foreseeable harm would infer a moral blame. Finally, the imposition of a duty on the college would further a policy of preventing harm and would not place an intolerable burden on the defendant.

Assuming the plaintiff was able to prove the facts as alleged, the school as a private property owner would clearly owe her a duty of due care. The question then becomes whether that duty is affected by the fact that the defendant is a public entity, the liability for which is controlled by California's Tort Claims Act which statutorily limits the liability of public entities in certain situations. Under this act, a public entity is not liable under the common law duty to protect invitees, but may be liable for injuries caused by a dangerous condition (a condition of property that creates a substantial risk of injury when such property or adjacent property is used with due care in a manner in which it is reasonably foreseeable that it will be used) on its property if:

> The plaintiff [establishes] that the property was in a dangerous condition at the time of the injury; that the injury was proximately caused by the dangerous condition; that the dangerous condition created a reasonably foreseeable risk of the kind of injury which was incurred; and that either (a) a negligent or wrongful act or omission by an employee of the public entity within the scope of his employment created the dangerous condition; or (b) the public entity had actual or constructive notice of the dangerous condition. ...

The majority of cases interpreting this statute have held that the conduct of third parties, alone, is not enough to constitute a "dangerous condition" for which a public entity may be held liable, unless that conduct was related to the condition of the property itself.

Although a public entity is clearly immune from liability for failure to provide adequate police protection under the law, Peterson's complaint goes further to allege that the school's failure to warn of her a known danger, to remove excessive foliage around the stairs, or to institute protective measures to prevent foreseeable harm to students, constituted a dangerous condition which caused or contributed to her injuries. The Supreme Court of California concluded that although a common law duty of care was insufficient to overcome the school district's immunity, the school was not immune under the Tort Claims Act for failure to warn its invitees of a known danger. Peterson was therefore entitled to pursue this action and to attempt to prove the allegations at trial.

PSCLR–JAN 89 (4)

Private University May Be Liable For Death Of Student Shot In Dormitory If It Failed To Provide Adequate Security Against Reasonably Foreseeable Danger:

George H. Nieswand and Patricia A. Nieswand, Administrators and Administrators Ad Prosequendum of the Estate of Erin Nieswand, Deceased; and George H. Nieswand and Patricia A. Nieswand, Individually v. Cornell University, a corporation organized under the laws of the State of New York v. Su Yong Kim, *No. 86-CV-1234, in the United States District Court for the Northern District of New York, decided August 17, 1988, reported at 692 F. Supp. 1464.*

Erin Nieswand, a freshman at Cornell University, was shot and killed in her dormitory room by a rejected suitor of her roommate. Nieswand's parents sued Cornell University for the wrongful death of their daughter, claiming that Cornell negligently failed to provide adequate security at the dormitory and that Cornell breached its implied contract regarding dormitory security based upon brochures mailed to prospective and accepted students.

Erin Nieswand shared a room in a North Campus dormitory at Cornell University in Ithaca, New York, with two other students, Jane Niehaus and Young Hee Suh. During the fall term, Suh received love letters and presents from Su Yong Kim, a non-student. Suh did not return Kim's feelings and, when he appeared at her room to visit, she asked her advisor to tell Kim that she was not around. The resident advisor offered to have Kim removed by campus security, but she rejected this offer and subsequently talked to Kim.

A few months later, Kim somehow gained access to North Campus 7. The dormitory doors were supposed to have been locked for the weekend, and it was not known how Kim gained entry to the dormitory. Suh talked with Kim, who remained in the corridor, through an open door, and then went to her room to talk with him. Kim attempted to close the door behind them, but Nieswand and four other students entered the room.

Kim suddenly pulled out a hidden rifle and announced that he intended to kill all six people in the room. Kim later allowed everyone but Suh and Nieswand to leave. Shortly after, two series of gun shots sounded.

Suh, who was shot three times, was found dead. Nieswand, who was shot twice, was found alive, was taken to a hospital, and subsequently died.

Nieswand's parents sued Cornell University for the wrongful death of their daughter, alleging negligence and breach of an implied contract based on the brochures that Cornell had sent to prospective and accepted students. The Nieswands alleged negligence on two different theories of liability: (1) Cornell, as a landowner, failed to provide adequate security; and (2) Cornell, by virtue of its relationship with its students, was obligated to provide adequate security.

To prove a case of negligence, a plaintiff must establish: (1) the existence of a duty on the part of a defendant to plaintiff; (2) defendant's breach of that duty; and (3) an injury proximately caused by the breach. The issue was whether Cornell owed Nieswand a duty to undertake security or protective measures in the dormitories it operated.

In general, a landowner has a duty to exercise reasonable care under the circumstances to maintain the property in a safe condition. The landowner, however, is not "an insurer of safety", and

> cannot be held to a duty to take protective measures unless he knows or has reason to know that there is a likelihood of conduct on the part of third persons which would endanger the safety of the visitor.

A landowner can be put on notice of such danger "if past history of criminal activity indicates that a criminal incident is a significant, foreseeable possibility." To avoid liability for negligence, a college must take minimal precautions to protect its students from the reasonably foreseeable acts of third persons.

Cornell argued that there was no evidence that it knew or should have known that Kim would endanger the life of any Cornell student. Student witnesses described Kim as a "well-dressed man who never caused a disturbance." Furthermore, Suh had rejected her resident advisor's offer to contact the Department of Public Safety. The University's Department of Public Safety had no records of any reports concerning Kim prior to the incident.

Cornell also asserted that there was no evidence of an extensive history of criminal conduct on the campus sufficient to impose a duty to provide security for the dorms. Prior to this incident, there had not been a murder or an attempted murder in the history of the university.

There was, however, some evidence that Cornell was aware of the threat posed by Kim. Nieswand's father testified that he had been advised by a Residential Halls official that Kim had been escorted off of the campus by university security on another occasion. In addition, the resident advisor knew that Suh had attempted to avoid Kim by hiding in the advisor's room.

Statistical evidence showed that, during the three school years prior to Nieswand's death, four rapes, eight robberies, nine assaults, 42 "other assaults," 19 sex offenses, 793 burglaries, and 2,582 larcenies occurred on the Cornell University campus.

Therefore, a significant question existed as to whether Cornell University could have foreseen the criminal activity that occurred in this case so as to give rise to a duty on its part to provide adequate security measures. Further, if the duty did arise, a question of fact for a jury existed as to what safety precautions would have been appropriate. A jury would have to take into account such variables as the seriousness of the risk and the cost of the various available safety measures.

The court also held that Nieswand's parents were not required to show a "special relationship" between Erin Nieswand and Cornell, creating a duty to provide police protection, because their claims were based upon Cornell's status as a landowner in its operation, maintenance, and supervision of its dormitories. In this case, Cornell would be liable only if it had failed to exercise reasonable care to prevent or minimize reasonably foreseeable danger.

Nieswand's parents also sued Cornell University for breach of an "implied contract" created between Cornell and Erin Nieswand, based upon a series of documents, brochures, leaflets, and pamphlets that Cornell sent to prospective students and to students accepted for enrollment. A brochure entitled "Living on Campus, Cornell University" contained a representation that:

> Concern for personal safety and security of property is shared by all segments of the University community. The Department of Resident Life provides facilities and programs that reflect this concern. Each building is locked during the evening and night, and periodic personal security programs are offered within the residence areas.

In addition, the "Residence Hall Policy and Safety Procedures" made similar representations, noting that "the ultimate responsibility for personal security is that of each resident," and including procedures for students to follow to keep the dorms and their rooms secure.

The court concluded that in this case, in which a student is seeking to enforce the provisions of the documents distributed to it by the university, "the rights and obligations of the parties as contained in the University's bulletins become a part of the parties' contract." The court rejected the argument that this rule would make Cornell "a guaranty of security." Contrary to this suggestion, the provision only required Cornell to lock each building at night and offer periodic security programs. In addition, the students had certain obligations to maintain adequate security, as provided in the brochure.

The court concluded that a significant question of fact existed as to whether either party had failed to meet the obligations under the contract established by the University brochures. The court, therefore, ordered that the case proceed to trial on both the negligence and breach of contract claims.

Employment Relationship

A. Liability to Employees

PSCLR – JUN 82 (2)

Owner Of Plant Liable To Widow Of Security Guard, Who Was Run Over and Killed While Performing Duties of Inspecting Truck, For Failing To Take Steps To Protect Guard From Danger:

Evelyn L. Zrust, as personal representative of the estate of John A. Zrust, deceased v. Spencer Foods, Inc., *Case No 81-1453, in the United States Court of Appeals for the Eighth Circuit, decided January 11, 1982, reported at 667 F.2d 750, on appeal from the United States District Court for Nebraska.*

John Zrust was a security guard employed by Per Mar Security and Research Corporation. Per Mar had an agreement with Spencer Foods to provide Spencer with security guards at Spencer's meat packing plant in Colfax County, Nebraska. Pursuant to this agreement, Zrust was assigned to work at the plant under the direction and supervision of Spencer Foods.

The guardhouse at the plant is located right at the exit from the plant, and among other duties, the security guards are responsible for checking each truck loaded with meat before it leaves the premises. Specifically, the guards record and adjust the readings of the thermostat and the temperature control for the refrigerated trailers, check the seals on each trailer, and copy certain information from the truck's bill of lading into a logbook.

On January 13, 1978, Zrust was run over and killed during his inspection of a truck driven by Donald Mathis. Mathis, who thought the inspection had been completed, started his truck forward while Zrust was reading the thermostat control for the trailer. Because this control is located behind the truck cab, and in front of the trailer, the driver could not see Zrust.

Zrust's widow, Evelyn Zrust, filed a wrongful death action on behalf of his estate against the driver Mathis, and against the lessor of the truck driven by Mathis, and against Spencer Foods.

At trial, the plaintiff contended that Spencer was negligent in failing to provide Zrust with a safe place to work. Specifically, she argued that Spencer failed to provide an electric barrier which could be controlled only by the guard from inside the guardhouse and which would ensure that each truck remained motionless during the check-out procedure. She offered evidence that similar plants had installed such automatic barriers as safety devices, but that the Spencer plant had only a manual gate which was not controlled from a place of safety because the guard had to leave the guardhouse in order to move the gate blocking the truck. The guards in fact usually used the gate only at night for plant security and not as a barrier for the trucks during inspection.

Spencer countered that the danger was open and obvious and that its supervisors' general warnings to all guards of the dangers of working around the trucks satisfied its duty to care.

The trial court returned a verdict for the decedent guard's estate in the amount of $175,000. The Eighth Circuit Court of Appeals, on review, affirmed the verdict on the basis that a landowner is responsible for open and obvious dangers if it is unreasonable to expect the invitee/employee to protect himself or if the duties of employment require unavoidable exposure to danger.

COMMENTARY:

A landowner clearly is required to keep its premises in a reasonably safe condition for its invitees/employees, but generally an owner has no duty to protect an invitee from known and obvious hazards from which the invitee can protect himself.

However, here we see that the law also states that the owner may be liable even for an obvious danger if it is unreasonable to expect the invitee to protect himself, as for instance where his attention may be distracted or the duties of his employment may require his unavoidable exposure to it.

Therefore, care should be exercised not only to keep the premises safe by inspecting for defects and dangerous conditions, and not only by warning invitees/employees of known dangers, but also by actually taking whatever measures are reasonably necessary to correct the defects and to protect people from dangers conditions, even the obviously dangerous conditions.

PSCLR – FEB 83 (3)

Female Employee Denied Recovery from Employer for Injuries Received in Attack by Male Co-Worker in Female Bathhouse Facility:

Patsy Gaskin v. Republic Steel Corporation, 81-9, *in the Supreme Court of Alabama, decided September 24, 1982, reported at 420 So. 2d 37.*

Patsy Gaskin, an employee of the Gadsen, Alabama, Republic Steel plant, stopped at the plant's bathhouse as she was leaving work on October 4, 1979. She was followed into the shower by a co-worker, Gerald Faulkner, who attacked her and tried to rape her. She frightened him away by screaming, and he was apprehended while still on company premises.

Gaskin sued Faulkner in a separate case, and sued Republic in this action. She claimed that Republic was negligent in placing the women's showers adjacent to the men's showers and near where male employees worked. Furthermore, she felt that Republic was negligent in failing to properly protect or guard the area, or to provide locks on the shower facility.

The court affirmed the dismissal of her claim, stating the test for imposing liability on an employer for assaults by third parties on employees:

> ... an action such as this will not lie except in the most extreme case where it is clearly shown that the employer in some manner, greatly and unreasonably, increased the risk to the employee without taking reasonable safety precautions.

The court ruled that the shower maintained by Republic did not greatly and unreasonably increase the risk of harm to its female employees.

PSCLR – FEB 83 (4)

Employee Shot by Intruder in Secured Area of Plant Could Not Recover from Employer or Security Company for Failure to Protect Her:

Madonna York v. Modine Manufacturing Co., a corporation, and R.R.S., Inc., a corporation, *No. 482-0190, in the Appellate Court of Illinois, Fourth District, decided November 8, 1982, reported at 442 N.E.2d 284.*

Modine Manufacturing Company and its contract security service, R.R.S., Inc., were sued by Modine employee Madonna York after she was shot on company property by Dennis York, who was not a Modine employee. York had been threatened by Dennis York previously and she had told Modine management that she did not want to have contact with him while on the job. Nonetheless, on November 8, 1979, Dennis York was allowed into a "secured area" of the plant and stayed there without supervision for three and one-half hours until 12:05 AM when he shot York.

York sued Modine for failing to exercise reasonable care in undertaking to provide security service for the protection of its employees. She alleged that her employer was liable because of its (1) failure to inform the security company R.R.S. of the threats on York's safety, (2) failure to inform York that she could not rely on the guard service to protect her and, (3) failure to hire a security service which would prevent the unauthorized entry of nonemployees into secured areas of the plant.

York also sued R.R.S. for its alleged negligence, complaining that it (1) failed to prevent the unauthorized entry by her assailant into a secured area of the plant, (2) failed to discover that Dennis York was in possession of a deadly weapon, (3) failed to keep Dennis York under observation while he was in a secured area, (4) failed to evaluate security and to advise Modine of the increased risk to its employees caused by inadequate security service, and (5) failed to inform York and other employees that they could not rely on the guard service for protection.

The court, which described this case as a "premises liability" case, stated that in Illinois, the extent of a landowner's duty to people on his property is determined by the status of the person injured. The court went on to say that the basis for the premises liability doctrine is to give "some redress to persons who would not otherwise possess it." In this situation, however, York's status as an employee gave her the opportunity for redress through the Workmen's Compensation Act, and in some instances, the Federal Employer's Liability Act. According to the court, "Any extension of an employer's duty to an employee beyond the ambit of these statutues should be approached with great caution."

Aside from her status as an employee, the court also said that it could find no facts which would raise some duty by Modine to protect York. Furthermore, there were no facts to show that the security company was hired to protect employees. Accordingly, York's suit against her employer and the security company was dismissed.

COMMENTARY:

Illinois remains a "status" jurisdiction; that is, as the court stated, the degree of care owed by a landowner to those on his property is determined by the "status" of those who come on the property. There are varying degrees of care owed depending on such status: great care, ordinary care, or slight care. This rule has been much criticized, even in Illinois (a strong dissenting opinion was written by an Illinois Supreme Court justice in 1976—see 361 N.E.2d 282), but it remains the law in that state.

One thing to be noted in the above case is that the plaintiff York failed to allege any facts to support her allegation that the security service was hired to protect employees as well as plant property. The court deemed this a critical factor and it is doubtful that her suit would have been dismissed had she been able to do so. All contracts with security companies should set out in detail the specific services to be provided and the protection to be accomplished. If the personal safety of employees is not to be covered, then the contract should be clear in this regard.

PSCLR – MAY 83 (2)

Retail Store Manager Cannot Recover From Security Company for Injuries Received When Shot by Robber in Store Protected by Security Company's Guards:

Joseph M. Paradiso v. Apex Investigators & Security Co., Inc., *in the Supreme Court of New York, Appellate Division, First Department, decided January 18, 1983, reported at 458 N.Y.S.2d 234.*

A manager for a Pathmark supermarket in Whitestone, Queens, in New York City, sued Apex Investigators & Security Company after he was shot during an armed robbery of the supermarket. He claimed that his injuries resulted from the negligence of the supermarket's security guard who had been furnished by Apex under an agreement with Pathmark. The suit was dismissed and the supermarket manager, Joseph Paradiso, appealed.

Paradiso was working in the counting room of the Pathmark store with another employee just after the 10 P.M. closing time on the night of June 4, 1978. Two armed men wearing clown masks forced their way into the room and demanded "the key." When Paradiso said, "I don't know what key," the nervous gunman replied, "Don't fool around, I want the key," put the gun to Paradiso's arm and it fired.

Shortly before the robbers entered, the store's security guard had gone to the bathroom with Paradiso's permission. He did not return until the robbers left. Paradiso's suit alleged that the security guard had neglected to lock the supermarket's doors and had not properly inspected the store before leaving his post. In addition, Paradiso felt that the guard could hear the robbery taking place and intentionally stayed away until it was over. Paradiso contended that the guard's presence could have deterred the robbers. He pointed out that the rules and regulations of Apex, applicable to all members of its uniformed security force, included directions never to permit "unauthorized persons ... to enter the premises [being guarded] or to loiter in unauthorized areas during a Security Officer's tour of duty." According to Paradiso, this placed an obligation on the security guard to secure the area, or at least to inspect for "unauthorized persons" before leaving his post.

With the exception of one of the court's five justices, the court held that the agreement between Apex and Supermarkets General did not indicate any obligation on the part of Apex security to protect the store's manager. Nor did the majority of the court feel that the unarmed guard was under any duty to prevent armed robberies or that his absence during the robbery could be considered the cause of Paradiso's injuries. In addition, the majority ruled that there was no negligence on the part of Apex which precipitated the injury to Paradiso. His case was dismissed without ever reaching jury trial.

COMMENTARY:

The dissenting justice's opinion was strongly worded, and concluded that the case should not have been dismissed prior to being presented to a jury. The dissenting judge felt that a jury should decide the nature and extent of the services to be rendered by the security guard, whether he should have inspected the store, and so on. In any event, this case reflects the necessity for continuing review and possible revision of contractual arrangements between security companies and their clients. New case decisions such as this repeatedly point out the legal consequences of the failure to include specific provisions.

PSCLR – MAY 83 (9)

Employer Has Duty to Establish Policies to Require Youthful Job Applicants to Present Proof of Age:

John Romano v. Henry M. Lomasney, John Lomasney, Eddie Borne, Commercial Union Insurance Company, and others, *No. 13164, in the Court of Appeal of Louisiana, Fourth Circuit, decided December 2, 1982, reported at 423 So. 2d 787.*

John Romano filed suit in New Orleans claiming damages for injuries he sustained while working for Southern Imperial Coatings Corporation, a paint manufacturer. His hair, pulled back into a pony tail, became caught in the shaft of a high speed mixing machine. His hair and scalp were ripped from his eyebrows to the back of his neck, leaving him permanently disfigured.

At the time of the accident Ramano had worked at Southern for several months, but he had lied on his employment application by stating that he was 18 years old, the minimum age for employment according to company policy. Southern did not, as a matter of course, require that applicants show proof of age. The Child Labor Law states:

> No minor under the age of eighteen years ... shall be employed, permitted or suffered to work: (1) In oiling, cleaning, or wiping machinery or shafting, or in applying belts to pulleys; ...

The court held that Southern had not fulfilled the duty imposed by this statute when it employed the 17-year-old Ramano to clean and wipe machinery. It was further pointed out that employers have been held to have breached this duty in situations where the minor had been hired inadvertently.

As a matter of policy, Southern did not employ persons younger than 18 years. As a matter of policy, however, neither did it ask for proof of age from employment applicants. A corporation has a duty to establish and enforce policies that require youthful person who apply for employment to present proof of age before they are employed.

The court stated that Southern was guilty of negligence in hiring Ramano to perform industrial work and that, like any corporation, it was under an obligation to "establish and enforce policies that require youthful ... [applicants] ... to present proof of age before they are employed. This is especially true when they are to be employed in industrial work."

Verdicts of liability were directed against Southern and its executive officers by the trial court. A jury, presented with the issue of damages alone, awarded Ramano $500,000.

The liability verdicts and damages award were affirmed on appeal.

COMMENTARY:

Be certain that you not only familiarize yourself with the child labor laws in your state, but that you also establish and enforce company hiring policies to insure that the laws are followed! Violations of these laws can not only carry stiff criminal penalties but, as seen in the above case, severe civil liabilities as well. Applicants for positions as armed guards or couriers should be required to present proof of their experience and training, in addition to proof of their age, because of the higher standards and increased liabilities relative to the use of weapons.

PSCLR—MAR 84 (2)

Motel Which Leases Space To Gift Shop Operator, Whose Employees Are Injured By A Known Hazardous Condition On Premises, May Be Liable For Failure To Protect Such Employees:

Gail Keller v. Holiday Inns, Inc., a Tennessee corporation, Great Western Investment Co., and Winegardner & Hammons, Inc., and Gene Burman and Joan Keller Burman, husband and wife v. Holiday Inns, Inc., a Tennessee corporation, Great Western Investment Co., and Winegardner & Hammons, Inc., *No. 14270, in the Court of Appeals of Idaho, decided November 4, 1983, reported 617 P.2d 1112.*

Is a property owner who leases space to a business operator, and who allows the business operator to bring unsafe appliances on the premises, liable to the operator's employees when they are injured by the appliances? This case addresses that troublesome issue.

Sisters Gale and Joan Keller were injured when a heavy iron security gate collapsed on them while they were working for a gift shop located in the Boise, Idaho, Holiday Inn. The sisters sued, among others, the Holiday Inn's corporate owner, even though the motel merely rented the space to another entity which operated the gift shop and employed the Kellers. An Idaho District Court granted the Holiday Inn summary judgment, and the Keller sisters appealed.

The gift shop was located inside the Holiday Inn, and was rented from the hotel on a month to month basis. The gift shop operator installed a portable iron security gate which surrounded the shop when it was not open for business. As part of their opening duties, the Kellers (usually with the assistance of Holiday Inn personnel) folded the gates, moved them on wheels, and stored them behind some doors in the motel hallway between the lobby and guests' rooms. The gates were known to be unstable and had collapsed several times, injuring gift shop and hotel employees. The Holiday Inn manager knew of and allowed the gates to be used by the gift shop proprietor and permitted them to be stored in the hotel hallway during operating hours.

The Idaho Court of Appeals began their review of this case by defining the Kellers, though employees of the motel's lessee, as "invitees" of the motel at the time of the accident. The gift shop clearly served a business purpose to the benefit of the motel, and by definition invitees are those who enter land "for a purpose connected with the business conducted there." Furthermore, the fact that the Kellers were employees of the lessee did not affect their status as invitees of Holiday Inn, particularly because the actual injury occurred while moving the gates through the motel.

The court then examined the motel's duty of care to the Keller sisters for the accident by applying two different theories concerning the scope of a landowner's duty: the "traditional notion," that a landowner cannot be held liable for an accident resulting from an obvious danger or one that was known to the plaintiff, and the "modern view," that the landowner's liability can be limited, but not excused, because the danger was obvious or the plaintiff had prior knowledge of the danger. The court chose the modern view, relying on two recent decisions by the Idaho Supreme Court which upheld this stand. One of the decisions held that liability may be limited where the risk was known and obvious to the injured party, except where "the land possessor should anticipate the harm despite such knowledge or obviousness."

The Idaho Court of Appeals also noted that regardless of whether the gates were considered "a product of a third party's conduct" or a condition of the property, a jury could find that the gates presented a foreseeable danger to invitees, that Holiday Inn knew or should have known that the gates might injure invitees, and that Holiday Inn should have taken some action to protect its invitees, which included the Kellers. Since the court found that certain questions of fact were yet to be resolved, including the extent of the Kellers' contributory negligence (if any), the summary judgment granted to Holiday Inn was reversed and the case remanded for trial.

COMMENTARY:

For readers who are interested in an in-depth analysis of the legal issues presented by this case, a review of the actual opinion would be extremely helpful. The opinion is too lengthy to be summarized here in any detail, but the following is an outline which indicates the comprehensive treatment given this subject by the court:

Part I is an overview of negligence liability, including the elements of negligence and the general grounds of showing a "duty of care" to exist. Part II examines the duty of landowners to protect others on their property from dangerous conditions and from hazardous activities of third person, and distinguishes between the "traditional" rule and the "modern" rule (which expands the owner's potential liability). Part III of the opinion applies the adopted rule to the facts of this case.

This case indicates once again the developing trend across the country to increase the property owner's liability for injuries to invitees. In view of this trend, property owners should be aware of their duty to protect even their lessee's employees from hazards on the premises, including even known and obvious hazards. They should also take care that their insurance coverage includes liability protection against such claims.

PSCLR – APR 84 (2)

Former Owner Of Business Will Not Be Held Liable For Employee's Death Where New Owner Had Ample Opportunity To Know Of And To Correct Design Deficiency And Unsafe Practices By Employees:

Raymond R. Martin, Sr., and Marianne Martin, Co-Administrators of the Estate of Laura Martin, Terese Dudek and Maureen Kincaid v. 1727 Corp., and John Portman, *No. 82-2104, in the Appellate Court of Illinois, First District, Fifth Division, decided December 23, 1983, reported at 458 N.E.2d 990.*

The liability of a former McDonald's restaurant licensee for the death of an employee killed during a robbery at the restaurant was explored in this case. The plaintiffs argued that the former licensees, John and Anne Portman (Mr. Portman was also president of 1727 Corp., a corporate entity created in 1974) sold the restaurant in 1979 to McDonald's Restaurants of Illinois, Inc. ("McDonald's), without informing the buyer of a hidden, defective and dangerous condition of the property. Specifically, the plaintiffs (Mr. and Mrs. Raymond Martin, parents of an employee who was killed at the restaurant) referred to the safety and adequacy of the restaurant's closing procedures and the construction of its back door.

The Martins' lawsuit was based on the following facts: The License Agreement under which the Portmans and 1727 Corp., operated the restaurant from 1971 to 1979 (when it was purchased by McDonald's), included an operations manual, "Cash Register Procedures and Security." This manual stated that every side and back door of the restaurant should contain a "security window." Further, the manual prohibited use of the back door after sundown and required that any trash accumulated after sundown be removed the next morning by employees who opened the restaurant. However, Mr. Portman and 1727 Corp., required employees to remove trash every evening through the rear door which was equipped with a peephole and not a security window.

The restaurant was never robbed from 1971 to 1979; however, only 29 days after McDonald's purchased the restaurant, an armed man entered through the rear door as an employee was removing the evening trash. He robbed the store and killed the Martin's daughter, Laura.

According to the Martins' lawsuit, the type of back door and the improper closing procedures were the result of Mr. Portman's negligence and presented "latent or concealed conditions which defendants knew would create an unreasonable risk of harm, and that defendants should have realized that the new owner, McDonald's, would fail to discover or appreciate these risks." The Appellate Court of Illinois, however, agreed with the trial court and held that even if the conditions described by plaintiffs were dangerous and latent conditions of the property, the new owner in this case could hardly be considered a naive victim operating under the mistaken belief that the property was safe: McDonald's itself constructed the restaurant and installed the peephole door. When it purchased the restaurant in 1979, the McDonald's operations manager and security manager met to take an inventory of the store's assets and to review procedures with the restaurant's manager, including the rule, "After closing, the store doesn't get opened. Everybody leaves together through the front door." New locks were installed on the doors at that time, keys were distributed, and a new combination was set on the safe. Within the 29 days between the purchase of the restaurant and the robbery, McDonald's security manager recommended that a closed hatch window be installed on the back door.

The Martins also urged the court to adopt the "New York rule," under which a vendor's liability for known defects extends until enough time has passed for the vendor to remedy the defects. This Illinois Appellate court, however, declined to follow that rule because the Illinois Supreme Court had previously declared that it was not generally accepted and should not be generally applied to excuse a new buyer from liability for a known defect.

Thus dismissing all of the Martin's contentions, the Appellate Court of Illinois affirmed summary judgment to 1727 Corp. and John Portman, the former owners of the McDonald's restaurant. The case against the new owners will proceed to trial.

COMMENTARY:

The general rule of law provides that a landowner's liability for injuries occurring in connection with the property is terminated at the time the landowner ceases ownership, possession, or control of the property. There is, however, an exception to this general rule which applies when the seller conceals or fails to disclose to the buyer some condition on the property which poses an unreasonable risk of harm to persons on the property. Such a seller is liable if the buyer does not know or does not have reason to know of the unsafe condition and if the seller does know or should know of the condition and also has reason to believe the buyer will not discover the condition or realize the risk involved. If the seller actively conceals the unsafe condition, then his liability will continue until the buyer discovers the unsafe condition and has reasonable opportunity to correct it.

The "New York rule" cited by the plaintiffs in the above case provides that the seller's liability continues to exist, even in the case of a known defect, until a sufficient time has passed to permit the buyer to remedy the defect. The Illinois appellate court declined to adopt the New York rule.

PSCLR – MAY 84 (5)

Franchisor May Owe Duty To Franchisee's Employee To Require Franchisee To Maintain Safe Place Of Employment:

Richard A. Clem, as administrator of the Estate of Janet L. Portman, deceased, and Thomas M. Portman v. Steveco, Inc., and the Southland Corporation, *No. 4-782A192, in the Court of Appeals of Indiana, Second District, decided June 29, 1983, reported at 450 N.E.2d 550.*

The Court of Appeals of Indiana here ruled that a franchisor (the Southland Corporation) may owe the employees of its franchisee store (Steveco, Inc.) an independent duty to require the franchisee to maintain a safe place of employment. This would be required where the franchisor retained sufficient control over its franchise. Depending on the degree of control or direction which the franchisor retained or exercised over its franchisee under the terms of the franchise agreement, the franchisor might be liable to an employee who was injured or killed due to the franchisor's failure to maintain such safe work environment.

Richard A. Clem, as the administrator of the estate of Janet L. Portman, and Thomas M. Portman, the husband of Janet Portman, jointly brought this negligence action against Steveco, Inc., and against the Southland Corporation. Steveco was the employer of Portman and the Southland Corporation was the franchisor of the retail store in which Portman was employed by the franchisee Steveco. Portman worked the night shift at a 7-Eleven Store in Indianapolis, Indiana. She was working alone on March 19, 1980, when she was abducted and murdered. The lawsuit claimed that Steveco and Southland failed to take reasonable steps to provide Portman a safe working environment and that such failure was the cause of her death.

In support of this claim, the plaintiffs assert that the convenience store had no camera surveillance or monitoring equipment, no security guards, no alarm system, and "no other ostensible means of security to protect the store premises or the people."

The trial judge dismissed the action against the employer Steveco because, the court found, the exclusive remedy for recovery against the employer for Portman's death is under the Workmen's Compensation Act (which limits death benefits to $1,500). The appellate court agreed with the trial court, finding that Portman's death was the result of and arose out of her employment as a cashier at the convenience store. Therefore, the Workmen's Compensation Act provided the sole and exclusive remedy.

However, the liability of a franchisor to the employees of its franchisee posed a more difficult question for the appellate court. This is, the court noted, a "developing area of the law." The court noted that the results of previous cases have been contradictory, even when confronted with identical franchise agreements. After reviewing the conflicting cases, the appellate court concluded that it would be necessary to examine the degree of control or direction the franchisor Southland retained over the operations of its franchisee Steveco. The case therefore was remanded to the trial court for further evidentiary hearings on this issue.

COMMENTARY:

Courts around the country have attempted to analogize the franchisor-franchisee relationship to other relationships under the law such as landlord-tenant, parent-subsidiary corporation, employer-contractor, lesser-lessee, principal-agent and the like. These analogies, however, are not directly on point because, with the sole exception of the employer-independent contractor relationships, all the other relationships are based to some extent on agency principals. Only the employer-independent contractor relationships carries with it some sort of connection which is similar to the franchisor-franchisee's employee relationship.

The franchise agreement itself should be the starting point for an investigation into the degree of control which the franchisor has the right to exercise over its franchisee's operations. Here again we see an area of contractual dealings which should be handled by well-versed professionals conversant with statutory and case law requirements in the particular jurisdiction(s) con-

cerned. Such professional insight is particularly needed in dealing in areas of the law, as noted by the court above, which are still "developing" and which have resulted in differing decisions in different jurisdictions.

PSCLR – JUN-JUL 84 (2)

Employer Liable For Injuries To Employee Who Continued Employment After Inducement By Employer's False Representations That It Would Provide Security Service:

Sandra Childress v. Church's Fried Chicken, *A021166, in the California Court of Appeal, First District, Division 5, decided November 10, 1983, reported at 404 Cal. Rptr. 404.*

In this case Sandra Childress, an employee of Church's Fried Chicken, brought an action against her employer for mental distress, emotional suffering and personal injuries due to its failure to provide sufficient security services at the restaurant location where she was employed. The trial judge dismissed her case and she appealed. The Court of Appeals reversed her dismissal, finding that her allegations, if true, would support a cause of action against her employer, would warrant an award of damages in her favor, and should therefore be presented to a jury at trial.

Childress was employed as a bookkeeper at a Church's outlet in Oakland, California. The Church's corporation, the owner and operator of the restaurant, had knowledge that the particular location was the scene of certain criminal activity and that people on the premises were subject to criminal attack by others. The restaurant had been robbed on a number of occasions, and such criminal activity was uncontrolled by Church's. The restaurant owners also knew that their own employees were in constant danger of robbery or assault or other criminal attack while at work; in fact, one of its employees had been kidnapped and murdered and Childress herself had been robbed. Childress advised her employer that she wished to quit because of the constant threat of criminal danger. The employer, in an effort to persuade her to continue as its bookkeeper, told her that it would employ security services to protect her and other employees while at work. These allegations by the plaintiff Childress, which must be accepted as true in determining whether or not she stated a cause of action for a lawsuit and whether or not her suit should be dismissed, include the claim that the defendant employer "falsely and fraudulently represented to [appellant] that security services, guards and undercover guards were present at or about the Church's [restaurant]." Relying on Church's representations, and believing that security services had in fact been put in place, she continued her employment as the restaurant location's bookkeeper. No such services were in fact hired, and on November 23, 1980, the restaurant was again robbed. Childress was shot and suffered serious injuries, including the loss of her right eye.

The defendant Church's contended that the Workmen's Compensation Statutes prevented any recovery other than those provided for in the Statutes. The court rejected this defense, concluding that in some exceptional circumstances, such as are present in the instant case, an employer cannot be freed from liability for his intentional acts even if the resulting injuries are compensable under Workmen's Compensation laws.

The case will proceed to jury trial.

COMMENTARY:

There are five elements which a plaintiff must plead and prove in order to sustain a cause of action for deceit: (1) a false representation by the defendant, (2) knowledge by the defendant that the representation is false, (3) an intention to induce the plaintiff to act in reliance on the false representation, (4) actual reliance on the false representation by the plaintiff and (5) and damage to the plaintiff resulting from such justifiable reliance.

PSCLR – OCT 85 (4)

Rape Of Nurse By Escaped Mental Patient Is "Accident" Arising Out Of Employment And Workers' Compensation Is Bar To Inadequate Security Tort Claim:

Jane Doe v. South Carolina State Hospital, *No. 0428, in the Court of Appeals of South Carolina, decided April 1, 1985, reported at 328 S.E.2d 652.*

The plaintiff, referred to in this case as "Jane Doe," was a nursing supervisor employed by the South Carolina State Hospital. She was assaulted and raped by a mental patient who had escaped from the hospital and she suffered serious physical and mental injuries as a result. She brought an action in tort against the hospital, but the defendant hospital was granted summary judgment by the trial court. On appeal, the Court of Appeals of South Carolina affirmed.

Nurse Doe's lawsuit alleged that the hospital's negligence allowed the rape to occur "because of the nature or setting of the work." Among the specific allegations of negligence, she listed the hospital's failure to secure the premises from easy access by unauthorized persons, to provide employees with a system to alert security personnel in the event of a problem, and to provide adequate security personnel. She also alleged that she was especially susceptible to an attack because her work area was not sufficiently visible to other areas of the building and because there were an inadequate number of supervisors on duty in the building.

Nurse Doe's hospital and doctor's bills totaled about $800 and were paid by the State Workers' Compensation Fund on a claim filed by the hospital. The hospital contended that this was the sole avenue of recovery for the nurse's injuries.

The Workers' Compensation Act in South Carolina states:

> The rights and remedies granted by this Title to an employee when he and his employer have accepted the provisions of this Title, respectively, to pay and accept compensation on account of personal injury or death by accident, <u>shall exclude all other rights and remedies of such employee...as against his employer, at common law or otherwise, on account of such injury, loss of service, or death.</u> (emphasis added)

The three criteria for determining applicability of the Act to an injury is whether the injury (1) resulted from an accident, which (2) arose out of the employment, and (3) occurred in the course of employment. The court stated that an employee whose injuries are covered by this Act does not have the option to accept Workers' Compensation or to bring an action in tort. Recovery for injuries meeting the criteria is "restricted to his right of claim for compensation" under the Act. In granting summary judgment to the hospital, the court ruled that the intentional assault by a third party upon Nurse Doe was an accident, which arose out of and in the course of her employment.

Doe appealed, arguing that the attack did not "arise out of" her employment because none of her duties were related to the escaped mental patient, and because every woman in the area was a potential victim of the escapee because of the hospital's negligence. She further argued that the Act did not provide recovery for the mental trauma of the rape and that bringing this tort action was therefore necessary.

The appellate court affirmed the trial court's decision that all of the criteria were met and Nurse Doe's right to recover damages for her injuries was restricted to the Workers' Compensation claim. Therefore, the summary judgment granted to the South Carolina State Hospital was affirmed.

COMMENTARY

This case reflects the marked distinction between the rights enjoyed by a tenant when injured on the premises of the landowner, as opposed to those of an employee when injured at his place of employment. And this case is reported in **PSCLR** *in order to remind our subscribers that the Worker Compensation laws of the various states can be an effective defense to inadequate security tort claims.*

PSCLR – FEB 86 (1)

Hotel Owner's Liability For Wrongful Death Of Moonlighting Security Guard (Independent Contractor) Is Upheld Where Owner Negligently Failed To Repair Defective Lock:

Marilyn Pucalik, Administratrix of the Estate of Lawrence J. Pucalik v. Holiday Inns, Inc., *No. 85-1382, in the U.S. Court of Appeals, Seventh Circuit, on appeal from the U.S. District Court for the Northern District of Indiana, decided November 13, 1985, reported at 777 F.2d 359.*

Lawrence Pucalik, an off-duty police officer working at a Holiday Inn as a security guard, was shot and killed during a robbery of the hotel. His wife brought a wrongful death action against Holiday Inn, alleging that the hotel's negligence caused Pucalik's death. After a jury trial, Pucalik's estate was awarded one million, two hundred fifty thousand dollars, a verdict and an award which the U.S. Court of Appeals affirmed.

Pucalik was a full-time patrolman with the Hammond, Indiana, police department, but worked a second job as a plainclothes security guard for a Holiday Inn. Since the hotel was located in a high crime area, the hotel management hired several plainclothes guards as independent contractors to provide protection for the hotel, and installed security locks on all the doors. At night, entry to the hotel could only be gained through the front door. Although the front door was locked, a guest could ring the front door bell and the lock could be released from the registration desk. The front door was designed to automatically relock.

On the night of the robbery, Pucalik locked and secured all of the hotel doors at 2:00 AM; the hotel could then be entered only as described above. At 3:00 AM Pucalik was working in an interior office, and heard the front bell. He went to the registration desk, observed a man at the door, released the front door, and allowed the man to enter. The door then closed, and should have automatically relocked. As the desk clerk prepared to register the guest, Pucalik returned to his office. At this point a second man entered through the front door without the clerk releasing the lock. He joined the first man and they demanded that the registration clerk put the money from the cash register and the cash boxes into a bag. Pucalik suddenly returned and saw that a robbery was in progress; he was not in uniform, did not draw his gun, did not identify himself as a police officer or attempt to make an arrest. He immediately began to retreat but the second man who had entered the hotel shot and killed him.

Mrs. Pucalik's wrongful death action alleged that Holiday Inns had breached its duty to her husband by failing to install and maintain certain security devices. There was evidence that the plainclothes guards were each responsible for different aspects of security at the hotel. Further, there was evidence that the guards had complained to the Holiday Inn management of unsafe conditions and defects in the security system. Although the management assured the guards that the problems would be corrected, there was no evidence that they were.

Holiday Inn's argued in its appeal that it could not be held liable for Pucalik's death at the hands of the robbers because he was "an independent contractor who assumed the risk that he would be injured or killed because of unlawful acts of third parties." However, the jury decided that the hotel had taken on the duty of maintaining the security locks and that Pucalik's death was due to the hotel's failure to maintain the front door's automatic locking mechanism. The jury concluded that Pucalik had not voluntarily assumed the risk that Holiday Inns would fail to repair a defective security lock.

Holiday Inn also submitted on appeal that Pucalik was a police officer and had died "in the line of duty." The defendants argued that their duty to protect Pucalik was therefore reduced under the "fireman's rule," which requires that a property owner "merely refrain from willful and wanton disregard for the safety" of a policeman or fireman. The hotel contended that certain documents filed with the Law Enforcement Assistance Administration by Mrs. Pucalik to gain federal benefits should be admitted into evidence, because they indicated that Mr. Pucalik had been killed while "in the line of duty." Nonetheless, the appellate court affirmed that the documents were not admissible, and that Mr. Pucalik was on the Holiday Inn premises as a business invitee.

Holiday Inn also argued that the amount of the jury award, $1,250,000, was excessive. The appellate court affirmed that the award was not excessive, noting that Pucalik was a 33-year-old man holding two jobs to support his wife and to pay child support and tuition expenses for a 16-year-old son by a previous marriage.

Therefore, the U.S. Court of Appeals for the Seventh Circuit affirmed that Holiday Inn had owed a duty to use ordinary care in maintaining reasonably safe premises, and its breach of that duty had resulted in Mr. Pucalik's death.

PSCLR – MAR 86 (8)

Business Owner/Operator Has No Duty To Warn Or Protect Contract Security Guard From Criminal Assault By Third Parties:

Andra Carter and Lilly Carter, his wife v. The Mercury Theater Company, Eugene Sloan and Richard Sloan, individually and as co-partners d/b/a United Motion Theater Company and Suburban Detroit Theaters, Inc., jointly and severally, *Docket Nos. 77773, 78453, in the Court of Appeals of Michigan, decided June 20, 1985, released for publication January 22, 1986, reported in 379 N.W.2d 409.*

Andra Carter and his wife sued the Mercury Theater in Detroit to recover for injuries which rendered Mr. Carter a paraplegic after being shot while on duty at the theater as a security guard. The defendants sought a summary judgment; the motion was denied by the trial court, but the Court of Appeals of Michigan reversed.

Carter was an employee of ACE Patrol Security Guard Service which was contracted to provide security at the Mercury Theater. While working on Christmas Eve, Carter and another ACE guard ejected two patrons due to their unruly and disruptive behavior. The guards instructed the other theater employees not to readmit the patrons, but the patrons did reenter and shot Mr. Carter rendering him a paraplegic. Carter's case put forth the theory that the theater's negligence allowed the shooting to take place, either by the patrons being permitted to reenter by the theater employees or by their use of the rear fire doors which did not have operative locking mechanisms.

The appellate court reviewed the allegations of the theater's negligence in the plaintiff's suit and found the facts to be undisputed. It quoted a previous Michigan case which had held:

> [The defendant] recognizing a duty to safeguard, protect and secure its ... business invitees, hired an independent security guard company for that purpose. What happened to [plaintiff] was the very reason [plaintiff] and his employer were hired, i.e., to safeguard against criminal acts of violence. It would be ironic to hold [defendant] liable to an employee of the very security guard company it hired for protection.

The Court of Appeals of Michigan concluded that this prior ruling was applicable binding precedent and therefore reversed the trial court's decision, and granted the Mercury Theater's motion for summary judgment.

COMMENTARY:

Can you conceive of any circumstances which would cause a court to hold that a business owner/operator could be liable to a contract security guard which the owner/operator had hired to protect its property and patrons? We believe that the above case should not be read as conclusive proof that no such circumstances could ever exist. Surely, for instance, if an assailant had called an owner/operator and threatened to kill or assault a security guard, and the company failed to notify the guard of such threat, it would then at least be conceivable that the court may rule that the owner/operator had a duty to warn the guard of the possibility of an assault or an attempt on his life.

Nevertheless, the general proposition applicable to most situations is correctly set forth in the above decision.

PSCLR—JUN-JUL 86 (9)

Security Guard Cannot Recover Damages From Supervisors Or Co-Employee For Unforeseeable Criminal Assault During Performance Of Duties, Except In "Most Extreme" Cases:

Craig A. Simpson v. Wolf Ridge Corporation, et al., *84-139, in The Supreme Court of Alabama, decided March 28, 1986, reported at 486 So. 2d 418.*

Craig Simpson was employed by Security Engineers, Inc., as a security guard. He was assigned as a guard for a Security Engineers client, Wolf Ridge Corporation. A tenant of Wolf Ridge had originally hired Security Engineers to provide security guards on the property; when the tenant moved, Wolf Ridge retained the tenant's contract with Security Engineers to maintain 24-hour security guards for the empty buildings. The property consisted of a large manufacturing building and a small warehouse inside a fence, and a small office building outside the fence.

Security Engineers notified Wolf Ridge Corporation that Alabama Power Company had discontinued electrical services to the premises after the tenant moved out. Wolf Ridge arranged to resume services to the buildings inside the fence, but not to the small office building outside the fence.

While on duty alone one evening, Simpson was patrolling the buildings on the Wolf Ridge property and he was accosted by a man with a shotgun. The man came out of the dark as Simpson was leaving the small office building and he forced Simpson to lie face down on the ground. He asked Simpson what the buildings contained and when Simpson answered that they were empty, he heard the click of the shotgun's hammer and Simpson immediately rolled his body away. Although he was shot in his right arm, Simpson broke away and escaped.

Simpson filed a personal injury suit against Wolf Ridge Corporation and five of his fellow employees of Security Engineers, Inc. The trial court granted summary judgment to Wolf Ridge, and dismissed the suit against Simpson's co-employees; Simpson appealed.

Although the general rule in Alabama is that employers are not liable for injuries caused by the criminal acts of third parties, Simpson alleged that Wolf Ridge should be held liable for his injuries under the exception to the general rule that an employer may be liable if he

> (1) employs the plaintiff under conditions that invite attack (i.e, an attack would be foreseeable); (2) greatly and unreasonably increases the risk of harm to the employee; and (3) fails to take adequate safety precautions. [The Alabama appellate court cited the case of *Henley v. Pizitz Realty Co.*, which was previously reported in the January 1985 **PSCLR**, case nine, page nine.]

Simpson argued that the circumstances of his assault were such that the exception applied: the assault was foreseeable by Wolf Ridge because the property was located in a high crime district and Wolf Ridge failed to take adequate security precautions by failing to have the electricity resumed and the lights on in the small office building where Simpson was shot.

The court reviewed a number of Alabama cases cited by Simpson in which a property owner was held liable for third party criminal assaults on invitees and employees under the exception to the general rule: *Law v. Omelette Shop, Inc.* (see the April 1986 **PSCLR**, case eight, page eight); *Ortell v. Spencer Companies, Inc.* (see the February 1986 issue, case nine, page six); *Latham v. Aronov Realty Co.* (see the August 1984 issue, case eight, page seven); and *Gaskin v. Republic Steel Corp.* (see the February 1983 issue, case three, page three). *Berdeaux v. City National Bank of Birmingham* (reported in the August 1983 **PSCLR**, case three, page three) was cited in contrast to the other opinions here cited. The cases providing the exception to the general rule

> emphasize that an action will not lie 'except in the most extreme case where it is clearly shown that the employer in some manner, greatly and unreasonably, increased the risk to the employee without taking reasonable safety precautions.

The Alabama Supreme Court ruled that the trial court did not err in granting the summary judgment to Wolf Ridge because there were no factual disputes and there was no legal basis to hold the defendants liable. Wolf Ridge couldn't be held liable for an unforeseeable assault by a third party and it was the former tenant not Wolf Ridge who had the electrical service terminated. Moreover, the court ruled that "Failure to return all of the lights to service does not make Wolfe Ridge's safety precautions unreasonable."

The appellate court also affirmed the trial court's dismissal of the suit against Simpson's supervisors. Simpson alleged that the Security Engineers' supervisors had "failed to properly instruct and warn him and failed to supply adequate personnel to guard an open and unlighted premises at night in a high crime district." The court affirmed that the supervisors should not be liable for Simpson's injuries for failing to predict an unforeseeable danger, to warn him of such a danger, or to instruct him on how to avoid such a danger. In the same way, the supervisors did not act negligently or unreasonably in assigning only one guard to fulfill the security contract with Wolf Ridge since the guard's purpose was to protect the unoccupied buildings from fire, theft, and vandalism. The trial court's dismissal of Simpson's personal injury suits were therefore affirmed by the Alabama Supreme Court.

PSCLR – JUN-JUL 88 (5)

Rape Of Waitress Was Caused By Employer's Negligence In Failing To Advise Supervisor Of Robbery Policy And Supervisor's Refusal To Cooperate With Robbers, Who Then Raped Waitress In Retaliation, Entitling Waitress To Recover Damages:

Mary E. Massie v. Godfather's Pizza, Inc., *Nos. 86-1080, in the United States Court of Appeals, Tenth Circuit, decided April 14, 1988, on appeal from the United States District Court for the District of Utah, reported at 844 F.2d 1414.*

Mary Massie sued her former employer, the defendant Godfather's Pizza, alleging that she was entitled to an award of money damages as compensation for a rape and assault which she suffered at the hands of robbers who attempted to rob her place of employment in Salt Lake City.

She based her allegations of negligence on the part of her employer on her claim that the employer had failed to communicate its written robbery policy to her shift supervisor, James Head. She further alleged that Head had failed to cooperate with the robbers' demands, which was contrary to company policy, and which, she claimed, caused the rape.

The case was tried to a jury, and verdict was returned in favor of Mary Massie in the amount of $200,000 general damages, plus an additional $36,000 in special damages. The trial judge, however, reduced the special damage award to $10,000, and both parties filed appeals from the trial judge's final judgment. Massie appealed the reduction of her special damage award, and Godfather's Pizza appealed the judge's denial of its motion for a judgment notwithstanding the jury's verdict, or for a new trial.

Godfather's also argued on appeal that the jury's damage awards were inappropriate and were not supported by sufficient evidence. Godfather's also claimed that the plaintiff's only remedy was under the state worker's compensation law, and that Godfather's did not owe any duty to Mary Massie to comply with the demands of robbers. Therefore, any negligence on its part was not the cause of the injuries suffered by plaintiff. Finally, Godfather's argued that the damages awarded to Massie were excessive.

The Court of Appeals reviewed the evidence presented at trial, in the light most favorable to the plaintiff (which the appellate court is required to do in reviewing the defendant's arguments on appeal) and found that the jury was justified in concluding that: Mary Massie had been employed at Godfather's Pizza, working at minimum wage, taking orders, running the cash register, and cleaning tables. She had nothing to do with the safe in the back of the restaurant, and had no supervisory responsibilities. She wore a uniform of blue pants, striped blue and tan shirt, with an apron, hat, and name tag, none of which bore the "Godfather's Pizza" name or logo.

James Head had been to shift supervisor several months before the incident which forms the basis of the lawsuit. As such, he was considered a part of the management team and was supposed to have a "thorough working knowledge of company policies and procedures," according to the company manual and to the restaurant manager, Carl Satchell. He had the keys to the restaurant, and he had the combination, as well as the key, to the safe. He wore a tie and vest, signifying his special position of authority as a shift supervisor.

According to the company manual, Head's job description required him to "maintain positive community relations at all times, both on and off the job" and he was required to be representing Godfather's even when "he was off the clock."

The manager of this particular Godfather's Pizza, Satchell, had not checked into the prior employment or army background of Head, and Satchell testified that Head had "acted macho and cocky" while at work. Satchell also testified that Head had brought a gun that was wrapped up and hidden on the restaurant premises.

Mary Massie knew that there was a safe on the premises located in the employee break room. She understood that Head had access to it because he was able to get change from it for her whenever she had requested it.

While the closing time for the restaurant was usually 11 PM, on the date of the robbery the shift ran later than usual because the restaurant was busy late into the evening. Massie did not have a car with her that day, so she made arrangements to walk home with Head after closing. She had often walked home, but never after dark.

After all the customers had been served that night, Massie, Head, and Satchell began preparing to close. Satchell left at approximately 12 midnight, leaving Head and Massie at the restaurant. Before leaving, Satchell "clocked" her out at 12:30 AM, although she did not actually finish working until about 1:15 AM (She was never paid for that additional time.)

While she was cleaning the restaurant, Massie saw someone try to open the locked door of the restaurant, but was not able to see the person well enough to identify him. There was no "open" or "closed" sign on the door, and people frequently would pull on the door after closing to see if the restaurant was open. After they finished cleaning up, Massie played one of the video games with Head, Head locked the outside door behind them, and they then left. They did not see anyone around the restaurant area as they were leaving.

At that time, Massie was dressed in her Godfather's uniform and a long navy coat. She had rolled up her apron and had put it in her pocket, and had left her hat and name tag in the restaurant. Head had taken off his tie, and he was wearing a black jacket over his clothes. The two employees were stopped by two assailants about two blocks from the restaurant, who ordered them to return to the restaurant. Head saw a knife in the pocket of one of the assailants, so he agreed to cooperate. On the way to the restaurant, Head lied to the men, telling them that he was unable to open the safe, when he was in fact able to do so. Subsequent evidence showed that Head also knew that there was only $500 in the safe at the time.

While returning to the premises, the assailants rifled Head's wallet and also asked for Massie's money. She only had a nickel in her possession, which they did not take. Upon arriving at the store, the robbers made Head give his keys to Massie. She unlocked the door, and the four entered the restaurant. The robbers then locked the door behind them, keeping the keys.

The cash registers were found to be empty, and the four then went into the break room. The assailants tied Head's hands together and searched the manager's office for money. They also tried to open the safe, and when they failed, demanded that Head open the safe. He repeated that he was not able to open it.

The robbers became frustrated at their inability to obtain money, and one of them said that they would rape Massie if they could not get any money (Head later denied hearing this statement). The assailants then took Massie to another part of the restaurant where they both raped and sodomized her. Head could hear Massie crying and screaming and finally agreed to open the safe after the second assault on her began, but not until the first assailant returned to where Head was tied up, zipped his pants in front of him, and threatened him with a roll of cellophane wrap.

After the assailants obtained the money from the safe, the two left Head tied up in the restaurant and took Massie with them. She was forced to run through an alley and to climb an eight foot fence thereby injuring her hands and a foot. When a policeman saw them, she was able to escape. The two assailants were never apprehended.

Head later testified that he had told the robbers that he was not able to open the safe because he felt that, as a supervisor and employee of Godfather's, he was in charge of the safe and it was his responsibility. He told an investigator that he did not want Godfather's to lose any money and he did not want to lose his job.

Before this incident occurred, Godfather's had devised a written policy calling for immediate cooperation with robbers, and the company manual set out the required procedure:

> No amount of money is worth the safety and well being of Godfather's Pizza employees. The criminal is usually interested only in money and, if not provoked to any other violence, will 'take the money and run.' The real key to safety of the victim is to allow the criminal to get in and get out as quickly as possible.
>
> Godfather's Pizza, in cooperation with law enforcement agencies, has established the following standard procedures in the event of a hold up or robbery.
>
> . Above all, remain calm.
>
> . Do not create a hostile situation.
>
> . Do exactly what the criminal tells you.
>
> . Do not provoke the criminal.
>
> . Offer no resistance.
>
> ***
>
> Above all, cooperate. Do not try to overcome the criminal. Do not resist. Do exactly what the criminal says.

The restaurant manager Satchell had previously told Head that he should obey any demands by any robbers, but Satchell had not reviewed the written policy, and it was not posted anywhere in the restaurant. Godfather's did not have an organized training program to teach employees how to act during a robbery. The manual setting out the policy was locked inside the manager's office and was not accessible to employees. According to Head's testimony, he was not aware of Godfather's written policy, and he had never received any training whatever concerning how to act during a robbery.

The court also noted that there had been no indication that either of the assailants intended to sexually assault the plaintiff until one of them said that they would attack Massie if they could not get any money.

The plaintiff called two expert witnesses who testified that Godfather's procedures for hiring and training its personnel were below standard and that Head's response to the robbers' demands were also substandard. These experts also testified that the robbers probably would have not harmed the plaintiff if Head had reacted properly by immediately turning over the money to them.

Concerning the issue of damages, the evidence was that prior to the rapes, Mary Massie had enjoyed her contacts with the public and life in general, she had an outgoing personality and a good marriage, with a "satisfying sex life." After the rapes, however, her sex life has been significantly impaired, her marriage has suffered, and she feels insecure, threatened, and frightened. She has sought counseling from several professionals and was diagnosed by Dr. Victor Kline as suffering from "post-traumatic rape syndrome," from which she will never completely recover. Dr. Kline also testified that her memories of the incident had been "permanently etched into her recollection by a chemical produced in high stress situations, locking vivid memories in place to recur from time to time."

Massie testified that she had attempted to return to restaurant work, but quit after her employer asked her to work nights. When the case reached trial, she had a job cleaning houses for two or three hours each day.

The jury, at the direction of the judge, returned a special verdict making specific findings as follows: Godfather's Pizza and/or James Head, while acting as an employee of Godfather's, was negligent, and that such negligence proximately caused the injury to Mary Massie; that Massie was not negligent and that the negligence of defendants was 100 percent at fault, and that Massie was not acting as an employee at the time of the incident.

The appellate court first rejected Godfather's argument on appeal that Massie's sole remedy was provided for under worker's compensation acts. The court held that when an assault against the employee is due, as here, to motives unrelated to employment, then the injury cannot be considered to have arisen out of the employment situation.

The rape of Massie was neither personal to her, nor distinctly associated with her employment, but rather was an act of wanton brutality directed at her merely because she is a woman.

> It is clear that she was not raped because of the nature of her duties or the nature of her workplace environment, nor because or any incident or quarrel growing out of the work.

The jury, the appellate court decided, was well within its authority in concluding that Mary Massie was not acting within the scope of her employment at the time of the assault. Accordingly, she should not be limited to a recovery merely under the worker's compensation act.

With regard to the defendant Godfather's argument that it had no duty to comply with demands of robbers, the court noted that Godfather's does have a duty to act reasonably under all of the circumstances attendant upon the situation. In addition, as the owner and occupier of the premises, Godfather's is liable when harm is caused by third parties, even when due to criminal conduct, if and when such conduct is foreseeable and Godfather's fails to exercise reasonable care to discover that such acts are being, or are likely to be, done.

The trial court decided, and the evidence presented was sufficient to establish, that Godfather's did in fact recognize that there was a foreseeable harm to its employees, including Massie, and in fact had created a duty to the public by establishing "robbery policy" which was set forth in its manual.

Furthermore, the two experts presented by the plaintiff established that such negligence on the part of Godfather's was the proximate cause of the criminal conduct resulting in the plaintiff's injuries, since their opinion was that the sexual assault would not have occurred if Head had promptly opened the safe at the robber's request. Again, the jury was well within its authority to conclude that proximate cause was established.

And, finally the court found that the $200,000 awarded as general damages was "not shocking or outrageous," and therefore allowed it to stand. However, as to the $36,000 awarded for special damages, the court noted that at the time of trial, Massie's medical expenses included a hospital bill for $98 plus $450 for six sessions with Dr. Kline, for a total of about $550 for past medical expenses. Without direct evidence as to future expenses, beyond Dr. Kline's testimony that Massie would need some counseling for the duration of her life, the appellate court ruled that the trial court did not abuse its discretion by reducing the special damage award to $10,000.

With this reduction approved by the appellate court, the verdict in favor of the plaintiff in the amount of $210,000 was affirmed.

PSCLR – AUG 88 (4)

Movie Theater Usher Assaulted By Patrons Can Maintain Action Against Contract Security Company For Failing To Provide Adequate Security At Theater And Failing To Follow Its Own Security Procedures:

Dina Cruz, individually and as parent and guardian of the infant plaintiff Moises Cruz v. Madison Detective Bureau, Inc. v. RKO Century Theaters, *in the Supreme Court of New York, Appellate Division, First Department, decided May 12, 1988, reported at 528 N.Y.S.2d 372.*

Moises Cruz, an usher at RKO Century Theaters 86th Street Theater, was attacked by patrons of the movie theater shortly after the close of the movie. Cruz sued Madison Detective Bureau, Inc., for negligence in failing to provide adequate security at the theater and in failing to follow its own security procedures. At the close of the plaintiff's case and before the defendant had presented its evidence, the trial judge directed a verdict in favor of the defendant Madison and dismissed the action.

On appeal, the court held that there was sufficient evidence for a reasonable juror to find that Madison's negligence caused injury to Cruz, and therefore reversed the trial judge's decision and remanded the matter for a new trial.

Brian Camp, the Madison security guard present on the night of the incident, testified that the only formal training provided to him by Madison was instruction in the proper way to hold a nightstick. Prior to his four months of employment by Madison Detective Bureau, he had no experience as a security guard. Camp testified that on the night of the incident, three security guards, including himself, were on duty. According to Camp, all three security guards were downstairs near the box office, leaving no one on the upper floor where the doors to the different theaters were located. That evening, the ticket taker at the theaters on the upper floor called out for Camp to come upstairs. When Camp arrived with another guard, three or four men were assaulting Cruz. After a brief altercation, the assailants were scared off when someone yelled that the police were coming.

Salvator Parete, the manager of the theater who hired Madison to furnish the guards, testified that he trained and directed the guards in crowd control. The two movies showing on the night of the incident were "blockbusters," requiring extra guards on duty. Parete testified that he instructed Camp to place two guards on the upper level at the close of the movies when crowds were exiting the theater.

The assistant manager of the theater testified that when three guards were on duty, at least one of them was supposed to be upstairs patrolling the theater. He stated that Cruz and another theater employee were attacked when they attempted to prevent certain patrons from leaving one theater and entering the other without paying a second admission. According to Tony Howell, the candy stand attendant who was also attacked, the two movies "broke" simultaneously that evening and, when the attack occurred, there was no guard upstairs, although the theater was crowded. Another employee testified that Camp was outside of the theater when he was summoned to the scene of the assaults.

Plaintiff Cruz testified that he and Howell were surrounded and punched and kicked by some men who had attempted to go from one theater to another without paying. Cruz stated that he had witnessed similar incidents, and that he and the other employees of RKO had been instructed to seek assistance from the guards in the event of any trouble.

In reversing the trial court, the appellate court stated the general rule that a case should not be dismissed at the close of the plaintiff's evidence "unless it is clear that by no rational process could the jury find in favor of plaintiff."

According to the appellate court, the plaintiff first had to prove a prima facie case sufficient to raise a jury issue in a negligence action: (1) the existence of a duty owing by defendant to the plaintiff; (2) defendant's failure to discharge that duty; and (3) an injury to plaintiff proximately resulting from such failure.

The appellate court held that Cruz had clearly made out a prima facie case for negligence under the above rule. The court noted the testimony that Madison Detective Bureau was hired to furnish security at the RKO theater and that part of its responsibility was to provide a guard at the upper level theaters. Further testimony established that none of the guards were on duty upstairs at the time of the incident.

In addition, there was enough evidence to suggest that Cruz's injuries resulted from Madison's breach of its duty to provide protection on the upper level. A jury could further find that Camp's failure to follow directions and take precautions in the face of a foreseeable risk constituted negligence for which his employer could be held liable.

Moreover, the jury could have determined that the contract security company itself, as the employer of the security guards, was negligent in failing to train its employees in proper security techniques and procedures.

Although the injuries to Cruz were directly caused by the acts of the assailants, the court concluded, a jury could reasonably have found that the attack should have been foreseen by the Madison security guards and, therefore, could be attributed to Madison's failure to provide adequate security and its failure to adequately train its employees. In other words, the negligence of Madison Detective Bureau could reasonably be seen by a jury as the "proximate" or legal cause of the plaintiff's injuries.

The evidence presented by plaintiff was sufficient to raise an issue concerning Madison Detective Bureau's negligence, and the evidence on that issue should have been presented to the jury. The appellate court, therefore, unanimously remanded the case for a new trial.

PSCLR—SEP 88 (9)

Security Guard's Employer Held Not Liable For Failing To Provide Safe Work Place, When Employer Had No Control Over Client's Premises On Which Injury Occurred:

Richard Gomes, et al. v. Revere Sugar Corp., and McRoberts Protective Agency, Inc., *in the Supreme Court of New York, Appellate Division, Second Department, decided May 23, 1988, reported at 528 N.Y.S.2d 646.*

Richard Gomes, a security guard with McRoberts Protective Agency, Inc., fell and was injured on the premises of Revere Sugar Corp. Revere had employed McRoberts to provide security at its plant, and Gomes was assigned to the job. There was no written contract between Revere and McRoberts.

Gomes sued Revere, asserting that his fall was caused by an accumulation of ice and/or syrup leaking from machinery. Revere then sued McRoberts, claiming that McRoberts was liable for failure to provide a safe place to work as required by Labor Law § 200, that McRoberts should contribute to any recovery by Gomes from Revere, and finally, that McRoberts should indemnify Revere for any amount recovered by Gomes against Revere. McRoberts moved to dismiss all of the above claims except for the claim for contribution. McRoberts submitted an affidavit from its president denying that the parties had entered into an indemnification agreement, and argued that it could not be held liable under Labor Law § 200 because it did not have control of Revere's premises and was not responsible for cleaning.

The court, finding that McRoberts lacked control over Revere's premises, dismissed Revere's claim that McRoberts should be held liable to Gomes under Labor Law § 200 for failure to provide a safe workplace. To impose liability on an employer, the statute requires that the party charged with the duty to provide a safe place to work "have the concomitant authority and degree of control over the activity which produces the injury to enable it to take the action necessary to correct or avoid an unsafe condition." Since Revere did not allege that McRoberts had control over the premises or any duty to inspect or clean the floors or machinery, the claim under federal labor law was dismissed.

In addition, since Revere admitted that there was no written indemnification agreement making McRoberts responsible for any harm to its employees on Revere's premises, and Revere did not contradict McRoberts' assertion that there was no oral indemnification agreement, the claim for indemnification was also dismissed.

PSCLR—OCT 88 (2)

Guard Service Not Liable For Injuries To Bank Employee Assaulted By Third Party For Reasons Unrelated To Bank's Business Activities:

Robert R. Shaffer, Jr., and Jill R. Shaffer, his wife, v. Wells Fargo Guard Services, a Subsidiary of Burns International Security Services, a Subsidiary of Baker Industries, Inc., *No. 86-3132, in the District Court of Appeal of Florida, Third District, decided March 1, 1988 on rehearing June 7, 1988, reported at 528 So. 2d 389.*

Robert Shaffer, Jr., an employee of Citizens Federal Savings & Loan, was assaulted in the bank by a man who was under the mistaken belief that Shaffer was romantically involved with the man's wife, also a bank employee. The bank had hired Wells Fargo Guard Services to protect the bank and its employees. The contract with Wells Fargo provided, in part, that Wells Fargo would "help provide the [bank] with a system of protection for its assets and employees against certain hazards."

Shaffer sued Wells Fargo for negligence and breach of contract in failing to protect him from the attack by a fellow bank employee's irate husband. The case was dismissed by the trial court, and Shaffer appealed.

The appellate court held that Shaffer's only recourse against Wells Fargo was for breach of contract, because Wells Fargo would have no affirmative duty to protect Shaffer against any harm, except for its contract with the bank. By naming the bank's employees as those for whom protection was to be provided, the contract created a right to protection by Wells Fargo for each bank employee. As an employee of the bank, Shaffer could sue Wells Fargo for breach of contract as a "third party beneficiary" of the contract between the bank and Wells Fargo.

Having determined that Shaffer could sue Wells Fargo as a third party beneficiary of its contract with the bank, the appellate court addressed the issue of whether Wells Fargo had contracted to provide protection from the kind of harm that Shaffer experienced. The court interpreted the term "certain hazards" in the contract to mean "certain hazards associated with carrying on a bank's business." Therefore, the court held that Shaffer could not sue Wells Fargo for "hazards totally unconnected to the activities or business of the bank," such as the attack that occurred in this case.

A dissenting judge would have allowed a jury to determine the scope of the "certain hazards" in the contract between Wells Fargo and the bank. For the court's purposes in deciding whether or not to dismiss the case prior to trial, the court should view any factual issues in the light most favorable to Shaffer. Viewed in the light most favorable to Shaffer, the term "certain hazards" could include any assault on an employee of the bank, regardless of the assailant's motivation.

The appellate court then granted a rehearing of the case, in which Shaffer raised the absurdity of using an after-the-fact determination of an assailant's motivation in assaulting a bank employee to determine the guard service's liability. The appellate court nevertheless upheld its earlier decision to dismiss the case, rejecting the argument that security guards would delay action to protect a bank employee until first determining whether the assailant's motive was connected to the bank's business.

After the rehearing, the dissenting judge stated his agreement with Shaffer's argument, reiterating his opinion that the security firm's legal obligation should not be determined by analyzing the assailant's motives and that the intent of the parties who entered the contract should be determined by a jury at the trial level, not by the appellate court. The dissenting judge noted that the appellate court's opinion would shield Wells Fargo from liability for failure to protect customers from people whose misconduct is not motivated by monetary gain, such as mentally unstable individuals or those under the influence of drugs.

Nonetheless, the majority's opinion stands, and the case against Wells Fargo was dismissed.

PSCLR–AUG 89 (1)

Hospital Is Liable To Employee Raped In Parking Deck Despite Absence Of Previous Violent Crimes In The Area, Based Upon Inadequate Security Established by Expert Witness:

Donovan Small, as the Special Administrator of the Estate of Teresa Small, Deceased v. The McKennan Hospital, *No. 16110, in the Supreme Court of South Dakota, decided March 8, 1989, reported at 437 N.W.2d 194.*

At about 7:30 p.m. one evening, Rocky Blair, a recent parolee, abducted Teresa Small from the McKennan Hospital parking ramp, raped her, and murdered her. Teresa Small's husband, Donovan Small, sued McKennan Hospital for negligence in failing to prevent Teresa's wrongful death and for causing Teresa's pain and suffering prior to her death. At trial, the jury found in favor of Small and awarded him $535,000 in damages. The Hospital appealed.

Teresa Small and her mother, Kathy Smith, were employed at McKennan. After finishing her shift, Teresa borrowed her mother's car, promising to return it to the parking ramp. Donovan Small followed his wife in his car when she was returning the car to the ramp. Small parked near the entrance/exit and waited for Teresa. After about five minutes, a white Buick driven by a single male left the parking lot.

Shortly thereafter, Small became concerned about Teresa and entered the parking ramp to investigate. There he observed the car parked about 30 to 35 feet from the elevator. When Teresa's mother appeared, they both proceeded to look for Teresa. On the fourth level of the parking ramp, Small noticed four young men parked in a vehicle, smoking marijuana and listening to loud music.

At trial, Small described the lighting in the parking lot as "kind of yellowish, hazy, spooky-like, real dim." The lights were recessed, creating shadows on the ramp. He noticed some cigarette butts in one area of the parking deck stairwell and human feces in one of the elevators. During his search of the ramp, he never observed any security personnel. Around 8:30 p.m., Small decided to talk to security at McKennan's front desk. Willard Myer, who was in charge of security that night, advised Small to notify the police, who then joined the search and located Teresa's dead body elsewhere.

McKennan is a large hospital located in a residential area in Sioux Falls, South Dakota.

The hospital covers approximately six square blocks and employs between 1,300 and 1,400 people.

McKennan's four-story parking ramp had opened approximately three years prior to Teresa's murder. McKennan had anticipated charging a fee for parking, and so a guard booth was built at the point of entry/exit. An intercom system connected to the front desk was installed in the guard house. The guard house was never staffed, due to budget considerations. McKennan employed one person for each eight-hour shift to secure the hospital, its grounds, and the parking ramp.

There were problems on the ramp with vandalism and theft. In addition, several people testified to using the ramp simply to park, drink alcohol, and smoke marijuana. Such parties were not interfered with by security personnel at the hospital.

Teresa's killer was caught, and he testified at trial that he had been released on parole 11 days prior to Teresa's murder. He had driven his 1965 Buick to McKennan at approximately 10:00 or 11:00 the morning prior to the killing. He and a friend drank most of the day. That evening he returned to the ramp, where he parked, drank whiskey, and smoked marijuana. He played loud music on his stereo with the windows down. He never saw a uniformed officer on either visit to the parking lot. Blair chose the parking ramp for no particular reason except that it was "quiet, dark, and lonely." When he noticed Teresa getting out of her car proceeding toward the elevator, he grabbed her, put her in his car, and drove out.

At trial, Charles A. Sennewald testified as an expert on security. He testified that parking ramps are generally dangerous because of the accumulation of vehicles, inviting theft, vandalism, and damage to vehicles. In Sennewald's opinion, McKennan's security was deficient in four major areas. First, there was no access control in place at the McKennan ramp. Second, the recessed lighting failed to provide the necessary illumination of the parking ramp area. Third, there was no closed-circuit television in use. Fourth, there was no systematic, routine, or managed patrolling by the McKennan security staff. Sennewald concluded that all these factors contributed to the incident leading to Teresa Small's death.

Sennewald indicated that in Sioux Falls, there were approximately 52 crimes for every 1,000 people during the year that Teresa was murdered. The average crime rate in South Dakota for that year was 26.4 crimes per 1,000 residents. Sennewald testified that crimes such as theft and vandalism can escalate into violent crimes, and that if illegal activities combine with alcohol and drugs, the risk of violence is increased. Sennewald stated that, in his opinion, the security at the McKennan hospital ramp was not adequate to protect the public against serious assault.

Howard Cone, McKennan's chief security officer, testified that over several years he had requested additional security personnel to better patrol the parking area at the hospital, as well as surveillance cameras for the ramp area. Most of these requests were denied, and the number of security personnel decreased in the year before Teresa Small's death. Cone was aware of the problems with people using the ramp for parties and of the prior incidents of vandalism and theft. Security personnel were to visit the ramp area two or three times a night.

In response to a "parking lot rapist" article appearing in the *Sioux Falls Argus Leader*, James G. Ward, assistant administrator in charge of security at McKennan, informed the media that employees had been instructed about such danger and asked to report any unusual activity. He was quoted as saying, "We've been fortunate up to this point." He also expressed his concern about the possibility of rape in the parking area.

In a prior appeal of this case, the appellate court held that the lack of prior similar incidents were not determinative of foreseeability of criminal activity, opting instead for a "totality of the circumstances" test. The court held that foreseeability must be analyzed in light of all the circumstances, recognizing that foreseeability depends on the facts of each individual case. Prior incidents, whether similar or not, were properly held to constitute evidence of foreseeability.

Under South Dakota law, two essential elements must be proven in order to impose liability for unlawful acts of a third party: (1) the existence of a special relationship between the landowner and the injured party, and (2) a finding that the intentional criminal acts are foreseeable. In this case, the court held that the "special relationship" element was proven in that Teresa was an "invitee" of McKennan by virtue of the fact that she was a "business visitor," which is "a person who is invited to enter or remain on land for a purpose directly or indirectly connected with business dealings with the possessor of the land." Because Teresa's mother was employed by the hospital, and because Teresa had returned to hospital property for her mother's convenience, Teresa qualified as a business visitor of the hospital.

In addition, the appellate court held that the evidence at trial was sufficient to submit the issue of foreseeability of Teresa's rape to the jury. The court noted that McKennan was aware of people using the area for parties. McKennan had warned its employees of the danger of sexual assault and admitted that it had been "very fortunate" that no assault had taken place on its ramp. In addition, it conducted rape awareness seminars, specifically warning its employees that 30 percent of all rapes began or occur in parking lots. There was no control of any kind at the entry/exit area of the ramp, despite the fact that a guard booth had been put in place and an intercom system was installed. The chief of security had specifically requested for several years an increase in personnel and equipment. Most of these requests were denied due to budgetary considerations.

The court held that "a long history of good fortune does not relieve a party from exercising ordinary and reasonable care in the operation of his particular business enterprise." Furthermore, the court noted that in its operation of a parking structure, McKennan may be said to have created a special "temptation and opportunity for criminal misconduct, thus increasing the foreseeability of the attack." The court concluded that failure to prove any criminal activity in the area is not fatal to the submission of the foreseeability issue to the jury, because criminal assaults occur in all neighborhoods, regardless of the frequency of criminal activity. "A violent criminal activity can be foreseeable simply upon common experience." Under the "totality of the circumstances test," the appellate court concluded that the evidence of foreseeability was sufficient to submit the question to the jury.

Finally, the court upheld the jury's award of $535,000 to Small for the injuries, pain, suffering and mental anguish that Teresa suffered prior to her death, in addition to his own economic loss from her death.

COMMENTARY:

This case can and should be used by those security directors who are experiencing difficulties in convincing management that known security hazards must be remedied and that "budgeting considerations," while relevant, will not excuse the failure to provide adequate security. Such failure can lead to very expensive budgetary consequences. More importantly, of course, it can also lead to tragic loss of life, as in the above case.

In the classic case where a person sues for injuries caused by a vicious dog, one of the traditional defenses has been that "each dog is entitled to one free bite." The theory holds that, unless the dog has bitten someone before biting the plaintiff, the dog's owner could not have known or foreseen that his dog was vicious and would bite the plaintiff, and therefore is not liable for the dog's "first bite."

The more enlightened view is that the owner's foreseeability should be tested by the "totality of the circumstances" to see whether he should have known from other evidence that the dog was vicious and may bite someone. Analogously, the above court ruled that no "free first murder" theory should allow the defendants to escape liability, just because no prior murder had occurred in the parking deck, but that the "totality of the circumstances" should be examined in determining the foreseeability of this first killing.

Therefore, even if you have been "fortunate," in that no assaults have occurred as yet on your property, you are still legally obligated to take reasonable steps to provide adequate security to protect against losses which you can or should foresee based upon all the circumstances.

PSCLR–AUG 89 (7)

Security Guard May Not Recover From His Employer, Or From Lessor Or Owner Of Property, For Injuries Sustained By Criminal Attack While Guarding Premises:

Atlanta Braves, Inc., et al. v. Leslie, et al., *No. 77548, in the Court of Appeals of Georgia, decided January 12, 1989, rehearing denied January 26, 1989, certiorari denied February 23, 1989, reported at 378 S.E.2d 133.*

Nathaniel Leslie was shot by an armed robber while guarding concession receipts at the Atlanta-Fulton County Stadium facility leased to the Atlanta National League Baseball Club, Inc. (the "Braves"). Leslie, a full-time Atlanta city police officer, also worked as a security guard for ARA Leisure Services, Inc., with whom the Braves contracted to provide concession services at the stadium facility at home games. When performing his duties for ARA, Leslie wore his police officer's uniform and carried a gun. It was while engaged in guarding a portion of the stadium premises where ARA's employees deposited their daily receipts, that Leslie was shot by the armed robber.

Leslie sued the City of Atlanta-Fulton County Recreation Authority, the owner of the stadium; the Atlanta National League Baseball Club, Inc., the lessor of the stadium; and his employer, ARA Leisure Services, Inc.; alleging the negligent breach of the duty to keep the stadium premises safe for "invitees." The defendants asked the trial court to dismiss the case without trial. The trial court refused, but allowed the defendants to appeal the order.

Where a property owner, by express or implied invitation, induces or leads others to come upon his premises for any lawful purpose, he is liable in damages to such person for injuries caused by his "failure to exercise ordinary care in keeping the premises and approaches safe." In guarding ARA concession receipts, Leslie was engaged in a facet of the business for which the Authority had leased the stadium premises to the Braves. "It is well to observe that where, as in this case, an owner of property leases it to be used in the conduct of a business, those coming upon the premises in connection with the conduct of the business are invitees of the owner and proprietor alike."

In this case, however, Leslie was on the premises as ARA's armed security guard and was shot by an armed robber while acting in that capacity. In the absence of contrary evidence, Leslie is assumed to be of ordinary intelligence and capable of knowing and appreciating the dangers incident to his employment. The ordinary risks of employment are assumed by an employee, especially when the dangers are obvious. "The business invitee of private premises assumes the risk of danger which he knows about and fully comprehends, or which is sufficiently obvious." According to the court, the possibility of a face-to-face encounter with a criminal is an "ordinary and obvious risk incident to employment as an armed security guard." Accordingly, Leslie, by accepting employment as ARA's security guard, assumed the risk that he would have such encounters on the stadium premises. The court noted that Leslie was, in effect, hired to provide an element of the very security that he alleges was negligently lacking. Accordingly, the appellate court held that the trial court should have dismissed the case against the Authority and the Braves.

As to the claim against ARA, the court held that Leslie was acting as an employee of ARA and not an independent contractor and, therefore, his exclusive remedy was under the Workers' Compensation Act. The question of whether an individual is an employee or an independent contractor is determined by whether or not the employer has the right to direct the time, manner, methods, and the means of the execution of the work. In this case, ARA's security coordinator controlled Leslie as to when to report to work, when to leave work, and what areas of ARA's operations to guard. Since Leslie acted as an employee of ARA, he could not sue ARA for damages outside of the remedy provided by the Workers' Compensation Act.

Furthermore, by accepting employment as an armed security guard, Leslie had no cause of action against ARA for injuries resulting from the ordinary and obvious danger of encountering a criminal. Therefore, the court dismissed Leslie's claims against all parties.

PSCLR–AUG 89 (9)

Bartender/Waitress May Recover Workers' Compensation Benefits For Sexual Assault By Customer On Employer's Premises:

Deborah Pharr Culpepper, Employee v. Fairfield Sapphire Valley, Employer, Aetna Casualty & Surety Company, Carrier, *No. 8810IC419, in the Court of Appeals of North Carolina, decided April 4, 1989, reported at 377 S.E.2d 777.*

Deborah Pharr Culpepper worked as a bartender and part-time waitress at Fairfield Sapphire Valley, an extensive resort community located in a remote region of the western North Carolina mountains. She was instructed when she was hired "to be very cordial and friendly and nice and [to] offer any assistance that [she] could" to members and guests since "[m]ost of the people coming up there were looking at buying property" at the resort. She was directed to serve only members and guests and those people were pointed out to her, one of whom was Ralph Harvey Henry. Culpepper had served Henry a number of times at the country club and at the inn within the resort. He telephoned her once while she was on duty to ask for a date. Culpepper declined. Culpepper and Henry had had no contact outside of her serving duties.

At about 11:00 PM one summer evening, Culpepper finished her work at the country club and drove to the inn, crossing both the resort's private roads and public Highway 64 to get there. Access to the resort's recreational, dining, lodging, and other facilities was by a system of private roads branching off of Highway 64. Highway 64 ran the length of and through the middle of the resort community, and provided the only means of entering or exiting the resort. Culpepper parked in front of the inn and went in to turn in the day's proceeds and paperwork. She noticed Henry sitting at the bar. When Culpepper left the inn, she discovered Henry had pulled his car beside hers and was standing at her car door. He asked her to go out with him. She again declined.

Culpepper waited for Henry to leave before she drove out of the parking lot. When he was out of sight, she drove down the resort road leading away from the inn. She turned on Highway 64, but before reaching the boundary of the resort, exited the highway and turned onto the unlit resort road which led to the employees' quarters.

As she drove down the road, Culpepper noticed Henry standing by his car and motioning her to stop. Culpepper "stopped to see if he needed help or wanted [her] to tell someone . . . to get him." Because "he was a guest and since he was on th[e resort] property, . . . [she] felt like it would be a good thing to stop and see if he needed help."

Henry walked over to Culpepper's car, leaned down to talk to her, then yanked the door open, forcing his way into the driver's seat. Culpepper jumped out the door on the passenger side and ran down the road. Henry caught her, hit her in the face, and forced her into his car. He drove her to a secluded area, where, after repeatedly threatening to kill her, he raped her.

Culpepper talked Henry into returning to her car, pretending that she would go out with him. As they approached Culpepper's car, they noticed people standing around it apparently concerned because the door was ajar and headlights were still on. Henry accelerated the car to get away. Culpepper leaped head first from his car, catching her ankle in the door. Her body struck the pavement and was dragged along the roadway until her foot came free.

Culpepper sustained a skull fracture and injuries to her back, leg, and face. As a result of her head injury, she suffered partial hearing loss, partial loss of the sense of taste, and the complete loss of the sense of smell.

Culpepper filed a claim for workers' compensation benefits. Following a hearing, the deputy commissioner awarded Culpepper $17,000 for her permanent injuries, temporary total disability for a two-month period, and all medical expenses incurred for treatment of her injuries. Fairfield Sapphire Valley appealed to the full commission. The full commission reversed the interlocutory opinion and award to Culpepper. The commission denied Culpepper's compensation, finding that she "diverted from her direction home on a private mission . . .." The commission also found that Culpepper was directed to be cordial at all times <u>while on duty</u>, since guests were prospective home buyers. The commission concluded that Culpepper "was acting on a personal mission . . . and was not under any <u>employment duty</u> to travel upon that road or <u>to stop to assist anyone being located on that road</u>." Culpepper appealed.

An injury is compensable under the Workers' Compensation Act only if the injury (1) is an "accident" and (2) "aris[es] out of and in the course of the employment." Injuries resulting from an assault are caused by "'accident' within the meaning of the Act when, from the employee's perspective, the assault was unexpected and was without design on her part."

The appellate court held that the commission correctly concluded that Culpepper's injuries stemming from the sexual assault were caused by "accident." Thus, the question on appeal was whether her injuries "arose out of" and were "in the course of" her employment. An injury arises out of employment if the employment was a "contributing cause" of the injury and if the injury was a risk inherent or incidental to the employment, and one to which the employee would not have been equally exposed apart from the employment. Finally, an injury arises out of employment if the employee's acts on behalf of a third person are of "appreciable benefit" to the employer.

The appellate court concluded that the only reason Culpepper stopped on the resort road was to offer a guest assistance, as her employer instructed her to do. The court found no evidence in the record to support the commission's finding that Culpepper was told to be helpful and cordial to guests only while <u>on duty</u>. Because Culpepper's decision to stop had its origin in her employment, the court held that her actions were sufficiently "work-connected" to warrant a conclusion that her injuries arose out of the employment.

The appellate court also found that the nature of Culpepper's job put her at an increased risk of sexual assault not shared by the general public. The nature of her job as a cocktail waitress and bartender subjected her to unwelcome advances from male customers, while at the same time requiring her to be "cordial and friendly and nice." In addition, Culpepper was required to work late at night in a remote mountain area. These work-related factors created a unique risk of sexual assault.

The appellate court also found that Culpepper's stopping to assist a guest bore a clear relation to her employer's interest and, therefore, that her resulting injuries arose out of her employment.

The phrase "in the course of employment" refers to (1) the time, (2) the place, and (3) the circumstances in which the injury occurred. The "course of employment" continues for a reasonable time after work ends and may include time spent going to or from work. The court noted that the assault on Culpepper occurred only moments after Culpepper's work day ended. The time lapse did not exceed a reasonable time in which to leave work. With respect to place, the "course of employment" includes the employer's premises and may extend into adjacent premises used as a means of ingress and egress to the employer's premises. In particular, injuries sustained on premises controlled or owned by the employer while going to or coming from work generally are considered "in the course" of employment. The court held that Culpepper was still protected by the "going and coming" rule when she stopped for Henry, even though she had left the employer's private road leaving from the inn and entered the public highway. The court found that Highway 64 was an integral part of the resort complex, and that the portion of Highway 64 within the boundaries of the resort constituted the employer's premises for the purpose of the going and coming rule.

The court rejected the employer's contention that, because Culpepper deviated from the direct route home on a "personal mission," she was not covered by the going and coming rule, and therefore was no longer in the course of employment. The court found that Culpepper returned to the course of employment when she decided to stop and help a known guest, even though she deviated from her trip home to see a friend. The court found that, although Culpepper drove on the road for a personal reason, the important fact was that she stopped for a reason related to her employment.

With respect to circumstances, an employee "engaged in activities that [s]he is authorized to undertake and that are calculated to further, directly or indirectly, the employer's business" acts within the course of employment. The court concluded that Culpepper returned to the course of employment when she stopped to assist a guest. Because Culpepper was engaged in a work-related activity reasonably calculated to benefit her employer, Culpepper's injuries did occur in the course of employment.

The appellate court, therefore, reversed the commission's decision and ordered the commission to award Culpepper the compensable consequences of her injury.

Employment Relationship

B. Liability for acts of Employees

PSCLR–JUL 82 (11)

Floor Guard Who Is "Off Duty" But Acting Within Scope Of Employment Subjects Employer To Liability For His Negligence:

Faye Q. Hansen v. Skate Ranch, Inc., *Docket No. 5237, Court of Appeals of New Mexico, decided February 4, 1982, reported at 641 P.2d 517.*

An employer is liable for the actions of its off-duty employees who cause injuries to patrons on the company premises if the employee was acting within the scope of his employment. The Court of Appeals of New Mexico affirmed an award to a skating rink patron whose injuries were caused by a "floor guard," who claimed to be off-duty but was employed by the rink.

While roller skating at the Skate Ranch in Albuquerque, Faye Hansen fell and seriously injured her wrist when Sparky Barks, a floor guard employed by the rink, executed a jump turn in front of her and she fell over him. Hansen sued Skate Ranch for negligence and failure to maintain its premises or to correct or warn of dangerous conditions, and claimed Skate Ranch was liable because she was injured by its employee. The jury at the trial agreed with the plaintiff Hansen and awarded her damages in the amount of $25,606.35.

On appeal, Skate Ranch argued that Barks could not be acting within the scope of his employment because he was not wearing the red vest worn by floor guards on duty, and because floor guards are not allowed to jump or do tricks on duty.

The Court of Appeals stated that, although Barks was not wearing his vest at the time, nor was he officially off-duty, Barks was acting within the scope of his employment. The evidence clearly showed that he was "helping out" his employer that night and was even being paid for his help, according to Barks' own testimony. Since the only difference between being on-duty and off-duty was that Barks was not wearing his red vest, the Court of Appeals affirmed the damages award.

COMMENTARY:

Employees should be cautioned that their responsibilities may extend past regular working hours if they continue to undertake to perform their duties.

Removing the uniform does not necessarily remove liability of the employer when the employee continues to perform duties.

PSCLR–SEP 82 (11)

Jury Awards Bystander Damages From Guard's Employer For Mental Distress Caused By Gunshot Fired By Guard Chasing Felon:

Giant Food, Inc. v. Geraldine Sherry, et al., *Docket No. 1258, Court of Special Appeals of Maryland, decided May 7, 1982, reported at 444 A.2d 483.*

Giant Food hired William Joyner as an armed security guard at its grocery store in the Blair Plaza Shopping Center in Silver Spring, Maryland. At approximately 7:30 PM on the evening of November 11, 1978, Joyner was on duty at the store when he saw a man brandish a pistol and rob one of the store's cashiers. Joyner followed the robber out of the store and began chasing him across the parking lot. Joyner himself, in his testimony at trial, described the incident:

> As soon as he [the robber] set foot on the parking lot he got into a slight trot across the parking lot. At this time I had already unbuckled my holster, and I jumped up and followed suit. I unholstered my revolver.
>
> I was running to the parking lot, and I called for the subject to 'halt, police officer.' He looked and kept trotting.
>
> I am still in pursuit as he reached the car. I noticed that he did not have an accomplice, and he reached in the car, he reached down into his inside coat and pulled his revolver back and brandished it up in the air like this. He did it as if he was going to turn and fire. He did not fire.
>
> He put his weapon back in his pants...and got in the car. This is when I hollered for him to halt again. At this time the subject got in the car and proceeded to pull off, and I fired my first shot which hit the rear panel of the car.
>
> He was pulling off in kind of a hurry. He burned a lot of rubber on the parking lot getting off the parking lot. I fired another shot in the back window of the car.

It was this second shot which sparked this litigation. Instead of hitting the fleeing felon's car, the bullet blasted through Geraldine Sherry's fifth floor apartment window across the street.

Ms. Sherry was just walking into her apartment living room when the bullet shattered the window and threw broken glass all over the room. While she was not struck by the bullet or by the debris, she thought someone was in fact trying to kill her and she became hysterical.

The event so frightened her that she was nauseated, and she continued to suffer for some time from insomnia, headaches, and general mental and emotional distress. She underwent psychotherapy, one period of hospitalization, and lost time from work.

Ms. Sherry and her husband sued Giant Food for damages on two theories of negligence: They alleged (1) that the security guard Joyner was negligent in firing the second shot and that Giant Food was responsible as his employer; and (2) that Giant Food was itself negligent in hiring Joyner and in giving him a firearm without proper training to use it.

It is clear that Joyner, as a private citizen, had the right to arrest the fleeing felon. A private person, such as a security guard, may make an arrest if he has reasonable grounds to believe that a felony was committed and that the person being arrested committed it. And, it is clear that Joyner could use reasonable force to accomplish the arrest.

What is not clear is the amount of force which is considered to be reasonable under the circumstances of a particular situation, and therefore this question was left for the jury, which returned a verdict for the Sherrys of $15,000.

In reviewing the case on appeal, the appellate court noted that a security guard's use of deadly force to arrest a fleeing felon is not privileged or justified in all cases. In attempting the arrest, a person has a double responsibility – one to the person arrested not to use unreasonable force against him, and one to the public at large to use reasonable force in a reasonable manner.

COMMENTARY:

Unfortunately, for our purposes, the appellate court did not discuss the issue of the employer's negligent employment of the guard. Nevertheless, since the employer was held liable for the negligence of its employee, this case reiterated the absolute necessity of proper training of all security employees, but especially of those being entrusted with firearms. While the verdict in this case was comparatively small, it can be easily seen that the same act by the guard could have resulted in a monumental verdict if the stray bullet had killed or paralyzed a bystander.

PSCLR – OCT 82 (1)

Employer's Responsibility For Negligent Hiring Of Ex-convict, Who Committed Assaults On Company Property, Is Limited To Business Invitees And Does Not Extend To Public Generally:

Carol Baugher and Genise Tumbleson v. A. Hattersley & Sons, Inc., *Docket No. 3-1281A338, Court of Appeals of Indiana, decided June 3, 1982, reported at 436 N. E.2d 126.*

The concept of negligent hiring was the basis of this damages suit brought against a business for employing an ex-convict who assaulted two women on company property.

Although on parole for armed robbery, Ronald Osborne was hired by A. J. Hattersley & Sons, an electric contracting firm, as a laborer. Over a period of four years, Osborne's job performance was exceptional and he was rewarded with increased responsibilities. Osborne was placed in charge of the shop in 1979, and was given keys to the premises and the use of a company car.

On September 28 of that same year, Osborne used his company car to pick up a cocktail waitress and her sister whom he had invited to join him for breakfast after the waitress got off work. Instead of going to a restaurant, however, Osborne stopped by the Hattersley shop and used a key given him by another employee to obtain a gun locked in the employee's desk. He then brandished the weapon, ordered the two women inside the building, raped, beat, and shot at the women, wounding one of them. Subsequently the women sued A. Hattersley & Sons to recover for their injuries which they claimed were caused by the company's negligence in hiring Osborne and in providing him with the car, the shop keys, and the gun – all of which aided in the women's assaults.

At trial, judgment was granted in favor of Hattersley. Although the court noted that Indiana law recognizes a cause of action based upon the negligent hiring of an employee, the court felt this implied that a business was required not to expose its customers, invitees, and patrons to intentional assault by its employees. The women assaulted by Osborne were

> ... neither customers, patrons, nor invitees of A. Hattersley & Sons, Inc. As such they were beyond the scope of the duty owed by the company to the public. ... no persuasive policy consideration [was presented] which would justify the expansion of that duty.

Although the company had supplied Osborne with a company car and shop keys, it had no prior notice that this "stable and competent employee" of four years would use his privileges to commit a crime. The court felt the company was not negligent in this respect, nor was it responsible for supplying Osborne with a gun since the company proved that it was completely unaware that the weapon was on its premises. Judgment for the company was therefore affirmed.

COMMENTARY:

While the employer here was exonerated because the employee's victims were not included in those particular groups to whom the employer owed the duty of care, it is quite clear that the result would have been otherwise if the employee had similarly assaulted a customer or patron who had come to the employer's property to conduct business as an invitee. At the same time, it should be seen that the law does not make the employer an insurer of the safety of the public. Furthermore the concept of negligence still includes the presence of foreseeability.

PSCLR–OCT 82 (7)

Police Officer's Misconduct Renders Both City Employer and Private Employer Liable When Officer Acting for Both Employers Simultaneously:

John A. Faust and Ingrid Piller v. John Mendoza, et al., *No. 14687, Court of Appeal of Louisiana, First Circuit, decided May 25, 1982, reported at 415 So. 2d 371.*

When a police officer also working as a private security guard used excessive force in an arrest, the officer and his employers – public and private – were held liable for personal injury damages incurred by the arrested persons.

The subject incident occurred around 10 PM on February 9, 1975, during the Mardi Gras celebration in the French Quarter of New Orleans, Louisiana. A couple who had been enjoying the festivities and drinking all day stopped at the ice cream parlor in the Royal Sonesta Hotel. Apparently the man, John Faust, rested his head on the parlor's counter and ignored requests that he move. At this point, Officer John Mendoza entered to wait the forty-five minutes until 11 PM when he was to begin work as a security guard for the parlor. He was to work until 3 AM, in his police uniform, for the parlor, after completing an 11 AM to 11 PM shift on a police assignment – controlling crowds around Mardi Gras parades.

After Mendoza approached Faust, testimony on what followed conflicts greatly. Although the particular details are unclear, it appears that Mendoza struck both Faust and his female companion, Ingrid Piller, with a billy club, smashed the ice cream parlor window (either accidentally or by throwing Faust against it), and arrested Faust and Piller for assault upon a police officer.

After being found not guilty on reduced charges of battery, Faust and Piller sued Mendoza, the Royal Sonesta Hotel, and the city of New Orleans for their injuries and for malicious prosecution. Awarding Faust $3,315 and Piller $1,578 for their injuries, the trial court rejected the malicious prosecution charge.

On appeal by defendants Mendoza and the Royal Sonesta Hotel, the Court of Appeal of Louisiana affirmed the award even though the officer had not yet gone on duty as the parlor's guard:

> While the tort occurred approximately forty-five minutes prior to the time he was scheduled to go on duty, Mendoza could give no explanation as to why he was at the ice cream parlor at that time. Moreover, forty-five minutes is not so long a period of time as to, by itself, remove Mendoza from the scope of his employment with Sonesta.

Although the trial judge stated that Faust's and Piller's conduct led to their attack by Mendoza, the Court of Appeal affirmed that the officer's excessive reaction warranted a damages award to the plaintiffs.

PSCLR—OCT 82 (11)

Private Business Not Liable for Death Resulting From Officer's Negligently Directing Traffic from Business Premises, Even Though Officer was Off-duty and Working for Business:

Glenmar Cinestate, Inc. v. Robert T. Farrell, Administrator, etc.; Ray Bentley, d/b/a Ray Bentley Productions and Neighborhood Theatre, Inc. v. Robert T. Farrell, Administrator, etc., *Nos. 800304, 800315, Supreme Court of Virginia, decided June 18, 1982, reported at 290 S.E.2d 366.*

Should a private business be held liable for the actions of off-duty public police officers hired by the business to perform police-type duties such as directing traffic out of the business' premises? A Virginia trial jury said "yes," granting a damages award of $52,911.30.

In this case, a drive-in theater had routinely used local deputy sheriffs "to quell disturbances, to watch for gate-crashers, and primarily to direct traffic" after late movies. The Patterson Drive-In Theater arranged for the deputies through the Goochland County Sheriff's office and paid each deputy $25 per night. In compliance with the sheriff department's policy, the deputies worked only in pairs, wore their uniforms, and were "technically on duty 24 hours a day," when working outside their regular duties.

On the night of June 25, 1978, one of the deputies on his way to work at the theater was delayed when he stopped to issue a speeding ticket. The other deputy, James Lilly, proceeded to the theater and immediately began directing traffic out of the theater and onto the highway, as was the regular practice. Lilly negligently directed exiting patron Herbert Rae onto the highway, and Rae's car was hit broadside by an oncoming car when he entered the highway's westbound lane. The driver of the oncoming car, John Farrell, died as a result of injuries suffered in this accident. His family sued the theater owner and operators on the claim that the theater, as Lilly's employer at the time, was responsible for his negligence.

Appealing the damages award granted the Farrell family, the defendant theater argued that the police officer acted as an "independent contractor" in performing traffic control for the theater and not as an "employee" of the theater. The theater argued that the trial court should have dismissed the case because a business cannot be held liable for the negligence of an independent contractor.

The Supreme Court of Virginia agreed and, quoting from an earlier case, defined an independent contractor as "one who undertakes to produce a given result without being in any way controlled as to the method by which he attains the result."

Since Lilly's private employer had no control over his method of directing traffic, and since the function of directing traffic is a public police function, the theater could not be held liable for his conduct, negligent or otherwise. Therefore, the court overturned the damages award granted to the Farrells and dismissed the case.

COMMENTARY:

It is frequently difficult to distinguish between an "employee" and an "independent contractor," as is illustrated by the conflicting opinions of the trial judge and the appellate court in this case. However, this issue had been addressed by the Virginia courts as far back as 1936, when it was held that

> *... a special police officer appointed by public authority, but employed and paid by a private party, does not subject his employer to liability for his torts when the acts complained of are performed in carrying out his duty as a public officer. The test is: in what capacity was the officer acting at the time he committed the acts for which the complaint is made? If he is engaged in the performance of a public duty such as the enforcement of the general laws, his employer incurs no vicarious liability for his acts, even though the employer*

> *directed him to perform the duty. On the other hand, if he was engaged in the protection of the employer's property, ejecting trespassers or enforcing rules and regulations promulgated by the employer, it becomes a jury question as to whether he was acting as a public officer or as an agent, servant, or employee.*

Norfolk Bus Term. v. Sheldon, 188 Va. 288, 49 S.E.2d 338 (1948).

PSCLR – NOV-DEC 82 (6)

Store Which Uses Contract Guards But Controls Their Conduct Is Liable for Guard's Use of Excessive Force in Lawful Arrest of Customer:

Safeway Stores, Inc. v. George I. Kelly, *Nos. 80-474, 80-659, 80-686, in the District of Columbia Court of Appeals, decided July 1, 1982, reported at 448 A.2d 856.*

The corporate owner and operator of a grocery store was held responsible by a trial court jury for the actions of its contract security guard who used excess force to arrest a customer. The jury awarded the customer $25,000 for assault and battery and $40,000 for false arrest, but the trial judge reduced the awards to $2,000 and $13,000 respectively. On appeal, the false arrest finding was reversed and dismissed.

Safeway Stores, Inc. had an arrangement with Seaboard Security Systems, Inc. for Seaboard to provide security guard services for Safeway's 16 Washington, D. C., stores. Seaboard was responsible for hiring and firing the guards, but each store's managers had "operational control" over the guards, could have a guard replaced by Seaboard, and set the guards' working hours. Also, the managers could and did direct the guards in the tasks they performed.

On a February night in 1976, customer George Kelly complained to a Safeway manager about a broken automatic exit door. During the ensuing conversation, an argument broke out between the two men. Kelly refused to leave when asked, so the manager requested that a security guard stand by while another employee was directed to call the police.

A police officer arrived, spoke to the manager, then told Kelly that the manager wanted Kelly to leave. The evidence as to what then occurred was in conflict. Kelly testified that he was grabbed by the throat from behind by the guard and was attacked by the officer before he could answer. Other witnesses testified that Kelly shouted at the police officer which prompted the guard and the officer to use force to subdue and to arrest him. At any rate, Kelly was handcuffed, taken into custody, and charged with unlawful entry. The criminal misdemeanor charge was later dropped.

Kelly's suit against Safeway resulted in a judgment of $15,000 in compensatory damages for false arrest and for assault and battery. Safeway appealed the judgment, contending that they could not be held responsible for the conduct of the guard who was an employee of an independent security service.

On review, the District of Columbia Court of Appeals affirmed that although the security guard was an employee of Seaboard, he nevertheless was the "servant" of Safeway because of the extent of control exercised over the guard by Safeway management. The court also affirmed the assault and battery award against Safeway because there was sufficient evidence to indicate that the guard used excessive force in assisting in the arrest of Kelly.

However, the court reversed the judgment against Safeway on the false arrest charge. The court stated that "[t]o prevail [in a false arrest action], the arresting officer need not prove probable cause in the constitutional sense, but rather must prove that he had a reasonable good faith belief that the suspect committed the offense." There was a reasonable basis for Kelly's arrest for unlawful entry because all of the evidence clearly showed that he had refused to leave the business premises when asked to do so by someone in charge. Therefore, Safeway was released from liability for the false arrest.

COMMENTARY:

As the opinion in this case correctly states, the determination as to whether a "master and servant" relationship exists depends on the particular facts of each case. It may be helpful to review the definitions:

> *MASTER— One who employs another to perform service in his affairs and who controls or has the right to control the physical conduct of the other in the performance of the service.*
>
> *SERVANT— One who is employed by a master to perform service in his affairs and whose physical conduct in the performance of the service is controlled or is subject to the right to control by the master.*

Therefore, the following factors are to be considered:

> *(1) the selection and hiring of the employee;*
>
> *(2) the payment of his wages;*
>
> *(3) the power to discharge;*
>
> *(4) the power to control his conduct; and*
>
> *(5) whether the work is part of the regular business of the employer.*

It is not necessary that a company meet all five of these criteria in order to be held liable as a "master" for the acts of a "servant." The most important criteria are the right to control and direct the employee in the performance of his work, and the right to discharge the employee. If a company exercises these rights over a person, it will be held liable for the acts of that person regardless of who selected, hired, trained, or paid him.

PSCLR—JAN 83 (5)

New Trial Ordered for Wrongful Death of Alleged Shoplifter Who Died During Arrest by Security Guard:

George C. Taggart, as Administrator, etc. v. Alexander's, Inc., et al., *in the Supreme Court, Appellate Division, Second Department, decided October 25, 1982, reported at 455 N.Y.S.2d 117.*

Michael, a 17-year-old boy, and two friends entered Alexander's Department Store in Queens just before closing time. Each of the three boys went to different floors to browse. One of the friends later testified that as he was riding up an escalator, he turned around and saw a security guard (Mark Farmer) fighting with Michael. He stated that the guard struck Michael, "threw him over his leg onto his face, and twice thereafter picked him up and threw him down again, first on his face and then on the back of his head."

A witness who was not involved in the struggle testified that the security guard then conferred with someone who appeared to be a superior, and subsequently "flipped" the boy while frisking him and handcuffed the youth behind his back even though he was unconscious and bleeding. Another disinterested witness confirmed this description of the event, and added that the guard "flipped the boy's head from side to side as he lay on the floor."

Michael died as the result of a fractured skull and resulting brain injuries.

The guard Farmer testified that he was locking the doors when the three boys came in the store after the closing announcement had been made. He saw them split up, and he said he saw Michael remove a coat from a rack, look around, and move quickly toward the entrance. After Michael stepped outside, Farmer ordered him to stop and showed his store identification. Farmer testified that the boy then hesitated and turned around, uttered an obscenity, and hit Farmer in the face. The two began to struggle and moved back inside the store entranceway. The guard said he struck Michael in the jaw just as he saw Michael's friends rushing toward them; however, two police officers arrived and stopped the boys.

The guard also testified that Michael had fallen backwards from the blow to his jaw, that he sat up and that one of the officers advised Michael to lie down again. Farmer called his superior and, after having the boy roll over, handcuffed him. Farmer stated that he had not noticed Michael bleeding and could not tell if he was badly hurt. He denied causing the boy's head to hit the floor after he fell, and denied that there was a clear line of vision from the escalator to the scene of the arrest as Michael's friend had testified.

The administrator of Michael's estate brought suit to recover damages for battery, false arrest, and wrongful death. The jury found in favor of the defendants, entering judgment against the plaintiff-estate.

However, the appellate division in a split decision (3-2) reversed the judgment because of improper arguments made by the defense attorney in his summation to the jury. A new trial will have to be held.

PSCLR – FEB 83 (1)

Where Guard's Activities Were Jointly Controlled by Both Contract Security Company and the Retailer, Both Held Liable for Guard's Mistake in Shooting Customer – $94,719.77 Verdict:

Gulf Oil Corporation, Empire Security Services and Empire Security Agency, Inc. v. Thomas Williams, *No. 9023, in the Court of Appeals of Texas, Texarkana, decided September 28, 1982, reported at 642 S.W.2d 270.*

When Thomas Williams stopped at a Houston, Texas, Gulf station for gasoline, he never expected to be mistaken for a robber and to be shot by a security guard. Nonetheless he was, for no apparent reason, and Gulf Oil Corporation ("Gulf") and Empire Security Agency, Inc. ("Empire") were held liable to Williams for $94,719.77 in actual damages and $50,000 in punitive damages.

It appeared from the evidence presented that Gulf had a written contract with Manpower, Inc., by which Manpower furnished guards for security at Gulf stations. Under this agreement, Gulf paid Manpower certain fees. Empire then agreed with Manpower under a separate contract to furnish security guards for Manpower's clients, such as Gulf. Empire hired and fired these guards, and furnished them to Gulf stations in accordance with its contract with Manpower. Gulf had no contractual agreement whatever with Empire, only with Manpower.

Empire hired Robert Gory as a guard, furnished him with his uniform and weapon, provided him training and instruction, and paid him for his work. Empire assigned him to provide security for Gulf.

At the trial in the lower court, the jury found that the security company Empire was guilty of negligence for hiring the guard Gory and for entrusting him with a deadly weapon. The Court of Appeals declared that Gory was obviously "incompetent and unfit as a security guard" – witnesses testified that Gory had simply "walked up to Williams' car, stuck a gun through the window, and shot him." Yet, Empire was able to prove that it had checked into Gory's past employment and that it did not have any knowledge, prior to the shooting, that Gory was incompetent. The court agreed that Empire was indeed innocent of negligence insofar as its own conduct was concerned, but this fact alone was insufficient to excuse Empire from vicarious liability for the guard's conduct.

The lower court also found that Gory was negligent in shooting Williams and that both Empire and Gulf should be held liable for his misconduct. Empire and Gulf took exception to the lower court's finding that Gory was "Gulf's borrowed servant" jointly controlled by the two companies.

The evidence conflicted on many points: It tended to show that Empire was responsible for supervising guard assignments and for instructing guards in the amount of force to use. However, other evidence indicated that Gulf hired and fired the guards, evaluated guard performances, required the guards to carry guns, scheduled the guard's hours and locations of work, and "directly supervised and controlled the guards." From this information, the court drew the conclusion that the companies exercised joint control over Gory, that Empire had "ratified" Gory's conduct, and that the "evidence was sufficient to sustain the joint liability ... for actual damages."

One of the justices on the Court of Appeals, a Justice Bleil, filed a dissenting opinion because he disagreed that Gulf should be held liable for Gory's actions. Justice Bleil's opinion stated in part:

> Regardless of any disputed testimony certain matters are conclusively established: (a) that Gulf had no agreement with Gory or Empire; (b) that Gory was engaged in the distinct occupation of being a security guard; (c) that Empire, not Gulf, furnished Gory his uniform and firearm; (d) that Empire, not Gulf, paid Gory for the work done; (e) that Empire, not Gulf, provided training and instruction for the security guards; and (f) that Empire, not Gulf, regularly engaged in the business of security services.

Applying these facts to Texas agency law, which provided the basis for determining whether or not an individual acted as the servant for another, Justice Bleil expressed the opinion that Gulf's relationship to Gory lacked the basic element of authoritative control over "the manner and details of Gory's work." In spite of conflicting evidence, it seemed to Justice Bleil that Gulf's control over Gory was more suggestive and sporadic than authoritative and therefore Gulf should not be subject to liability for Gory's actions.

The majority ruling of course controlled and the verdict for the plaintiff was modified to set aside the $50,000 punitive damages award, and affirmed against all defendants for $94,719.77 actual damages.

COMMENTARY:

This case illustrates the fact that it is not always a simple matter to determine who is the responsible master of a negligent servant. Earlier cases reported in **PSCLR** *have set out tests applied by the courts in attempting to make this determination. Suffice it to say here that each fact* *situation must be evaluated on its own merits, and when it appears that both the security company and the retailer exercise joint control over the guard, then the courts will rule that both are liable for the guard's misconduct.*

PSCLR–JUN-JUL 83 (8)

Employer of Gardener May Be Liable for Shooting of Trespasser If Gardener's Duties Included Security Functions, Or If Employer Negligently Supervised Gardener:

Dermot Murphy v. The Army Distaff Foundation, Inc., *No. 82-5, in the District of Columbia Court of Appeals, argued September 30, 1982, decided March 11, 1983, reported at 458 A.2d 61.*

The Army Distaff Foundation operated a retirement home in Washington, D.C., for the widows of army officers and it had employed Ernst Flink as the resident gardener since 1957. In October of 1977, at about midnight, Flink was strolling the grounds and observed a group of young people in the parking lot of the Army Distaff Hall. It was not the first time he had caught youths trespassing on the grounds, and using the parking lot as a place to congregate and drink alcohol, use drugs, and engage in sexual activities.

The gardener approached one of the boys, Dermot Murphy who was there with a girl friend, and ordered him and his friends to leave. A heated argument arose between the two; Murphy ended up hitting Flink, who then drew a gun and shot the youth six times in the arm and torso. Murphy subsequently sued the foundation to recover for the injuries he suffered in the shooting incident. The court granted the defendant's motion for summary judgment, stating that it had not found any issues of fact in the suit which should be presented to a jury. Murphy appealed.

The D.C. Court of Appeals examined the facts of the shooting incident and of Flink's employment relationship with the foundation, after first noting that summary judgment was "an extreme remedy, entitling the moving party to judgment as a matter of law only when no genuine issue of material fact is present at the time the motion is made." According to the court, the key to the foundation's liability to Murphy would lie in whether Flink had acted within the scope of his employment when he shot the trespassing youth, and whether the foundation knew or should have known that Flink regularly removed trespassers while armed with a gun.

Flink's supervisor testified that checking the parking lots on the grounds was one of the Flink's duties, and the manager of the Army Distaff hall testified that, "He lives in Knoll House as part of the conditions of employment, and he does secure and lock up the Knoll House." Although Flink might have had personal reasons for assaulting Murphy, the court felt that this evidence was sufficient to infer that Flink's duties included at least some security functions, therefore raising a jury question as to whether the encounter could be considered job related.

The evidence further indicated that the employer knew that Flink frequently strolled the grounds at night and that he had had previous encounters with trespassing youths. The court determined that there was a conflict in the evidence concerning the foundation's possible negligent supervision of Flink, particularly on the question as to whether it was aware or should have been aware that the gardener carried a gun with him on his evening walks.

Because the court felt that these questions of fact were raised by the conflicting evidence on these issues, the dismissal of the suit was reversed and the case was remanded to the lower court for trial.

PSCLR–OCT 83 (2)

Patron Assaulted By Security Guards Recovers From Security Company For Its Negligent Equipping, Training, And Supervising Of Guards:

Raymond J. Gonzales and Linda Gonzales v. Southwest Security and Protection Agency, Inc., f/d/b/a Southwest Investigative and Protective Agency, Inc., a New Mexico corporation, and The City of Albuquerque, a municipal corporation, Fundamentals, Inc., a New Mexico corporation, Mike Sedillo, Denny Sanchez, Alex Sedillo, Ysidro Victor Vigil, aka Victor Vigil, and David Chavez, *No. 5845, in the Court of Appeals of New Mexico, decided June 7, 1983, reported at 665 P.2d 810.*

This lawsuit was initiated by a patron at a wrestling match who was injured by security guards standing watch during the match. The plaintiff, Raymond J. Gonzales, sued for damages resulting from an alleged false imprisonment, battery, and negligent hiring, training, supervision, and retention of the security guard employees. He was awarded a verdict of $15,000 at the trial, and the defendant security company filed this appeal.

The plaintiff Gonzales sued the city of Albuquerque, New Mexico, and Fundamentals, Inc., and Southwest Security and Protection Agency, and five security guard employees of Southwest. Gonzales was injured while attending a wrestling match sponsored by Fundamentals and held in the city's civic auditorium which had been rented to Fundamentals. Southwest Security was hired by Fundamentals to provide the necessary security services for the wrestling match.

Gonzales alleged that he was handcuffed by the security guards and taken to a small room in the auditorium where he was beaten. He claimed that he was a bystander, merely observing a disturbance caused by his friend and a woman. Nevertheless, according to Gonzales, the guards threw him to the ground and handcuffed him. The evidence showed that Gonzales apparently remained calm and did not provoke the guards.

Judgments were entered in favor of Gonzales against Southwest employees Mike and Alex Sedillo, David Chavez, and Denny Sanchez, as well as against Southwest Security itself. The City of Albuquerque, Fundamentals, Inc., and Southwest employee Ysidro Victor Vigil were dismissed without liability. Only Southwest Security appealed the adverse judgment against it, in the amount of $15,000. The company argued that it should not be held responsible for the intentional torts of its employees and claimed that the evidence did not show that it was negligent.

It is well settled law that "an employer is liable for the intentional torts of its employees if the torts are committed in the course and scope of employment." The court then reviewed the evidence presented at the trial to determine whether the security guards' actions in the instant case were within the course and scope of their employment by the security company. More particularly: "The question is whether the security guards' imprisonment and battering of Gonzales was within the course and scope of their employment." In order to answer this question, the court looked to the nature of Southwest operations and its security guards' work. The court found from the evidence that Southwest Security had provided the security guards with their uniforms, handcuffs, guns, nightsticks and instructions to maintain order. The court then concluded that the evidence was sufficient to indicate that the guards' misconduct was naturally incident to their employer's business, even though it was not specifically authorized by the employer. It was enough that the guards' conduct substantially arose out of an effort to perform Southwest's business of maintaining order at the arena. Accordingly, Southwest was held liable for the intentional acts of its employees.

With regard to Gonzales' claim that Southwest Security itself was negligent, the court reviewed the evidence and determined the following facts:

(1) Southwest Security did not adequately investigate the background and character of the individual security guards prior to hiring them;

(2) Southwest Security did not adequately supervise the guards in general, especially on the evening in question;

(3) Southwest Security failed to adequately train the security guards in the use of nightsticks and handcuffs which were issued to them;

(4) Southwest Security failed to adequately instruct the security guards in the proper method of restraining and arresting individuals;

(5) Southwest Security was on notice of one other beating of an individual by one of the security guards and failed to take appropriate action to avoid further such incidents.

Therefore, the appellate court affirmed the award in favor of Gonzales against Southwest Security and Protection Agency and its five security guards.

COMMENTARY:

It cannot be said too often or too strongly: It is imperative that those exercising supervisory authority over security guards (whether contract or proprietary) initiate affirmative measures to ensure that the guards are physically, mentally, and morally capable and qualified to perform their assigned duties, and that they are properly trained in the use of their weapons and in the basic legal principles applicable to their work, and that their conduct on the job is then monitored and supervised adequately.

PSCLR—JAN 85 (1)

Restaurant Which Employs Security Guard To Protect Premises And Patrons Is Liable For Injuries Due To Guard's Negligence In Performing His Duties:

Henry Harris, et al. v. Pizza Hut of Louisiana, Inc. and St. Paul Fire and Marine Insurance Company, *No. 84-C-0304, in the Supreme Court of Louisiana, decided September 10, 1984, reported at 455 So. 2d 1364.*

In March of 1979, during an armed robbery of a New Orleans, Louisiana, Pizza Hut Restaurant, one patron was shot and killed, and her daughter wounded as they sat in the restaurant with other family members and a family friend. The Harris family sued Pizza Hut to recover for their injuries, the loss of their mother, and damages suffered by the survivors, because, they alleged, the negligence of the restaurant's security guard was the actual cause of the shootings.

The trial court jury agreed with the allegations and awarded $50,000 to the injured Harris child, Zina Harris, $50,000 to each of six Harris children for the death of Mrs. Harris, and two other Harris children were each awarded $65,000 and $80,000 respectively. On appeal, the Louisiana Court of Appeal reversed the judgment, finding no negligence on the part of the security guard, but this judgment is now reversed by the Louisiana Supreme Court on the most recently published appeal.

The armed robbery in question occurred at a Pizza Hut which was located in a high-crime neighborhood. The robbery occurred on the evening of March 17, 1979. The security guard who was present that night was New Orleans Police Officer Maxie Walker. Officer Walker had worked for the Pizza Hut in the past though on this night he was substituting for another police officer who then worked for the restaurant. He was scheduled to work from 9 PM to 1 AM, and the evidence showed that Pizza Hut had never given him any specific instructions concerning his duties.

Officer Walker arrived at the Pizza Hut after leaving a supermarket where he worked from 2 PM until 9 PM as a security guard. Upon entering the restaurant, Walker first prepared a salad from the salad bar and got a soft drink. He sat near the front of the restaurant and, after eating, remained there to talk with an off-duty employee. His view of the front door was obscured by a partition.

At about 10 PM a masked armed robber entered the crowded restaurant through the main door. The robber did not see Walker until the officer stood and reached for his gun. Another robber at that point entered and pointed his gun at Walker and ordered him not to move. However, Walker did move slightly, and the off-duty employee standing with him dived to the floor. According to several witnesses, the robber fired at Walker (according to the robber, he fired after Walker shot at him), wounding Walker, hitting Zina Harris, and killing Mrs. Harris. As he fell to the floor, Walker returned fire, killing one robber and wounding another. A third robber escaped unharmed.

The Harris's lawsuit alleged twenty claims of negligence on the part of Pizza Hut and its employees, which can be summarized in essentially three claims surrounding the security guard Walker:

> (1) the security guard was negligent in drawing his gun and firing it before the shotgun blast, (2) in moving and provoking the robber to fire the shotgun, (3) in being seated inside the restaurant, being inattentive, visiting with other employees and perusing a book, rather than being at the door of the restaurant or outside the door of the restaurant in a position where he would have been seen by potential robbers and thereby would have deterred the crime.

Security experts testified that the ability of a security guard to serve the purpose of deterring crime is dependent on his/her visible presence. An expert testifying for the plaintiffs emphasized that Walker was very poorly positioned at the time of the robbery, that he should have been sitting near the main entrance or patrolling the premises on foot and not sitting at a table, eating a salad. One expert testified that if a guard were temporarily off-duty or on a break, he should stay out of sight since he would not be able to provide effective deterrence to crime. Security experts also testified concerning the appropriate behavior of a police/security officer who is confronted by an armed individual. All agreed that the officer should not move, resist, or draw his gun, particularly when there is a substantial risk of harm to innocent by-standers.

There was also evidence that the Pizza Hut had been robbed or burglarized more than 20 times since opening, and that there had been seven criminal incidents in the three years before this incident, most of which were armed robberies occurring between 9 PM and 10 PM About five months before this incident, the restaurant had begun to employ security guards and there was an apparent decrease in criminal occurrences.

In reviewing the case, the Louisiana Supreme Court noted that the question of a business's liability to its patrons for third party criminal assaults where the business employed security guards had never before been addressed in Louisiana. However the court discussed similar cases decided in other jurisdictions (including *Banks v. Hyatt Corporation*, reported in the April 1984 **PSCLR** at page one; *Butler v. Acme Markets, Inc.*, a New Jersey case reported in the November-December 1982 issue at page six; *Mullins v. Pine Manor College*, a Massachusetts case reported in the August 1983 issue at page seven; and *Tucker v. Sandlin*, a Michigan case reported in the October 1984 issue at page nine). The court concluded that a business basically owes a duty to provide a reasonably safe place to its invitees. Where it additionally undertakes to provide protection for itself and its patrons through a security guard, it will be liable for physical injuries occurring due to the negligent acts of the guard.

Since Louisiana relies on a "duty-risk analysis" in tort claims, the court then applied this to the case. A duty-risk analysis seeks to discover whether (a) the conduct which is under question was a "cause-in-fact" of the injury, (b) the defendant had a duty to protect against the risk of the injury, (c) the duty was breached by the defendant, and (d) any damages were suffered by the plaintiff. Here, the jury found that Officer Walker's negligence was a cause-in-fact of the Harris's injuries. The court ruled that Walker had a duty, once employed as the Pizza Hut security guard, to fulfill his responsibilities in a "reasonable and prudent manner, ... not to contribute by his act or omission to an escalation of violence during an armed robbery." And, finally, the court found that the "failure of Walker to be alert and visible at the time of the crime constituted a breach of his duty to patrons of the Pizza Hut."

Reinstating the trial court's judgment for the plaintiffs and the associated damages awarded, the Louisiana Supreme Court reversed the decision by the Court of Appeal.

However, two justices on the court expressed their qualms with this decision:One justice concurred, agreeing with the decision of the majority based on the ordinary negligence of the security guard. He disagreed however with much of the discussion surrounding its decision, which included other theories this justice felt were unnecessary.

Another justice dissented, because he felt the decision could be interpreted to require a restaurant, once it had hired a security guard, to require the guard to stay on duty in clear view to those who enter, without taking a break. Further, such a requirement would unreasonably assume a guard would be able to remain perfectly motionless when threatened by an armed individual and, failing that, the guard would be negligent and liable to those injured as a result. He stated that to require such from a restaurant and the security guards it may employ was to place the restaurant in the unreasonable position of having to insure the safety of its patrons, even where a third party intentionally commits a violent act.

COMMENTARY:

The importance of this case is accentuated by the fact that amicus curiae (friend of the court) briefs were filed in the Supreme Court by the Louisiana Restaurant Association and by the Louisiana Trial Lawyers Association (an organization consisting primarily of plaintiffs' trial attorneys). The Restaurant Association argued that a decision in favor of the Harrises would require every restaurant in the state to provide a security guard.

The court, however, noted that this issue, as to whether a restaurant which had suffered a large number of armed robberies would have such a duty, is not the true issue before the court in this case inasmuch as the defendant Pizza Hut had already concluded that it was in fact necessary for the protection of its customers and its property to hire a security officer.

The Restaurant Association also argued that having to provide a security guard would create an undue economic burden on the restaurants of the state. However, the Supreme Court made it clear that by recognizing the duty provide a safe restaurant, the court is not requiring that all restaurants must have security guards, or that an undue economic burden must be placed on the restaurants throughout the state.

In this regard the court pointed out that there are many restaurants which do not require any more protection than that which is available from the public police authorities. At the same time, the court noted, many restaurants do feel the need for security officers to reassure their customers and to protect their property.

The inherent difficulties in this case are also highlighted by the fact that we have two appellate courts disagreeing as to the result of the case, and the Supreme Court itself could not reach a unanimous conclusion.

These are not simple issues and, as noted by the Supreme Court in its majority opinion, great consideration was given to the testimony of various expert witnesses, who also disagreed as to the proper duties of the Pizza Hut in retaining a security officer, and as to the duties of the security officer himself. They did agree, however, that the primary purpose of a security guard is the deterrence of crime by his visible presence to would-be violaters. They all also agreed that an officer with a shotgun pointed at him should not move and should follow the instructions of the perpetrator.

Indeed, there is an obligation to innocent citizens not to create a substantial risk of harm to them by opening fire or by causing another to open fire. The security officer in this case acted contrary to the experts' opinions as to what his trained reaction should have been: According to the expert testimony, there is a probability that no one will get injured in a commercial robbery if there is no resistance, but Officer Walker testified that he would still "try to pull [his] gun" if a gunman caught him by surprise. (As a footnote, the court noted that Officer Walker conducted himself bravely and heroically once the shooting began, and in fact received a written commendation and medal of merit from the police officers' award committee).

However, the court held there were various derelictions of duty by Walker as a security guard prior to the shooting, and noted that the jury could well have found that the officer's conduct was the responsible cause of the resulting harm to the plaintiffs. The experts all agreed that (a) Officer Walker should have maintained a high visibility, in that the evidence was that the hiring of a security guard had resulted in an interruption of criminal activity at the Pizza Hut, and (b) this particular armed robbery would not have occurred if the robbers had known that a security guard was present.

Furthermore, the court noted that "while it is difficult to fault a man in Walker's perilous situation, there was certainly sufficient evidence from which a jury could have concluded that Maxie Walker's reaction to the leveled shotgun was negligent and resulted in the death of Valeter Harris and the injury to Zena Harris."

In general, there is no duty to protect others from the criminal activities of third persons. However, as we see here, when a duty to protect others against criminal misconduct has been assumed, a concomitant liability may be created by the negligent breach of that duty. Once the decision has been made to employ a security guard, the security guard is then under a duty to discharge his obligations "in a reasonable and prudent manner, and especially not to contribute by his act or omission to an escalation of violence during an armed robbery which resulted in patrons being wounded and killed."

For further discussions and examples of the problems presented by these types of cases, please review the earlier decisions reported in **PSCLR** *as noted in the case above.*

PSCLR – JAN 85 (3)

Employer Of Rapist Held Not Liable For Negligent Hiring For Failing To Ascertain Employee's Criminal History:

Carolyn W. Cramer v. Housing Opportunities Commission of Montgomery County, *No. 2. Sept. Term, 1984, in the Court of Special Appeals of Maryland, decided October 9, 1984, reported at 482 A.2d 156.*

Carolyn Cramer was allegedly raped by an employee of the Housing Opportunities Commission of Montgomery County, Maryland, which owned the housing development in which she rented a townhouse. She sued the Commission on the basis that it had been negligent in hiring the employee, George Slater, as a housing inspector, because it had failed to verify that Slater had no criminal record. At trial, the jury found the Commission not negligent and the judge entered judgment for the defendant Commission. (The jury did award the plaintiff $500,000 against the employe.) On appeal, the Court of Special Appeals of Maryland agreed with the trial court's decision.

Slater was employed by the Commission through a Comprehensive Employment Training Act program administered by the Montgomery County Office of Family Resources (OFR). OFR obtained background information and employment history from Slater, and conducted personal interviews. Slater indicated to OFR that he did not have any record of criminal convictions. OFR then referred him to the Commission for an entry level position as a housing inspector. The Commission's own interview process was restricted to inquiries concerning his employment history and skills. Slater was hired, and his training and supervision were conducted by the Commission's director of maintenance, who accompanied him on several inspections. On one such inspection, Slater was assigned to the home of Cramer. She was present, along with a male friend, during the 45 minute inspection.

A few weeks after Slater inspected Cramer's home, she was awakened by a male intruder who blindfolded her and raped her repeatedly. Though she could not see the man, she noticed and remembered certain physical features. Slater was subsequently arrested and charged with the rape. Although he had not been convicted of the rape at the time of the trial of this case, the Commission conceded for the trial that Slater was the rapist. (It appears that he may have entered into a plea bargain which resulted in his not being prosecuted for this offense.)

Cramer based her negligent hiring claim on the fact that Slater had been convicted of burglary, robbery, and assault, and had been arrested on a variety of other criminal charges, including sexual offenses, for which he had not been convicted. She alleged that the Commission was negligent in failing to verify Slater's criminal record, which might have prevented his being hired; she claimed this negligent act resulted in her rape and assault by the employee.

The appellate court stated that Cramer was attempting to make new law, to "declare that failure by some employers to ascertain the existence of their prospective employees' prior criminal record is negligence per se." Ordinarily, the question of an employer's negligence in hiring is a jury question, a question of fact. In most situations, an employer does not have a duty to inquire into a job applicant's employment history, unless the nature of the conviction would indicate a lack of fitness for the position. For example, someone convicted of armed robbery would be considered an inappropriate choice for a position as armed security guard. In order for a plaintiff to prove a negligent hiring, he must have proof that the employer knew or should have reasonably known that the employee was potentially dangerous.

In this case, the Commission was able to prove to the satisfaction of the jury that it was not negligent in hiring Slater. The appellate court refused to set aside the jury's determination of this question of fact to hold the Commission negligent for failing to determine whether Slater had a criminal record. Therefore, the decision in favor of the Commission was affirmed.

COMMENTARY:

The Maryland court in this case based its decision on practicalities:The court noted that if employers were required to investigate each prospective employees' criminal record, a positive finding unquestionably would mean that the applicant would not be hired. The costs of employing a person would therefore also rise due to the cost of the investigation. And, the court noted that "it takes no social scientist or behavioral expert to recognize that such a policy will scuttle a society's concept of rehabilitation of criminals in favor of the idea of once convicted, forever condemned."

We believe that it would be dangerous for employers, however, to take the position that they have no obligation whatsoever to investigate the criminal history record and background of applicants who are being hired for sensitive positions. The concept of negligent hiring is in fact finding favor with many courts around the country, and many employers are finding that they have an increased obligation in this regard. Even the Maryland court above noted that, "employing an armed guard ... may require the employer to assure that the guard is not an irresponsible person who, once armed, will readily cause injury and damage to others."

PSCLR–FEB 85 (1)

Hotel May Be Liable For Negligent Hiring And Retention Where Habitually Intoxicated Employee Sexually Assaulted Minor Invitee:

Q. Lee Pittard, as father and next friend of Cody Pittard, and Kim Pittard, individually v. The Four Seasons Motor Inn, Inc., *No. 7323, in the Court of Appeals of New Mexico, decided April 19, 1984, certiorari quashed September 5, 1984, reported at 688 P.2d 333.*

An employee of the Four Seasons Motor Inn in Albuquerque, New Mexico, sexually assaulted a young boy who was visiting, with his parents, his grandparents who were registered guests of the motel. The child's parents, Q. Lee and Kim Pittard, sued the hotel to recover damages under several causes of action, including (1) respondeat superior; (2) breach of duty to care for the safety of guests and invitees; (3) failure to provide adequate security; (4) negligent hiring; (5) negligent retention; and (6) inadequate supervision. Judgment was entered for the hotel at trial, and the Court of Appeals affirmed the judgment except as to the negligent hiring and retention claims.

David Leroy Perales was working for the Four Seasons Motor Inn as a steward assisting in banquet preparation. He was intoxicated when he reported for duty that day, and continued to consume alcohol as he worked. The hotel was conducting several promotional activities on the day in question (a Super Bowl Sunday), and there were many customers, guests, and invitees present, including the Pittards and their son. While still on duty, Perales walked to the swimming pool area and encountered the Pittard boy whom he enticed into a hotel bathroom. He locked the door and assaulted the boy. He was later apprehended in the hotel kitchen and admitted to committing the assault.

The Pittard's lawsuit ended in directed verdicts to the hotel on the negligent hiring and retention claims and adverse jury verdicts on the other counts. They appealed, claiming that the trial court erred in its instructions to the jury on the hotel's duty of care to its invitees, and in granting the directed verdicts. The instruction in dispute concerned the liability of a hotel to its guests. The Pittards insisted that the hotel should be held strictly liable for their son's injuries, but the court ruled that negligence was the proper standard to be applied to hotels: "Innkeepers are held to a 'reasonable care commensurate with the quality of the accommodations offered' standard ...; no matter how strict the standard, however, the hotel is not an insurer." The court noted that other states had imposed a higher standard of care for assaults by hotel employees under a theory of "implied breach of a legal or contractual obligation of an innkeeper to treat guests with due consideration for their safety and comfort," but it declined to follow the higher standards imposed by those cases since New Mexico does not recognize "degrees of negligence."

However, the court agreed with the plaintiff that there were questions of fact involved in the negligent hiring and retention claims which had been dismissed on directed verdicts. In this case, Perales' employers were aware that he had a drinking problem and a propensity for violence. In fact, Perales had been fired from a job as dishwasher because of his drinking. He had returned to the hotel after his termination to plead for his job. He was drunk and became violent when asked to leave the hotel. Hotel security personnel forcibly subdued him and he left only when threatened with criminal prosecution. He was later rehired as a steward which meant contact with customers and invitees and easy access to alcoholic beverages which he made use of regularly. The appellate court held that "notice of an employee's alcoholism and tendency toward violent behavior may make sexual assault by the employee foreseeable to the employer." Because a jury rather than a judge should resolve this factual question, the case was remanded for a new trial on the negligent hiring and retention claim.

PSCLR – MAR 85 (3)

Security Company And Its Employee/Guard Retained By Restaurant To Protect Patrons, Can Be Held Liable To Patron Injured On Restaurant Premises If Guard Failed To Act "Reasonably":

Kevin Marois v. Royal Investigation and Patrol, Inc., et al., *D000914, Civ. 31050, in the California Court of Appeal, Fourth District, Division I, decided November 28, 1984, reported at 208 Cal. Rptr. 384.*

Security guards James Prince and Derek Miller were employed by Royal Investigation and Patrol, and were assigned to provide security at a Jack-in-the-Box restaurant owned by a client of Royal, Foodmaker Inc. The security company, Prince, Miller, and the restaurant itself were sued by a customer who was attacked by a third person on the restaurant premises one evening while the guards were on duty. The customer, Kevin Marois, alleged that the guards were under a duty to prevent assaults on customers of Jack-in-the-Box, had failed in that duty and were therefore liable, along with their employer, for his injuries. Foodmaker, the owner of the restaurant, settled with the plaintiff before trial. The guards Prince and Miller were granted a motion for nonsuit by the trial court after the plaintiff's evidence was heard, and Marois appealed. The California Court of Appeal reversed, holding that substantial issues of fact existed that preclude the judge's dismissal of the case.

Royal had agreed with Foodmaker, Inc., to provide security at a Jack-in-the-Box restaurant in San Diego, California, which had been the site of problems due to persons loitering and drinking in the parking lot. The security service was intended to reduce vandalism and other problems in the parking lot. The guards, who were trained "on the job," were responsible for patrolling the parking lot, asking "rowdy or boisterous persons" to leave the premises, "effecting citizen's arrests" if "absolutely necessary for the protection of property and safety of employees and customers upon the premises." The guards were further encouraged by Royal to emphasize dealing with incidents in such a manner that physical action would be used only as the last resort.

On the evening Kevin Marois was assaulted, his attacker, Terry Hunt, had earlier entered the restaurant with his face and clothing bloodied. Prince asked him to leave the restaurant, and he did so immediately though he remained in the parking lot. Twenty minutes later, Hunt and Steve Gracelli used a baseball bat to vandalize a photo processing booth located 25 yards from the restaurant's drive-up window. A crowd gathered to watch, and a restaurant employee called the police. Marois and Bruce Stingle were waiting in a car for an order at the drive-up window. Stingle left the car and went over to the photo booth, and got into a fight with the vandals. Marois, arming himself with an empty beer bottle, joined the fight. When he saw that Hunt was holding a baseball bat, Marois dropped the beer bottle and retreated. Prince and Miller were observing the incident and attempted to verbally dissuade Hunt from attacking Marois with the bat. Nonetheless, Hunt struck Marois with the bat, injuring him. By this time the two were just outside the drive-up window. Prince and Miller went into the restaurant and asked an employee to call for emergency medical assistance. They recorded the make and license tag number of the vehicle in which Hunt and Gracelli drove away from the premises. This information was subsequently turned over to the police.

The defendants argued that their duty to Marois to prevent his injury was merely to conduct themselves in a reasonable manner. Moreover, even if they had failed to act reasonably, they were under no duty to "affirmatively prevent" Marois' injury. In addressing this, the court determined that there were <u>no</u>

> established policy considerations which warrant a rule that a business establishment's security guard may never be liable to a patron of that establishment for injuries caused by the intentional tortious acts of third parties, regardless of the guard's negligence.

The appellate court stated that a business has a duty to take reasonable steps to protect its customers from harm and may be liable for failure to do so. The court reasoned that this rule applied likewise to a security company which is under contract with a business to provide protective services, reasoning that the security company thereby creates a "special relationship" between its guards and the client's customers. The mere existence of this type of relationship would be sufficient to require the guard to act affirmatively to protect the client business's customers. Indeed, the court stated:

> a security guard hired by the business should be liable to an injured customer where the guard fails to act as would a reasonable security guard under similar circumstances and that failure causes the customer's injury.

However, the court questioned the validity of Marois' charges that Prince and Miller were negligent in allowing Hunt to remain on the Jack-in-the-Box property after he walked into the restaurant with a bloody face. It would be unreasonable to expect that the guards could or should focus their attention on one person, particularly where the person appeared to cooperate, as with Hunt. Up until the time Stingle approached the photo booth, the court held that there was no question that the guards acted reasonably and were not negligent. From that time, however, many questions of fact could be raised to support charges of the guards' negligence: Should Prince and Miller have attempted to warn or physically prevent Stingle from approaching Hunt? Should the guards have attempted to restrain Hunt and Gracelli once they began fighting with Stingle? When Hunt picked up the baseball bat and approached Marois, in the face of an apparent risk of serious physical harm to Marois, were the guards then required to do more than simply attempt to verbally dissuade Hunt from attacking Marois? Because answering these questions would require extensive scrutiny into the physical layout and the sequence of events leading up to the attack, the court ruled that such questions should be determined by a trier of fact (jury) and not dismissed by the judge on a summary judgment.

Therefore, the issue yet to be determined is:

> whether the security guards, ..., acted reasonably with respect to the risk which obviously confronted Kevin Marois. ... The only question in this case is whether the security guards responded reasonably to this clearly foreseeable risk.

The trial judge's decision to dismiss the case against the guards and the security company was reversed.

PSCLR – OCT 85 (5)

Owner Of Service Station Is Not Liable For Acts Of Lessee's Employees Unless Owner Has Power To Control The Employees' Conduct:

Deloris Giles v. Shell Oil Corporation, *No. 83-1450, in the District of Columbia Court of Appeals, decided February 8, 1985, reported at 487 A.2d 610.*

Deloris Giles's son was shot and fatally wounded by a service station attendant at a District of Columbia Shell Oil station. She brought suit against the owner of the station, Shell Oil Corporation, alleging it was liable under the theory of respondeat superior or the theory of negligent hiring or supervision. The case was dismissed by summary judgment granted in favor of the oil company, and Giles appealed.

The District of Columbia Court of Appeals affirmed the trial court's ruling that Shell Oil was not liable for the shooting as a matter of law under the theory of respondeat superior. In order to prove Shell Oil's liability, Giles would have to show that a master-servant relationship existed between Shell and the service station attendant who shot her son. A master-servant relationship exists where a person or entity (1) selects and hires the servant, (2) pays the servant's wages, (3) has the power to discharge the servant, (4) has the power to control the servant's conduct, (5) and employs the servant to work at part of the "master's" regular business. Previous court rulings have held that the power to control an employee's conduct and to terminate that employee at will are the most crucial elements to the master-servant relationship. (The court specifically cited an earlier District of Columbia case, *Safeway Stores v. Kelly*, which was reported in the November-December 1982 **PSCLR**, case 6, page 4.)

Giles argued that there was a factual dispute over the extent of Shell's control over the station's operation because Shell set the standards of operation and had the right to terminate the agreement if the operator failed to meet those standards, or if the operator failed to stock Shell Oil products for a specified period of time. Giles also claimed that Shell Oil was liable because it had knowledge, prior to her son's death, that the station operator had a gun on the premises which was used carelessly. In fact, the attendant who shot her son had previously shot a hole into the ceiling of the station while he was "playing with the gun." Notified of the damage and how it occurred, Shell Oil arranged and paid for the repairs to the ceiling.

Although there was proof that Shell Oil owned and maintained the service station premises, arranged for and paid for repairs, had the right to approve of alterations to the premises, and was entitled to inspect the premise and set operating standards, these factors do not prove that Shell had the right of control over the station. The management and operations of the Shell service station were granted to the station operator by Shell under a "Motor Fuel Service" lease and a "Dealer's Agreement." Each of these contracts included a clause stating that "nothing in this [contract] shall be construed as reserving to Shell any right to exercise any control over or to direct in any respect the conduct or management" of either the leased premises or the dealer's business. The station operator confirmed in his deposition that Shell Oil had no control over the station's employees. There was no evidence to indicate that Shell Oil exercised control over any of the station's day-to-day operations. As the station's landlord and products distributor, Shell had a reasonable interest in maintaining and inspecting the premises, and therefore in repairing the premises when the ceiling was damaged in the accidental shooting.

The District of Columbia Court of Appeals also affirmed the summary judgment to Shell Oil on the alternate theory of liability for negligent hiring or supervision because of the lack of a master-servant relationship between Shell and the attendant who fired the fatal shot.

PSCLR – JAN 86 (2)

Grocery Store Owner Is Vicariously Liable For Assault Of Customers By Store Employee During Performance Of Employee's Duties:

Timothy Phillip Wattik and Charles Wattik on behalf of minor David J. Wattik v. The Lewis Grocer Co dba County Market and the Travelers Insurance Company, *No. 17174-CA, in the Court of Appeal of Louisiana, Second Circuit, decided September 25, 1985, reported at 476 So. 2d 444.*

Timothy and David Wattik, brothers, were parked outside the County Market Grocery Store while their mother was inside. Shortly before their mother exited the store, they moved the car into the fire lane in front of the store. An employee of the store, Kenneth Woods, who was then sweeping off the sidewalk, approached the car and asked them to move out of the fire lane; a heated argument ensued, during which the Wattiks, who are white, "hurled racial slurs at Kenneth Woods," who is black. Eventually, an assistant manager intervened and Woods went inside the store. However, Woods then clocked out, got a machete from his own car, went to the Wattiks' car, and jabbed the machete at the boys through an open window. After Woods ran away, the Wattiks went inside the store and the police were called. Woods was arrested for aggravated assault. The Wattiks subsequently sued the grocery store owner to recover for personal injuries and emotional distress allegedly resulting from the assault by the employee Woods.

After the trial of the suit, the court did not award damages for any physical injuries but found The Lewis Grocer Co., to be vicariously liable for the acts of its employee, granting awards of $600 and $900 to Timothy and David Wattik for emotional distress. The Wattiks appealed, contending the court erred in denying the physical injuries award and requesting a higher award for emotional distress. The defendants appealed the finding of vicarious liability, and requested that the award for emotional distress be decreased.

An employer is vicariously liable for the employee's tortious act if it was committed within the "course and scope of his employment at the time of the tort." In an earlier decision, the Louisiana Supreme Court set out the relevant factors to determine whether an employer should be held vicariously liable for an employee's tort. The factors include whether the tort was "primarily employment-rooted," whether the tort was "reasonably incidental to the performance of the employee's duties," and whether the tort occurred at the employment location and during employment hours. The state's supreme court has also indicated that each case must be analyzed individually; the factors are not required to be present but are only guidelines to determine:

> whether the tortious conduct of the employee was so closely connected in time, place, and causation to his employment-duties, as to be regarded as a risk of harm fairly attributable to the employer's business, as compared with conduct motivated by purely personal considerations entirely extraneous to the employer's interest.

Applying this standard to the present situation, the trial court found, and the Louisiana Court of Appeal affirmed, that Kenneth Woods was performing employment-related duties when the altercation began, and the subsequent assault upon the Wattiks was not separated either by time or location from Woods' employment. There was no evidence that the altercation between Woods and the Wattiks was based on any pre-existing personal dispute, but that it clearly resulted from Woods' performance of employment duties. Accordingly, the appellate court affirmed the finding that the employer, The Lewis Grocer Co., should be held liable for damages inflicted by the employee Woods.

The appellate court also affirmed the trial court's decision to deny any award for physical damages and affirmed the amount of damages awarded by the court for the Wattiks' mental distress injuries. Timothy Wattiks claimed that he suffered physical injuries, not in the machete attack itself, but in a fall as he left his car after the machete attack. His testimony concerning the alleged fall and resulting injuries was inconsistent, however and the trial court chose not to accept it as true. The appellate court could find no error in this decision. Likewise, the court chose not to disturb the trial court's award for mental distress damages, because of the "overwhelming evidence that the Wattiks actually provoked the assault." Therefore, the appellate court affirmed the trial court in every respect.

PSCLR–FEB 86 (6)

Former Owners Of Shopping Center Face Lawsuit By Parent Of Minor Who Was Raped And Assaulted By Shopping Center Security Guard:

Mark Heindel, Individually and as parent and natural guardian of Donna Heindel, an infant v. Lawrence Wein et al., doing business as Newburgh Mid-Valley Associates, *in the Supreme Court of New York, Appellate Division, Third Department, decided November 27, 1985, reported at 495 N.Y.S.2d 295.*

This appeal was brought by the former owners of a shopping mall named as defendants in a lawsuit by the parents of an individual who was raped by a security guard on the mall property. The former owners asked the court to dismiss the complaint against them and the trial court refused.

Lawrence Wein and Peter Malkin, doing business as Newburgh Mid-Valley Associates, owned the Mid-Valley Mall and its adjoining parking lot which is located in Newburgh, New York. The Bowery Savings Bank held the mortgage on the property and in 1980, when the mortgage went into default, the bank started a foreclosure action. The court appointed a receiver of the property to receive rents and profits, and to maintain the property while the foreclosure was being settled. The court eventually granted the foreclosure and the mall was sold at a public auction to the Bowery Bank in September 1982. In November 1983, the court granted Bowery's motion to vacate the final judgment of foreclosure in order to add additional party defendants. In December 1983 Donna Heindel was raped and assaulted by Robert Turner, a security guard employed by Interstate Security North who was hired by the court-appointed receiver to police the mall premises.

The defendants here contend that the complaint against them should be dismissed because the order vacating the foreclosure judgment reinstated the property to the control of the receiver. They claimed that this relieved them of any liability. The appellate court disagreed and stated that there was a question of fact as to who had actual control of the property at the time of the assault on Donna Heindel. The decision by the trial court to deny the defendant's motion to dismiss is therefore affirmed.

PSCLR – MAR 86 (5)

Tort Action Against Employer Alleging Negligent Hiring Was Cause Of On-Premises Rape By Co-Employee Is Not Barred By Colorado Workers' Compensation Laws:

Deborah S. Tolbert v. Martin Marietta Corporation, *Civ. A. No. 83-C:1465, in the U.S. District Court for the District of Colorado, decided November 15, 1985, reported at 621 F. Supp. 1099.*

Deborah Tolbert was assaulted and raped by Arthur Martinez while on her way to lunch in the Martin Marietta Aerospace facility in Denver, Colorado. Tolbert was working as a secretary at the facility and Martinez was employed as a janitor at the time of the rape; Tolbert and Martinez had had only minimal contact before the incident. Tolbert brought this action against Martin Marietta to recover for her injuries, claiming that it had negligently hired Martinez and had failed in its duty to make the facility safe for employees. She also claimed that Martin Marietta knew or should have known that Martinez had a record which would indicate that he was capable of rape or assault. Martin Marietta filed a motion to dismiss Tolbert's claim, on the basis that her sole avenue of recovery lies in worker's compensation, but the District Court ruled otherwise and denied summary judgment to the defendants.

The District Court quoted the relevant Colorado statute which states that workers compensation applies when these conditions are met:

> (a) Where, at the time of the injury, both employer and employee are subject to the provisions of said articles and where the employer has complied with the provisions thereof regarding insurance;
>
> (b) Where, at the time of injury, the employee is performing service arising out of and in the course of his employment;
>
> (c) Where the injury or death is proximately caused by an injury or occupational disease arising out of and in the course of his employment and is not intentionally self-inflicted.[Emphasis supplied by the court.]

In this case, it is only condition (c) which is disputed by the parties, and indeed the District Court could not find a consistently applied test used by the Colorado appellate courts to decide whether an assault by an employee on a co-employee is an injury "arising out of and in the course of employment." The District Court found that two tests have been applied in cases of this nature in Colorado: the positional risk test of causality and the increased risk test of causality.

The increased-risk test requires that workers compensation be awarded only where the employment increases the employee's risk of injury above that of the general public. If applied to this situation, Tolbert's rape would not be covered by workers' compensation since her employment as a secretary within a secured defense plant would not increase her risk of a rape above that of the general public.

The positional risk test provides a broader coverage of injuries in that it allows a workers' compensation award where the employee's injury occurs "when one is reasonably required to be at a particular place at a particular time" while in the course of employment. Tolbert argues that this test also excludes her from a workers compensation recovery because she was on her lunch break when she was accosted and raped by Martinez and her job did not require that she have lunch at any particular place.

The District Court held Tolbert's rape to be "a non-employment motivated act of wanton brutality directed at the plaintiff because she was a woman." The court continued, "Considering the policy and purpose of workers' compensation, the rape of a worker is simply not the type of injury intended to be covered," and concluded that Tolbert is entitled to seek recovery in a tort action:

> Adopting this position has the additional advantage of providing employers an incentive to make reasonable efforts to screen prospective employees so as to avoid hiring rapists or those having the identifiable characteristics of potential rapists. Tort law does not impose strict liability—far from it; the plaintiff has the burden of showing negligence, causation, and damages. Women in the workplace are surely entitled to at least the protection of reasonable care where a prospective employee's available past record indicates a substantial foreseeable risk that, if employed, he may rape some woman he encounters on the employee's premises.

Martin Marietta's motion for summary judgment was therefore denied and the case will proceed to trial.

PSCLR—JAN 87 (3)

Employer Is Not Liable For An Employee's Unprovoked Use Of Physical Violence Against A Subordinate Employee After Reprimanding Subordinate:

Hester v. Church's Fried Chicken, *No. C-850505, in the Court of Appeals of Ohio, Hamilton County, decided May 21, 1986, reported at 499 N.E.2d 923.*

Hester was employed by a Church's Fried Chicken restaurant in Cincinnati. She was physically attacked by her supervisor, Al Howard, after he had reprimanded her for unsatisfactory job performance. She sued the restaurant under the doctrine of respondeat superior, but the trial court dismissed the suit based on the court's conclusion that the assault had not taken place either in pursuit of the employer's business or within the scope of Howard's employment. Hester appealed.

The confrontation between Hester and Howard began as a reprimand, but began to change when the supervisor made "several disparaging remarks" about the employee's personal life. The confrontation ended when Howard became enraged; he grabbed Hester by her collar, threw her down, and kicked her in the back.

Although there was evidence that Howard's supervisory authority extended to disciplining subordinates in keeping with Church's established policies, there was no evidence that his authority included acts of physical violence against a co-employee. There was also no evidence that Church's ratified or approved of Howard's actions. According to the appellate court, the evidence supported the trial court's finding that Howard's assault of Hester was motivated "solely out of feelings of personal ill will and malice harbored against his victim." The appellate court affirmed the trial court's ruling that the assault took place outside the scope of the supervisor's employment, and held that as a matter of law Church's could not be held liable for its employee's assault of a fellow employee. The appellate court affirmed the trial court's decision for the defendant Church's.

One judge dissented from the majority holding. This judge expressed his belief that there was a factual dispute concerning whether Howard acted within the scope of his employment duties when he attacked Hester or whether he acted purely out of personal malice and vindictiveness toward the plaintiff. The dissenting judge concluded:

> There is no presumption that the wrongful act of the agent was the act of the principal, [cites] but on the other hand, there is no presumption that a wilful, malicious act of an employee is beyond the scope of employment. [cites] Because the incident under scrutiny started with an attempt to enforce the appellant's performance of her job ... , I cannot say that the assault was beyond the supervisor's scope of employment as a matter of law.

Nonetheless, the majority ruled, and the summary judgment granted to Church's was affirmed.

COMMENTARY:

There is a fairly widespread but erroneous perception that an employer is per se responsible and liable for any and all acts of an employee performed by the employee while "on the job," even including acts of misconduct by the employee. The above case is included in this issue as a reminder to the security professional that one of the first questions which should be discussed when attempting to determine a company's liability for the acts of an employee is whether the employee's acts were within the scope of the employee's employment, and in pursuit of the employer's business. This relates not only to intentional acts of misconduct by an employee but also to negligent acts by the employee. If the facts establish that the employee was not acting within the scope of his employment and in pursuit of the employer's business, then, as is seen by the holding in the above case, the employer is not liable or such acts. The employee is deemed to be "on a frolic of his own." Sometimes, however, the question as to whether the employee is or is not within the scope of his employment cannot be answered definitively. As we see above, one of the appellate court judges would have left this conclusion to be answered by the trial jury.

PSCLR—APR 87 (5)

Retailer Liable For Battery Where Security Guard, Acting Within Scope And Course Of Employment, Touches Customer's Breasts:

Joy Clark v. Skaggs Companies, Inc., *No. WD 38081, in the Missouri Court of Appeals, Western District, decided December 6, 1986, motion for rehearing and/or transfer to the Supreme Court of Missouri denied February 3, 1987, application to transfer denied March 17, 1987, reported at 724 S.W.2d 545.*

Joy Clark sued Skaggs Companies, Inc., for false imprisonment and battery after she was detained by a Skaggs security guard as a shoplifting suspect, and the guard "squeezed" her breasts while searching her.

Joy Clark went shopping at the defendant's store in Kansas City, Kansas. She first saw Johnny Rogers, a uniformed security guard, as she was browsing in the cosmetics department. At one point, she picked up a bottle of nail polish and took it with her into another section of the cosmetics department to compare the nail color with the lipsticks. She decided against buying any cosmetics; she did not return the nail polish to its area, but left it in the lipstick section. Approximately 15 minutes later, she tried to leave the store without having bought anything. It was then that she saw that the security guard was following her, so she asked him if he knew her from somewhere. The guard said, "No, but I was wondering where you put that nail polish." Clark responded that she left it somewhere in the store, and the guard asked her to help him find it.

At the trial of Clark's allegations, it was then that she realized that the security guard suspected her of shoplifting. She also testified that she believed that she had no choice in the matter and was required to accompany him back into the store. As Clark led Rogers back to the lipsticks section, he gestured toward some doors that said "Employees Only" and suggested that they go "through those doors and see if it's in your purse." Rogers dumped the contents of her purse out, and the nail polish was not among her things.

Clark was wearing blue jeans, a plaid blouse, and a sweater-jacket. She removed the jacket and handed it to Rogers, saying, "It's not in my sweater pockets either." He declined to search the sweater. Clark then patted her jeans pockets and said that the nail polish was not in there either. Rogers also did not search the jeans pockets. Clark stated, "It's not in my shirt pockets," indicating the breast pockets of her blouse and saying, "See" and unbuttoned the pockets. She stated that although it was abundantly clear that nothing could be in her blouse pockets because there was no bulge or lump of any kind. However, she also testified that Rogers stated, "Let me check," and that he then stepped forward, reached out, placed his hands upon her breasts, and squeezed them. Clark testified that his action made her feel "offended, repulsed, embarrassed, violated, and terrified." She told officer Rogers that she did not think that his action was "legal."

The security officer then apologized, and requested that Clark return with him into the store in order to locate the nail polish. A Skaggs clerk joined them, and with Clark's assistance, the three did locate the nail polish in the lipstick section. The clerk stated, "Well, I guess we have to take your word for it." Clark then asked if she could leave, but neither Skaggs employee responded. Rogers then apologized once more, and she repeated her request to leave. The clerk replied, "Yeah, go on, get out of here."

At trial, Clark stated that she felt she was not free to leave during the time she was with the security guard, which was a period of approximately 15 minutes. She also testified that she now feels paranoid whenever she goes shopping, she does not pick up small objects while she is shopping, and she does not feel free to browse as she had before this incident.

The jury awarded Clark $7,500.00 on the false imprisonment count, and an additional $20,000.00 for the battery (improper touching).

The defendant Skaggs first ground of appeal was that the plaintiff had failed to present substantial evidence that the guard in fact restrained her against her will. The law is clear that anyone who attempts to recover for a false arrest or false imprisonment has the burden of proving that he or she was unlawfully arrested by the defendant, who either instigated the arrest or assisted, directed, or encouraged it. It is clear from the defendant's own evidence that the security guard, Johnny Rogers, was in fact an employee of the defendant and that his instructions were to conduct surveillance and to apprehend and/or detain customer suspected of shoplifting. While there was no evidence in the record that officer Rogers used any physical force or threats, nor that he ever actually accused Clark of the offense of shoplifting, and that she did in fact voluntarily accompany him back into the store, nevertheless the jury was authorized to accept her testimony that she did not believe that she was free to depart. The words and conduct of the officer caused her to believe that resistance or attempted flight on her part would be useless. Therefore, reasonable people could infer from that evidence that she was intimidated or coerced to the extent that she had reasonable grounds to believe that she could not leave and that any effort to do so would be futile. Accordingly, the appellate court rejected the department store's arguments.

The defendant store next claimed that the security guard Rogers was not acting within the scope and course of his employment when he touched Clark's breasts. In this regard, the court first noted that the law is settled that an employer is in fact liable for assaults by his employee if the assults are committed within the scope of the employment.

In support of its argument, the department store introduced testimony from two witnesses who were familiar with the security policies at the store. Both of the witnesses testified to the effect that it was

> against policy for security personnel to make sexual advances to a suspected

shoplifter and if such incidents should occur, and be brought to the attention of the management, the employment should be terminated.

In further support of its argument, the store cited the plaintiff's own testimony that it was obvious that she had nothing in her breast pockets, thereby indicating that the security guard was acting for his own gratification and not to benefit the business of the employer, Skaggs.

The court concluded that "reasonable minds could in fact differ on this issue," and since the jury had decided that Rogers was acting within the scope of his employment when he touched Clark, the employer should be held liable.

The court also concluded that the award in the amount of $7,500 for false imprisonment, and $20,000 for battery was not excessive.

PSCLR – AUG 87 (10)

Company Can Recover From Union Costs Of Reasonable And Necessary Security Measures Taken To Protect Property From Unlawful Conduct During Labor Strike:

Carter-Glogau Laboratories, Inc. v. Construction, Production & Maintenance Laborers' Local 383, No. 1, *CA-CIV 8107, CA-CIV 8128, in the Court of Appeals of Arizona, Division 1, Department A, decided October 30, 1986, reconsideration denied March 5, 1987, review denied May 12, 1987, reported at 736 P.2d 1163.*

The plaintiff company sued the defendant union, and local labor council, claiming that they were liable for the illegal acts of union members during a strike against the company which took place at its manufacturing facilities in Glendale, Arizona. The company sought to obtain reimbursement for the costs incurred for implementing certain security measures, as well as payment for property damages incurred by certain of its employees who had not participated in the strike, plus an amount for punitive damages due to wrongful activity during the strike on the part of union members.

After a trial by jury, a verdict was rendered against the defendants in the amount of $118,672.62 compensatory damages, plus $600,000 punitive damages against the union itself (and not against the labor council). The trial judge, at the request of the union, reduced the punitive damages award to $200,000, but declined to grant the defendants' motion for a new trial. The union therefore appealed, and the company cross-appealed, claiming that the trial judge should not have reduced the punitive damages award from $600,000 to $200,000.

The defendant union began picketing the plaintiff company's premises on the morning following the expiration of its collective bargaining agreement. The picketing continued for several months, and during the course of the strike, various actions on the part of the picketers included threats of violence, sexual harassment, and property damage against non-striking employees of the company. The evidence showed that bent nails were repeatedly thrown into the employee parking lot, acid was thrown onto their cars, several of the non-striking workers were followed home from work, one of them was spit upon, and there were numerous express threats of physical violence, and incidents of sexual harassment. The evidence also support the conclusion that union stewards, a business agent of the union, as well as picket captains approved of such activities and on a number of occasions even participated.

Because of the nature of these incidents, the company hired security guards and expert security consultants, and installed security cameras on the property. They also reimbursed the non-striking employees who suffered damage to their automobiles.

The union argued that the damages award should be set aside because the cost incurred were caused by protected strike activity and the security measures undertaken by the company did not result from activity on the picket line which is not protected by law. However, referring to a Connecticut case, the appellate court concluded:

> We hold that if 'spoken threats and moral intimidation' reach a point where a trial court may reasonably concluded that 'physical violence and property damages is feared, then actions of picketers have gone beyond a protected level.'

The appellate court therefore concluded that the evidence presented shows that a trial jury could reasonably conclude that the union activities were unlawful and therefore not protected by the country's labor laws.

In addition, the appellate court concluded that the amount expended by the company was supported by appropriate expert testimony. Mr. William J. Overman of R.H. Investigation presented testimony relating to the reasonableness and the necessity of the particular security measures employed in view of the misconduct of the picketers as outlined above during the strike. The company had hired Overman within the first couple of days of the strike to assess security requirements for the company. He testified that he first had recommended the use of a car patrol, but when the reports of the harassment and intimidation of employees began coming in, and the nails were thrown into the employee parking lot, and acid was thrown on the employees' cars, then he decided that a foot patrol would be necessary to prevent such acts. Overman also indicated that the need for various security measures was reevaluated on a daily basis, and as the incidents of misconduct declined, the security measures were accordingly reduced. He presented evidence as to the reasonableness and necessity of the costs incurred for the security measures which were employed as a result of some of the picketers' unlawful conduct. The appellate court therefore concluded that the award of compensatory damages to repay the company for the costs incurred was proper and was supported by the evidence.

The court also concluded that the punitive damages awarded, although reduced by the trial judge was also proper, in order to deter similar future misconduct and to punish the misconduct which did occur.

Therefore, the trial court's judgment in favor of the plaintiff for compensatory damages and reduced punitive damages was affirmed, and the denial of the defendants' motion for a new trial was likewise affirmed.

COMMENTARY:

One sure way for a security manager to endear himself to his or her superiors would be to manage the security measures employed in a strike situation in such a way that there is a basis for recoupment of the expenses incurred in protecting the company's property and employees from illegal union activity as in the case above. In order to be able to do this effectively, daily assessments of the need for increased and/or decreased security measures must be made and must be adequately and fully recorded, giving reasons and justifications for the additions or deletions of certain security measures.

The costs of various alternatives must also be explored, and the reasons for those chosen or not chosen should be explained in a daily log kept for this purpose. Furthermore, the non-striking employees should be asked to submit written reports of any and all illegal union conduct, including documentation as to any and all property damage. Extensive record keeping is a must.

The addition of one or more expert security consultants may also be considered, depending upon the security manager's own degree of expertise and the perceived need for additional assistance.

The records should also include a distinction between those costs incurred as a result of proper and legal union activity as opposed to those incurred due to illegal activities of union members and picketers.

While the record keeping procedure outlined above may seem and actually be onerous, the rewards (as can be seen from the result in the above case) can be significant. Without such records, such recoveries would not be possible.

PSCLR – NOV-DEC 87 (1)

Security Guards Breached Duty Of Care To Hotel Guest By Requiring Intoxicated Guest To Drive Car Off Premises, In Spite Of Others' Offers To Hire A Taxi, Or Pay For A Room For Intoxicated Guest:

Lonnie Akers v. Irvine Marriott Corporation, *G003268, in the Court of Appeal of California, Fourth District, Division 3, decided March 5, 1987, reported at 235 Cal. Rptr. 92.*

Lonnie Akers' suit for negligence against the Irvine Marriott Corporation, and for intentional infliction of emotional distress was dismissed by the trial judge, and Akers appealed. The appellate court reversed.

The facts are that Lonnie Akers, an 18-year-old male, attended a party at the Irvine Marriott Hotel in Irvine, California, given by Akers' employer. When he arrived at the hotel, he parked his car in the hotel parking facility. At the party he was served a number of alcoholic beverages by Marriott employees in violation of state statutes because he was too young under California's minimum drinking age.

Ronnie Akers' co-workers saw that he was intoxicated, and they therefore "fixed" his car so that it would not start and so that he would not be able to injure himself or others by driving while drunk. However, certain security guards employed by the Marriott saw Akers in his intoxicated condition, and had a conversation with him and with several of his co-workers. The co-workers told the guards that Akers' car had been rendered inoperable, and that they would be willing to hire a taxi for Akers or to pay for a hotel room for him to spend the night. The security guards, however, rejected the co-worker's suggestions and directed them to fix the car so that it could be driven, which they did reluctantly. While the car was being fixed, the security guards notified the Irvine police department, and afterward the guards required Akers to drive the car off the hotel property. Acting on these instructions, Akers did drive a short distance, but he was immediately apprehended in the hotel parking lot by a local police officer, at which point he was arrested and charged for driving under the influence of alcohol.

Akers subsequently pleaded guilty to the charge, incurred expenses for attorney's fees, a fine, court costs, plus the suspension of his driver's license, the entry of a drunk driving conviction on his permanent record, increased insurance costs, loss of his freedom as a result of the arrest, and loss of driving privileges as a condition of probation.

Akers sued the Marriott, alleging that the actions of the Marriott security guards were

> intentional, deliberate, and done to satisfy their sadistic urges ... at the financial, physical, and emotional expense of the plaintiff, and for the malicious purpose of causing Akers to suffer humiliation, mental anguish, and emotional and mental distress ....

Akers argued that the hotel security guards breached their duty to exercise reasonable care for his safety while he was a guest of the hotel, by insisting that he drive his car off the property while he was obviously intoxicated.

The appellate court concluded that the proximate cause of Akers' losses was not his own consumption of alcohol, but rather was the security guard's requirement that he operate his automobile while drunk. Obviously, if he had not been required to drive his car, he would not have been arrested for driving under the influence of alcohol. The issue was therefore whether the guards owed Akers, a hotel invitee, any duty of protection. The court, in attempting to answer this issue, noted that if Akers was, as claimed in his lawsuit, "visibly intoxicated," then the guards could have foreseen that Akers would be arrested for driving under the influence, particularly since the guards themselves had apparently summoned the police for that specific purpose. The court also noted that the guards should likewise have foreseen that Akers could have hurt himself or others by driving while he was intoxicated. While Akers was stopped by the police before any such injuries took place, nevertheless, the security guards' conduct in requiring him to drive while intoxicated showed their indifference to the possibility of such harm to him or others, according to the court.

The Marriott's defense relied upon the holding set forth in the case of *DeBolt v. Kragen Auto Supply, Inc.*, 182 Cal. App.3d 269, 227 Cal. Rptr. 258 (1986), in which the California Court of Appeals affirmed the dismissal of a wrongful death action against an employer who had furnished alcohol to a minor guest/employee at an office party. The suit was brought by the survivors of two victims of a car crash caused by the intoxicated employee after she had left the defendant/employer's party. The plaintiffs had alleged in their lawsuit that the defendant had supplied and served alcoholic drinks to the minor employee to such an extent that she became intoxicated and then required her to leave the party because she became drunk. Furthermore, although the employer knew that she would drive her car and it was obvious from her condition that she would not be able to drive safely, he had nonetheless failed to provide any alternative or safer means of transportation, or even to call the minor employee's parents. The plaintiffs had urged the court to hold that the employer had a legal duty not to create the unreasonable risk caused by forcing the minor employee into her car and onto the highway in her drunken state. Therefore, having failed to uphold that duty, the employer/defendant should be held liable for the resulting and reasonably foreseeable deaths. However, the appellate court concluded that "social host immunity" shielded the employer from liability.

In the present case, however, the California appellate court found that the circumstances in the *DeBolt* case, while somewhat similar to those of the present case, were distinguishable in one important respect, to wit: Akers was compelled to drive his automobile in spite of reasonable alternatives suggested to the guards by his co-workers, which were to allow the co-workers to send Akers home in a taxi or to get a hotel room for him to spend the night. The appellate court noted:

> One who knows or has reason to know that a third person is giving or is ready to give to another aid necessary to prevent physical harm to him, and negligently prevents or disables the third person from giving such aid is subject to liability for physical harm caused to the other by the absence of the aid which he has prevented the third person from giving.

In conclusion, the Court of Appeal of California found that Akers does have a viable lawsuit under "the unique facts alleged." The court found that ordinarily the ejection of an intoxicated person would not actionable under the law even where the person's only apparent means of transportation is by driving an automobile. What is actionable, according to the court, is conduct designed to prevent a third person from rendering aid or assistance to the intoxicated person, or preventing such third person from providing a safe alternative for the intoxicated person.

Accordingly, the appellate court reversed the trial judge's dismissal of Akers's suit and the case is currently pending trial.

COMMENTARY:

It should be clearly understood that, according to the Court of Appeal of California's decision, the security guards would have owed Akers the same duty of care even if he had arrived at the hotel in an intoxicated condition, as opposed to becoming intoxicated on the premises. Further-

more, the duty the security guards owed to Akers was not dependent upon their status as security guards: any and all employees of the hotel would find themselves in the same legal position under similar facts and circumstances.

Furthermore, the appellate court also noted in its decision that Akers had a cause of action for intentional infliction of emotional distress, and that he may be entitled to an award of punitive damages for the conduct of the security guards if he is able to prove his factual allegations.

There was also included in the published opinion of the appellate court's decision a scathing dissent by one of the three judges who heard Akers' appeal. The dissent noted that Akers had pleaded guilty to committing the crime of driving drunk, and was here attempting to recover the damages caused by his conviction from the security guards who had in effect aided and encouraged him in committing the crime. The dissent compared it to

> *a suit by one highwayman against another for an accounting of their plunder. The bill was dismissed with costs to be paid by the defendant; the plaintiff's solicitors were attached and fined 50 pounds each for contempt. Both plaintiff and defendant were subsequently hanged. In short, contribution was not allowed.*

He then concluded: "These highwaymen (and their counsel) are doing considerably better."

PSCLR – NOV-DEC 87 (6)

While Property Owner Cannot Be Liable To Party Injured By Independent Contractor Guard Service Based On Agency Principles, Owner May Be Liable For Negligent Retention Of Guard Company:

Robert W. Savinsky v. The Bromley Group, Ltd., a Canadian corporation, d/b/a Mediterranean Garden Apartments, and Amos Cason and Lino Gonzales, d/b/a Sierra Security Service, *No. 8923, in the Court of Appeals of New Mexico, decided June 9, 1987, certiorari denied August 12, 1987, reported at 740 P.2d 1159.*

The plaintiff Robert W. Savinsky brought suit against The Bromley Group, Ltd., (Bromley) in its capacity as owner of an apartment complex where Savinsky was injured. Savinsky alleged in his suit that his injuries arose out of a confrontation with a security guard, Cason, who was on duty at the Mediterranean Garden Apartment, owned by The Bromley Group, Ltd.

Cason had originally been hired by Security Protection Services, which was subsequently merged with, or acquired by, the defendant Sierra Security Service (Sierra), by whom he was employed at the time of plaintiff's alleged injuries. Savinsky alleged that Bromley, as the property owner, was negligent in its selection and retention of the independent contractor security guard service, and that Bromley was also liable to him under the doctrine of agency, or respondeat superior.

Bromley moved for summary judgment, arguing that Sierra was an independent contractor and Bromley therefore could not be held responsible for the acts of Sierra's employee Cason, and also on the ground that Bromley was not negligent in contracting with Sierra to provide security services at the apartment complex. The trial court granted the motion and dismissed the plaintiff's case and he appealed.

The appellate court affirmed the trial judge's decision to the effect that Bromley could not be held responsible for the acts of the independent contractor security guard services' employee. To determine whether an employer/employee relationship exists, as opposed to an independent contractor relationship, the primary test is whether the employer has the right to control the details of the work to be performed. Other secondary tests as to the existence of the relationship includes evidence as to the right of the employer to control the employee, the method of compensation, the furnishing of equipment to do the job, and the legal right to terminate the relationship. The evidence here was to the effect that the security guard Cason reported directly to Sierra's management, and the security service set his rate of pay and issued the payroll checks. The guard neither reported to, nor was supervised by the apartment managers, although the guard would investigate anything suspicious reported to him by them. The guard would call Sierra for assistance with any problems he encountered.

Furthermore, the resident manager of the apartments testified that it was the security company which hired and assigned Cason to Mediterranean Garden Apartments, and this was done without consulting with or obtaining the approval of Bromley, or its employees. It was the security service which decided which guard to assign to patrol the apartments, determined the guard's working hours, and patrol schedule, and directed the manner in which the guard was to respond to incidents.

Accordingly, the court concluded that The Bromley Group, Ltd., had maintained the right only to direct the result to be achieved by the security guard and the security company, but had no right to control the manner in which the work was to be performed. There was therefore no error on the part of the trial judge in granting summary judgment for the apartment owners on this count of the complaint.

The appellate court next reviewed the negligence count of the complaint, first noting the general rule that a plaintiff cannot recover for the negligence of a defendant's independent contractor. At the same time, the court noted that there are "numerous exceptions to the rule." Savinsky alleged in this count of his suit that the apartment complex owner owed a duty to the members of the public

> to ensure that security guards employed to protect its premises were adequately trained and properly licensed in regard to carrying weapons, using weapons, use of force, and detaining citizens.

Furthermore, Savinsky claimed that the security guard Cason had not been properly trained and licensed, and suggested that Bromley may have breached its duty of care to the public by using security guards who were illegally armed. In an effort to counter these allegations by the plaintiff, Bromley submitted evidence showing that its manager at the apartments had contacted several security agencies listed in the telephone book, had advised them of the security needs of the apartment complex, and had learned that the Security Protection Services, with which the owner subsequently contracted was licensed and bonded to act as a private security office. However, the court noted that Security Protection Services was not a party to this case (it was acquired by the defendant Sierra Security Services). And, the court also noted that there was no evidence in the summary judgment record that the defendant Bromley took any action whatsoever in connection with the retention of the successor Sierra. Nor had the apartment owners presented any evidence to show that the independent contractor guard service was not engaged in inherently dangerous work, in fact the evidence indicated that the work was in fact inherently dangerous.

Further, the defendants had argued that it could not be liable for the plaintiff's injuries, because it had not owed any duty to keep the premises safe for the plaintiff who was a trespasser on the premises. The court agreed that while as a general rule a property owner would not owe a duty to exercise reasonable care to make the premises safe for the uninvited visitor, the property owner would be required to use ordinary care to avoid injuring even a trespasser when the owner engaged in activities which involve an unreasonable risk of death or great bodily harm. The court noted that there was no evidence to prove Savinsky had been trespassing, and even if there had been, the fact that Cason was engaging in the dangerous activity of carrying a loaded weapon would require him to use ordinary care to avoid injuring the trespasser.

For these reasons, among others, the appellate court concluded that the apartment complex owner had failed to submit sufficient evidence to permit the court to conclude that the apartment owner could not be liable to the plaintiff for negligence in the selection and retention of the defendant guard company, and therefore the trial judge's decision to grant the summary judgment as to this count was reversed, and the case was remanded to the trial court to permit the plaintiff to attempt to establish his allegations by proof.

COMMENTARY:

This is a very significant case and carries major implications for property owners and others who believe that they are immune from liability for the acts of security guards who are employed by an independent contractor guard service and who are not the agents or employees of the property owner!

As we have discussed at various seminars on this subject in the past, while a defendant cannot be held liable for the acts of the employee of the independent contractor according to the legal principles of agency and respondeat superior, nevertheless the defendant may be held liable to a party injured by such an employee if the party can prove that the defendant had a duty to protect the injured party, and was negligent in his selection and retention of a guard service to provide that protection. This ruling puts an affirmative burden on the property owner who contracts with a guard service company to make an independent investigation and determination, based upon findings to the effect that the independent guard service company is competent, qualified, capable, properly trained, properly licensed, and otherwise able to fulfill the duty of care owed by the defendant to the public (recognizing that the degree of care owed to certain members of the public may vary depending such members' status as invitee, licensee, or trespasser). In addition, it should be said that, in the event of litigation, it may be helpful that the defendant property owner is able to show the written memoranda, correspondence, and other physical evidence which was generated by its investigation into the competency and qualifications of the various guard companies upon which it based its decision, as well as the reasons for the decision, to select and retain and contract with the particular guard service in question.

PSCLR—JAN 88 (8)

A Corporation Is Liable For Punitive Damages For Intentional Misconduct By Its Employees Only If Management Authorizes, Participates In, Consents To, Or Ratifies Such Misconduct Or Intentionally Retains An Unfit Employee:

Bruce Neal v. C.F.M. Enterprises, Inc., *in the Supreme Court of New York, Appellate Division, Third Department, decided October 22, 1987, reported at 520 N.Y.S.2d 656.*

The plaintiff Bruce Neal had received injuries at the defendant C.F.M. Enterprises, Inc.'s (CFM) store in Schenectady, New York, when he was assaulted by two other customers of the store after making his purchases. In fact, on the night Neal was assaulted, he had come into the store for the purpose of buying cold cuts and had a problem with one of the store employees. It appears that just before closing time, the defendant's employees at the store were required to clean the slicing machine which was used to slice cold cuts of meat, that the employees had previously complained to their supervisor that Mr. Neal often entered the store immediately before closing for the purpose of buying cold cuts, and that Neal frequently harassed the employees. On this occasion, Neal's disagreement with the employees caused two other customers, who wanted to make their purchases and leave, to begin yelling at Neal. As Neal left the store, the two other customers followed him into the parking lot and assaulted him. After he was attacked, a friend of Neal's helped him to Neal's house where they called the police. Neal and his friend then returned to the CFM store to retrieve Neal's car and to wait for the police whereupon the plaintiff Neal once again was assaulted by the same two customers.

The evidence which was introduced during the trial showed that the defendant's employees in the store witnessed both attacks on Neal, but they did nothing to assist him. At trial, after presentation of the evidence to the jury, the jury concluded that CFM was 60 percent at fault, and the plaintiff Neal was 40 percent at fault for his injuries, and therefore awarded Neal both compensatory damages for his injuries as well as punitive damages. The defendant appealed.

The defendant's only argument on appeal was that the evidence showed on its face that the plaintiff was not entitled to an award of punitive damages. The appellate court agreed since an award of punitive damages against a corporation would not be appropriate unless the management had

> authorized, participated in, consented to or ratified the conduct giving rise to such damages, or deliberately retained the unfit servant.

This was not the case here, as the court pointed out, for the following reasons:

The transcript of the trial record showed that the defendant's store manager was not working on the night of the assaults, and therefore did not know of the assaults on Mr. Neal until after they had taken place. Prior to the assaults, she had been informed by the store employees that Neal had often bought cold cuts immediately before closing time, but she had still instructed the employees to clean the slicing machine immediately before closing. Neal argued that the store manager had thereby contributed to a "potentially explosive situation" by requiring the employees to clean the machine even though she knew that Neal regularly bought cold cuts "at approximately the same time."

Neal also argued that the store manager's actions constituted a "ratification" of the employees' conduct when Neal was assaulted. The appellate court rejected this argument, concluding that, "Surely, no one had any idea that cleaning a slicing machine would lead two unidentified patrons to assault plaintiff." Since the store clerk's supervisor had not instructed the clerks to refuse to slice cold cuts for Neal, and had not instructed them to refuse to call the police in the event of an assault outside the store, the store manager could not be deemed to have ordered, ratified, or participated in the employees' conduct, and the defendant CFM therefore could not be subjected to punitive damages.

Furthermore, the appellate court concluded that the supervisor had not delegated her managerial duties to the store employees and the employees therefore could not be held to have acted in the manager's place.

Lastly, the plaintiff argued that the defendant should be held liable for the punitive damages for deliberately retaining an unfit employee. However, the evidence showed that both of the store clerks had adequately performed their responsibilities prior to the assault on the plaintiff, and at best, the evidence indicated only that the defendant company was negligent in failing to properly train the clerks. However, such failure on the part of the company would not be sufficient to support the award of punitive damages to the plaintiff. In conclusion,

> Since the evidence failed to show that the defendant's management authorized, ratified, or participated in its employees' malicious conduct or that management duties had been delegated to the employees wrongfully, or that defendant had deliberately retained unfit employees, plaintiff was not entitled to an award of punitive damages.

Therefore the appellate court set aside the award of punitive damages to the plaintiff and otherwise affirmed the award of compensatory damages to him.

PSCLR—FEB 88 (7)

Where Security Guard Uses Weapon In Horseplay To Scare Third Party, Resulting In Accidental Gunfire And Shooting Of Plaintiff, Conduct Is Outside Scope Of Employment And Employer Is Not Liable:

Kensington Associates v. Harry W. West, *Record No. 841524, in the Supreme Court of Virginia, decided November 25, 1987, reported 362 S.E.2d 900.*

The plaintiff Harry W. West was injured when he was shot by the defendant Willis Chittum, a security guard employed by the defendant Kensington Associates. At trial, the jury returned a verdict for the plaintiff West in the amount of $200,000 against both Kensington and Chittum. The defendant/employer Kensington appealed, claiming that Chittum was beyond the scope of his employment at the time he shot the plaintiff, and that Kensington therefore cannot be held liable to the plaintiff.

The evidence showed that United Services Industries had contracted with Kensington Associates to renovate the former Johnston/Willis Hospital building in Richmond, VA, which was owned by Kensington. Harry West was one of the construction workers employed by United Services, and United Services provided living quarters on the premises for the workers. Kensington employed security guards on the premises, and hired Willis Chittum as a security guard with the responsibility of protecting the hospital property and preventing vandalism on the premises. Chittum was armed and carried a pistol in a holster while he was on duty, and Kensington management was aware that he carried the weapon, and acknowledged that it was for Kensington's benefit. Chittum had been told by Kensington management to notify the police if there was any trouble on the property.

On several different occasions before the incident in question, Willis Chittum had engaged in horseplay with a United Services construction worker by the name of Willie Archie. At those times, Chittum had taken his weapon from his holster and "waved it around" to scare Archie. On the night in issue, Chittum completed his rounds through the hospital building, and stopped in the hallway outside the construction workers recreation room, in which he thought Archie was located. Chittum did not know that West in fact was in the room. Chittum removed his weapon from the holster, and the pistol accidentally discharged, shooting Harry West in the foot. According to Chittum's testimony, he did not pull the pistol in order to protect any of Kensington's property, but only to have "fun" with Archie. Chittum also admitted that he had had a "couple of" beers that night, although Kensington had prohibited drinking while on duty. Additionally, Kensington had ordered Chittum not to enter the construction workers' recreation room or otherwise to bother the construction workers.

It is well-settled agency law that an employer can be held liable for the misconduct of his employee only if the employee was acting within the scope of his employment and was pursuing his employer's business interests at the time of the misconduct. The test for making this determination is as follows:

Generally, an act is within the scope of the employment if

> (1) it was expressly or impliedly directed by the employer, or is naturally incident to the business, and (2) it was performed, although mistakenly or ill-advisedly, with the intent to further the employer's interest, or from some impulse or emotion that was the natural consequence of an attempt to do the employer's business, 'and did not arise Wholly form some external, independent, and personal motive on the part of the [employee] to do the act upon his own account.'

The court felt that from the evidence it was clear that the horseplay by Chittum which resulted in the shooting of West was not done to further any business interests of Kensington, but was an independent effort on Chittum's part to satisfy his own desire to play around by scaring Archie; this independent effort suspended temporarily the employer-employee relationship between Chittum and Kensington. Accordingly, the appellate court reversed the trial court's judgment in favor of the plaintiff West, and entered judgment in favor of the employer Kensington.

PSCLR – MAY 88 (5)

Employer Cannot Be Held Liable For Its Security Guard's Conduct Which Was Outside The Scope Of Employment, Where Guard Participated In Robbery Of Employer's Store And Shot And Killed Store Manager:

Angela Vallejo, Elizabeth Vallejo, Lewis L. Vallejo, and Jacqueline Vallejo v. Osco Drug, Inc., *No. WD 39429, in the Missouri Court of Appeals, Western District, decided October 27, 1987, motion for rehearing and/or transfer to supreme court denied December 29, 1987, application to transfer denied February 17, 1988, reported at 743 S.W.2d 423.*

During 1986, Advance Security provided armed guards for Osco Drug, Inc. In June of that year, one of the guards supplied by Advance Security, Marvin Duncan, along with an accomplice, robbed the Osco Drug Store he had been hired to protect. At the time of the robbery, Paul Vallejo was employed as manager of that store. Duncan shot and killed Vallejo during robbery.

Vallejo's survivors sued Osco Drug, Advance Security, and Marvin Duncan for Vallejo's wrongful death. As to Osco, the survivors claimed that Duncan acted for Osco when he shot and killed Vallejo, and that Osco was thus responsible for Vallejo's death. Osco denied responsibility and moved for summary judgment. The trial court entered summary judgment in Osco's favor, and the survivors appealed.

The plaintiffs claimed that Osco owed Vallejo the duty to provide a safe work place and that this duty was nondelegable. They asserted that hiring Advance Security, and thus Duncan, to fulfill this duty towards Vallejo, transformed Duncan, through Advance, into Osco's alter ego. Since Duncan was performing an act for which Osco was responsible, Duncan's act became Osco's act. The plaintiff's argument then followed that since Duncan's act in killing Vallejo was intentional, and should also be imputed to Osco, workers' compensation laws would not preempt the survivors from suing Osco in tort for Vallejo's death.

While the court agreed that Osco owed Vallejo a duty to furnish a safe work place, the court stated, "However, an employer is not an insurer of safety of the employee and the failure to exercise ordinary care to provide a safe work place constitutes negligence." Duncan's act, even if attributable in some way to Osco, would only be perceived by the court as a violation of ordinary negligence standards.

The court next noted that Osco's duty to provide a safe place to work was indeed nondelegable, but nonetheless discounted the importance of this duty since workers' compensation laws had eliminated the necessity of determining whether a fellow employee or the employer had performed the negligent act which had injured the plaintiff. Since workers' compensation laws had abolished the negligent fellow servant defense, the importance of nondelegability of the duty to furnish a safe place to work had virtually disappeared.

The court, finally, held that an employer will not be responsible for an employee's acts outside the scope of that employee's duty which cause the work place to become unsafe and which cause injury to another employee. The court, in this portion of the opinion, relied on *Sherrill v. American Well and Prospecting Company.* In that case, a group of employees manufactured a cannon to celebrate New Year's Eve. The foreman knew of and acquiesced in their decision to make the cannon. The cannon exploded in a work room while employees were pounding gun powder into it, and an employee was injured. The court held that since the act by the employees which created a dangerous work place lay outside of the scope of their employment, the employer could not be held responsible for that act.

Turning to the present case, the court ruled that in robbing the store, Duncan definitely acted outside of the scope of his employment. Since Osco had hired Advance and Duncan to prevent the occurrence of acts such as robberies, Duncan's own robbery of the Osco Drug store could hardly be said to be within the scope of his duties. Because Duncan's unlawful act in robbing the drug store, rather than Osco's act in hiring security guards for the drug store, proximately caused Vallejo's death, Osco could not be held liable for the intentional act which caused Vallejo's death.

PSCLR – AUG 88 (8)

Employee May Sue Employer For Attack By Co-Employee Motivated By Purely Personal Animosity, When Employer Is Directly Negligent In Hiring Or Retaining The Co-Employee, Knowing Of His Dangerous Propensities:

Lindsey v. Winn-Dixie Stores, Inc., *No. 76200, in the Court of Appeals of Georgia, decided April 19, 1988, reported at 368 S.E.2d 813.*

Brett Allen Lindsey, produce manager at Winn-Dixie Stores, Inc., sued Winn-Dixie and its part-time security guard, Charlie Whitehead, for injuries sustained when Whitehead choked him unconscious and dropped him to the floor. Lindsey suffered a fractured jaw, a lacerated chin, and the loss of three teeth.

Prior to the incident, Lindsey and Whitehead, who was a city policeman as well as a Winn-Dixie security guard, had maintained a pleasant and friendly working relationship, except for one incident several months previously in which Whitehead had unjustly accused Lindsey of stealing merchandise. Lindsey reported the matter, and Whitehead later apologized.

The day before the assault, Whitehead had confronted Lindsey, claiming that he had heard that Lindsey had called him stupid. Lindsey denied ever having made such a statement. That evening, Whitehead announced to the store's acting manager that "he was going to take [Lindsey's] face and mop the back room with it." In addition, other store employees had reported threats of physical violence against them by Whitehead to Lindsey's superiors at Winn-Dixie. No disciplinary action had been taken against Whitehead in response to those reports.

On the day of the incident, Whitehead had approached Lindsey while Lindsey was in the back of the store unpacking produce and told him, "I'm off the clock . . . . if you want to call me stupid, you can call me stupid." Lindsey replied, "Charlie, I told you yesterday that I didn't call you stupid, but if you insist, I will call you stupid," which he did. Whitehead grabbed him from behind in a choke hold, continued choking him until he passed out, and dropped him to the floor.

The trial court dismissed Lindsey's action against Winn-Dixie on two grounds. First, the trial court held that the State Workers' Compensation Act provided Lindsey's exclusive remedy and, second, that the "fellow-servant doctrine" precluded Winn-Dixie's liability for injuries to an employee caused by the negligence or misconduct of a co-employee. The appellate court reversed the trial court on both grounds. The appellate court held that the State Workers' Compensation Act does not cover injuries caused by Whitehead's attack on Lindsey because the attack was due to purely personal animosity, rather than animosity that arose from Lindsey's performance of his work-related duties. The Georgia Workers' Compensation Act specifies that "'[i]njury' and 'personal injury' shall not include injury caused by the willful act of a third person directed against an employee for reasons personal to such employee . . . ."

However, the court stated that the Act does cover injuries occurring on the job that stem from personal animosity "if the animosity arose from reasons related to the employee's performance of his work related duties."

According to the appellate court, the evidence of record so far did not suggest any causal relationship between Lindsey's performance of his duties as produce manager and his attack by the security guard. In fact, the evidence suggested that the attack was purely personal and thus not covered by the Workers' Compensation Act.

The appellate court, in sum, held that Lindsey was not limited to remedies under the Workers Compensation Act for his injuries inflicted by the willful acts of a third person, in this case, the store security guard. Those acts were spurred by animosity totally unrelated to his duties at work. If, however, the security guard had beaten him because of the way he performed his duties as produce manager, he would not be able to sue Winn-Dixie for his injuries, but would have to proceed under the Workers' Compensation Act.

The appellate court next addressed the issue whether the fellow-servant doctrine barred recovery from Winn-Dixie. The court distinguished between vicarious liability (the liability of an employer for the acts or omissions of an employee) and direct liability of an employer for its own negligence in hiring or retaining the employee.

Under the fellow-servant doctrine: "[T]he employer shall not be liable to one employee for injuries arising from the negligence or misconduct of other employees about the same business." This doctrine would protect Winn-Dixie from vicarious liability for Whitehead's acts of negligence or misconduct, but would not protect it from liability for its own negligence in hiring or retaining an employee with knowledge that the employee poses a danger to co-employees.

Thus, in order to succeed at trial, Lindsey will have to show that Winn-Dixie knew of the danger posed by retaining Whitehead as its security guard. This fact is clearly suggested by the evidence that Lindsey's superiors had been notified of threats of physical violence by Whitehead to other workers and by the evidence that, the day before the incident, Whitehead had told the assistant store manager that he intended to mop the floor with Lindsey's face.

This evidence, if proved at trial, would allow a jury to conclude that Winn-Dixie's negligence in retaining Whitehead in its employ proximately caused Lindsey's injuries when he was attacked by Whitehead.

PSCLR – NOV-DEC 88 (9)

Punitive Damages Award Against Casino For Abusive Acts of Security Personnel Upheld When Acts Were Authorized By Casino's Managing Agent:

Craig Cerminara v. California Hotel and Casino d/b/a Sam's Town Hotel Gambling Hall and Bowling Center, *No. 18058, in the Supreme Court of Nevada, decided August 25, 1988, reported at 760 P.2d 108.*

Craig Cerminara, a regular patron of Sam's Town Hotel and Casino, owned and operated by California Hotel and Casino, won a jury award in the amount of $5,000 for past and future compensatory damages and $100,000 for punitive damages due to the casino employees' abusive treatment of him after a brawl broke out in the casino. The trial court reversed the punitive damage award, and Cerminara appealed.

Craig Cerminara, his brother-in-law Mike Morabito, and friends Randy and Lisa Vickrey, spent an evening together at Sam's Town Dance Hall. The four decided to leave at about 1:30 a.m. Near the exit, Randy Vickrey met a man referred to as Dudley Do Right and engaged him in friendly conversation. Without warning, Morabito approached the two and took a wild swing towards Do Right, missing him. Cerminara and Randy Vickrey immediately separated the two and attempted to escort Morabito out of the dance hall.

According to the plaintiff's version of the incident, as Cerminara and Vickrey were attempting to remove Morabito from the dance hall, Don Lovett, the dance hall manager, standing six feet tall and weighing over two hundred pounds, charged at the group "like a maniac." Lovett pushed or "stiff-armed" Cerminara, knocking him over a table. Lovett then choked Randy Vickrey until Vickrey was unconscious. He also picked up Lisa Vickrey by the neck, choking her. The casino security guards then arrived and handcuffed Randy Vickrey.

Lovett and the guards led Cerminara and Morabito to a small security office. The guards dragged Vickrey, his hands handcuffed behind his back and his feet dragging, down the stairs to the office. In the security office, Louis Hines, the casino shift manager, instructed the security guards, Fitro, Odom, Rude, and Puentes, to "86," or permanently oust, the men.

When Odom attempted to take Vickrey's photograph for future reference in keeping him out of Sam's Town, Cerminara put his hand in front of the camera, demanding an explanation for the way the men were being treated. One of the guards pushed Cerminara down on a chair, and Morabito told the guards to leave his brother-in-law alone, causing the guards to wrestle Morabito to the ground. Cerminara then tried to pull one of the guards off Morabito, and another security guard grabbed Cerminara's arm, forcing it behind his back and causing Cerminara to fall to his knees. Cerminara yelled that the guards had broken his arm but, despite his complaints, the guards proceeded to handcuff Cerminara's hands behind his back. The arm began to swell badly, and Cerminara continued to complain of the pain.

Security guard Fitro noticed the swollen condition of Cerminara's arm and told the other guards to uncuff him and cancel the request for police assistance. The guards told the men they were "86-ed" and led them off the casino premises.

According to Vickrey, "Craig couldn't even straighten his arm out. He couldn't put it down. He had to hold it up from the pain. It was already swollen twice its size in the elbow and it was already turning black." Vickrey also testified that when he rejoined his wife, Lisa, her neck was bruised and discolored.

Lisa and Randy Vickrey immediately reported the incident to the police, who declined to charge Lovett with assault and battery. Cerminara went to a doctor and found that his elbow had been dislocated. The doctor who examined his arm testified that Cerminara's ulnar nerve was torn lose, causing continuing pain and discomfort, and that calcification had occurred.

The Sam's Town employees' version of the evening's events was significantly different than the plaintiffs' version. Don Lovett testified that he used only reasonable force to quell the fight that had broken out among the men. He testified that the security guards had handcuffed Vickrey because Vickrey had struggled, protested, and would not peacefully cooperate. He denied choking Vickrey or stiff-arming Cerminara.

Odom testified that when he had tried to take Morabito's picture, Morabito "smashed the camera into" Odom's face. When the guards had tried to restrain Morabito, Cerminara jumped on one of them and turned the commotion "into a major free-for-all." He admitted that one of the guards had somehow twisted Cerminara's arm, either in attempting to get Cerminara off the guard or in the process of handcuffing him.

Cerminara sued Sam's Town for assault and battery, false arrest and false imprisonment, intentional infliction of emotional distress, and negligence. The jury found Sam's Town negligent in its treatment of Cerminara. The jury also found Sam's Town liable for assault and battery and intentional infliction of emotional distress. However, on the questionnaire submitted to the jury, the jury indicated that Cerminara suffered "$0" in damages due to battery, "$0" from intentional infliction of emotional distress, and left blank the amount of damages suffered due to Sam's Town's negligence. As to "the total amount of [past and future] compensatory damages," the jury awarded Cerminara $3,000 past damages and $2,000 future damages. The jury also awarded Cerminara $100,000 in punitive damages.

In an effort to overturn the award of punitive damages, Sam's Town argued that the punitive damages award was improper, without an accompanying award of compensatory damages for either intentional infliction of emotional distress or battery, both of which required a finding of intentional misconduct on the part of Sam's Town. The trial court agreed with Sam's Town and overturned the punitive damages award.

Cerminara appealed the trial court's decision, arguing that Nevada law allows punitive damages to be awarded when a defendant "[h]as been guilty of oppression, fraud, or malice, express or implied." The jury in this case had concluded that Sam's Town assaulted, battered, and intentionally inflicted emotional distress on Cerminara. The jury also awarded $100,000 in punitive damages for Sam's Town's malicious or oppressive behavior.

The appellate court found that there was substantial evidence to support the jury's findings as to assault and battery, intentional infliction of emotional distress, and malicious or oppressive behavior on the part of Sam's Town employees. The court concluded that the jury's responses to the questionnaire could be interpreted to provide Cerminara a lump compensatory damages award of $5,000 for harm suffered from all three torts found to have been committed by Sam's Town. Therefore, the court did not need to decide whether a jury must award compensatory damages for an intentional tort before it may award punitive damages for the same behavior. The court held that the trial court should not have reversed the punitive damages award because, under its interpretation of the jury's verdict, the jury did make an award of compensatory damages for the underlying tort.

The court also addressed the issue of whether Sam's Town could be held liable for punitive damages for the acts of its employees. The jury was instructed as follows:

> Punitive damages may be awarded against principal for acts of the employee only if:
>
> * * *
>
> The principal authorized or ratified the conduct which is found to be the basis for punitive or exemplary damages.
>
> If the principal is a corporation, the... authorization, ratification, or act of oppression, fraud, or malice must be on the part of any officer, director, or managing agent of a corporation.

The trial court concluded that there was no evidence of manager or director participation, ratification, or approval of the intentional torts committed by the Sam's Town employees. The appellate court disagreed.

In determining who is a managing agent, the court must "look to what the individual is authorized to do by the principal and to whether that agent has discretion as to both what is done and how it is done." In this regard, "[j]ob titles...should be of little importance."

The appellate court stated that Fitro, Rude, Puentes, and Odom were security guards and clearly did not possess the necessary discretion and authority to be considered managerial agents. However, Don Lovett, the dance hall manager, testified that he "hired employees, set schedules, [made] sure the room was up set [sic], ... [and] order[ed] liquor." He also stated that he "was in charge of making sure that no minors were allowed in the bar and that all the patrons were taken care of and no one was harassed or hassled."

The court held that, given Lovett's responsibilities, discretion and authority concerning the dance hall, Lovett could be reasonably considered a "managing agent" of Sam's Town.

Because Lovett, as a managing agent, incited the brawl at the casino by a violent overreaction to the conflict between Morabito and Do Right, he was involved in and authorized the oppressive and malicious acts of the security guards leading to the jury's award against Sam's Town. According to the court, Lovett "established a tone or pattern of unnecessary and excessive force that continued until the guards realized they may have seriously injured Cerminara's arm. Moreover, Lovett's pushing Cerminara over a table, in and of itself, could have been the basis for the jury award." In sum, the court held that whether Lovett merely authorized the security guards' acts or actually participated in the malicious and oppressive behavior, the jury's award of punitive damages against Sam's Town should be reinstated.

PSCLR—NOV-DEC 88 (11)

Bar Owner Is Liable To Injured Customer For Negligent Retention Of Bartender With History Of Violent Acts:

Nicholas A. Foster v. The Loft, Inc. et al., *No. 87-513, in the Appeals Court of Massachusetts, Middlesex, decided August 12, 1988, reported at 526 N.E.2d 1309.*

Late one evening, Nicholas A. Foster was having a drink at a bar in Worcester, Massachusetts, known as The Loft when a fight broke out between his friend and the bartender. In the ensuing melee, The Loft's doorman pinned Foster's arm behind his back, and the bartender, Gino Rida, punched him in the face, fracturing Foster's cheekbone and injuring his right eye. Foster sued The Loft, Inc., for negligent retention of Gino Rida as its bartender. The jury found in favor of Foster, and The Loft appealed from the jury's verdict.

The Loft is a complex of five different bars, providing music for dancing on the premises. The complex has a seating capacity of 500 people. The Loft employs two uniformed police officers to control disturbances, in addition to four individuals to check identifications and to help maintain order. The manager also watches over the premises to prevent or to quell any outbreaks of violence. There had been disturbances at The Loft prior to the evening in which Foster was injured.

The Loft, Inc., had purchased the bar only one month prior to the incident. It retained in its employ Gino Rida and a number of other former employees. In his role as bartender, Gino Rida was regularly in contact with members of the public. He was in charge of taking orders for drinks, making drinks, and handling customer complaints.

Prior to the evening in question, but after The Loft had purchased the bar, Ian Stuart, a Worcester police officer and part-time employee of The Loft, informed co-manager James Flynn that Gino Rida had a criminal record. Apparently, neither Flynn nor anyone else in The Loft's management made any attempt to check Rida's background, made any inquiry of him of job or character references, or asked him to fill out an employment application. Rida's criminal record, which was presented to the jury, revealed that Rida had pled guilty to charges of assault and battery by means of a dangerous weapon (a knife); assault with intent to commit rape; and kidnapping. He was sentenced to one year on the assault and battery charges and given two years probation on the other charges, to commence after serving his prison term. Rida was on probation at the time he struck Mr. Foster.

On the evening in question, Mr. Foster went with friends to The Loft. Because there was standing room only on the dance floor, Mr. Foster and his friends moved to the third floor, where Rida was the bartender. Foster and his friend, Koler, ordered drinks. When Koler complained to Rida that his drink was improperly mixed and requested a new drink, Rida responded with obscene language and gestures. Koler then threw his drink at Rida, and Rida lunged over the bar and struck Koler. It was at that time that the doorman pinned Foster's arm behind his back, and Rida proceeded to punch him in the face, causing Foster's fractured cheekbone and right eye injury. Foster sued The Loft for negligence in hiring or retaining Rida in its employ.

The doctrine of negligent hiring or retention provides that

> an employer whose employees are brought in contact with members of the public in the course of the employer's business has a duty to exercise reasonable care in the selection and retention of his employees.

An employer must be careful to avoid the selection or retention of any employee whom he knows or should know is unsuitable to deal with the persons invited to the premises by the employer. The employer's knowledge of a history of violent or inappropriate acts is generally sufficient to forewarn the employer that the employee may eventually commit an assault.

> Negligent retention...occurs when, during the course of employment, the employer becomes aware or should have become aware of problems with an employee that indicated his unfitness, and the employer fails to take further action such as investigating, discharge or reassignment.

Mr. Foster contended that The Loft was negligent because The Loft knew, through its co-manager, that Rida had a criminal record yet did nothing to determine the nature of the record or its extent in order to assure itself that Rida's past criminal activities did not pose a present danger to its customers. The Loft countered that the only evidence of its negligence was its knowledge prior to the incident that Rida had a criminal record, and that this alone could not support a finding of negligence.

The appellate court held that The Loft's mere knowledge that Rida had a criminal record would not, by itself, establish The Loft's negligence. Rather, The Loft's negligence in retaining Rida depended upon the nature of the criminal record and all of the circumstances. The court stressed that it was not holding that an employer can never hire a person with a criminal record or retain such a person in its employ without risking liability for negligence if that employee commits an assault. Rather, the employer must consider all of the circumstances, including the nature of the criminal record and the position for which the employee is retained.

The court noted that The Loft was not a quiet cocktail lounge, but a large complex of bars, serving alcoholic beverages to a large, young crowd. The Loft had employed several persons, including police officers, to maintain order. Therefore, the jury could have inferred that Rida's work environment was volatile, with a high potential for violence. In addition, Rida had the responsibility of handling customer complaints, involving situations that could easily lead to heated confrontations. Thus, the jury could have found that The Loft was negligent in retaining Rida as its bartender, and that such negligence caused Foster's injuries. Assuming that The Loft knew of the nature of Rida's criminal past, The Loft took an unreasonable risk that Rida would injure one of its customers. Having taken such an unreasonable risk, The Loft was liable to Foster for the injuries thereby sustained.

The appellate court emphasized that its decision did not mean that an employer cannot hire or retain a person known to have a criminal record. Further, the court agreed with the general rule that an employer is not required to make an inquiry with law enforcement agencies about an employee's possible criminal record, even when that employee is to deal regularly with the public. In this case, The Loft knew that Rida had a criminal record and acted unreasonably in allowing him to remain as the bartender, given the nature of his past criminal acts and the duties of a bartender.

PSCLR—APR 89 (8)

Employer May Be Liable For General Damages For The Wrongful Act Of An Employee And For Negligently Hiring The Employee, But Cannot Be Liable for Punitive Damages For Both Torts:

K-Mart Corporation and Nathan Christian v. Minerva Martinez, Elsa Nichols and Marta Mata, *No 13-87-362-CV, in the Court of Appeals of Texas, Corpus Christi, decided November 23, 1988, rehearing denied December 30, 1988, reported at 761 S.W.2d 522.*

Minerva Martinez, Elsa Nichols, and Marta Mata sued K-Mart Corporation and Nathan Christian, K-Mart's loss prevention district manager, for actual and punitive damages for slander, assault, false imprisonment, and negligent hiring.

At trial, the jury found K-Mart and Christian to be jointly and severally liable to the plaintiffs for Christian's slanderous statements in the following amounts: (1) Martinez—$37,000; (2) Nichols—$35,000; and (3) Mata—$36,000.

For false imprisonment, the jury found Christian and K-Mart jointly and severally liable to the plaintiffs in the following amounts: (1) Martinez—$10,000; (2) Nichols—$10,000; and (3) Mata—$10,000. Although the jury found that Christian acted maliciously in making slanderous statements and in his acts resulting in the false imprisonment, the jury did not assess any punitive damages against Christian.

The jury found that K-Mart ratified the acts of Christian, and that K-Mart negligently hired Christian. The jury assessed punitive damages against K-Mart for slander, false imprisonment, and negligent hiring as to each plaintiff.

The jury found that K-Mart was liable for punitive damages to the plaintiffs as follows: (1) Martinez—$15,000 for slander, $10,000 for false imprisonment and $30,000 for negligent hiring; (2) Nichols—$15,000 for slander, $10,000 for false imprisonment and $30,000 for negligent hiring; and (3) Mata—$15,000 for slander, $10,000 for false imprisonment and $30,000 for negligent hiring.

On appeal, K-Mart contended that the $30,000 punitive award to each plaintiff should be reversed as a double recovery for the same injury already compensated by the award of punitive damages for false imprisonment and slander by Christian.

The plaintiffs presented two different theories of recovery of punitive damages against K-Mart. Under the first, an employer can be vicariously liable for punitive damages for an employee's misconduct if one of the following is shown:

(1) the employer authorized the doing and the manner of the act;

(2) the employee was unfit and the employer was reckless in employing him;

(3) the employee was employed in a managerial capacity and was acting in the scope of employment; or

(4) the employer or a manager ratified or approved the act.

Since the jury found that K-Mart ratified the acts of Christian and that K-Mart was reckless in employing Christian, the plaintiffs were entitled to recover for Christian's reckless and malicious conduct constituting slander and false imprisonment under the theory of vicarious employer liability.

An employer can also be directly liable for punitive damages for its own negligence in hiring an employee. Since the jury found that K-Mart acted with reckless disregard in hiring Christian, and that the negligent hiring caused the plaintiffs' damages, K-Mart could be properly assessed punitive damages for its own act of negligence in hiring Christian.

The issue before the court in this case was whether each plaintiff could recover punitive damages from K-Mart for both the malicious acts of K-Mart's employee and for K-Mart's own act of negligently hiring the employee. The court held that to allow recovery under both theories would amount to a double recovery because K-Mart's vicarious and direct liability is based upon K-Mart's single act of negligently hiring the employee, Christian, and because both theories of punitive damages are predicated upon the same findings of actual damages.

Therefore, the court found that the plaintiffs should have been required to elect between the amount of punitive damages for negligent hiring of Christian, $30,000, or the slander and false imprisonment, $25,000. Since the plaintiffs were not given the opportunity to elect between the alternative measures of damages, the court rendered judgment for the plaintiffs in accordance with the greater recovery. The court, therefore, reversed the awards of $15,000 and $10,000 to each plaintiff for slander and false imprisonment and affirmed the $30,000 to each plaintiff for K-Mart's own act of negligent hiring.

PSCLR – MAY 89 (2)

Store Owner Not Liable For Injury To Customer During Armed Robbery Of Store, If Employee's Acts Were Reasonable, Considering Urgency Of Situation:

Schultz v. Elm Beverage Shoppe, *No. 88-313, in the Supreme Court of Ohio, decided December 30, 1988, reported at 533 N.E.2d 349.*

Two men robbed the Elm Beverage Shoppe in Youngstown, Ohio at gunpoint. At the time of the robbery, the only person present in the store was the clerk, David Wildes. During the course of the robbery, David E. Schultz, a regular customer, entered the store. Schultz stopped just inside the door when he noticed Wildes' disheveled appearance and the presence of another man behind the counter. When he saw Schultz, Wildes shouted either, "Run, run, call the cops," or "Call the cops, call the cops." As Schultz turned and attempted to run out of the store, the robber shot him in the left leg and left arm, causing him to fall to the pavement. The robbers then shot and killed Wildes and shot Schultz again as they fled the store. Schultz stated that he knew a robbery was taking place soon after entering the store and did not need Wildes' advice to leave the store.

Schultz sued Wildes for the injuries he incurred during the robbery and contended that Elm Beverage Shoppe should be liable for the negligent acts of its employee during the robbery. The trial court dismissed the case against Elm Beverage prior to trial; Schultz appealed.

The appellate court reasoned that certain conduct is "privileged" when circumstances justify or excuse conduct that would ordinarily subject the actor to liability. The court held that because the conduct of the store clerk, David Wildes, was reasonable in light of the urgency of the situation, his acts were privileged and therefore the store could not be liable for the injuries sustained by Schultz during the robbery. Therefore, the court upheld the trial court's dismissal of the action against Elm Beverage Shoppe.

COMMENTARY:

The law recognizes that conduct undertaken to prevent a felony is privileged under certain circumstances. For example, a private person, such as a private security officer, is privileged to use force which is not intended or likely to cause death or serious bodily injury against another for the purpose of preventing any felony which the security officer reasonably believes the other is committing or about to commit if the officer reasonably believes that the commission of the felony cannot otherwise be prevented.

In addition, a security officer (or any private person) may use force intended or likely to cause death or serious bodily harm if the person reasonably believes that the felony cannot otherwise be prevented, and the felony which the person attempts to prevent threatens death or serious bodily harm or involves the breaking and entry of a dwelling place. Robbery is expressly included among the felonies which may be prevented by deadly force. This privilege has been applied to protect stores from liability where customers have been injured as a result of actions taken by store owners, store operators or their agents against armed robbers.

The above privilege is limited where the force, or tortious conduct, is directed towards a third party rather than the felon. As the court noted in the above case:

> *An act which is privileged for the purpose of protecting the person from harmful or offensive conduct or other invasion of his interest or personalty subjects the person to liability to a third person for harm unintentionally done to him only if the person realizes or should realize that his act creates an unreasonable risk of causing such harm.*

Thus, the acts of a store owner, store operator, or his agent while coping with an armed robbery are "privileged" if they do not create an unreasonable risk of harm to customers or other third parties on the premises. An important factor in determining the reasonableness of such actions is the urgency created by the situation.

Financial Institutions

PSCLR—AUG 82 (1)

First Impression Case Finds Bank Has Affirmative Duty to Protect Customers From Criminal Acts of Third Parties:

Marian Stalzer v. European American Bank, *Civil Court of the City of New York, Queens County, Small Claims, decided March 15, 1982, reported at 448 N.Y.S.2d 631.*

"What duty, if any, does a bank owe a 'business invitee' who, after cashing a check, was the victim of a robbery by a third party while still inside the bank premises?" It is a novel question, without precedent, according to the New York Civil Court.

Customer Marian Stalzer was robbed of $502.02 in the European American Bank located in the Gulf & Western Building in New York City, after she cashed her payroll check and stepped away from the teller to a customer table a few feet away. While she was counting her money, a man approached her from behind and grabbed her cash. Screaming that she had been robbed, Stalzer received no assistance from bank employees, although she reported the theft to the. Stalzer returned to her office and informed her employer, who called the bank and the employer's security director.

Stalzer subsequently sued the bank, alleging that "lack of security or inadequacy of security" was the proximate cause of her loss, and that the bank had failed to anticipate and protect against a foreseeable danger in light of the history of criminal activity around the building.

The bank denied the inadequate security claim, noting that it had "standard barriers" to crime—teller windows with protective glass, cameras, alarms, and concave mirrors. The bank claimed "guards are basically ineffective in the greater majority of the time where there is a holdup," and therefore it only used security guards in branches without "standard barriers."

The bank offered evidence that its plain clothes investigators periodically visit branch offices to make sure that each branch adheres to Regulation P of the Federal Reserve Board's "Minimum Security Devices and Procedures for Federal Reserve Banks and State Member Banks." The bank contended that it owed Stalzer "no special duty of care" beyond its regular security precautions and Stalzer "was on her own" once she stepped away from the teller's window.

Neither the court, nor the attorneys for either side, could find a case from New York, or any other jurisdiction, directly on the issue of a bank's duty to make its premises safe for the public.

The court quoted the New York Court of Appeals' established "rule of thumb" regarding a bank's handling of customer money and other typical banking transactions:

> A banking corporation occupies a different relation to the public than do ordinary corporations, and its transactions frequently are subjected to a closer scrutiny and tested by a higher standard than that which applies to ordinary commercial affairs.

This "public trust" aspect carries over into the question of foreseeability of the possibility of robbery or other related harm to its customers by a third person:

> This court believes that the general public assumes that when one is present in a bank premises to transact business that he or she is present in a safe and security place; that the public reaches this conclusion of confidence by reason of the very special character and purpose of a bank, namely, that it receives, keeps, and distributes money, money being a most convertible commodity. This special characteristic in public perception makes a bank quite different from any other store or commercial premises.

After considering the "very nature of a bank and the business it does and considering the many recent media-documented robbery attempts and completions, it seems reasonable to conclude that banking invites certain very real risks to the public at large." The court noted that bank robberies are an everyday occurrence in New York City, with an average of 750 committed in the city every year for the past three years.

Applying these principles, the court ruled:

> ... where, as here, the bank either knows or has reason to know from past experience that there is a likelihood of conduct on the part of third persons which is likely to endanger the safety of a bank visitor, the bank is obliged to take all necessary protective measures and to provide a reasonably sufficient number of servants to afford reasonable protection.

The court determined that the elements of reasonable security measures will depend on "current specific and general crime statistics, geographic location of the bank, physical arrangements of the bank in relation to traffic and transportation, predictable customer use, and the perceptive deterrent effect of bank guards and various other anti-crime paraphernalia." The court concluded by awarding plaintiff Stalzer judgment in the amount of her loss, finding that "inadequate security precautions were a proximate contributing cause of the claimant's loss."

COMMENTARY:

This case should prompt management of banks and other lending institutions to perform an immediate and thorough evaluation of existing security precautions and employee in light of the specific factors mentioned by the New York court.

For example:

. Have you recently reviewed which days are the heaviest payroll check-cashing days for local business which might necessitate additional security personnel?

. Is the physical arrangement of the bank's interior calculated to discourage criminal behavior, including aspects like appropriate lighting, arrangement of customer tables relative to location of security, and visibility of security personnel at the doors?

. Are your employees instructed to report suspicious activity in the customer area? Have they been provided with police, fire, and other emergency numbers?

. Do you regularly review reports of incidents for possible methods to prevent a repeat of problems?

PSCLR– NOV-DEC 82 (7)

Bank Customer Injured by Criminal Attack in Bank Parking Lot Cannot Recover from Bank for Its Failure to Employ Armed Guards:

Charles B. McKinney, et al. v. Louisiana National Bank and Aetna Life & Casualty Company, *No. 14765, in the Court of Appeal of Louisiana, First Circuit, decided May 25, 1982, reported at 416 So. 2d 948.*

A Louisiana court refused to award damages to a bank customer who was the victim of an attempted purse-snatching in the bank's parking lot.

Customer Ruth McKinney and her husband Charles McKinney sued Louisiana National Bank and its insurance carrier, claiming that the bank owed a "stricter duty" to protect its customers and potential customers because the bank's "product" was money and because the presence of money increased the risk of attack on the bank's customers. The McKinneys asserted that the bank was negligent in failing to employ armed security guards, and that their negligence precipitated the attack on Mrs. McKinney.

The Court of Appeal of Louisiana disagreed with the McKinneys' contention that a bank owed its customers extended protection against criminal attack, pointing to the substantial security measures practiced by banks and the trend toward paper transactions over cash transactions in banking.

Reviewing past Louisiana appellate court decisions, the Appeals Court stated that the operator of a business could not be held liable for an unexpected criminal attack on customers except in three situations:

First, bar proprietors may be held liable to patrons attacked by other, intoxicated patrons because intoxication in bars is "expectable and sometimes encouraged," and proprietors could or should foresee such attacks.

Second, innkeepers and landlords impliedly warrant the safety of their premises by encouraging guests and tenants to stay for an extended time period; therefore, a higher degree of security is required of this class of business operators.

Third, liability may arise if a business's employees or management either participated in the incident or had knowledge of prior criminal incidents on or near the business premises such as would or should cause them to foresee the incident in question.

In the present case, however, none of these situations existed. Mrs. McKinney was on the bank property only briefly, during daylight hours, and the evidence indicated that very few such incidents had occurred prior to the attempted theft of Mrs. McKinney's purse. Moreover, the court was satisfied that the bank's decision not to have armed guards present was based on its assessment that "the presence of an armed person can cause more injury in a robbery than without the person." Their absence was not the result of poor planning nor of an effort to cut expenses. There was no showing that the presence of guards could have prevented the attempted purse snatching, and the court concluded that the McKinneys' lawsuit was without basis.

COMMENTARY:

This case should not be construed as absolving banks (or any property owners or business proprietors) from responsibility for criminal attack on their patrons. In fact, a number of cases have been reported in which owners and/or operators have been held liable for injuries sustained in such incidents. (See the next two cases below.)

In order to fully understand the limited applicability of the above case, it should be noted that the case was decided on a pre-trial motion for summary judgment in which the McKinneys' attorney stipulated and agreed that there were no issues of fact to be decided by a jury. Why the plaintiffs' counsel did so is open to conjecture. The attorney also did not obtain (and perhaps could not obtain?) any counter-affidavits showing that the presence of guards could have prevented the attack. Such an affidavit from an expert security consultant would almost certainly demand a different result in this case.

It is still the rule that a bank (indeed, every proprietor) must exercise reasonable care to protect business invitees from risks which it knows or should know will pose an unreasonable hazard to their physical well-being. In the McKinney case, there simply was no showing made by the McKinneys to suggest that the bank violated its duty in this respect.

PSCLR – MAR 83 (6)

Mere Licensee on Bank Property Cannot Recover From Bank for Injuries Caused by Unknown Hazard:

Daniel Wiegand, a minor, by his Parent and Natural Guardian, James Wiegand, and James Wiegand, in His Own Right, v. Mars National Bank, *in the Superior Court of Pennsylavania, decision filed December 17, 1982, reported at 454 A.2d. 99.*

When a property owner allows his vacant lot to be used by the general public for a variety of community recreational activities, what are his responsibilities to people using his property? This case explored that question in a Pennsylvania suit brought by a boy and his parents after the boy was injured while playing ball in an empty lot.

Mars National Bank in Pittsburgh, Pennsylvania, owned a vacant lot across the street from its bank building. For many years, the bank allowed local residents to use the property for community activities. In 1974, the Kiwanis Club was given permission to erect a large Christmas tree on the lot. In doing so, club members had driven a large wooden stake into the ground. Unable to remove the stake after the tree was taken down, a club member used a sledge hammer to drive the stake below ground level and covered the spot with dirt. Even though a bank employee regularly mowed the grass in the lot that spring, the stake was not discovered. In May of 1975, however, Daniel Wiegand and some friends were playing ball in the lot and Daniel tripped over the stake, injuring his bare foot.

Daniel, 16 years old at the time, and his parents sued the bank to recover for Daniel's injuries, claiming that Daniel was the bank's "invitee." As an "invitee," Daniel would be owed the highest duty of care by the property owner – the owner would be obligated to make the land safe for his use, to discover any hazards, and to warn of or to repair any existing hazards.

The court disagreed with the Wiegand's argument. The court defined an invitee as " ... one invited to enter or remain on the land for a purpose for which the land is held open to the public. ..." and stated that liability would have to be based on the owner "encouraging others to enter the land for a purpose beneficial to the owner." The court felt that evidence had more than sufficiently established that the land was used for many different community activities, but that the land had not been intended specifically as an athletic field. There was no evidence that organized games were played or that the field had ever been marked off for games of any kind.

The court agreed with the bank's designation of Daniel as a "licensee." Therefore, the bank was obligated only to exercise "reasonable care to warn of dangerous conditions known to [the bank], and this is distinguished from the obligation to an invitee who has an implied assurance that the land has been made safe. ... " In the context of this relationship, the bank had no duty to look for unknown hazards, but to only correct or to warn of hazards of which it was aware. Bank employees had neither driven the stake into the ground, nor had they subsequently discovered the stake. Daniel himself admitted that after hurting his foot, he had trouble finding the stake. Therefore, the bank could not be held liable for Daniel's injuries.

The Wiegands also attempted to persuade the court that Daniel could be considered a "trespassing minor" and would therefore be owed a greater duty of care than an adult. A landowner is required to warn of or to protect even uninvited children he knows to be on his property, because children are less likely to be aware of potential hazards. Although this duty usually would be applied to children younger than 16, the court noted that there was no age limit on the rule. At any rate, this rule would still rely on the property owner's superior knowledge the existence of a hazard, which was not the situation here.

Finally, the court noted that the bank would be completely protected from liability in this instance by a Pennsylvania law entitled "Recreation Use of Land and Water." This act provides that owners of land and water areas made available for public recreation would have "no duty of care to keep premises safe.

PSCLR – MAY 83 (3)

Bank Ruled Not Liable to Customer Who Was Shot and Injured by Bank Robbers During Holdup:

J. L. Berdeaux v. City National Bank of Birmingham, Alabama, *No. 81-885, in the Supreme Court of Alabama, decided December 22, 1982, reported at 424 So. 2d 594.*

Although the question of a bank's duty to protect its customers has been raised before and such cases have been reported in **PSCLR**, this case presented the Alabama Supreme Court its first opportunity to decide whether banks should have a special duty to protect customers from criminal attack.

On October 5, 1979, the Oxmoor Road Branch of the City National Bank of Birmingham, Alabama, was held up by three armed robbers. The bank did not have a security guard working on the premises at the time of the robbery. The bank did have "sequence" video tape cameras and an alarm system to the local police in operation. One of the bank's customers, a Mr. Berdeaux, was in the bank making a loan payment at the time of the robbery and was shot by at least one of the robbers. He later filed this suit against the bank, alleging that his injuries were a result of the bank's negligence in operating, maintaining and controlling the bank premises.

Berdeaux, considered an invitee of the bank at the time of the robbery, admitted that the law does not require that private persons protect invitees or others from criminal attack by a third person unless some special relationship or circumstances are shown. However, he contended that a financial institution is under a special obligation to protect customers from criminal attack because of the nature of its business. Criminal intruders, Berdeaux argued, would be particularly drawn to branch banks such as the one where he was attacked, because of the remote locations and the large amounts of cash available. He based his arguments on the proliferation of branch banks and automatic "tellers" which should compel the banking industry to practice greater security to deter criminals who are "indirectly invited" by the establishment of these largely remote and unsecured modern banking methods. For this reason, he asked that the Alabama Supreme Court adopt a rule which would impose "a special duty on the banking industry to protect its customers from injury at the hands of criminal intruders who ... are drawn to banks ... ."

The Alabama Supreme Court expressed sympathy with Mr. Berdeaux's injuries, but declined to adopt a rule which would impose a special duty on the banking industry to protect its customers from criminal attack. The court stated that the issue "presents a policy matter peculiarly within the province of the legislative as opposed to the judicial branch of the government." Therefore, summary judgment which had been granted in favor of the bank by the trial judge was affirmed by the appellate court.

COMMENTARY:

It is most important to note here that Mr. Berdeaux had conceded that there was no liability on the part of the bank for the criminal act of the armed robbers when tested by the traditional rule governing the duty of a business owner/operator to a business invitee. In other words, Mr. Berdeaux apparently had determined that under the particular fact situation of his case, he would not be able to prove that the bank had had a previous problem with criminal activities likely to harm customers and had simply ignored such dangers and failed to protect its customers from such known dangers. Had he been able to prove that, he would have been entitled to a recovery because banks do have the same duty to protect its invitees from known foreseeable dangers as do all business owners/operators.

(See the following cases reported in earlier issues of **PSCLR** *for situations where this traditional standard of care was applied to banks: Stalzer v. European American Bank,* **PSCLR** *August, 1982, a New York case in which a customer who was robbed on the bank premises did recover from the bank; McKinney v. Louisiana National Bank,* **PSCLR** *November-December, 1982, where the court denied a recovery to a customer who was victimized on the premises.)*

Since Berdeaux realized at the outset that he would not be able to recover under this traditional theory, he argued that the court should judicially impose a higher standard of care on banks because of the very nature of the banking business in having large sums of cash on hand and thereby attracting more criminals. This the court declined to do, stating that only the legislature has the right to impose higher standards of care on one industry as opposed to others.

PSCLR–OCT 83 (5)

Police Officer/Bank Teller's Conduct Increased Danger Inherent In Armed Robbery And Violated Bank's No-Provocation Rule:

Leonard Shaw v. Manufacturer's Hanover Trust Company, et al., *in the Supreme Court of New York, Appellate Division, First Department, decided June 28, 1983, reported 464 N.Y.S.2d 172.*

Leonard Shaw claimed to have suffered permanently debilitating spinal injuries, when he was shot by Donald Summers, an off-duty New York City police officer, who was working as a part-time teller at defendant's branch office when an armed robbery occurred.

Shaw was standing in line at a teller's window when two armed men entered the bank and demanded money from a bank officer. The bank officer gave one of the robbers the money. When the other robber saw Summers in a teller's case, he said, "I know he's a cop. If he touch the gun, I'll shoot him." The other robber then escaped with the money. The remaining robber stood with his gun pointed at Summers, when Summers, identifying himself as a police officer, ordered him to drop his gun.

Telling everyone in the bank to get on the floor, Summers took a position behind a column and fired two shots at the robber. The robber moved toward the exit, and grabbed Leonard Shaw to use him as a shield as he made his escape. Summers, from his position behind the column, fired two more shots at the escaping bandit; one shot struck Shaw, causing the spinal injuries.

Shaw sued Summers and Manufactuer's Hanover Trust; the defendants won a jury verdict, and Shaw appealed.

At trial, the judge permitted the jury to examine the defendant Summer's conduct under the "Emergency Rule." The appellate court ruled that this was improper, because the defendant bank was not entitled to the Emergency Rule instruction, since the possibility of a robbery at a bank would be an anticipated. A situation could only be considered a true emergency if it could not be anticipated. A bank, and a police officer working as a teller in a bank, must be prepared for the possibility of an armed robbery, since the banking business is an activity in which such robberies are likely to occur.

Furthermore, defendants would be entitled to the Emergency Rule instructions only if an emergency occurred which was not precipitated by the defendants' conduct. In this case, bank employees had been specifically instructed to do nothing to provoke bank robbers. Summers admitted that he had been so instructed, and further testified that he knew customers were in the bank at the time of the robbery. It therefore appeared that his response to the robbery was in direct violation of the bank's instructions.

The defendant Manufacturer's Hanover argued that Summers should not be considered an employee of the bank because he had announced that he was a police officer. As such, he should be considered as having acted solely in his role as a police officer, and not as an employee of the bank, according to the bank's argument.

The court made short shrift of this argument, pointing out that Summers was paid at a higher rate than the other tellers, and there was evidence that the bank valued the "extra protection" of his presence at the bank.

The court concluded that the police officer/teller had aggravated the situation and had increased the danger inherent in a bank robbery:

> Had Summers abided by the bank's instruction not to provoke armed robbers, it is possible that both robbers who never returned [gun]fire would have left the bank without harming anyone.

Accordingly, because the Emergency Rule instructions were improperly given at trial, the verdict for defendants was reversed and the case was returned for a new trial.

PSCLR—JAN 84 (3)

Bank Faces Negligence Suit Brought by Customer Shot by Robber at Bank's Night Depository:

Jose J. Torres v. United States National Bank of Oregon, a corporation, *No. A8007-03990; A 25563, in the Court of Appeals of Oregon, decided October 12, 1983, reported at 670 P.2d 230.*

On July 30, 1978, at about 6:30 P.M., Jose Torres stopped at a branch of the United States National Bank of Oregon in Portland to make a deposit for his employer. Although the branch was closed, a depository was available for after-hours transactions. As Torres approached the depository, he was shot and wounded by a robber hiding in bushes nearby.

Torres brought suit against the United States National Bank of Oregon, alleging that the bank's own negligence created a dangerous situation—one which resulted in his injuries. Torres' suit was dismissed by the court and he appealed.

The bank claimed that Torres failed to show that it had a duty to protect him against the criminal acts of third persons. The court noted that, indeed, an individual would not ordinarily be under an obligation to protect another individual from criminal attacks by third parties, and that a possessor of land would not be required to insure his visitor's safety. That is, unless the individual or landowner knew or should have known from past experience that there was a likelihood that the conduct of third persons in general could endanger the safety of the visitor. If the location, type of business, or prior incidents were such that the proprietor had reason to expect that third persons, in particular or in general, could endanger the safety of his visitors, then the proprietor might be under a duty to take reasonable precautions to protect his visitors.

Torres' suit accused the bank of inviting its customers to use the night depositories, although it knew of the hazards inherent in the use of night banking facilities. According to Torres, the bank was under an obligation to take special precautions to protect customers, but had not done so. More particularly, Torres contended that the design of the bank created a dangerous situation because it was such that the depository could not be seen by passersby, including police patrols. He stated that it was "common knowledge that visibility deters criminal activity" and the bank had a duty to its night customers to make that area visible, or at least not to completely eliminate visibility. In addition, Torres pointed out that his employer had, in fact, moved his account to this branch of the bank because of robbery hazards at another branch, and had been assured of greater safety by the bank.

The bank countered with the argument that Torres had not alleged or proved that the bank knew or should have known that criminals could be hiding in the bushes near its night depository, or that criminal conduct was likely to occur there and endanger the safety of a bank customer.

The Court of Appeals of Oregon noted that it was not necessary for the plaintiff to state specifically that "the defendant knew, or should have known, that customers using the depository in this particular bank had been robbed or assaulted in the past or that potential criminals were lurking around its night depository." The court continued,

> For defendant to breach its duty to plaintiff, the harm to plaintiff must be foreseeable. It must be within the general class of harm a reasonable person would anticipate might arise from defendant's conduct, and the person injured must be within the general class threatened by the conduct.

In this case, the court held that Torres had sufficiently stated a claim. The bank should have foreseen that customers might use the depository, as invited, after regular banking hours; the design of the bank building obscured the view of police and passersby; "the general class of harm" which might befall a customer in making a deposit included assault and robbery; and Torres, as an employee of one of the bank's customers, was among those threatened by such conduct. The court therefore reversed the judgment dismissing Torres' negligence claim and remanded his suit for trial.

Hospitals

PSCLR – JUN-JUL 83 (6)

Hospital Is Not Liable to Invitee Injured in Fall Without Proof of Hospital's Negligence in Maintaining Floors and Elevators Where Fall Occurred:

Hughes v. Hospital Authority of Floyd County, Georgia, *No. 65670, in the Court of Appeals of Georgia, decided February 23, 1983, reported at 301 S.E.2d 695.*

While visiting the Floyd Medical Center in Rome, Georgia, in February of 1978, Mr. Hughes entered an elevator, slipped, and fell. He later sued the hospital to recover for injuries he suffered in the fall, alleging that the hospital had failed in its duty to keep the elevator in a safe condition. A verdict was directed for the hospital at trial and Hughes appealed.

The evidence showed that there were two sets of elevators in the hospital; although testimony indicated that some coffee had been spilled in one of the elevators earlier that same night, it was not the one in which Hughes fell. There was no evidence to show that there was any liquid in the floor of that elevator.

On review, the Georgia Court of Appeals pointed out that a landowner/possessor can be held liable for injuries to invitees which are caused by a failure to practice ordinary care in maintaining the property. If the owner knew or should have known (through regular inspection of his property) that a dangerous condition existed on his proerty and failed to warn invitees, the owner will be considered to have acted negligently. An owner will not be held liable if the invitee was aware of a hazard, but ignored it.

In this case, however, there was no indication that the hospital was actually aware that the elevator floor was slippery. Nor was there proof that the hospital should have known – there were no hospital employees in the immediate vicinity of the elevator before Hughes fell, and no evidence was offered to show that the hospital had neglected to inspect the elevators.

Because Hughes could not present any evidence to prove that the hospital had been negligent in any way, the court affirmed the judgment granted at trial. The court found no basis for Hughes' argument that a higher standard of care – one of extraordinary diligence – was owed him because he was injured while using an elevator. The court noted that only in cases where injury was caused by the mechanical failure or improper use of an elevator had courts required a higher standard of care to invitees.

PSCLR – MAY 84 (3)

Hospital Owes Duty To Parents To Protect Their Infant From Abduction While In Hospital's Custody:

Cynthia Johnson, et al. v. Jamaica Hospital, *in the Supreme Court of New York, Appellate Division, Second Department, decided October 17, 1983, reported at 467 N.Y.S.2d 634.*

Mr. and Mrs. Johnson filed this suit seeking money damages for psychic injuries suffered by them due to the disappearance of their infant daughter from the Jamaica Hospital nursery. They alleged that the disappearance resulted from the hospital's negligence.

The infant child, Kawana, was left in the care and custody of the hospital for additional treatment following her birth and her mother's discharge from the hospital. When the child was nine days old, the mother visited the hospital nursery ward and learned her infant was missing. She had been abducted by an unrelated third party (not an employee of the hospital), she was recovered through the efforts of the police approximately 4 and 1/2 months later. The hospital moved to dismiss the complaint, and this motion was denied by the trial judge. The hospital filed this appeal to the Appellate Division of the New York Supreme Court, which affirmed the trial judge. However, as will be seen below, the issue is complex and the trial judge's decision was affirmed by a mere three to two majority with strong dissents being voiced by the minority.

In the opinion of the two judges who dissented, they felt that the direct injury resulting from the abduction of the child was sustained by the child herself; the parents' emotional injuries were only an indirect result of the abduction. In support of this, the dissenters noted that a separate lawsuit had been brought on behalf of the child herself for negligent care. They further stated that the "fundamental underpinning of any cause of action founded upon negligence is the existence of a duty, the breach of which is the proximate cause of the damage suffered by the injured party." They examined the policy considerations and concluded that the defendant hospital's duty, and concomitant liability, should be limited to the care and treatment of its patient, the child. They also felt the extension of this doctrine to find that the hospital owed a duty to the parents would lead to a further extension on the part of the hospital to all persons who might foreseeably suffer harm or psychic injury as a result of injury to the child and this would create virtually unlimited liability on the part of the hospital. It has long been the law, noted the dissenters, that there is no duty to protect a mere bystander from emotional injury due to the trauma of observing injuries or death of another. "Since there was no duty of care flowing directly from the hospital to the parents of the patient, the emotional harm sustained by the parents solely as the result of the injury inflicted upon their child is not actionable." Furthermore, they argued, even if the hospital's duty of care was extended to both the child and her parents, a recovery by the parents for their emotional injuries would still be barred, because their emotional trauma was not the direct result of the child's injury; as such the parents' injury was a consequential rather than a direct result of the hospital's breach of duty.

These arguments were not persuasive to the other three judges on the appellate court. The majority ruled that the defendant hospital clearly owed a duty to the parents themselves to protect their child in the hospital's custody, and the parents, as "direct victims" of the breach by the hospital, do have a cause of action for their emotional injuries. The majority further stated that in view of the parent-child relationship, and the direct and certain emotional impact of an abduction as occurred, the risk of harm to the parents resulting from the hospital's alleged negligent management of the nursery was reasonably foreseeable to the hospital. While the law does generally deny the recovery for negligent infliction for emotional harm unless there has also been some physical injury, there are certain exceptions to this general rule of law, such as where recovery was permitted by next of kin for emotional injuries incurred as the result of a hospital's mishandling of a corpse, or for the negligent communication of false information to a daughter that her mother had died. Accordingly, the trial judge's denial of the hospital's motion to dismiss the lawsuit was affirmed by the appellate court and the case will proceed to trial.

PSCLR – JUN-JUL 86 (4)

Hospital Liable For Inadequate Security Which Permitted Trespasser To Assault Patient In Hospital Room; No Expert Testimony Is Need To Establish Inadequacy of Security:

Thomas C. Roettger and Diane D. Roettger v. United Hospitals of St. Paul, Inc., *No. C4-85-1283, in the Court of Appeals of Minnesota, decided February 4, 1986, reported at 380 N.W.2d 856.*

United Hospitals of St. Paul appealed from a verdict awarding Diane Roettger $300,000 and awarding her husband Thomas $22,500 for a violent criminal assault on Mrs. Roettger in one of its patient rooms. The Court of Appeals of Minnesota affirmed the verdict and damages awards.

While recuperating from giving birth to a baby girl, Mrs. Roettger was staying in a second floor room at United Hospitals, a large urban medical complex consisting of 700,000 square feet and over 21,000 rooms located in downtown St. Paul, Minnesota. As she lay awake in bed at about 2:00 AM on April 19, 1983, Charles Brown was loitering in the fourth floor visitor's lounge, pacing about nervously and smoking cigarettes. Two women were also in the lounge trying to sleep, and Brown apparently attempted to lure one of the women into a tunnel leading to another hospital building by telling her that a doctor had asked him to get her. The woman smelled liquor on his breath, became suspicious, and did not go into the tunnel; she and her companion reported Brown's behavior to the personnel at the nurse's station.

Brown left the fourth floor via an elevator before a hospital security officer arrived at the nurse's station. As the officer was speaking to the two women, he was told over his radio that there was a problem on the second floor.

The "problem" on the second floor proved to be Mr. Brown. He had taken an elevator to the second floor and walked into Mrs. Roettger's room. After talking to her a few moments, she asked him to leave and began to pick up the telephone. As she did, the phone made a ringing sound and Brown suddenly threw his body onto Mrs. Roettger's body, and the two of them fell to the floor. Brown hit her in the face several times and when she screamed, he held her throat until she had difficulty breathing. Brown sat on top of Mrs. Roettger with her back to the floor; he held her head and struck it repeatedly against the floor. She believed he was going to kill her. Two nurses entered the room and pulled Brown off Mrs. Roettger. He fled the room and the hospital. Mrs. Roettger was hysterical, her left eye was badly bruised and swollen, her face was cut on her eyebrow and required stitches, and there were marks on her neck.

Brown was eventually caught, arrested, and committed to another hospital. There was evidence that Brown had been seen inside United Hospitals on at least three occasions prior to his assault of Mrs. Roettger. On the last of the three occasions, he was arrested for trespassing by a security officer whose report stated, "This is the party mentioned by Children's [another division of the United Hospitals] as sleeping in the hospital the past few nights."

United Hospitals' security chief, Jerry McDonald, testified that in the 17 years he had been associated with United Hospital he couldn't recall a single incident involving a trespasser assaulting a patient. McDonald and a security chief at another area hospital testified that they were of the opinion that the security measures practiced at United were adequate and reasonable.

United's security practices consisted of two security officers being stationed outside the hospital buildings and one officer patrolling the hospital's eight floors. On the night of Mrs. Roettger's attack, the inside security officer did not go above the first floor between midnight and two a.m. The women who had reported Brown to the fourth floor nurses' station testified that they had spent nearly every night for several weeks in the visitor's lounge and neither could recall ever seeing a security officer patrolling that part of the hospital.

In addition to Mrs. Roettger's physical injuries, there was evidence she suffered from extreme anxiety after the attack; upon returning home from the hospital, she felt frightened and insecure and was unable to be alone. According to medical testimony by experts for both parties, she suffered from post-traumatic stress disorder.

United Hospitals' appeal of the trial court's verdict and damages awards was based on its contention that its security procedures were reasonable and adequate in light of the circumstances, and that the evidence did not support the jury's finding that its negligence was the direct cause of Mrs. Roettger's injuries. The appellate court noted that whether or not the defendant's security procedures were reasonable and adequate was a question for the jury to decide (the court cited *Butler v. Acme Markets, Inc.*, reported in the November-December 1982 **PSCLR**, case nine, page six, and *Isaacs v. Huntington Memorial Hospital* , reported in the May 1985 **PSCLR**, case three, page three), and that there was sufficient evidence produced to enable the jury to determine that the security was inadequate.

On this issue the Hospital argued that the plaintiffs had failed to introduce any expert testimony to the effect that the security measures employed by the Hospital were inadequate. (The plaintiffs had called a witness who testified about the availability and use of certain security devices which should have been employed but the trial judge had ruled that the witness was not an "expert" on the issue of the hospital's negligence and whether it had or had not provided adequate security.)

The appellate court noted that "expert testimony is not always necessary to establish negligence." Expert evidence is necessary only if the issue to be presented to the jury is not within the common knowledge and experience of the average juror. The jurors in the Roettger's case were presented extensive evidence regarding the security measures which were employed by the hospital, and heard evidence concerning the trespasser's presence above the first floor of the hospital. Under these circumstances, the court concluded that the jurors were competent to determine whether the security provided by the hospital was reasonable under the circumstances, and they were entitled to reach a different determination from that of the hospital's security experts based upon their own independent evaluation of the situation.

The appellate court then reviewed United Hospitals' argument that there was insufficient evidence to support a finding that its negligence directly caused Mrs. Roettger's injuries. The court relied in part on the recent Minnesota case of *Rullman v. Fisher* (reported in the November-December 1985 **PSCLR**, case four, page three) which also dealt with the issue of inadequate security as the cause of a criminal assault. In *Rullman*, this court stated:

> On the issue of causation, the plaintiff must produce evidence which affords a reasonable basis for the conclusion that it is more likely than not that the conduct of the defendant was a substantial factor in bringing about the result. A mere possibility of such causation is not enough; and when the matter remains one of pure speculation or conjecture, or the probabilities are at best evenly balanced, it becomes the duty of the court to direct a verdict for the defendant.

In this case there was substantial evidence to show that Brown had been seen in the hospital without legitimate reasons for being there not only on the night of the assault of Roettger, but also on three previous occasions. In spite of the previous sightings and in spite of the testimony by United Hospital's security expert that people are critical to effective security, the security department had not alerted the hospital staff that Brown or any trespasser had been seen on the patient floors. Moreover, the hospital's only interior security guard did not patrol above the first floor on the night of Mrs. Roettger's assault. The appellate court here concluded that the jury was justified in its conclusion that United Hospital failed to provide adequate security on its patient floors and this failure was a substantial factor in bringing about Diane Roettger's injuries (the court referenced *Virginia D. v. Madesco Investment Corp.*, reported in the September 1983 **PSCLR**, case one, page one).

Finally, the appellate court affirmed that the amount of damages awarded to Mrs. Roettger and her husband was appropriate in light of the brutality of the assault; the timing of the assault, only two days after she gave birth; and the severity of her physical and emotional injuries.

COMMENTARY:

This case once again illustrates the need for continuing assessment and reassessment of the security needs of a particular business or enterprise. When one or more individuals has been found to breach the existing security measures, then that in and of itself should give rise to a reevaluation of the effectiveness of the security measures being utilized. Repeated breaches should alert security personnel that the measures being employed are insufficient and immediate steps should be taken to in order to prevent future breaches of security.

In the above case if it had not been for the fact that Brown had previously been found by the hospital security personnel to have effectively avoided the then-existing security measures and if he had assaulted Mrs. Roettger during his first intrusion into the hospital proper, then the court may well have concluded that the security measures were in fact reasonable and adequate under the circumstances then existing.

However, when the hospital failed to improve or increase its security after it had become aware of Brown's repeated trespasses, the circumstances changed to the extent that the jury and court could conclude, and did conclude, that the hospital should have improved its protection of its patients.

Therefore, it is not sufficient for a security manager to merely read incident reports received from his field personnel, but he must also exercise judgment and evaluate the effect of the reports. Do the reports justify or indeed mandate an increase or perhaps decrease in the level of security being provided at a particular location? What should the reasonable security manager do when confronted by the particular facts and circumstances set forth in the reports you receive from your personnel?

And, finally, the case points out the necessity for reminding field personnel to file accurate, complete, and timely reports of incidents in order that the security manager can have the appropriate information on which to base his decisions.

PSCLR – NOV-DEC 86 (11)

Suit Alleging Inadequate Security Resulting In Assault And Murder Of Hospital Patient Is Not A Medical Malpractice Action And Need Not Be First Presented To Medical Review Panel:

Dalton Reaux, Sr., et al. v. Our Lady of Lourdes Hospital, et al., *No. 85-529, in the Court of Appeal of Louisiana, Third Circuit, decided July 21, 1986, rehearing denied August 19, 1986, reported at 492 So. 2d 233.*

The relatives of Nola Reaux brought suit against Our Lady Of Lourdes Hospital of New Orleans, alleging that she was assaulted, raped, and battered while she was a patient in the hospital, and that her subsequent death resulted from the attack. The suit also claimed that Mrs. Reaux's death was caused by the hospital's failing to provide Mrs. Reaux with proper security against intruders at a time when she was helpless and in the hospital's care and custody.

The hospital and its insurance company, filed pleadings alleging that the Reaux family's suit was premature because they had failed to present their claims to a medical review panel which, the hospital argued, was a statutory prerequisite to filing the lawsuit against the hospital. The Reaux family, however, responded that the hospital's alleged misconduct did not constitute "medical malpractice" and that the statute requiring presentation of the claim to a medical review panel was therefore inapplicable. The trial judge agreed with the hospital and sustained its position, thereby effectively dismissing the lawsuit. This appeal by the Reaux family ensued.

The Louisiana Court of Appeals noted that the Louisiana Medical Malpractice Act does in fact require that any malpractice claim against a health care provider must first be reviewed by a medical review panel before it may be filed in a court of law. Furthermore, the definition of "health care provider" in the statute does specifically include hospitals.

The statute itself defines malpractice as follows:

> Malpractice means any unintentional tort or any breach of contract based on health care or professional services rendered, or which should have been rendered, by a health care provider, to a patient ....

Tort means any breach of duty or any negligent act or omission proximately causing injury or damage to another.

Health care means any act or treatment performed or furnished or which should have been performed or furnished by any health care provider for, to, or on behalf of a patient during the patient's medical care, treatment, or confinement.

The Court of Appeals concluded that the act complained of here was not based on "health care services or professional services rendered, which should have been rendered" to Mrs. Reaux. The court noted that it had previously ruled in an analogous situation that a slip and fall suffered by a hospital patient did not constitute an actionable wrong based upon services which should have been rendered by a health care provider and likewise was outside the provisions of the Louisiana Medical Malpractice Act. And, the court examined other cases indicating more typical acts of medical malpractice, such as a hospital employee tripping over a patient's IV tube, a hospital's failure to provide sterile and clean conditions thereby causing a staph infection, and the malfunctioning of a life support system. Therefore the appellate court reversed the trial judge, and held that the Louisiana Medical Malpractice Act does not require that the claim in question be presented to a medical review panel. The case will be returned to the trial court for further proceedings where the jury will be allowed to hear the case.

PSCLR – MAR 87 (9)

Liability of Non-Profit Hospital For Damages To Hospital Visitor Received In Criminal Assault in Hospital Parking Lot Is Limited To Statutory Amount:

Flora Katz and Paul Katz v. Rahway Hospital, John L. Yoder, Director of Rahway Hospital, Robert Pohamus, Director of Security of Rahway Hospital, et al., *in the Superior Court of New Jersey, Appellate Division, decided December 19, 1986, reported at 519 A.2d 895.*

This case involves an appeal taken by the Rahway Hospital from a verdict and judgment in favor of Flora and Paul Katz. They sued the hospital for injuries arising out of a criminal attack of one of the Katz, which took place in the hospital parking lot after visitor hours.

A New Jersey statute provides that:

> ... any non-profit corporation ... organized exclusively for hospital purposes shall be liable to respond in damages to such beneficiary who shall suffer

> damage from the negligence of such corporation, ... to an amount not exceeding $10,000, together with interest and cost of suit ... and to the extent to which such damage ... shall exceed the sum of $10,000 ... to which non-profit corporation ... organized exclusively for hospital purposes shall not be liable therefor.

The hospital argued on appeal that the total amount payable to the plaintiffs should be limited to $10,000 under the above-quoted statute, while the plaintiffs argued that the interest was also payable under the statute even though it may result in a recovery exceeding $10,000. After reviewing the legislative history of the enactment of the statute, and the status of the common law of the state as interpreted by previous court decisions, the appellate court concluded that the pre-judgment interest as well as the post-judgment interest on the damages of $10,000 should be paid by the hospital to the plaintiffs in addition to the $10,000 in damages incurred by them.

Another interesting ruling contained in the appellate court's opinion was its affirmance of the trial judge's ruling prohibiting the hospital from presenting evidence on the extent of its security-related expenditures as compared to the expenditures by other hospitals in the country as a whole, and to average expenditures by other hospitals in the state of New Jersey. The trial judge had refused to admit this evidence on the grounds that it was irrelevant and did not indicate the extent of parking lot security expenditure; as such, it was of such dubious value that the information could not assist the jury in reaching its conclusion.

In another part of its ruling, the court also agreed with the trial judge's decision to permit an expert in the field of security to testify, ruling though that he was not qualified as an expert on hospital security and even though he had not consulted certain security guides which had been expressly designed for hospitals.

The verdict and judgment in favor of Mr. and Mrs. Katz were affirmed.

PSCLR—JAN 88 (4)

Security Director At State University Hospital Held Immune From Negligence Claims Alleged In Suit By Survivors Of Victims Killed By Former Psychiatric Patient:

Janice C. Beck et al. v. Alfredo Calvillo; Myrtle A. Williams et al. v. Alfredo Calvillo, *Civ. A. Nos. 83-2094-S, 83-2095-S, in the United States District Court for the District of Kansas, decided May 19, 1987, reported at 671 F. Supp. 1555.*

One of the defendants in this case is Jack C. Pearson, the director of the University of Kansas Medical Center Police Department. This decision deals only with his motion to dismiss and/or for summary judgment in his favor, against claims made by the family survivors of victims shot and killed in the medical center by a former psychiatric patient of the medical center. The victims of the tragedy were Dr. Mark Beck and Ruth Rybolt.

The plaintiffs alleged that Pearson, in his capacity as administrative manager of security at the medical center, was negligent in failing to provide proper security measures in the medical center emergency room, thus causing injury and damages to the plaintiffs.

The plaintiffs also alleged that their fundamental right of liberty and their due process interest in having a continuing family relationship and enjoying life together as a family were interfered with and cut short by the defendant's negligence. The court, however, reviewed the record presented to it and determined that it would not support a finding of the necessary intent on the part of Pearson to deprive the plaintiffs of the continuing family relationship with the deceased victims. Even if Pearson were determined to be grossly negligent or that he carried out his duties with a reckless disregard of foreseeable consequences, the court concluded that such conduct would not constitute an unconstitutional deprivation of the plaintiffs' rights, since the plaintiffs had not alleged that Pearson directed any such intent or reckless disregard toward the plaintiffs. The court therefore granted Pearson's motion for summary judgment as to these claims of constitutional deprivation by the plaintiffs.

The court then addressed the defendant's argument that the plaintiffs' allegations against him in his individual capacity should also be dismissed. Pearson argued that the complaints against him related to his conduct as an official of the state government and that he therefore would be entitled to immunity. Plaintiffs responded to this argument by claiming that Pearson's alleged failure to take action in spite of a known danger presented in the hospital emergency room was outside and beyond the scope of his official position and authority as security manager of the state facility and that he therefore was not entitled to any immunity. The court proceeded to examine, without ruling on the merits of the plaintiffs' suit, the evidence which the plaintiffs indicated would show that security personnel in the hospital knew that mentally unstable patients were entering the emergency room of the hospital, and particularly had knowledge that Bradley Boan, the psychiatric patient responsible for the murder of the plaintiffs' family members, had previously presented a danger one the medical center property.

The particular facts upon which this claim by the plaintiffs is founded are revealed in the record as follows: Dr. Beck was in the Kansas University Medical Center Emergency Room, fulfilling his duties as a resident of the department of medicine. The deceased Rybolt had brought her mother to the emergency room for medical care. Thus they were both present when Bradley Boan entered the emergency room with a gun and began firing. Rybolt and Dr. Beck were both shot and died as a result.

The evidence did show that security officers of the Kansas University Medical Center knew and were concerned that mentally ill and sometimes violent patients were being permitted to enter the emergency room. The medical center police department also was aware that Bradley Boan specifically had been a threat to individuals at the medical center in the past. In fact, they knew that Boan had been convicted and jailed because he had previously pulled a gun on someone in the medical center. At the same time, the court noted that the university police department, while aware of prior incidents of violence, still did not become "an absolute insurer" of the public's safety. Furthermore, the medical center personnel did not know that Boan had been released from Kansas State Penitentiary, nor that Boan had any intent to seek revenge at the medical center before Boan actually entered the emergency room and began shooting.

Furthermore, although the record of the case does support plaintiffs' contentions of negligence and insufficient security measures at the medical center, Pearson's conduct could not be considered as having created the threat which Boan posed to the deceased victims, nor could Boan's conduct (the actual cause of the injuries) be deemed "under color of state law" (state action). The court therefore concluded that Pearson should be granted his summary judgment motion opposing plaintiff's allegations that he had deprived them of certain constitutional entitlements because the required element of "state action" was not present.

In addition to the constitutional claims, the plaintiffs also sued Pearson under the Kansas Tort Claims Act, and the defendant argued that he is immune from suit because of the "police protection" exception provided by the Kansas statute:

> A governmental entity or an employee acting within the scope of the employee's employment shall not be liable for damages resulting from: ... failure to provide or the method of providing police or fire protection.

The allegations made by the plaintiffs in an effort to support this portion of their suit were that Pearson had failed to

> (1) disarm individuals in the emergency room with weapons, (2) restrain an individual in the emergency room who posed a danger to others, (3) place security provisions in the emergency room, and (4) establish other protocol that requires employees to observe people entering the facility or have other type of security officers on guard in the emergency room.

After reviewing the applicable statute, and case law precedent, the federal district court judge concluded that Pearson's conduct came within the police protection section provided in the statute, and therefore ruled that he was immune from liability to the plaintiffs as provided in the statute.

The court went on to express its "shock and horror" at the tragic events which resulted in the deaths at the Kansas University Medical Center, and stated,

> The court in no manner seeks to condone the conduct of Jack Pearson or the lack of security at the Kansas University Medical Center at the time of the tragedy,

but the court also explained that "this court is without the power to create a remedy where none exists." The motion for summary judgment by the defendant Pearson was granted.

PSCLR–FEB 88 (1)

Hospital Liable For Employee's Misconduct Only If Hospital Was Itself Negligent For Employing Or Retaining Employee, Or If Employee's Conduct Is Within Scope of Employment:

G.L. v. Kaiser Foundation Hospitals, Inc., v. Kaiser Foundation Hospitals, a California Corporation, *A8309-05835, CA A40470, in the Court of Appeals of Oregon, decided December 9, 1987, reported at 746 P.2d 731.*

This lawsuit arises out of the sexual assault of the plaintiff, "G.L.", by an employee of the hospital in which the plaintiff was a patient. The suit was grounded on three theories:

(1) That the plaintiff had an implied contract with the hospital whereby the hospital was to provide "adequate care, safety, treatment and security for her while she was an in-patient";

(2) That the hospital is "strictly liable" for the intentional misconduct of its employee;

(3) That the hospital itself was negligent in that the hospital management either knew or should have been aware of the dangerous propensities of the employee who assaulted the plaintiff and the hospital failed to take the necessary protective action to prevent the employee's misconduct against the plaintiff-patient.

The trial judge dismissed the first and second claims of the plaintiff, and the case went to trial on the third claim alone. A verdict was returned by the jury in favor of the defendant on the third claim of negligence. The plaintiff filed an appeal alleging that the trial judge had erred in dismissing her implied contract and strict liability claims.

It appeared that the Oregon appellate courts had never been required to answer the question as to whether a hospital could be liable for the intentional misconduct of an employee against a patient when the employee is acting beyond the scope of his employment and the hospital itself was not negligent in any respect. The court therefore had to look at other prior cases to determine what the law should be on this issue. In reviewing these prior applicable cases, the court noted that they fell into two categories: cases dealing with an employer's liability for crimes and assaults by employees, and cases dealing with the duty that employers such as hotels, common carriers, and hospitals owe to the public. With regard to the first category of cases, the Oregon Supreme Court had ruled that any employer who hires employees and assigns them job responsibilities which require them to have contact with members of the public during the course of their employment should be held responsible to exercise reasonable care in hiring and retaining such employees. Further, such employers should be held liable for any negligence in hiring or retaining an employee who has dangerous propensities which are known to the employer or which could have been learned by the employer had he made a reasonable inquiry concerning the employee. This requirement to use reasonable care in the hiring and retention of such employees is imposed because that it is reasonably foreseeable that such an employee could cause an unreasonable risk of injury to the public.

On the other hand, the second category of cases mentioned above includes the case where a senior citizen, who was a resident of a home for the aged brought a suit against the home for injuries incurred as a result of an attack on her by an intruder. Her suit was based on a city ordinance which required homes for senior citizens to have personnel "adequate to ensure proper protection and care for all guests at all times." She claimed that the home had clearly violated this ordinance, that the violation resulted in her injury, and that the home should therefore be liable for her injury. However, the court in that case rejected this theory, stating that the ordinance merely reiterated the defendant-home's common law duty to provide reasonable protection to the residents, and did not impose an absolute duty on the home to protect residents from criminal attack. (*Laurie v. The Patton Home*, 267 Or. 221, 516 P.2d 76 (1973))

The plaintiff in the present contended that the public policy of the state required that hospitals "be held to an absolute standard in safeguarding its patients from intentional injurious conduct by its employees." But the defendant contended that the legislature had not mandated such a policy and only the legislature should properly determine whether such was in fact the public policy of the state of Oregon. The Oregon Supreme Court agreed with the defendant:

> Although neither the Supreme Court nor we have expressly decided whether an employer's liability for its employee's intentional torts may be grounded on something other than the employer's negligence or culpability, our decisions have emphasized that the employer's responsibility only arises when it is negligent or when the employee's act is within the scope of his employment.

Accordingly, the appellate court ruled that the trial judge was correct in striking the plaintiff's claims based on strict liability and implied contract theories.

COMMENTARY:

In the above case, the plaintiff tried to extend to hospitals the same degree of responsibility to its invitees that the legislature had imposed on hotels, arguing that a hospital should not be held to any lesser degree of responsibility than a hotel by attempting to show that a hospital is indeed "a special kind of hotel." The plaintiff noted that the legislature had provided for innkeepers to be liable to guests for injuries caused by the innkeeper's employees and servants. However, the court noted that while the legislature had indeed imposed such a responsibility on innkeepers, it had not chosen to do so with respect to hospitals.

PSCLR—OCT 88 (3)

Hospital Is Not Strictly Liable, Nor Liable For Breach of Implied Contract, For Injuries To Unconscious Patient Who Was Sexually Assaulted By Staff Member:

G.L. v. Kaiser Foundation Hospitals, Inc., *Nos. CC A8309-05835/CA A40470/SC S34799, in the Supreme Court of Oregon, En Banc, decided June 7, 1988, reported at 757 P.2d 1347.*

Terry Daniel, a respiratory therapist employed by Defendant Kaiser Foundation Hospitals, Inc., sexually assaulted the plaintiff in this case (identified only as "G.L." in the court's decision), while she was unconscious in her hospital bed in Kaiser Foundation's Sunnyside Hospital, recovering from surgery in a semi-private room. The plaintiff's hospital roommate, an elderly, partially-paralyzed patient, reported the assault to her and to hospital personnel. The respiratory therapist later pled guilty to attempted rape in the second degree.

The plaintiff contended that Kaiser was liable for the acts of its staff member on three separate theories: (1) that Kaiser was negligent in retaining and supervising its employee, the respiratory therapist; (2) that Kaiser was strictly liable (liability unrelated to any fault by the hospital) for the acts of its employee; and (3) that Kaiser had breached its implied contract with the plaintiff.

The trial judge dismissed the second and third grounds, the case went to trial on only the first ground, negligence, and the jury concluded that Kaiser had not been negligent in retaining the guilty employee, nor in failing to supervise him properly. The plaintiff-patient appealed first to the Oregon Court of Appeals, which ruled against her; she then filed for a final review by the Oregon Supreme Court.

In this opinion, the supreme court reviewed the decision by the two courts below to dismiss the plaintiff's second and third theories of liability prior to the trial and not to permit the jury to hear them.

In asserting the strict liability theory, the plaintiff attempted to expand the traditional respondeat superior doctrine whereby employers are held liable for the acts of their employees when they are acting within the scope of their employment.

Numerous courts in the United States have held employers liable for the intentional torts of their employees. Often these cases involve employees, such as "bouncers," who are authorized by the employer to use force but who have exceeded the amount of force which they were authorized to use. One source states the law of respondeat superior liability as follows:

> A master is subject to liability for the intended tortious harm by a servant to the person or things of another by an act done in connection with the servant's employment, although the act was unauthorized, if the act was not unexpected in view of the duties of the servant.

Moreover, the court noted that courts have frequently held that the intentional act must have been "undertaken with the intent of furthering the business purposes of the employer, however misguided that intent might seem."

The court then reviewed the law of respondeat superior liability as requiring three elements:

1. that the act occurred substantially within the time and space limits authorized by the employment;

2. that the employee's motivation was, at least partly, to serve the employer; and

3. that the act is of a kind which the employee was hired to perform.

The appellate court here concluded

> where, as here, there is no allegation...that the employee was acting for the purpose of furthering any interest of the employer, plaintiff has failed to state a claim for tort liability, and dismissal of plaintiff's claims beyond theories of negligence was proper.

The court then reviewed the plaintiff's third contention: that Kaiser should be held liable on an "implied contract" theory. The plaintiff asserted that, although she had no actual contract with the hospital that the hospital would protect her safety, all hospitals impliedly contract with their patients to provide them a safe place for their recovery. The plaintiff did not offer any evidence of statements by the hospital which led her to rely on her safety while at the hospital. Thus, the plaintiff's theory is that all hospitals should be held "to have implied an offer to all its patients that the hospital is a place of safekeeping from the criminal acts of others."

The appellate court refused to take the "extreme step" of finding that all hospitals implicitly have made a contract to insure the safety of their patients. In the absence of special facts showing that the hospital had assumed such a responsibility, the court would not hold a hospital liable for the criminal acts of an employee. Further, the court, citing various regulatory statutes, reasoned that the legislature, if it chose to do so, could extend the liability already in place for long-term care facilities to include hospitals.

Because the jury had found no negligence on the part of Kaiser in retaining or supervising the respiratory therapist who assaulted the plaintiff, and because the court rejected the plaintiff's strict liability and implied contract theories, the hospital was not liable to the plaintiff for her injuries.

COMMENTARY:

An argument can certainly be made that hospitals should be held strictly liable for the safety of their unconscious patients. It is because travelers sleep in hotels that the law has for many years held innkeepers liable for failing to protect their guests from harm.

The innkeeper's duty has, for that reason, been recognized to be greater than the duty, for example, of the retailer to protect its business customers – they are not unconscious while on the retailer's premises doing business with the proprietor, whereas hotel customers are.

It is because innkeepers are held to this greater duty that legislatures have enacted statutes limiting the innkeeper's otherwise "strict liability" exposure. The Oregon Supreme Court believes that the state legislature, rather than the court, should be the body to extend hospitals' liability exposure to patients.

Note: the Oregon Court of Appeals' review of this case was reported in the February 1988 **PSCLR**, *case one, page one.*

PSCLR – OCT 88 (7)

Nursing Home Is Liable For Wrongful Death Of Unsupervised Patient Struck And Killed By A Vehicle Driven By His Daughter:

Helen H. Fields v. Senior Citizens Center, Inc. of Coushatta; Janet Fields DuPree and Randall Wayne DuPree v. Senior Citizens Center, Inc. of Coushatta, *Nos. 19,542-CA, 19,543-CA, in the Court of Appeals of Louisiana, Second Circuit, decided May 4, 1988, reported at 528 So. 2d 573.*

At approximately 7:00 PM one evening, William Fields, a 68-year-old resident of a nursing home owned by defendant Senior Citizens Center, Inc. of Coushatta, Louisiana, wandered off the premises of the home and was struck and killed by a van driven by his own daughter, plaintiff Janet Fields DuPree.

Fields' widow, Helen H. Fields, and her children, including Janet DuPree, sued the nursing home for damages caused by the wrongful death of Fields. The jury returned a verdict for the plaintiffs, awarding Mrs. Fields $200,000, Janet Fields DuPree $92,500, and each of the other children, $7,500. The nursing home appealed from the jury's verdict, contending that the nursing home was not negligent, that Janet Fields DuPree and her father were at least partially at fault for Fields' death, and that the awards to Mrs. Fields and Janet Fields DuPree were excessive.

The evidence showed that Fields had a history of mental illness: he was admitted to the VA Hospital in Shreveport for psychosis approximately six years prior to the accident; four years prior he was admitted to the hospital for mild dementia; and three years prior he was referred for psychiatric consultation while being treated for skin cancer. When Fields became progressively more confused and his wife was no longer able to care for him, the family placed him in the Senior Citizens Center.

Approximately two months prior to the accident, Fields entered the nursing home, confused and disoriented as to time and place. He was at the home on a six-month VA contract, which provided for room and board in the amount of $1,350 per month.

Fields' doctor authorized the nursing home to use restraints as necessary for Fields' safety, at the discretion of the nurse in charge. The VA Hospital prescribed a major tranquilizer for him, which caused him difficulty with his speech, and blurred vision. According to testimony by Fields' physician, any patient taking this tranquilizer should be under constant supervision. Fields' physician also stated that four days prior to the accident, Fields was suffering from dizziness and vertigo.

The administrator of the nursing home, Lynn Stevens, testified that, on the date of the accident, 81 of the 83 beds in the nursing home were occupied. Six employees were on duty at the time of the accident, suppertime, described as a very busy time of the day. Fields was placed in a special room close to the nursing station in order to provide him enhanced supervision.

The front doors to the home were equipped with a security system, which the nurses were instructed to turn on after the evening meal and after all visitors had departed. Stevens testified that the security system had not been activated on the night of Fields' accident however, because of its loud, annoying noise.

Several members of the nursing home staff testified that Fields frequently wandered from the nursing home premises. They also testified that Fields suffered from confusion, hearing impairment and blurred vision.

Sally Johnson, the nurse in charge on the night of the accident, testified that she had seen Fields walking towards his room at about 7:00 PM She then went to the medicine room to prepare the evening's medications, but no one replaced her at the nurses' station that evening. In any event, the nursing home administrator conceded that a nurse sitting at the desk at the nurses' station could not observe the front door of the home. (After the accident, the home installed a mirror so that the door was visible from the nurses' station.)

Stephanie Miller, ward clerk at the nursing home, was also working at the time of the accident. The nurses had told her to watch Fields carefully because of his tendency to wander from the nursing home. Earlier on the day of the accident, Miller had stopped Fields on his way out the door. Fields told her that he was going home and pointed to a house across the street. She had informed Johnson of the incident, and Fields had been restrained.

That evening, when Fields wandered out just before the accident, Miller was tending to another patient. One of the residents reported Fields' absence, and Miller immediately left the building to look for him. She saw Fields across the street on the shoulder of the road walking toward Coushatta. Miller went to the parking lot and called out to him. When she got Fields' attention, she instructed him to stand where he was until she could get to him. It was at this time that the van struck and killed Fields.

An eyewitness to the incident testified that it was dark and that he had his lights on as he passed the nursing home on Highway 71. He was driving about 50-60 feet behind the van, and both vehicles were traveling at about 30-35 miles per hour. Traffic was moving slowly in both directions. Bamberg testified that he saw the van swerve to the left and then saw Fields lying in the highway, but that he had not seen anyone on the highway before the accident.

When police officer Scotty Hill arrived at the scene, he found DuPree (the driver of the van and Fields' daughter) lying in the corner of the parking lot screaming hysterically, "I killed my daddy, I killed my daddy, I ran over my daddy, I didn't see him, I didn't see he was there."

DuPree testified that she had had her lights on as she approached the nursing home, and that she had seen something moving to her left, felt the impact to the right of her van, and then felt her right back tire run over something. She pulled into the parking lot and discovered that she had run over her father.

DuPree testified that, since the accident, she experienced difficulty sleeping and often woke up in the middle of the night with nightmares; she had sought professional counseling regarding her emotional trauma. Almost a year after the accident, she did not feel that her life had returned to normal. She had difficulty driving and staying home alone at night with her children, and suffered frequent flashbacks of the accident. Her husband corroborated her testimony about her nervous condition.

The DuPrees' marriage and family therapist testified that DuPree's grief over her father's death was complicated by her feelings of responsibility and guilt related to the accident. The therapist testified that DuPree's emotional scars and traumatic memories would always remain and that her life would never completely return to normal.

The widow, Mrs. Helen Fields, testified that she and Fields had been married for approximately 37 years and had lived together until he was institutionalized. Mrs. Fields had warned the nursing home workers about her husband's tendency to wander and was advised that he would be closely supervised at the nursing home.

The reviewing court upheld the jury's verdict and damage awards to each of the plaintiffs. As to the negligence of the nursing home, leading to its liability for wrongful death, the court stated that the standard of care imposed upon a nursing home is that of reasonable care, considering the patient's known mental and physical condition. The court stated that a nursing home is required to take reasonable steps to prevent injury to an ambulatory but mentally confused patient. All of the Senior Citizens Center employees involved in Fields' care and supervision were aware of his confused mental state and his tendency to wander from the nursing home grounds. Further, the court noted that the Senior Citizens Center had an alarm system which it had failed to use properly. In addition, the nursing station was unattended when Fields wandered out, and it was not possible to observe the front door from the nursing station.

The appellate court also found that Janet Fields DuPree and the victim were without fault in his accidental death. The general rule applicable to accidents involving motorists and pedestrians is that "a motorist is under a duty to see what an ordinarily prudent driver should have seen under the circumstances, and to avoid striking objects in the road ahead." However, an established exception to this rule is that a motorist driving at night is not liable for striking an unexpected or unusual obstruction, which is difficult to see and which the motorist has no reason to anticipate encountering on the highway. A motorist is liable for striking a pedestrian only when the motorist has an opportunity to perceive the pedestrian's peril in time to avoid the accident.

In addition, Janet DuPree neither saw, nor had any reason to expect to see, her father in the roadway before she hit him. She was driving within the speed limit and had her headlights on. Furthermore, her attention may have been distracted by Stephanie Miller shouting and waving her arms in the nursing home parking lot (in an attempt to get Fields' attention).

As to the contributory negligence of Fields himself, the appellate court held that, like a child, an elderly person suffering from impaired sensory and mental facilities is subject to a "relaxed standard of care."

> Disability resulting from advanced age affects the test of care required. A person who suffers from impaired senses and mental facilities due to old age is not guilty of contributory negligence where his failure to use that degree of care which an ordinarily prudent person would use under the same or similar circumstances is due to such disability.

Because the record in this case was replete with testimony regarding Fields' mental confusion, inability to care for himself, poor vision, and impaired hearing, and because the accident was caused by these impairments, the reviewing court upheld the determination by jury that Fields was not contributorily negligent.

The appellate court further upheld the jury's damages awards to Helen Fields and Janet Fields DuPree. The $200,000 awarded to Mrs. Fields represented compensation for her loss of love and affection, in addition to lost income. Further, the jury was justified in awarding $92,500 to Janet DuPree, far in excess of the $7,500 awarded to her siblings, due to the evidence of her mental anguish, over and above that suffered by her siblings.

The appellate court also upheld the trial court's refusal to allow the jury to hear evidence of the Senior Citizens Center admissions document for Fields, purporting to release the home from liability for accidents of this nature. The appellate court concluded on this point:

> We agree with the trial judge, who concluded that an attempt to prospectively absolve oneself from liability for one's negligent acts is against public policy.

The reviewing court thus upheld all of the rulings of the trial court, as well as the jury's verdict and damages awards.

PSCLR – MAY 89 (12)

Hospital Patient Who Was Sexually Assaulted May Recover From Hospital For Breach Of Implied Contract To Provide Secure Environment For Care:

K.M.H. v. Lutheran General Hospital, *No. 87-742, in the Supreme Court of Nebraska, decided October 28, 1988, reported at 431 N.W.2d 606.*

K.M.H. entered Lutheran General Hospital as a surgery patient. While occupying a private room, K.M.H. received pain and sleeping medication late one evening. At about 3:00 a.m., she was awakened by Thomas Eckles, a male employee of the hospital. Eckles touched K.M.H. and announced that he was performing a "bed check." He then sexually assaulted K.M.H.

K.M.H. sued the Lutheran General Hospital and Lutheran Medical Center, alleging that the hospital entered into an implied contract with her whereby the hospital agreed to provide a "private, safe, secure environment for her care and to protect her privacy, safety, and security." K.M.H. alleges the hospital breached the implied contract in (1) neither providing for nor protecting her privacy, safety, and security by staff, facilities, or otherwise; and (2) failing to provide the required degree of care to protect plaintiff from sexual indecencies by male personnel, considering her weakened and stressful condition. The case was dismissed by the trial court, and K.M.H. appealed.

The hospital contended that it followed proper and acceptable procedures in hiring Eckles and that if K.M.H. was assaulted by Eckles, such act by Eckles was not in the performance of his assigned duties at the time, and, therefore, the hospital was not vicariously liable for Eckles' intentional bad act.

The appellate court found, however, that K.M.H. alleged sufficient issues of fact to allow her action, based on an implied contract with the hospital, to proceed to trial. In order to recover on an action for breach of contract, a plaintiff must plead and prove the existence of a promise, its breach, and damage. K.M.H. alleged an implied contract, imposing upon the hospital a duty and obligation to provide her a private, secure environment for her care and to protect her privacy, safety, and security.

The appellate court held that the facts and circumstances creating the implied promise, together with the duties and obligations to be performed, are material fact questions whether or not the same are oral, express, assumed, or imposed by law. In reaching its decision, the appellate court quoted a 1914 case stating,

> a patient is generally admitted to a hospital . . . under an implied obligation that he shall receive such reasonable care and attention for his safety as his mental and physical condition . . . may require.

The appellate court reversed the trial court's decision dismissing the case and ordered that the case proceed to trial on the implied contract theory.

Hotels/Motels

PSCLR—AUG 82 (2)

Sleeping Guard And Negligent Delivery Of Pass Key Render Motel Liable To Guests For Third Party's Assault, Battery And Theft:

Larry Kraaz and Joyce Kraaz v. La Quinta Motor Inns, Inc., *No. 81-C-0937, in the Supreme Court of Louisiana, decided March 1, 1982, reported at 410 So. 2d 1048.*

In spite of Louisiana statutes limiting innkeeper liability for damages, a motel corporation was hit with a jury verdict of $72,000: to Joyce Kraaz, $3,500 for physical injuries, $30,000 for traumatic neurosis, and $13,000 for future medical expenses; to Larry Kraaz, $2,500 for physical and mental pain and suffering; and $23,000 for cash taken in the armed robbery of the couple, which occurred while they were guests in the defendant's motel in Metairie, Louisiana. The pair were in Louisiana pursuing their business of buying, selling, racing, and betting on horses. Because race horses are generally purchased with cash, the Kraazes had $41,000 in their possession at the time.

In the early morning hours of December 31, 1978, the Kraazes were awakened by two men, who broke the chain on the motel room door and assaulted and robbed them. Mr. Kraaz was knocked out of bed, and held at gunpoint until tied up with wire and gagged. Mrs. Kraaz was mauled, hit in the face, and placed in the bathtub.

The thieves took $23,000 in cash from Mrs. Kraaz's purse, and a .25 caliber Smith & Wesson automatic weapon, but they overlooked the other $18,000 also in the room. After the assailants fled, Mr. Kraaz broke free, got a gun from his car, and chased the men. His chase took him by the motel office, where he spotted an elderly security guard sleeping soundly.

At trial, evidence showed that the assailants had earlier approached the 17-year-old desk clerk of the motel and one of them told the boy that he was Benson in room 233 and had lost his key. Because there was no extra key available for 223 and a Benson was registered to the room, the clerk disregarded motel policy and gave the man the pass key which opened all of the room doors.

The clerk later saw the men in the hall, one with a gun in his waistband, and became concerned. He then contacted room 223, and learned that Benson knew nothing about the key request.

A police detective confirmed that a key had been used to open the door to the Kraaz room before the chain lock was broken. The Kraaz robbery was not the first security problem the motel had experienced. Another guest, also involved with race horses, reported an attempted robbery of his room the night before the Kraaz incident. He had cracked open his door in response to a knock and a voice saying that it was the motel manager. He was able to shut it quickly when he saw that the man at the door was wearing a ski mask and gloves. The guest reported the incident to the desk clerk. Although the clerk claimed to have called the police, no police investigated.

After the robbery, Mrs. Kraaz became paranoid, lost weight and began relying on sleeping pills and tranquilizers. She refused to travel with her husband, although she had done so for ten months of every year prior to the robbery. Mrs. Kraaz was described by a psychiatrist who examined her a few days after the robbery as "extremely frightened, apprehensive, and essentially unable to function." The doctor also reported that Mrs. Kraaz had experienced several traumatic incidents as a child which she had suppressed until the robbery; it was his opinion that her life would have been normal had the robbery not occurred.

The Kraazes' marriage became impaired by Mrs. Kraaz's unstable condition. The couple consulted the psychiatrist, who recommended treatment at least twice a week for the next two years, and possibly longer.

On appeal from the judgment granted the Kraazes, the motel tried to rely on two statutes which limit innkeeper liability. One requires an innkeeper to make a safe available to guests for the deposit of valuables and to notify guests by a "conspicuously posted" notice of the availability of a safe. The notice also indicates a limitation of liability to $100 if the guest suffers a loss, except "through the fraud and negligence of the landlord, or some clerk or servant provided by him in such inn or hotel." A second statute exempts the innkeeper from responsibility "for what is stolen by force or arms, or with exterior breaking open of doors, or by any other extraordinary violence."

The motel manager claimed that a notice was posted on the door of the room which informed guests of the availability of a safe and included the applicable Louisiana statutes. The Kraazes testified that they were not aware of the notice, and several witnesses could not remember seeing any notice. The court was not satisfied that adequate notice had been given.

The Supreme Court of Louisiana affirmed the trial judgment and ruled the statutes did not apply because the forced entry was accompanied by gross negligence on the part of the innkeeper and its employees. The pass key was the critical element in the robbery, and it was given to the robbers by the desk clerk, while the guard slept nearby.

The court noted that a motel guest is due a "high degree of care and protection" by the innkeeper:

> The innkeeper has a duty to take reasonable precautions against criminals. Safeguarding the room keys is a minimum requirement. The duty to avoid handing a pass key to any stranger is even stronger. Breach of this duty was a direct cause of plaintiffs' physical, emotional, and financial damages....The [innkeeper] is liable for the resulting damages.

Two justices dissented, stating that the testimony supported adequate notice and agreeing with the motel's argument that plaintiffs were contributorily negligent for keeping such large sums of cash in the motel room. As frequent travelers, they should have been familiar with the availability of motel safes. Nonetheless, the majority affirmed judgment to the Kraazes, because "mere possession of money does not constitute negligence."

COMMENTARY:

The onerous responsibility associated with cautious delivery of keys applies not only to innkeepers, but to virtually every security guard, landlord, and other business person who has access to pass keys or other limited access keys.

. Are your personnel trained to make appropriate inquiries before relinquishing any keys, such as requesting and confirming identification?

. Do you have firm and clearly articulated procedures for dealing with lockouts, including alternatives of having the guard or clerk retain the key and unlock the door for the guest?

. Are our pass keys properly coded so that unauthorized duplications cannot be made?

The sleeping guard in this case compounded the error. Careful selection of personnel is obviously critical. A security guard working on a part-time basis could already be exhausted from daytime labors. Appropriate investigation before hiring could reduce that risk.

PSCLR–OCT 82 (5)

Hotel Not Liable for Causing Rape of Guest is Nevertheless Liable for Aggravation of Injuries Caused by Delay in Aiding Victim:

Jackie Bernice Boles v. La Quinta Motor Inns, a Texas corporation, *No. 81-2098, in the United States Court of Appeals for the Fifth Circuit, on appeal from the United States District Court for the Southern District of Texas, decided July 21, 1982, reported at 680 F.2d 1077.*

Although a motel was not found negligent when a motel guest was raped and robbed in her room, it was held liable for $78,000 in damages when motel employees refused to come to the guest's assistance after the attack.

Jackie Boles was raped when a man forced his way into her room at a La Quinta motel. Threatening to return and kill her, he left Boles tied and gagged in the darkened room. She was able to knock the telephone off the hook and alert the front desk clerk to her attack and to her attacker's threat to return and kill her. The clerk and other unidentified motel employees questioned Boles over the phone before calling the police. After she was finally told by the clerk that the police had been called, Boles was asked to "hang up the telephone." When she replied that she was unable to find the telephone to hang it up, the clerk retorted in a sarcastic manner.

The police arrived about twenty-five minutes later to find three La Quinta managers outside Boles' room door while her screams and cries could be heard from within the room.

Boles' subsequent lawsuit against La Quinta alleged that the motel failed in its duty to her as a guest by failing to provide a safe, well-lit hallway and by not coming to her aid when she called. Although the trial court agreed that the hallway was not safe, it decided the condition of the hall did not lead to the rape. However, the motel employees' behavior in cruelly ignoring Boles' cries for help was determined to be the cause of aggravating Boles' injuries. In fact, psychiatric testimony indicated that the time delay between the rape and her rescue led to "a greater impact and more injury and symptoms." Further testimony showed that helping Boles would not have put the employees in any danger.

Since the employees knew that Boles was in the room alone and was obviously suffering from both her injuries and fear, their lack of action in rescuing her was the cause of her damages. The judgment of $35,000 for present damages and $43,000 for future damages was therefore affirmed by the U.S. Court of Appeals.

PSCLR–OCT 82 (6)

Absent Special Facts Requiring Higher Standard of Care, Hotel Defeats Liability Claim by Proving Compliance with Safety Ordinance:

Helen Montgomery and Kenneth Montgomery v. Royal Motel, *No. 12884, Supreme Court of Nevada, decided June 3, 1982, reported at 645 P.2d 968.*

Because a Las Vegas motel met minimum security standards set forth in a city ordinance, it was able to avoid liability to motel guests who were assaulted in their room.

Helen and Kenneth Montgomery were attacked and robbed shortly after returning to their room in the Royal Motel in Las Vegas, Nevada. Although the Montgomerys testified they normally kept their door locked, the only lock – a deadbolt latch – was not engaged when their attacker entered the room. Because the door was not also furnished with an automatic door lock, the Montgomerys sued to recover for injuries they suffered from the assault. The Montgomerys claimed their injuries were a result of Royal Motel's failure to protect them from criminal attack.

The Supreme Court of Nevada agreed with the trial court's decision that Royal Motel had adequately complied with the Las Vegas ordinance concerning minimum hotel security standards. The ordinance requires motels to provide room doors with deadbolt latches; automatic locks are not required. Moreover, the testimony at trial indicated that no prior criminal attacks had necessitated tighter motel security than was currently being used, and that the Montgomerys knew that their door would not lock by itself. Royal Motel therefore was held not negligent in carrying out its duty to protect the Montgomerys.

COMMENTARY:

It is important to note that the hotel's compliance with the ordinance meant that the hotel had met only the minimum standards of care for protecting its guests. Therefore, if the hotel had failed to comply with the ordinance, it probably would have been held liable per se to the plaintiffs and the only question would have been how much damages should be awarded for the plaintiffs' losses and injuries.

On the other hand, as the court noted, since the ordinance prescribed only a minimum standard, the plaintiffs still may have been able to recover if they had been able to show that some special circumstance existed which required that the hotel be held to a higher standard of care. Mere compliance with local ordinances alone, therefore, will not always protect a defendant from liability; whereas a failure to comply will always subject it to liability.

PSCLR – NOV-DEC 82 (8)

Hotel May Be Liable For Failure To Warn Guest And To Prevent Criminal Assault Where Prior Criminal Activity In Area Should Have Made Attack On Motel Guest Foreseeable To Owner/Operator:

Tomas H. Urbano v. Days Inn of America, Inc.; Office Parks of Charlotte, Inc., d/b/a Days Inn Budget Luxury Motels; Commercial Management, Inc.; and Days Inn Budget Luxury Motels, *No. 8126SC1121, in the Court of Appeals of North Carolina, decided September 21, 1982, reported at 295 S.E.2d 240.*

Where it is reasonably foreseeable that criminal activity will occur on property, a possessor of land may be required to protect invitees from attack. The key to required security lies in the foreseeability of criminal attack occurring on the property. However, a possessor of land cannot be considered an insurer of invitees' safety.

In this case, motel guest Tomas Urbano was assaulted by unknown criminals in a Charlotte, North Carolina, Days Inn parking lot. Urbano had registered late at night and was getting his belongings out of his car when the attack occurred.

Urbano sued the motel and its owners to recover for his injuries and damages suffered as a result of the attack. His suit alleged that damages occurred because the lot was improperly lighted, was not enclosed by a fence or other protective measures, and was not guarded or monitored by motel personnel (the motel relied on routine police visits). Moreover, guests were not warned that the motel parking area was hazardous, in spite of the fact that 42 criminal incidents had taken place at the motel in the three years preceding Urbano's attack, including one armed robbery, and seven illegal entries (12 of these incidents had occurred in the three and one-half months immediately preceding this incident). The suit further alleged that Days Inn had provided guests with warranties of safety, through advertising and other representations. At trial, judgment was granted to Days Inn by the trial judge, who summarily dismissed the case and denied the claimant a jury trial.

On appeal, the Court of Appeals of North Carolina reviewed a motel's duty to protect its guests from attack and a motel's duty to warrant a guest's safety. The court, relying on past North Carolina appellate decisions, said that "foreseeability is the test in determining the extent of a landowner's duty to safeguard his business invitees from the criminal attacks of third persons," and that "a parking lot provided by a business owner for the use of his invitees is considered part of the business premises." Quoting from an earlier case, the court stated:

> [T]he conduct of hotel innkeepers in providing security must conform to the standard of ordinary care. In the context of the hotel-guest relationship, it is foreseeable that an innkeeper's failure to maintain adequate security measures not only permits but may even encourage intruders to rob or assault hotel patrons. *Peters v. Holiday Inns, Inc.*, 278 N.W.2d 208 (1979)

Because it saw questions of fact concerning the alleged deficiencies in the motel's security based on the criminal activities history, the court remanded the case to the trial judge for a jury trial. The court, however, denied Urbano's claim of an implied warranty of safety, stating that: "Under our decisional law, an innkeeper or other occupier of land is not the insurer of the personal safety of business invitees."

COMMENTARY:

Once again we note that the courts are requiring owners and occupiers of land, and operators of businesses, to take reasonable precautions to warn patrons and guests of possible criminal attacks, and to employ reasonable efforts to prevent such attacks, where a pattern of criminal activity has shown that attacks are likely to occur.

As was noted by the North Carolina appellate court in the above case, landowners and business proprietors may not be able to rely on the efforts of public law enforcement personnel to protect their guests from criminals. If the history of criminal activity in the area is sufficient to indicate that police efforts alone will probably not be successful in preventing attacks, then the owner/operator must undertake security measures on his own, or incur the consequences of liability verdicts against him.

With regard to such verdicts, it should be noted that many juries returning verdicts in favor of injured plaintiffs are beginning to include an additional award for punitive damages, and, while the negligent owner/operator may be able to cover some of his losses in these cases by liability insurance, some policies of insurance do not cover punitive damages but only actual damages (e.g., medical, property damage, pain and suffering, etc.).

PSCLR – NOV-DEC 82 (14)

Guest Who Deposits Valuables for Safekeeping Entitled to Recover Full Actual Value, and Limitation on Liability of Hotel Inapplicable for Noncompliance with Statute:

Sarah Zacharia v. Harbor Island Spa, Inc., *No. 611, No. 81-7558, in the United States Court of Appeals for the Second Circuit, on appeal from the United States District Court for the Eastern District of New York, decided June 25, 1982, reported at 684 F.2d 199.*

In Florida, a hotel can limit its liability for the negligent or accidental loss of a guest's valuables held for safekeeping by complying with a statute which states:

> [L]iability ... shall be limited to $1,000 for such loss, if the [hotel] gave a receipt for the property (stating the value) on a form which stated, in type large enough to be clearly noticeable, that the [hotel] was not liable for any loss exceeding $1,000 and was only liable for that amount if the loss was the proximate result of fault or negligence of the operator. Fla. Stat. § 111(1) (1979).

However, a United States Court of Appeals held that a hotel must comply "strictly" with the statutory requirements or be barred from claiming the benefit of the $1,000 limitation on its liability.

When Sarah Zacharia registered at the Harbor Island Spa Hotel in Miami Beach, she signed a registration card which included a notice which read: "HOTEL'S LIABILITY IS LIMITED AS PROVIDED IN POSTED 'IMPORTANT NOTICE TO GUESTS'." Later, she deposited her valuables with the hotel for safekeeping and was asked to sign two cards containing language certifying that the value of the deposited items did not exceed $1,000, that the items deposited would not exceed $1,000 in value at any time during the stay, and (with a cite to the Florida statute) that the guest waived any claim over $1,000 for a loss of deposited items.

Mrs. Zacharia filled in the blanks on the first card (her name, the hotel's name, and the date of deposit) and signed both cards. Similar blanks on the second card were not filled in, and the amount of limitation was marked out on the first card. Furthermore, the hotel did not give her a receipt for the deposited property, and the hotel clerk retained both cards.

During Mrs. Zacharia's stay at the hotel, a clerk absconded with the contents of several of the hotel's safety deposit boxes, including Mrs. Zacharia's valuables. When she submitted her claim for lost items in excess of $10,000, and her claim was rejected by the hotel, she sued the hotel's owner. The trial judge ruled that the hotel's liability was limited by the statute to only $1,000 and rejected her suit.

On appeal, a United States Court of Appeals reversed the dismissal and remanded the case for trial. The court based its decision on the requirements of the Florida statute and on recent Florida appellate court decisions which "emphasize that the burden of compliance is on the hotel rather than on the guest, since the hotel has 'superior position and knowledge ... with regard to the mandates of the statute'," and that compliance by the hotel must be "strict."

The court pointed out that evidence did not show that Mrs. Zacharia had been given a receipt for the items she deposited as is required by the statute. In addition, the hotel's casual attitude toward the filling in of the two deposit cards indicated that Mrs. Zacharia was not made fully aware of the content of the cards. Therefore, the $1,000 statutory limitation of loss would not apply in this instance, and the case was remanded to the trial judge for a trial to determine the actual amount of Mrs. Zacharia's loss.

COMMENTARY:

Not only hotels, but also hospitals, financial institutions, security houses, private vault operators, and others, routinely accept valuable deposits for safekeeping. In most instances, applicable state statutes limit their liability to a set amount, typically $1,000 as in the above case. However, as is seen here, this limitation on liability is completely worthless to those institutions whose personnel fail to comply strictly with the requirements of the statute.

Have you reviewed the particular requirements of your state's statute limiting your company's liability? Have you checked recently for any amendments to the statute, or case law interpretations of the statute, which may have changed the requirements? And, equally important, have all employees (including new employees) received adequate training and recent reminders to follow the specific procedures mandated by the statute?

PSCLR—JAN 83 (11)

Statutory Limitation of Innkeeper's Liability for Loss of Guest's Property Inapplicable to Theft of Guest's Car and Contents from Motel Parking Lot:

Charles T. Vilella v. Sabine, Inc., et al., *No. 57722, Supreme Court of Oklahoma, decided July 27, 1982, rehearing denied November 1, 1982, reported at 652 P.2d 759.*

On February 3, 1979, Charles Vilella's two sons stopped at an Oklahoma City motel operated by the defendant Sabine, Inc. They were moving their father's personal property to Pennsylvania in a truck and trailer and inquired about a safe place to park the vehicle. They were told that the motel had an open parking lot which was patrolled by a security guard at night. They left the truck and trailer there, and the vehicle and its contents were stolen during the night. Vilella sued the motel owners.

Oklahoma law provides that:

> An innkeeper ... is liable for all losses of, or injuries to, personal property placed by his guests ... under his care, unless [caused] by an irresistible superhuman cause, by a public enemy, by the negligence of the owner, or by the act of someone whom he [the owner of the property] brought into the inn. ...

In construing this statute, the court ruled that an innkeeper is made "a virtual insurer of the safety of property entrusted to his care by a guest." The law does so not only because of the traveler's vulnerability, but also because the innkeeper is in a better position to compensate for his losses by regulating his room rates. Also, in order to find the motel liable for loss, the court determined that it is not necessary that the property be placed under the motel's exclusive care, such as in its safety deposit box. It is only necessary that the property be brought into the motel in the normal way and that it not remain under the exclusive control of the guest; it is enough if the property is in the "general and implied control" of the motel. Since Vilella's vehicle and its contents were in the motel's guarded lot, the property was under the care of the motel, and the defendant is liable for its loss.

Another question for the court in this case was whether the $250 statutory limitation of an innkeeper's liability applied to this loss, or whether Sabine was liable for the full value of the loss. The applicable statute relied on by Sabine provides:

> [A hotel] proprietor ... shall not be liable for the loss of or damage to per-

sonal property brought into such hotel ... by any of the guests thereof exceeding $250 in value. ...

The statute goes on to provide limits of liability for loss of specific items such as trunks ($100), valises ($75), boxes ($25), and so on. But nothing is mentioned about vehicles and their contents. Therefore, the court ruled, since no limit on liability was set for the loss of vehicles, Sabine would be liable for the full value of Vilella's loss.

COMMENTARY:

This case does not involve any claim of negligence on the part of the hotel/motel owner. That would be a harder case for the plaintiff to prove. As the court here ruled, the motel operator is a "virtual insurer" of the guests' property in its care, and the plaintiff need only show the value of the loss incurred.

This case undoubtedly will come as unwelcome news to innkeepers and their insurance carriers, especially in Oklahoma. Not only are they held "automatically liable" for the theft of a guest's vehicle in their care, but also the $250 limitation of liability does not apply.

Are you familiar with the similar statutes in your own jurisdiction(s)? And, equally important, are you aware of the most recent court interpretations of those statutes? You may be under the impression that a statute limits your liability to a nominal amount only, when in fact a court may have ruled otherwise. If you have any doubt, consult your attorneys.

PSCLR – MAR 83 (5)

Corporation Cannot Recover From Hotel For Lost Profits And Other Damages Caused By Wrongful Death Of Its Corporate Manager In Hotel Fire:

Arrow Electronics, Inc. v. The Stouffer Corporation, et al., in the Supreme Court of New York, Special Term, New York County, Part I, decided December 20, 1982, reported at 458 N.Y.S.2d 461.

A fire broke out on the third floor conference wing of the Stouffer's Inn of Westchester, located in Harrison, New York, on December 4, 1980, shortly after 10 AM. The fire spread rapidly, and 26 people were killed. Among those who died were 13 Arrow Electronics, Inc., employees.

Arrow sued the owners of the inn, the Stouffer Corporation, and the William L. Crow Construction Company, which had been retained by Stouffers for the planning, design and construction of the inn, and a firm of consulting engineers who performed the mechanical engineering and electrical work on the inn.

Arrow alleged that in October 1980, Stouffers agreed to rent guest and meeting rooms to Arrow for its senior level management annual budget metings to be held between November 30, and December 6, 1980. Arrow also alleged that the agreement implied that the rooms were suitable and safe, reasonably free of fire hazards, and reasonably equipped with devices to minimize the danger of fire. Arrow sought $5,000,000 compensatory damages plus $5,000,000 punitive damages for the loss of its management; costs of recruiting new employees; loss of books, records, and papers; death benefits paid to the widows and estates of deceased employees; and lost profits.

The defendants all moved to dismiss the claims of Arrow, except the claim for loss of books and records, which defendants agreed was a viable claim.

The trial judge examined the status of the law on this issue in New York, and ruled that Arrow Electronics was in essence attempting to pursue a wrongful death claim. However, claims for damages for wrongful death belong exclusively to the personal representative of the decedent, according to a New York statute. Furthermore, limiting recovery to ther personal heirs of the decedent is the wise course, according to the court, "if recovery could be had by all who are injured in fact, the resources available to compensate...the survivors...would be diluted."

The trial court dismissed Arrow's contention that this was an action for injury to the corporation, and therefore dismissed Arrow's claims for damages or lost profits and other compensatory and punitive damages resulting from the deaths of the employees.

PSCLR—APR 83 (3)

Motel Is Not Liable For Guest's Injuries From Criminal Attack Unless Motel's Negligence Made Attack Possible:

Allen Zang v. Eddie Leonard, Jack C. Spence, and Helen P. Spence, Individually, and as Partners d/b/a Spence Properties, Spence Manor Motor Hotel, Spence Motels, Inc., and Motel Systems, Inc.; *in the Court of Appeals of Tennessee for the Middle Section at Nashville, decided June 1, 1982, permission to appeal denied by Supreme Court, October 4, 1982, reported at 643 S.W.2d 657.*

A resident of Spence Manor Motor Hotel in Nashville, Tennessee, Allen Zang, arrived home on May 30, 1977, and parked his car in the motel lot. While removing his luggage, Zang was approached by a man who robbed and shot him, and then stole his car. Because Zang felt the motel's security precautions were inadequate to protect guests from such attacks, he sued its owners, Jack C. Spence and Helen Spence personally, and as partners operating Spence Properties and Spence Manor Motor Hotel, as well as Spence Motels, Inc., and Motel Systems, Inc. At the trial, a jury awarded Zang $40,000 damages; the defendants appealed.

In reviewing the laws applicable to this situation, the appellate court first made it clear that a property owner is not an insurer of guests against injuries caused by third persons. Land owners/possessors would be liable for misconduct of third persons only if the owner/possessor's own negligence was the actual cause of an injury. To determine whether an owner/possessor was negligent in a particular case, the facts and circumstances of that case would have to be examined and a finding would then be made as to what steps a reasonable and prudent owner would have taken under those same facts and circumstances. If the actions of the owner in question conformed to the standard of the "reasonably prudent" owner, then the owner in question would not be considered negligent and therefore could not be held liable to a guest injured on the property.

The appellate court continued, stating that for an owner to be held liable to a guest who had been injured by the criminal act of another, proof that the owner had actual notice of a specific, threatened criminal act by a specific person would not be necessary. It would be sufficient to prove that the guest's injury at the hands of a third party was made possible by the failure of the owner to exercise reasonable care under the circumstances (i.e., to be negligent) in protecting the guest.

Furthermore, the court pointed out that the type of services and accommodations advertised to prospective guests affects to some extent the standard of care the motel must provide its guests. The quality of care represented to the public would have to be considered in a determination of a motel operator's negligence.

In an effort to prove that the Spence Manor defendants were indeed negligent, Zang's lawyers called the criminal assailant himself as a witness. The robber testified that he would not have attacked Zang if the lobby of the hotel had had better lighting and if other people had been present. However, the appellate court ruled that such evidence was not admissible because the issue was not whether the attacker would or would not have attacked, but what could and should have been anticipated by the defendant hotel owners.

In conclusion, the court dismissed the suit against Mrs. Spence since she was only a stockholder in the corporation and did not perform any management duties, and the suit against Spence Motels, Inc., since nothing in the record justified a verdict against that corporation. The case was reversed and returned to the trial judge for retrial against the other defendants.

COMMENTARY:

Cases against innkeepers for injuries or losses suffered by guests are not new to the law, but they do seem to be proliferating at a rapid rate.

The above case articulates a new element in the determination of whether an innkeeper was or was not negligent in a particular situation, and that is: the type of services and accommodations advertised to prospective guests. It seems that the court is saying that the quality of care impliedly promised to prospective guest by a hotel's (or hotel chain's) advertising agency can and will have a direct effect on the standard of care to which the courts will hold the innkeeper responsible in protecting the lives and property of the hotel's guests.

The standard remains the same: "that degree of care which would have been exercised by a reasonably prudent innkeeper under the same or similar circumstances." But what the court is saying is that the "same or similar circumstances" to be considered in determining what care the reasonably prudent innkeeper would have exercised will vary depending upon the type and quality of care that the particular innkeeper promised to the public through its advertising. That is to say, the "reasonably prudent innkeeper" who has implied to the public through advertising that he will "take better care of "guests than his competitors will be required by the law to do just that.

PSCLR—APR 83 (11)

Hotel Guest's $4,000,000 Suit Against Architect for Design Deficiencies Allegedly Resulting in Death of Guest's Wife Is Barred by Statute:

Dr. John W. Yarbro, surviving spouse of decedent Geraldine Yarbro v. Hilton Hotels Corporation, Third-Party Plaintiff v. I.M. Pei, individually, and I.M. Pei & Partners, a partnership, formerly known as I.M. Pei and Associates, Third-Party Defendants, *No. 80SA274, in the Supreme Court of Colorado, En Banc, decided December 13, 1982, rehearing denied January 10, 1983, reported at 655 P.2d 822.*

Drs. Geraldine and John Yarbro attended a medical convention being held at the Denver Hilton Hotel in May of 1977. Geraldine Yarbro presented one paper at the convention and was scheduled to present another on May 18. On May 17, the Yarbros enjoyed a cocktail party and retired to their fifteenth floor room at about 11 PM While John had his back to his wife and was removing his coat, Geraldine apparently tripped over the radiator which was located at the base of the room's window. She fell through the window glass and to her death.

Dr. John Yarbro sought to recover damages from the hotel for his wife's death in the amount of $4,000,000. His suit alleged that his wife's fatal fall was either caused by her coming in contact with the poorly situated radiator and losing her balance or by her tripping over the radiator and then through the window which could not withstand the impact of her body against the glass. Evidence indicated that the room's windows were one-eighth inch thick glass set in pre-cast concrete and were not equipped with any type of protective device, guards, or screens. Further, the radiator projected about 15 inches from the wall under the window, leaving very little room between it and the bed.

Hilton in turn sued I.M. Pei and his professional architectural firm which had provided design services for the building in 1960, stating that the architect and not Hilton was responsible for Dr. Geraldine Yarbro's death. Dr. John Yarbro also sued Pei directly, alleging certain design deficiencies on the part of the architects.

However, a Colorado statute prohibits lawsuits against an architect for "damages for injury to a person or property caused by the design [or] planning...of any improvement to real property brought more than ten years after the substantial completion of the improvement to the real property...." Because Pei's work on the Hilton was unquestionably completed nineteen years before this wrongful death action was commenced, judgment was granted in his favor.

Dr. Yarbro attacked the constitutionality of the statute, but without success. The appellate court affirmed the judgment dismissing the architects from the case, and the case by Dr. Yarbro against Hilton alone will now proceed to trial.

COMMENTARY:

Every astute property owner, developer, and manager knows that security problems include much more than security guards and burglar alarms alone. This case illustrates the fact that the design of a structure is itself a security concern. While this particular case relates more to an engineering or design defect, it presents an opportunity to emphasize that the skill and expertise of security professionals should be utilized in the design and construction stages of all new and renovated properties to insure that matters such as traffic flow patterns, adequate locks, and similar security concerns can and will be addressed during these stages.

PSCLR – MAY 83 (14)

Motel's Installation of Safety Devices Following Rape of Guest Generally Cannot Be Used to Prove Motel's Negligence, But May If Motel Denies Feasibility of Devices:

Linda Anderson, Derriel Anderson v. James E. Malloy, Peter Zes, Edward J. Gibson, *No. 81-1923, in the United States Court of Appeals for the Eighth Circuit, decided March 9, 1983, reported at 700 F.2d 1208.*

During the months of January and February, 1979, Linda and Derriel Anderson were staying in an apartment building which had been converted to long and short-term motel accommodations. On the night of February 7, Linda Anderson was alone in the room when she heard a knock at the door. She asked who was there, heard a muffled voice which she thought was her husband's, and opened the door to an unknown man who forced his way into the room and raped her.

The Andersons filed a suit which alleged that the motel owners (James Malloy, Peter Zes, and Edward Gibson) had been negligent by failing to provide safe lodging to them as guests and by misrepresenting the level of security provided to guests. The jury returned a verdict for the motel owners.

The Andersons appealed the verdict because they felt the trial judge had erred in refusing to allow certain evidence to be presented to the jury. The evidence that the judge would not admit was certain testimony (1) concerning the security measures which were used by other hotels and motels in the same area as the defendants' motel, (2) concerning the fact that the defendants had installed safety chains and peepholes in their motel's guestroom doors after Anderson's rape, (3) concerning the fact that an apartment complex next to the motel was a "breeding ground for crime," and (4) by a woman concerning her own rape at the same motel five months before Anderson was raped.

The U.S. Court of Appeals for the Eighth Circuit did not agree with the trial court's ruling that the evidence as to security measures taken by other area hotels and motels was not relevant because of the "unusual physical layout of the defendant's motel." The defendants' motel had been converted from an apartment building and did not have interior hallways, upstairs units, a restaurant or similar facility, unlike other motels/hotels in the area. The Court of Appeals held:

> The defendants' motel and other hotels and motels serve the same purpose in the same way, they are located in the same area of the same city, and customers rely equally on each hotel and motel to provide proper security. We do not find that the difference in physical layout between the defendants' motel and others [invalidates] the relevance of evidence concerning the security measures of others. ...

On this issue, the appellate court noted that evidence of the customary practices of similar businesses does not establish the legal standard of care which a motel owner owes his guests. The standard is "reasonable prudence" and a defendant may be found to have failed to meet the standard (i.e., to have been negligent) even though he conformed to the customary practices of the industry. Likewise, the fact that a defendant did not conform to customary trade practices does not establish conclusively that the defendant was negligent.

However, customary practices by similar businesses such as other motels in the area should have been admitted into evidence, according to the court, because that testimony would have some evidentiary value to the jury in determining whether the defendants met the standard of "reasonable prudence."

The Andersons also argued on appeal that they should have been able to show the jury that the defendant motel owners had installed door chains on and peepholes in the guestroom doors after the incident. The motel owners argued that the door chains and peepholes could not prevent crime and would actually create a false feeling of security which would actually make the guest less secure. The court noted that evidence of remedies which are put into place by defendants after an alleged act of negligence is generally not admissible to prove defendants were negligent prior to taking the remedial measures. However, in this case, the motel owners had taken the position, and had testified at trial, that the use of door chains and peepholes was not feasible, thereby indicating to the jury that such devices were not "capable of being ... utilized, or dealt with successfully." The court said:

> For the defendants to suggest that installation of peepholes and chain locks would provide only a false sense of security not only infers that the devices would not successfully provide security, it also infers that the devices would in fact create a lesser level of security if they were installed.

Therefore, the court said that the Andersons should have been allowed to show the jury that the safety devices of peepholes and door chains were in fact feasible by showing that, although the motel owners had testified that they had done everything necessary for a safe and secure motel and that peepholes and door chains would not be successful, they had in fact installed these exact devices after Linda Anderson was raped.

The Court of Appeals affirmed that the trial court's ruling that evidence concerning crimes at the adjoining apartment building was not "sufficiently relevant to justify admission." Moreover, the testimony by the earlier rape victim at the motel was not considered relevant to this suit.

Therefore, the Court of Appeals returned the suit to the trial court for another trial because evidence concerning the motel owners' subsequent security measures had been denied admission at the original trial.

COMMENTARY:

As pointed out by the court's decision, it is generally true that evidence of remedial measures is inadmissible. A party who is charged with negligence would not be inclined to fix his unsafe property or product (assuming he knows it to be unsafe and in need of fixing) if his own improvements could then be used aginst him to prove his negligence.

The Anderson case here should not be considered as a change in this longstanding legal doctrine. The law that evidence of remedial measures taken by a defendant after an alleged act of negligence is inadmissible is still good law. The only reason such evidence became admissible here was because the defendants' own attorney had elicited testimony from one of the motel owners indicating that the safety devices should not be installed and would not be successful. This testimony "opened the door" for the plaintiffs to prove the opposite.

Certainly any conscientious hotel/motel owner or operator would and should initiate remedial safety measures when the need for them becomes apparent, and the evidence of such will generally not be admissible to prove that he was negligent at an earlier time before the safety measures were in place.

PSCLR – JUN-JUL 83 (4)

Hotel Cannot Qualify for Statutory Limitation of Liability for Theft of Guest's Valuables Unless It Establishes Adequate Facilities for Safekeeping of Valuables:

Eliza M. Goncalves, also known as Eliza G. M. Salles v. Regent International Hotels, Limited, doing business as Mayfair Regent, et al. (Action No. 1), and Sara Cecconi v. Regent International Hotels, Limited, doing business as Mayfair Regent, et al. (Action No. 2), *in the Court of Appeals of New York, decided February 17, 1983, reported at 447 N.E.2d 693.*

The Mayfair Regent is a luxury hotel located in Manhattan in New York City and is owned by Regent International Hotels, Ltd. In November, 1979, two guests of the hotel, Eliza Goncalves and Sara Cecconi, were traveling with jewelry collections reportedly worth $1,000,000 each. Both women turned the jewels over to the management of the hotel for safekeeping and each signed and received a "Safe Deposit Box Receipt" which described the terms and conditions of the deposit:

> In consideration of the privilege herewith granted me by Mayfair House whereby I am allowed the sole use of an individual safe deposit box for the keeping of my valuables ..., I hereby agree that the liability of Mayfair House for loss or damage to the contents, at any time, of such safe deposit box, by theft, burglary, or otherwise, is and shall be limited at all times to a total amount of $500, and, for the purpose of determining any liability of Mayfair House for any such loss or damage to the contents of such safe deposit box, I hereby agree that the value of the contents of said box shall never exceed at any one time the sum of $500. ...

The New York statute on which this agreement was based limits the absolute liability of innkeepers for the loss of guests' valuables. (Prior to the enactment of this statute in 1855, an innkeeper was held totally responsible as an insurer of his guests' valuables, regardless of the amount.) According to present law, an innkeeper's liability can be limited to $500 for any loss of a guest's property in his possession (unless the guest and the hotel came to another agreement in writing), if he provided "safe in the office of such hotel ..., or other convenient place for the safekeeping of any money, jewels, ... " and if he had notified guests that such facilities were available by "posting a notice ... in a public and conspicuous place," as in the office and public rooms of the hotel. The statute also states that the hotel could not be held liable if the guest refused to deliver valuables to the hotel for safekeeping, and that the hotel could not be required to receive a guest's property for deposit if the value exceeded $500.

Notices were posted in the Mayfair Regent, as required by New York law, informing guests that safety deposit facilities were available for the storage of guests' valuables. The particular facilities provided by the Mayfair were rows of safety deposit boxes located inside a plasterboard room. Access to the room was gained through two hollow-core wood doors, only one of which was equipped with a lock – an ordinary tumbler type lock. Access to the boxes required two keys, one held by Mayfair and one held by the guest. Mayfair also kept a card file which contained information concerning guest use of the boxes, including dates of deposit and removal of valuables.

On November 25, 1979, the safe deposit room was broken into by thieves and some of the boxes, including those holding Ms. Goncalves' and Ms. Cecconi's jewelry collections, were emptied. The Mayfair's owner, Regent International, refused to cover the guests' losses beyond the statutory limited liability of $500 as stated in the "Safe Deposit Receipt" and the two women brought separate actions against the hotel owner for $1,000,000 in damages each. Their suits were later consolidated. The New York trial court entered summary judgment for Regent International and awarded the plaintiffs $500 each; Ms. Goncalves and Ms. Cecconi appealed.

The suits against the hotel contended that the hotel could not be protected by the statutory liability limitation because the hotel's negligence in providing security was the proximate cause of the theft. The plaintiffs argued that the hotel had not provided a "safe" within the meaning of and as required by the statute and therefore could not claim the benefits of the statute. The hotel defended itself by contending that the statute limited its liability to $500 regardless of the cause of a loss, and argued that it did in fact provide a "safe" as required, but that in any event the statute did not even require a safe as such, but allowed for "[an]other convenient place" to satisfy the condition.

The New York Court of Appeals noted that the statute must be strictly followed in order for a hotel to benefit from the limitation of liability. Limited liability would be granted to a hotel which had strictly adhered to the conditions of the statute even if some other negligent act caused the loss of guest property. However, negligence by failing to satisfy the conditions of the statute would remove the protection. In this regard the court noted that a hotel which provided no safe at all, or which failed to post the notices conspicuously, or provided a poorly-constructed receptacle as a safe, could not claim the benefits of the statute's protection and would be fully responsible for a loss, as innkeepers were before the passage of the statute in 1855.

Therefore, the court then addressed the question of the type of facility to be considered a "safe" within the meaning of the statute. Given the court's admitted lack of expertise in the field, it held that uniform technical specifications would be inappropriate: the security requirements of luxury hotels differ greatly from those of small, budget motels because different clientele and different areas are served by each. In consideration of this, the court determined that for a "safe" to fall within the meaning of the statute, it "should be a receptacle that, under the circumstances, provides adequate protection against fire, theft, and other reasonably foreseeable risks. In deciding this question, all aspects of a hotel's security may be considered."

The plaintiffs submitted an affidavit by a security expert with 29 years of experience in designing, installing, and selling safes and vaults. It was the expert's opinion that safe deposit boxes could be easily entered in less than 30 seconds, that safe deposit boxes should be housed in a locked vault and not a plasterboard room with one flimsy lock, and that the Mayfair's deposit facilities were inadequate. In light of this opinion, the court felt that a material question of fact was raised concerning whether the hotel had actually provided an adequate "safe" in accordance with the requirements of the statute, and ruled that a jury should decide whether the facility provided by Mayfair was or was not a "safe."

In sum, the court stated that a hotel owner must prove that it had provided a "safe" within the meaning of the law to invoke the protection of the statute limiting its liability in the event of the loss of a guest's property. If the hotel could not prove this, the statutory limitation of $500 per guest would be inapplicable and could not be used in the hotel's defense.

Concerning the hotel's reliance on the safe deposit box receipts signed by each of the plaintiffs, the court ruled that such reliance was ill-placed:

(1) If the hotel was found to have provided a "safe" as required by the statute, then the statutory $500 limitation would apply and the receipt-agreement would be immaterial and unnecessary to protect the hotel; and

(2) If the hotel was found to have failed to provide a "safe" as required by the statute, then the receipt-agreement would be void because it is against public policy and it would provide not protection to the hotel.

Therefore, according to the majority opinion, the defendant hotel's motion for summary judgment should have been denied. The lower courts' rulings were reversed, and the case was sent back for trial and for a jury to determine whether the Mayfair Regent's safe deposit facilities amounted to a "safe" as required by the statute. If so the hotel's liability will be limited to $500 per plaintiff; if not, the hotel will face a claim of $1,000,000 per plaintiff.

COMMENTARY:

The holding in this case constitutes a setback for the hotel industry in New York. Recognizing that the case would have important consequences to the industry, the counsel for the New York State Hotel and Motel Association filed a "friend of the court" brief on this appeal, but to no avail. Four of the seven judges sitting in hearing on the appeal agreed with the majority opinion summarized above.

The dissenting opinion, concurred in by the three remaining judges, points out the practical difficulties caused by the majority's ruling: By leaving it to a jury to decide whether a hotel's safe is in fact a "safe" as required by the statute, the court is opening the door for lay jurors to decide that question on a case-by-case basis, with possible widely different results. A jury in one town in New York could decide that an installation such was in the Mayfair Regent was a "safe," while another jury could find in another case that the identical installation did not amount to a "safe." As stated by the minority:

> *Indeed, the wholly unacceptable result of the majority's holding will be that, as a practical matter, in every case where the plaintiff alleges that no !qsafe' was provided within the meaning of the statute, the question will have to be submitted to a jury.*

The minority concluded that the court and not the jury should determine what constituted a safe, and further concluded that the Mayfair Regent's safe deposit boxes did meet the standard. These three judges would have affirmed the summary judgments for the hotel limiting damages to $500 per plaintiff.

It remains to be seen whether the majority opinion will stand, or for how long, before being changed by some subsequent appeal or perhaps by action of the legislature. But for now, it is the law of the state of New York, and it will undoubtedly be used in argument by plaintiffs seeking recoveries in other states with similar statutes.

See Zacharia v. Harbor Island Spa (November-December 1982 **PSCLR** *at page 9) for another case in which a New York hotel's defense of limited liability was refused by the court for noncompliance with the statute's requirements.*

PSCLR – JUN-JUL 83 (10)

Town Can Be Proportionately Liable to Motel for Injuries to Motel Guests If Town Breached Duty of Care Owed Motel:

Timothy S. Garrett, Individually and as Limited Administrator of the Estate of Nancy J. Garrett, Deceased v. Holiday Inns, Inc., et al., and Town of Greece, *(Four Main Actions and 20 Third Party Actions), in the Court of Appeals of New York, decided February 23, 1983, reported at 447 N.E.2d 717.*

After an extensive fire at a Holiday Inn in the Town of Greece, New York, killed or severely injured many of the motel's guests, several lawsuits were brought by those injured and the families of those killed (hereafter, "the plaintiffs"). Lawsuits were brought against Holiday Inns, Inc., which was lessor/operator of the motel; the corporation which had developed and built the motel; the past and present owners of the motel; the owner of the property, and the Town of Greece itself.

The Town of Greece was sued by the plaintiffs because the town allegedly failed to require that the motel owners comply with the town's fire and safety codes. The plaintiffs also contended that the Town of Greece had negligently issued a certificate of occupancy to the motel and had not properly inspected the motel during construction or since its completion, and that this negligence contributed to the loss of life and property in the fire. The trial court however dismissed the complaint against the municipality because it determined that the allegations by the plaintiffs suggested "no more than a violation of a general duty owed by the town to the public at large." Because the town did not owe a specific duty to the motel guests, the complaint was insufficient to support an action against the town.

Holiday Inn and the other remaining defendants then filed their own actions against the Town of Greece, claiming that while the town may not be directly liable to the plaintiffs (as the trial court had ruled), the town should nevertheless be held responsible to the other defendants for "contribution or indemnification to the extent the plaintiffs' losses were caused by the town's misconduct." A lower court, however, dismissed these claims against the Town of Greece, ruling that the town could not be liable to Holiday Inn and the other defendants without some showing that it owed some duty of care directly to the plaintiffs who suffered losses in the fire. Holiday Inn and the defendant motel owners appealed.

Stated simply, the issue of this appeal was whether the town, although previously held to owe no duty directly to the plaintiff motel guests, could nevertheless be held proportionately liable to the owners and lessee-operators of the motel for breaching an independent duty owed the motel. As a rule, proportional liability will be assigned among defendants where two or more wrongdoers "share in responsibility for an injury, in violation of duties they respectively owed to the injured person." The court noted that such joint liability could arise from a "nexus of duty between wrongdoers" which was independent of each wrongdoer's obligation to the injured parties.

The New York Court of Appeals was therefore asked to decide whether the town was obligated to the Holiday Inn's owners and lessee-operators to use reasonable care in issuing a certificate of occupancy and in approving construction alterations in order to guarantee that the building did not violate fire and safety laws. According to case law, the existence of a special duty, as opposed to a general duty owed to the public, is key to the liability of a municipality for the negligent performance of a government function. The court interpreted this principle as imposing liability

> ... where the municipality has violated a duty commanded by a statute enacted for the special benefit of particular persons ... ; where the municipality has voluntarily assumed a duty, the proper exercise of which was justifiably relied upon by persons benefited thereby ... ; or where it as-

> sumes positive direction and control under circumstances in which a known, blatant and dangerous safety violation exists ... .

The court determined that merely failing to uncover fire and safety violations during inspections of the motel was not sufficient to impose liability on the town. However, the defendants here had alleged much more against the Town of Greece. The appellate court stated:

> If, as is alleged ..., known, blatant, and dangerous violations existed on these premises, but the town affirmatively certified the premises as safe, upon which representation [the motel owners and operators] justifiably relied in their dealing with the premises, then a proper basis for imposing liability on the town may well have been demonstrated ... .

The court then concluded that in such a case the municipality clearly would be liable to the defendant motel for damages to the motel, but more importantly, the municipality would be liable to the motel for a proportionate share of the motel's tort liability to the plaintiff guests which was caused by the town's breach of duty to the motel. Therefore, Holiday Inn's claims against the Town of Greece will go to trial along with the injured plaintiffs' claims against Holiday Inn and the other defendants.

COMMENTARY:

This is one of the more procedurally-convoluted cases we have reported in **PSCLR**. *Rather than get overly involved in the procedural aspects, however, it is best to concentrate on the practical result of the case: A hotel/motel owner/operator who gets sued by an injured guest may be able to recover a share of its losses to the guest from the local government if it can be shown that the government knowingly permitted blatant safety code violations to exist on the property. Of course, as is often the case, astute legal counsel will be needed to successfully pursue such a recovery. And, as is always the case, the better course of action is to insure that your property is in compliance with all safety codes and that you not incur any losses to your guests in the first place.*

PSCLR—AUG 83 (5)

Constitutional Challenge of Statute Limiting Motel's Liability for Theft of Guest's Property Is Dismissed:

Sheldon S. Morris, Naomi J. Morris, and the Continental Insurance Company, a foreign insurance corporation, v. Hotel Riviera, Inc., a Nevada corporation, d/b/a Riviera Hotel, *No. 82-5734, in the United States Court of Appeals for the Ninth Circuit, on appeal from the United States District Court for the District of Nevada, decided April 26, 1983, reported at 704 F.2d 1113.*

The plaintiffs in this case attacked the constitutionality of Nevada's Innkeeper Statute. The statute limits a motel owner's liability to $750 (without a written agreement to assume greater liability) for the loss of or damage to a guest's property.

Sheldon and Naomi Morris along with their insurance company brought suit to recover the alleged $75,000 value of the Morrises' jewelry which was lost during the Morrises' stay at the Riviera Hotel in Las Vegas, Nevada. The defendant Riviera Hotels, Inc., ("Riviera") relied on the statute's liability limitation of $750 and requested that the district court dismiss the suit because the "amount in controversy" was less than $10,000, the minimum amount for a suit of this type brought in a U.S. district court. The Morrises challenged this statute by contending that the state could not validly limit the amount of recovery for property losses because that would violate the Fourteenth Amendment's equal protection clause by distinguishing between a suit for property loss and a suit for personal injury claims.

The court agreed with Riviera and dismissed the suit for lack of jurisdiction. The plaintiffs appealed to the U.S. Court of Appeals, which examined the constitutionality of the Innkeeper's Statute in Nevada and determined that the Nevada state legislature had passed the law for legitimate reasons. The importance of the tourism industry to Nevada's economy led the legislature to pass the law in order to encourage the operation of hotels and motels. Further, "[t]he likelihood of fraudulent claims for property loss, and the difficulty for innkeepers to prevent crimes involving personal property, are reasons justifying a distinction between victims of property loss and personal injury victims." Because there was "ample justification for the difference in treatment" between victims of the two classes of injuries, the Nevada Innkeeper's Statute withstood the equal protection challenge. Because the distinction was "rationally related to a legitimate state purpose in the sphere of local economic regulation," the court upheld the limitation of Riviera's potential liability to $750 which automatically dismissed the case from the jurisdiction of the U.S. District Court.

PSCLR – SEP 83 (1)

Hotel Found Liable to Rape Victim, Even Without Proof of Prior Such Assaults:

Virginia D. v. Madesco Investment Corp., d/b/a Bel Air Hilton Hotel, *No. 64183, in the Supreme Court of Missouri, En Banc, decided March 29, 1983, as modified on denial of rehearing April 26, 1983, reported at 648 S.W.2d 881.*

The Bel Air Hilton, located in downtown St. Louis, Missouri, was a seventeen-story hotel which has a parking garage with access to the hotel elevators. The hotel had the usual street-level lobby, including a small gift shop and two restaurants. A lower level lobby could be reached by stairs or by elevator and there were men's and ladies' restrooms on this level. This level also contained various banquet and meeting rooms.

The two restaurants in the hotel were a Trader Vic's and a Miss Hullings. On September 24, 1976, the plaintiff in this case, referred to as Virginia D., and several friends went to Miss Hullings restaurant for dinner. The restaurant had no public restrooms and patrons were directed to the restrooms in Trader Vic's or to those in the lower lobby. At approximately 5:30 PM, the plaintiff went to the lower level restroom. She saw no one in the lower lobby area or in the ladies' restroom itself. There were six to eight people setting up the banquet room for an evening function with the doors to the lower lobby open. As Virginia was about to leave the ladies' restroom, she was accosted from behind by an assailant, threatened, and then sexually attacked. A bellhop heard her screams and apprehended the attacker who was eventually convicted and sentenced for the assault. Virginia brought this suit against the hotel, alleging that its negligence had led to her injuries.

There was nothing about the assailant's appearance which would have prompted any hotel employee to detain or question him, if he had been seen entering the hotel. However, there was evidence that certain persons having no business in the hotel proper entered the hotel with some regularity and were a cause of concern to the hotel security personnel. Certain acts of vandalism and breaking into automobiles had occurred previously in the hotel garage.

But there was no evidence of recent incidents of rape, assault, robbery, or other violent crime in or about the hotel premises. Neither was there evidence of any previous criminal activity or violence in the lower lobby ladies' room area.

There was one security officer on duty 24 hours a day, supplied by a private security agency. The agency had previously recommended that additional guards be employed but this had not been done. The assigned security officer's rounds required him to be elsewhere in the hotel at the time of Virginia's attack, and he did not regularly survey the lower lobby level in any event.

The former operations manager of the hotel, Charles Powers (who had either been fired or forced to resign), testified that, during his tenure as operations manager, he had been concerned about security in the lower lobby. He had decided to lock the lower lobby restroom doors after 5:00 PM as a security measure, but this practice was discontinued after the manager of Miss Hollings restaurant complained. Powers also testified that cost factors figured prominently in decisions made about security measures at the hotel.

Mr. Baddock, who was the manager of the hotel at the time of the incident, testified that the security system could have been a bit more efficient. He recognized that anyone could walk into the lobby and that the potential for an assault was present. He had considered installing a T.V. monitor in the banquet preparation area in the lower level, but not near the restrooms.

In addition, Lt. Colonel Atkins Warren of the St. Louis police department testified on behalf of the hotel. He stated that the hotel was located in a low-crime area, as compared to the city as a whole. A security consultant testified for the plaintiff and criticized the design of the ladies' restroom, especially the corridor between the lavatory and the toilet area, and the absence of mirrors.

The plaintiff Virginia D's complaint alleged that the defendant hotel was negligent in failing to provide a restroom attendant, failing to provide uniformed security guards, failing to maintain internal surveillance systems, failing to train hotel employees in security measures, and failing to warn customers of the potential dangers. She further alleged that the hotel management knew or should have known that an assailant could conceal himself in the lower level ladies' restroom and attack a hotel customer such as herself. After considering the evidence presented at the jury trial (only a summary of which is provided above), the jury found in favor of the plaintiff and awarded her a verdict of $100,000.

The judge at the trial, however, set aside the jury verdict and entered a judgment in favor of the defendant hotel. The plaintiff first appealed to the Missouri Court of Appeals which affirmed the judge's decision by a vote of two to one. The plaintiff then took the present appeal, to the Missouri Supreme Court.

Preliminarily, the supreme court noted that the plaintiff in this case must first establish the elements of a negligence claim: existence of a duty, breach of that duty by the defendant, and a causal relationship between the breach and resulting damages. However, an innkeeper would not be held an <u>insurer</u> of his guests' safety. In this regard, the hotel's arguments were threefold: (1) The hotel owed no duty to take special precautions against an assault of the type which occurred, since nothing of this type had happened in the recent past; (2) there was no breach of duty by the hotel in failing to detect the assailant or to eject him from the property since nothing about his appearance or conduct prior to the attack indicated that he was dangerous; (3) the element of causation was absent because there was no way to be sure that the suggested security precautions, if in place at the time of the attack, would have prevented the attack.

With regard to the first defense, the supreme court found that "There is no requirement that there be at least one mugging or rape before the innkeeper is obliged to consider the possibility. The duty is one of the appropriate degree of care under the circumstances."

With reference to the defendant's second defense, the supreme court stated:

> We agree that the defendant could not be found negligent on the basis of the assailant's presence in the hotel, but only up to the point at which he

> entered the ladies' room. Had his entry there been observed by a hotel employee, an immediate duty to take action would have arisen. But the entry was not observed.

According to the court, the plaintiff then must prove that the suggested security measures could have increased the possibility that the assailant would have been discovered in time to prevent the assault. After reviewing the evidence, the court found that Virginia had presented sufficient proof to permit the jury to draw such a conclusion. The court found that her evidence presented a variety of possiblities for improving security on the lower lobby level. The plaintiff was not required to prove that the security measures would certainly have prevented the assault.

Therefore, the court concluded that the question of the defendant's negligence and the question of causation were issues to be decided by the jury, which had found in favor of Virginia and had awarded her a verdict. The supreme court ruled that the trial court should not have preempted the jury's findings by setting its verdict aside. The trial court's ruling, and the Court of Appeals' decision affirming that ruling, were reversed and the verdict for the plaintiff was reinstated

COMMENTARY:

The above decision was originally a five to two split decision by the Missouri Supreme Court. However, after one of the dissenting justices issued his opinion, one of the majority changed his position and adopted the dissenting opinion, thereby resulting in a split decison in favor of the plaintiff by only a four to three margin. The three dissenting justices criticized the majority opinion strongly, stating that it has the effect of imposing absolute liability upon the owners of hotels and motels. The dissenting opinion stated:

> *As a theoretical matter, anything is foreseeable. The law, however, must be tempered with a measure of practicality. ... Nothing about previous incidents at the hotel suggests that defendant had a reasonable basis for anticipating the attack of which plaintiff complains. In such a case defendant breached no duty owed to plaintiff.*

Thus it is seen that the plaintiff here prevailed by only the narrowest of margins.

It must be remembered that the law recognizes that there is a special relationship between hotel operators and their guests, which relationship imposes affirmative duties on innkeepers to protect the persons and property of their guests. Furthermore, the case law is clear that this duty owed by innkeepers is more extensive than the duty owed by the owner of general business premises to invitees on the property. In fact, the hotel owner's duty is owed not only to those guests who take rooms, but also to those who come onto the property for food and drink. And, this duty also extends to guests of those restaurants in the hotel proper which may be operated by lessees of the hotel and not by the hotel management itself.

PSCLR–OCT 83 (3)

Motel Ruled Not Liable For Assaults On Guests Without Proof Of Prior Incidents Or Proof That Such Incident Was Likely To Occur:

Charles Kveragas and Esther Kveragas v. Scottish Inns, Inc., Hsu Enterprises, Inc., and Tzu-Chi Hsu, *Civ No. 3-82-609, in the United States District Court for the Eastern District of Tennessee, Northern District, decided February 28, 1983, reported at 565 F. Supp. 258.*

Esther and Charles Kveragas were robbed and assaulted by three unknown assailants while staying at a Scottish Inns motel on February 16, 1982. The couple subsequently brought this suit against the owners and operators of the motel to recover damages.

The U.S. District Court heard the case, and relied on a ruling by Tennessee's Supreme Court in reaching its decision:

> There is no duty upon the owners or operators of a shopping center, individually or collectively, or upon merchants or shopkeepers generally, whose mode of operation of their premises does not attract or provide a climate for crime, to guard against the criminal acts of a third party, unless they know or have reason to know that acts are occurring or about to occur on the premises that pose imminent probability of harm to an invitee; whereupon a duty of reasonable care to protect against such act arises.

The district court held that this same rule should be applied to motel owners and operators.

The Kveragases were unable to prove that the motel's owners/operators had prior knowledge that an attack on them was likely to occur, according to the court. And, although past criminal activity in a neighborhood might make the area more dangerous, the court held that such evidence would be insufficient to meet "the degree of imminent harm required" under the law.

The court concluded by holding that the sudden and unexpected criminal acts which occurred to the Kveragas couple on February 16, 1982, were the "sole proximate legal cause of [their] injuries.... For this additional reason, plaintiffs have not made out their case."

This suit was therefore dismissed by the district court.

COMMENTARY:

We continue to urge all of our readers to review all of the cases reported in **PSCLR** *because many of the legal principles which apply to one industry are equally applicable to another—as is clearly seen in the above case. The judge here had no legal precedent in any Tennessee hotel case, but he did find applicable legal principles in a Tennessee shopping mall/retail case.*

Do not be misled into thinking that because a particular case involves another industry you need not consider its implications. Though the settings may vary, the issue and legal principles may be closely related to your own situation. Review them all!

PSCLR—OCT 83 (6)

Motel Guest Injured In Criminal Attack Recovers Compensatory Damages ($750,000) From Motel Owner And Operator, But Punitive Damages ($1,000,000) Is Disallowed:

Orlando Executive Park, Inc. v. Patricia Dale Robbins; Howard Johnson Co. v. Patricia Dale Robbins, *Nos. 61165, 61166, in the Supreme Court of Florida, decided March 31, 1983, rehearing denied June 29, 1983, reported at 433 So. 2d 491.*

Patricia Dale Robbins was attacked by an unidentified man while she was a guest of a Howard Johnson Motor Lodge which was owned and operated by Orlando Executive Park, Inc. She alleged that both defendants had violated their legal duty to exercise reasonable care for her safety as a guest on the premises.

The case was tried to a jury, and she was awarded $750,000 compensatory damages against both defendants jointly, plus $500,000 punitive damages against each defendant separately. The trial judge set aside the punitive damages awards but granted judgment to the plaintiff Robbins for the $750,000 compensatory damages. This judgment was affirmed by the District Court of Appeal, and the defendant Orlando Executive Park, Inc., then appealed to the Florida Supreme Court.

The evidence presented by plaintiff included testimony by a security expert, who testified that there should have been three guards on around-the-clock duty at the motel. Furthermore, the security service previously employed by Orlando Executive Park had also suggested the need for three full-time guards. Robbins also presented evidence to the effect that numerous episodes of criminal activity known to the motel management had taken place during the six-month period immediately preceding the attack on her.

The court also noted that the judgment against Howard Johnson should be affirmed since although Howard Johnson Company did not own or operate the hotel, it did represent to the public that a certain level of service could be found at the motel, and the Howard Johnson Company did operate the restaurant, lounge, and theater at the motel. The court concluded that the motel complex was an "integrated commercial enterprise" and that the Howard Johnson Company's direct participation was significant.

According, plaintiff's judgment for $759,000 in compensatory damages was affirmed.

COMMENTARY:

The decision by the appellate court in the above case was not unanimous, but was a four-to-one decision. The dissenting opinion took the position that there was no evidence to support the claim that defendant's negligence actually contributed to causing the plaintiff's injuries. The dissent argued that even if defendants had provided the security measures which plaintiff's expert witness had said were needed, there was still no evidence that such measures would have prevented the attack.

In support of this position, the dissenting justice called attention to the following facts:

> *The plaintiff testified that when she registered, the people at the desk told her that she could park in the lot which was adjacent to and could be seen from the office. She testified that after taking her luggage to her room she returned to her car for some papers. She said that while she was at the car, she saw a man standing by the office. As she walked back to her room, she saw the same man looking, she supposed, for an ice machine. The man came up and assaulted her when she was two doors from her room. On cross-examination, she testified that the man was neatly dressed and well-groomed. Before the assault she did not find anything suspicious about his appearance or behavior. The guard who was on security duty that night testified that he did not see any suspicious characters. He also stated that he had walked through and around*

> *the building containing the plaintiff's room not more than 15 minutes prior to the assault. The guard said that after touring the grounds, he got in his car. He noticed the plaintiff by her car and parked in a location where he could observe most of the premises.*

The dissent concluded that the attacker probably would not have been prevented from assaulting the plaintiff even if there had been two additional security guards on the property. According to the security expert's testimony, those two additional guards would have been patrolling other buildings the time the assault occurred in any event. Therefore, defendant's failure to have two additional guards did not cause the plaintiff's injuries.

While forceful and logical, the dissenter's argument did not persuade a majority of the appellate court and the plaintiff prevailed.

PSCLR – OCT 83 (10)

Hotel Guests May Recover Fair Market Value Of Property In Luggage Lost By Hotel Computer; Limitation Of Liability Statute Inapplicable:

Sukhminder Singh Bhattal and Mohinder Kaur Bhattal v. Grant Hyatt-New York and Hyatt Corporation, *No. 82 Civ. 7803-CLB, in the United States District Court for the Southern District of New York, decided May 18, 1983, reported at 563 F. Supp. 277.*

Residents and citizens of India, the plaintiffs in this case checked into defendant's Grand Hyatt Hotel (formerly known as the Commodore) in New York City on July 19, 1981, and were assigned to Room 2946. As is the usual practice, they turned over to the bell captain their personal luggage which was then transported by hotel employees to the Bhattal's hotel room.

The Bhattals did not request that any valuables be placed in the hotel's safe deposit box. They left their luggage in their hotel room, locking the door with the room key, and went to lunch with friends. Upon their return later that evening, they discovered their luggage and its contents were missing.

As noted by the court:

> All things in the modern world which go wrong for reasons other than the application of Murphy's Law, seem to go wrong because of a mechanical malevolence known as 'computer error.'

The defendant hotel had used a computer for the assignment of rooms to its guests. As the result of an error, the plaintiff's luggage was carried from their room to JFK International Airport, along with the luggage of aircraft crew nemeses from Saudi Arabia who had previously occupied Room 2946. It appeared that the computer or more accurately, its operator, failed to note that the room had been vacated by the air crew and had been rerented to the Bhattals. As may be guessed, the plaintiffs' luggage left for Saudi Arabia and has not been seen since.

The Bhattals claimed that their luggage contained a pearl valued at $250,000, for which amount they sued defendants. (The appellate court noted that "a missing pearl is always a pearl of the finest water." The hotel moved to dismiss the plaintiffs' claim and to have summary judgment entered for the hotel.

The hotel based its defense on the New York General Business Law which included a section purporting to limit the liability of innkeepers, the relevant part of which provided as follows:

> Whenever...any hotel...shall provide a safe...for the safekeeping of any money, jewels [etc.],...belonging to the guests of...such hotel,...and shall notify the guests...by posting a notice stating the fact that such safe is provided,...; and if such guest...shall neglect to deliver such property...for deposit in such safe, the...hotel shall not be liable for any loss of such property sustained by such guest by theft or otherwise; but no hotel...shall be obliged to receive property on deposit for safekeeping exceeding $500 in value; and if such guest shall deliver such property...for deposit in such safe, said [hotel] shall not be liable for any loss thereof sustained by such guest by theft or otherwise in any sum exceeding the sum of $500.

Another section of the statute provided that the innkeeper shall not be liable for damage to or loss of clothing for any sum exceeding $500 unless the loss occurred through the fault or negligence of the innkeeper.

The question then presented to the court by defendant's motion for summary judgment is whether the New York statutory law effectively limited the liability of the hotel in a case where the hotel took custody of the plaintiffs' luggage and, without plaintiffs' authorization, intentionally though inadvertently caused the luggage and its contents to be sent to Saudi Arabia.

The court decided that the above law did not protect the defendant in this case. The purpose of this law cited by the defendant would apply to a situation where there was a mysterious disappearance of property either as the result of a theft by a hotel employee or by a third party for whose acts the innkeeper was not responsible. The law was also intended to protect an innkeeper from a fraudulent claim made by a guest in a situation where the property claimed to be lost never existed or was taken or stolen by the guest himself. The court concluded that the law did not protect an innkeeper from liability which was based on the innkeeper's exercise of control over the guest's property.

The court then noted that the plaintiffs, in addition to claiming that they were entitled to claiming the fair market value of the lost property, also asserted a claim for extreme emotional distress, mental anguish and mental suffering. However, without proof of malicious intent on the part of the defendant hotel, the plaintiffs could not recover damages for mental suffering caused by the loss of property. Since there was no evidence to suggest that the hotel acted maliciously in causing the loss of the plaintiffs' luggage and it is clear that the loss was not the result of malice, no recovery for any mental suffering would be permitted.

The trial judge then denied the hotel's motion for summary judgment, and on his own motion, the trial judge granted partial summary judgment in favor of the plaintiffs, ruling that the hotel was liable to them for the value of their loss. The case will therefore proceed to trail at which the parties will present evidence as to the fair market value of the lost property at the time and place of the loss.

PSCLR – NOV-DEC 83 (1)

Motel Held Liable For Placing Guest In Situation To Be Assaulted And Failing To Come To His Aid After Assault:

Raymond Davenport v. George E Nixon, et al., d/b/a Admiral Motel, et al., *No. 82 CA 0954, in the Court of Appeal of Louisiana, First Circuit, decided June 28, 1983, reported at 434 So. 2d 1203.*

The duty of care owed by innkeepers to their guests is more stringent than that required of other types of business. In this case, the extent of an innkeeper's duty to protect and assist his guests is explored.

Ray Davenport, a 48-year-old man who was partially paralyzed due to two previous strokes, moved out of his apartment late during the evening of November 8, 1980, and spent about an hour in a local lounge. Because he had trouble seeing at night, he had two men take him and his truck to the Admiral Motel.

It was about 2:45 AM when Davenport rang the bell at the late night check-in window located outside the motel's front office. He was a little nervous – a stranger had approached him as he walked up to the window and, although the man had only invited him to his room for a drink and left, Davenport felt suspicious.

Mrs. Pearl Nixon, who along with her husband and sons owned and operated the motel, registered Davenport. He had stayed at the motel before, and Mrs. Nixon knew that he was somewhat paralyzed. She filled out the registration form for him. Davenport expressed his suspicion of the stranger, and asked Mrs. Nixon to allow him to come inside the office to pay for the room. He testified that he told her that he was uncomfortable about taking his money out while standing outside where the man might see it. She refused, saying that it was against the motel's policy. She did, however, comply with his request for the room directly adjacent to the front office. He later testified that he asked for this room because it was close to the office and also the outside lighting was better in the front of the motel. She then shut the window and returned to her family's living quarters located adjacent to the office. Davenport walked over to his truck which was parked in front of the check-in window.

When he reached his truck, the same man approached him again and demanded that Davenport hand over all of his money. He tried to resist the attacker, but the man slashed at him with a knife and reached directly into the pocket holding all of Davenport's money. It seemed obvious to Davenport that the man had watched him pull his money out to pay for the motel room.

The assailant continued to slash Davenport with the knife until the victim was able to grab the blade and break it. As the mugger fled, Davenport made it over to the check-in window and again rang for Mrs. Nixon. He asked her to call an ambulance, which she did. However, she offered no other help, and when the ambulance arrived, Davenport was found lying on the curb with bleeding wounds. He spent three hours in the hospital receiving treatment for the lacerations.

Subsequently, Davenport filed suit against the Nixons and the motel's insurance company. The basis of the suit was the Nixons' alleged negligence in taking reasonable precautions to provide for their guests' safety. Judgment was awarded to plaintiff, against all defendants, in the amount of $25,240. The defendants appealed.

The defendants' appeal was centered on their contention that the trial court's decision required of the Nixons an unreasonably "stringent duty" to protect Davenport. Moreover, the defendants contended that the verdict held them liable for a criminal act about which they had no warning and over which they had no control.

The Louisiana Court of Appeal examined the law concerning an innkeeper's duty to protect its guests and quoted from *Kraaz v. LaQuinta Motor Inns, Inc.*, (reported in the August 1982 **PSCLR**):

> An innkeeper does not insure his guests against the risk of injury or property loss resulting from violent crime .... a guest is entitled to a high degree of care and protection. ... The innkeeper has a duty to take reasonable precautions against criminals.

In its review, the court emphasized that a proprietor's duty to protect customers would not extend to unforeseeable/unanticipated criminal attacks, but only those of which the proprietor had some kind of forewarning. Moreover, the proprietor would only be liable if it was reasonably within his power to protect his guests.

The defendants disputed any implication of negligence on their part: there was no forewarning that a criminal attack might take place. Indeed, the attacker was later discovered to be another guest of the motel, Byron Fraser, who had been at the motel for over a week and had caused no trouble in that time. In addition, no other such attacks on guests had occurred on motel property. The court, however, pointed out that evidence showed that Nixon was suspicious of Fraser before his attack on Davenport. He had been hanging around the pay telephone near the office and it seemed to Nixon that he was only pretending to use the phone.

Based on the evidence and the nature of the duty owed to guests by innkeepers, the court held that Nixon could have taken certain precautions to protect Davenport. At the very least, she should have watched him until he got to his room. If he had still been attacked, she could have then sounded the siren alarm located in her office to scare the attacker away. The appeals court agreed with the trial court which stated:

> ... the duty of care required of an innkeeper was not met by the defendants. The defendants were aware of criminal activity in the area and consequently did take measures to insure safety from crime – but only for themselves and their property. In fact, the [motel owners] themselves inadvertently increased the possibility of attack on their guests. Because of the check-in method they devised for late night arrivers, their patrons were more likely to be the object of attack and therefore the defendants should have taken even greater precautions to safeguard their patrons safety.

The defendants's claim that the $25,240 award was an excessive amount was also dismissed. Davenport's injuries were quite extensive. In addition to the permanent and disfiguring scars, Davenport also suffered from nerve damage, a sprained ankle, a broken toe, and recurring nightmares. The court therefore affirmed the judgment and the award to Davenport.

COMMENTARY:

The court noted that prior to the *__Kraaz__* *decision on March 1, 1982, the law (in Louisiana) was clear that*

> *the duty to protect business patrons did not extend to the unforeseeable or unanticipated criminal acts of an independent third person. Only when the owner had knowledge, or could be imputed with knowledge of a third person's intended criminal conduct which was about to occur, and which was within the power of the owner to protect against, did such a duty of care towards a guest arise.*

However, in the *__Kraaz__* *decision, this duty was expanded: As noted above, a motel guest is entitled to a "high degree" of care, and an innkeeper must take reasonable precautions to protect guests from criminals* *__in general,__* *not just from specific criminal acts about to occur.*

The court in the *__Davenport__* *case noted however that Davenport would be entitled to a recovery* *in either situation since the Nixons failed to protect him from foreseeable criminal conduct of a known suspected fellow guest, and also failed to protect him from criminals generally.*

PSCLR – JAN 84 (4)

Hotel Has Duty to Exercise Reasonable Care to Protect Its Guests from Criminal Attacks by Third Persons:

Allen B. Morrison, et al. v. MGM Grand Hotel, et al., *No. CV-R-81-244-ECR, in the United States District Court for the District of Nevada, decided September 20, 1983, reported at 570 F. Supp. 1449.*

Allen Morrison was an invited guest of the MGM Grand Hotel in Reno, Nevada, at a New Year's party on January 1, 1981. Planning to leave the hotel on January 2nd, he cashed in his gambling chips and withdrew his jewelry and cash which he had stored in the hotel's safe deposit box at approximately 3:00 AM that morning. As he left the hotel desk, he was followed into the lobby and onto the elevator by a man who robbed him at gunpoint, immediately after the elevator doors closed. His valuables and money were stolen and he was knocked unconscious.

He brought suit against the hotel, contending that it was negligent in failing to provide adequate protection against such criminal acts. He alleged that the hotel in fact had knowledge that an identical elevator robbery had recently occurred. The hotel had failed to take any additional security measures as a result of that robbery, and had failed to warn guests of the known risk of riding the elevator without an escort. The hotel moved for summary judgment in its favor, taking the position that it was under no duty to protect the plaintiff from an unforeseeable criminal act by a third party. The hearing magistrate recommended that the motion be granted; however, the district court declined to adopt that recommendation, holding:

> Under Nevada law, a landowner must exercise ordinary care and prudence to render the premises reasonably safe for the visit of a person invited on his premises for business purposes. ***

Notwithstanding the existence of a special relationship [such as innkeeper and guest], the duty of a landowner to take affirmative action to control the wrongful acts of third persons arises only where he has reasonable cause to anticipate such acts and the probability of injury resulting therefrom and fails to take affirmative steps to control the wrongful conduct.

The district court determined that the plaintiff Morrison had specifically alleged that the MGM Grand could have reasonably foreseen the criminal misconduct in question. He had alleged that the hotel was on notice because of the similar recent elevator robbery. The court therefore concluded that a duty to protect its guests from loss due to such misconduct did exist and the question as to whether the hotel fulfilled this duty in a reasonable manner should proceed to a jury trial. The defendants' motion for summary judgment was denied.

PSCLR – JAN 84 (7)

Casino/Hotel Guests Cannot Recover for Property Stolen from Room Absent Hotel's Gross Negligence:

Anne S. Levin, a minor by her parents and natural guardians, Marvin J. Levin and Charlotte W. Levin and Marvin J. Levin and Charlotte W. Levin in their own right v. Desert Palace Incorporated t/a Caesar's Palace, *in the Superior Court of Pennsylvania, decision filed August 19, 1983, reargument denied October 11, 1983, reported at 465 A.2d 1019.*

The plaintiff Marvin Levin and his family checked into Caesar's Palace hotel in Las Vegas on September 7, 1976. They claimed that their room was burglarized as they slept and that they lost jewelry worth $18,550 and $1,500 in cash. They sued the hotel based on the allegation that the casino/hotel failed to provide adequate security measures to protect against guests' property losses.

While some of the facts were disputed, it was undisputed that each inner door to the Levins' suite had a warning to guests to use the night locks before retiring, as well as a notice that safe deposits boxes were available. The parties disagreed on whether or not the doors had working deadbolt locks. In any event, the plaintiffs agreed that they had no evidence of any gross negligence on the part of the hotel/casino.

The defendant moved for dismissal of the plaintiffs' claims, alleging that no cause of action had been stated against them under applicable provisions of the Nevada Innkeeper's Statute. The trial court agreed, the plaintiff's suit was dismissed and they appealed.

The appellate court first turned its attention to whether Pennsylvania (the Levins' resident state) or Nevada law should be applied to the case. This was important because the Nevada Innkeeper's Statute limits an innkeeper's liability for guests' losses to those instances where the innkeeper was guilty of "gross neglect." In contrast, the Pennsylvania statute places a higher degree of care on the innkeeper, making him liable for guests' losses unless each room is equipped with bolts and locks on all doors and windows. The court noted:

> The law of that state which has the greater interest in the application of its law governs. ***
>
> Obviously, Pennsylvania has an important interest in protecting the welfare of its citizens. We conclude, however, that this interest is outweighed by Nevada's interest in regulating the conduct in prescribing the liability of hotel owners within its jurisdiction. A hotel owner relies on the law of the state in which the hotel is located to determine the standard of conduct required of him. It cannot be expected that a hotel should comply with the laws of all the states of which its guests are citizens.

The plaintiffs also contend that Nevada common law rather than the Nevada Innkeeper's Statute should apply. Pursuant to Nevada common law, an innkeeper is almost an insurer of his guest's property. It was the plaintiffs' position that the Innkeeper's Statute controlled only when property was "left" in the hotel room and the guests were absent, but not when the guests were actually present in the room. The court, however, rejected this argument as being "illogical." The phrase "property left in the room" in the Statute merely distinguished between property turned over to the hotel for safekeeping as opposed to that property which guests kept in their rooms. (The court noted that the Fifth Circuit Court of Appeals had decided similarly.)

Since the Nevada Innkeeper's Statute did apply, and since that law placed liability on the innkeeper only on a showing of gross negligence, the defendant Caesar's Palace could not be held liable to the Levins for their property loss.

COMMENTARY:

We see from this case that the decision as to which state's law should be applied to a situation can be highly material. Had Pennsylvania law governed the conduct of the innkeeper here, the plaintiffs would have at least had an opportunity to present their case to a jury.

We also see that had there been no Innkeeper's Statute, the "common law" would have made the innkeeper almost automatically liable for any loss of his guests' property. The so-called "innkeeper's statutes" are in derogation of common law since they strictly limit the innkeeper's liability to guests. Therefore, those innkeepers who wish to avoid liability by relying on their state's statutes must strictly adhere to the requirements of the statutes.

For further reading, see these cases published in **PSCLR** *concerning innkeeper's limitation of liability statutes: Montgomery v. Royal Motel, October 1982, on page 4; Zacharia v. Harbor Island Spa, Inc., November/December 1982, on page 9; Goncalves v. Regent International Hotels, Ltd., June/July 1983, on page 3; Morris v. Hotel Riviera, Inc., August 1983, page 4; Bhattal v. Grand Hyatt-New York, October 1983, on page 9.*

PSCLR – APR 84 (1)

Hotel Operator Liable For Death Of Hotel Guest Shot By Armed Robber Outside Entranceway To Hotel:

Philip Aaron Banks, et al. v. Hyatt Corporation and Refco Poydras Hotel Joint Venture, *No. 81-3377, in the U.S. Court of Appeals for the Fifth Circuit, on appeal from the U.S. District Court for the Eastern District of Louisiana, decided January 9, 1984, reported at 722 F.2d. 214.*

This case discusses the question as to whether a landowner or an innkeeper can be held liable to a hotel guest for a criminal assault by a third person which occurs just outside the entrance doors to the hotel on property which is owned by the landowner but which serves as a public sidewalk. The facts are as follows:

Dr. Robert Banks and Dr. John Hakola were returning to the Hyatt Hotel in New Orleans, Louisiana, at 9:30 PM on October 12, 1979. They had almost reached the door at the mall and the hotel's Loyola Avenue entrance when they were approached by two men, one brandishing a handgun. Dr. Hakola was able to enter the hotel, but Dr. Banks was shot and killed on the sidewalk, only four feet from the hotel door.

A suit for wrongful death was brought by Dr. Banks' widow and children against both the Hyatt Corporation, as the lessee and operator of the hotel, and against Refco Poydras Hotel Joint Venture, as the owner and operator of the Poydras Plaza Mall and owner of the hotel. The suit charged the defendants with failure to provide adequate security to protect Dr. Banks from criminal assaults and failure to warn him of the danger of criminal activity near the hotel entrance. A jury returned a verdict in favor of the plaintiffs against both Hyatt and Refco in the amount of $975,000. The trial judge set aside the verdict as to Refco, but affirmed it as to Hyatt Corporation.

The appellate court took great pain to explain its reasons for affirming the trial court's decision. The evidence showed that the hotel/mall complex had been the scene of a number of crimes. There had been 11 armed robberies and five simple robberies within the immediate vicinity of the hotel during the three month period immediately preceding Dr. Banks' death. Also, Dr. Banks was the second person to be shot at the hotel entrance and the fifth victim of an armed robbery at the entrance since the hotel's opening in 1976.

The Hyatt Director of Security, Joseph Murray, reviewed 21 incident reports describing occurrences at the hotel entrance between 1976 and the date of Dr. Banks' death, and 12 of those occurrences involved weapons. Earlier in 1979, the defendants and three other property owners became concerned about the increase in crime in the vicinity, and had requested the New Orleans Police Department to assign additional police patrols to the area. The department denied the request. Off-duty, uniformed and armed police officers were then hired by the defendants to patrol the property perimeter.

The court ruled that

> the owner or operator of a business owes a duty to invitees to exercise reasonable care to protect them from injury. This duty does not extend however to unforeseeable or unanticipated criminal acts by an independent third person.

The court therefore affirmed the trial judge's decision to set aside the verdict against Refco.

The court went on to note that the Louisiana Supreme Court recently held in *Kraaz v. La Quinta Motor Inns, Inc.* (reported on page 2 of the August 1982 issue of **PSCLR**) that innkeepers owe their guests a duty of care higher than reasonable care, and this includes a duty to take reasonable precautions against criminal assaults on guests.

The court then concluded that, at least under Louisiana law,

> the duty of a business to protect invitees can extend to adjacent property, particularly entrances to the business premises, if the business is aware of a dangerous condition on the adjacent property and fails to warn its invitees or to take some other reasonable preventive action.

COMMENTARY:

We have here an example of a situation in which the hotel operator did in fact know of criminal activity in the area and did in fact take some steps to prevent assaults on its patrons. However, the jury found that the hotel simply did not go far enough in protecting its guests.

It is also interesting to know that the appellate court included in its opinion the statement that "court law has become increasingly concerned with placing liability upon the party that is best able to determine the cost-justified level of accident prevention." The court noted in this regard that holding a negligent innkeeper liable when there is a third party assault is "sensible," not because of some abstract conceptual notion about the risk arising within the course of the relationship, but rather because the innkeeper is better able to identify and carry out reasonable preventive measures.

Subscribers interested in this area of the law would be well advised to obtain a copy of the entire opinion in this case which is rather lengthy, and which has only been briefly summarized above. This is especially true of hotel/motel operators in the Louisiana jurisdiction, and in Florida where, as the court noted, the higher degree of care is imposed on innkeepers.

The following cases, all reported in previous issues of **PSCLR**, *provide some insight into different states' court rulings in this area of the law: Morrison v. MGM Grand Hotel (Nevada), January 1984, page 4; Davenport v. Nixon (Louisiana), November/December 1983, page 1; Kveragas v. Scottish Inns, Inc. (Tennessee), October 1983, page 3; Orlando Executive Park v. Robbins (Florida), October 1983, page 6; Anderson v. Malloy (Missouri), May 1983, page 9; and Zang v. Leonard (Tennessee), April 1983, page 3.*

PSCLR – OCT 84 (2)

Innkeepers' Limitation-Of-Liability Statute Protects Innkeepers Against Negligence Claims As Well As Contractual Claims:

Paulette Laubie, wife of/and Andre Laubie v. Sonesta International Hotel Corp., Archer Casbarian, C. Mark Stevenson, Sam F. Uchello, United States Fire Insurance Co., *Civ. A. No. 77-1584, in the United States District Court for the Eastern District of Louisiana, decided June 27, 1984, reported at 587 F. Supp. 457.*

Andre and Paulette Laubie are French citizens who were guests at the Royal Sonesta Hotel in New Orleans. While away from the hotel, their rooms were entered by unknown persons who stole jewelry valued at approximately $50,000. Without going into the details of their claim, suffice it to say that they filed suit in federal court claiming that the defendant hotel, and three of its executive officers, and its insurer, were negligent and that their negligence caused the loss.

To maintain an action in a federal district court, it must be shown that the amount in dispute is in excess of $10,000. The defendants won a dismissal of the case because of the Louisiana innkeeper limitation of liability statute. However, the case was caught up in a procedural quagmire before this result was finally obtained by the hotel.

When the limitation-of-liability defense was first raised, the statute made no distinction as to whether an innkeeper's liability was limited on contractual claims only, or on negligence claims as well. On an earlier appeal to the Louisiana Supreme Court, that court ruled that the limitation of liability statute as it then read did not protect innkeepers against negligence actions, and that since the loss claimed was in excess of $10,000, and negligence was alleged, the case should proceed to trial and the hotel would be held liable for the full amount of the loss if the plaintiffs could prove that the defendants' negligence caused the loss. However, the Louisiana legislature then passed an amendment to the statute which specifically included provisions protecting officers and employees of hotels from contractual or negligence actions by hotel guests resulting in the loss of the guests' property. Limitation was raised from $100 to $500, but still well below the $10,000 jurisdictional limit for federal court cases.

The question presented on the present appeal was whether the amendment to the statute should be given retroactive application to the Laubie's claim. The United States District Court determined that the Louisiana legislature's amendment was merely "interpretive" and did not make any substantive changes in the previous statute. Accordingly, the Laubie's negligence claim was limited to a total recovery of $500, and since this amount is below the federal court limit, their case was dismissed.

COMMENTARY:

As has been noted in earlier cases reported in **PSCLR**, *the innkeeper's limitation of liability statutes provide necessary protection to hotels and motels in the event of guests' property losses, but only if the innkeeper's employees comply strictly with the specific dictates and details of the statute in their respective jurisdictions. It is imperative that security operations officers insure that these procedures are taught to the appropriate hotel employees, and that they are scrupulously followed. Furthermore, since different states have different requirements, those innkeepers with properties in different jurisdictions must insure that the employees in each location are fully trained as to the requirements of the applicable statute in their location.*

PSCLR – OCT 84 (8)

Hotel Is Liable To Guest For Injuries Suffered In Fall Due To Dangerous Condition On Premises Where Hotel's Knowledge Of Danger Was Shown By Statement Of Security Officer:

Stouffer Corporation v. Henkel, No. 67138, in the Court of Appeals of Georgia, decided February 29, 1984, rehearing denied March 20, 1984, certiorari denied April 25, 1984, reported at 317 S.E.2d. 222.

While a guest of a Stouffer Hotel in Atlanta, Georgia, Mr. Henkel was strolling about the premises and happened upon a small bridge which appeared to be safe to cross. As he walked across the bridge, Henkel was faced with a large puddle of water over which a board had been placed. Instead of retracing his path or stepping into the water, he assumed that the board had been placed there to allow guests to cross over the water to get to the hotel and he stepped onto the board. The board "vibrated" and Henkel fell. After picking himself up, he returned to his room and called the front desk to report the incident. Shortly thereafter, a uniformed individual, who appeared to be a hotel security officer, arrived and announced that he was employed by the hotel and told Henkel that he believed "that spot should have been either roped off or closed down and nobody should have been able to go through there."

Henkel subsequently brought this lawsuit against Stouffer Hotels to recover for his injuries, claiming that the hotel was aware of the danger and had not taken steps to protect its guests from this danger. The trial ended in Henkel's favor and the hotel appealed. The hotel contended that the jury's decision that the hotel had had prior knowledge of the danger was based on hearsay testimony by the plaintiff concerning the unidentified individual's statement that the incident "never should have happened."

The court noted that the law clearly demonstrated that the basis for a property owner's liability for invitees' injuries on the property is the owner's superior knowledge of the danger which caused the injury. In this case the danger was presented by an "insecure makeshift plank bridge across the puddle." Henkel was obviously aware of the plank before he crossed it; however the evidence was sufficient to indicate that he was not aware of the danger presented by the plank. There was no evidence that the plank had been placed over the puddle by the hotel, and mere proof that a dangerous condition existed on its premises did not raise a "presumption of negligence" on the part of the hotel.

However, the court also ruled that the words of the "unidentified visitor" provided sufficient evidence to prove that the hotel was aware of the dangerous condition before Henkel's fall. Moreover, the visitor could reasonably be inferred to be a hotel employee by the timing of his arrival, his announcement that he was a hotel employee, the uniform and the badge he wore, and the fact that the hotel did not prove he was not an employee.

Dismissing other grounds of error presented by the hotel's appeal, the court affirmed the trial verdict in favor of Henkel.

COMMENTARY:

The facts of this case do not indicate that the security officer had reported the bridge/board defect to the manager, but simply that he had knowledge of it. A question arises as to whether he had been instructed to report problems of this nature to management or to his supervisor so that remedial measures could be taken. What instruction have your security officers been given concerning reporting dangerous conditions on the property in addition to reporting dangerous or suspicious activities.

PSCLR—NOV-DEC 84 (1)

Innkeepers Have "Special" Duty To Protect Guests, Which Differs From Retailer's Duty To Protect Invitees:

Charles Kveragas, et ux. v. Scottish Inns, Inc., et al., *No. 83-5197, in the U.S. Court of Appeals for the Sixth Circuit, on appeal from the U. S. District Court for the Eastern District of Tennessee, decided May 2, 1984, reported at 733 F.2d 409.*

Charles and Esther Kveragas sued the Scottish Inns corporation to recover for injuries sustained in a motel room assault and robbery during the couple's February 1982 stay in a Knoxville, Tennessee, Scottish Inns motel. The U.S. District Court for the Eastern District of Tennessee heard the plaintiff's case and directed a verdict for the defendants, ruling that the innkeeper did not have a duty to protect the guests, because "the sudden criminal acts of the assailants were the sole proximate cause of plaintiffs' injuries." (See the October 1983 **PSCLR** at page three for a full report of that decision.) The U.S. Court of Appeals here reviewed and reversed that decision, remanding the case for a new trial.

Briefly, the facts were that Mr. and Mrs. Kveragas were attacked in their motel room by three armed intruders. The assailants entered the room by breaking down the door which was secured by a door lock and chain. Mr. Kveragas was shot and $3,000 in cash was stolen. Evidence indicated that the hollow core door did not fit the door frame, was knocked in with a single kick, and was not equipped with a deadbolt lock which could have withstood the force used to break the door down.

The District Court based its verdict for the defendant on the *Compropst* decision in which a shopper's suit for a parking lot assault was dismissed because there was no actual notice to the defendant shopping center of a specific risk to the plaintiff. However, the U.S. Court of Appeals determined that the *Compropst* rule was not the proper standard to decide this case, since the court in *Compropst* addressed the duty of shopping center owners and operators and those of similar establishments to its invitees, whereas the relationship between an innkeeper and its guests has been long recognized by the courts as "special."

The Court of Appeals held that the rule established in the case of *Zang v. Leonard* should instead apply. *Zang* set forth a duty of care on the innkeeper to reasonably protect its registered guests from the reasonably foreseeable misconduct of third persons. (For a full summary and analysis of this case, see the April 1983 **PSCLR** at page three.) The relevant test would be to determine the protective measures which would be employed by a "reasonably prudent motel operator" under all the relevant conditions, including but not limited to the motel's advertising of facilities, the location of the motel, the history of criminal assaults at or near the premises, the costs vs. the expected benefits of protective measures, the motel's established security practices and compliance with internal security procedures, and any other factors relevant to the question of reasonable protective measures.

Applying the *Zang* standard to this situation, the U.S. Court of Appeals decided that the Kveragases's suit presented sufficient evidence to overcome the directed verdict granted the defendant. The dismissal was therefore reversed and the suit was remanded for trial.

COMMENTARY:

We see from this case that there is a different standard of care applied to innkeepers than to retail establishments and other commercial operations. While the latter must have "actual notice of a threatened specific act by a specific person" before liability attaches for criminal acts of others, liability may be imposed on an innkeeper for failure to exercise reasonable care to protect their guests from such attacks in general. At the same time, it is important to note (and the court so ruled) that innkeepers are not insurers; that is, they are not liable for any and all injuries to guests caused by the criminal acts of third parties.

The innkeeper can be held liable only when the negligence of the innkeeper is the proximate cause of the injury to the guest. As the court said, "some criminals are bent on theft or violence regardless of the obstacles placed in front of them; some crimes are therefore not foreseeable and others are not preventable, even with the exercise of utmost care."

Accordingly, reasonable protective measures, determined in light of all surrounding circumstances, must be employed by the hotel/motel operator in order to avoid liability.

As a matter of historical interest, the liability of an innkeeper to its guests did in fact approach that of an insurer in that innkeepers were responsible for almost all dangers encountered by the guest with the exception of "acts of God." While the standard of care is not as strict today, we see that it is stricter than the standard of care applied to the owners and operators of commercial establishments.

PSCLR – NOV-DEC 84 (2)

Guest Cannot Recover From Hotel For Theft Of Property Which Hotel Had Not Undertaken To Protect:

Jose Maria Berga De Lema v. Waldorf Astoria Hotel, Inc., *No. 83 Civ. 1075-CSH, in the U.S. District Court for the Southern District of New York, decided May 14, 1984, reported at 588 F. Supp. 19.*

Prior to the enactment across the country of "innkeeper's liability statutes," a hotel/motel operator was considered to be an insurer of property held by the hotel/motel operator for the guest, and the innkeeper therefore was absolutely liable for the loss of any such property. (This principle of strict liability on innkeepers was discussed in the *Goncalves v. Regent International Hotels Ltd.* case, decided in 1983 and reported in the June-July 1983 **PSCLR** at page three.) The innkeeper's exposure, however, has been restricted by statutes which have been passed to protect hotel owners from excessive liabilities. The New York statute sets a $500 maximum limitation of liability, unless the guest declares a greater value in which case the innkeeper is not required to accept the property for safekeeping, or can charge additional for undertaking the added responsibility.

Mr. De Lema arrived in New York City on October 30, 1982, and proceeded to the Waldorf Astoria Hotel where he had hotel reservations. In addition to his suitcases, he carried a case containing $300,000 in jewelry. Upon arriving at the hotel, he went straight to an assistant manager's desk, said that he had a reservation, was taken to the registration desk, and immediately inquired about a safety deposit box. He put his jewelry case on the floor beside him; while filling out the reservation form and paying an advance deposit for the room, he was jostled by an unknown woman. When he next looked down, he saw that the jewelry case was gone. It was never recovered, and De Lema sued the Waldorf Astoria for his loss.

The trial judge dismissed the case due to the fact that De Lema had failed to disclose the value of his property to any hotel employee prior to the time of the theft. Therefore, since the hotel had not had the opportunity to accept or reject the additional responsibility of safeguarding his property, nor to accept or reject the additional liability, the New York statute limiting the innkeepers' liability for lost property to $500 was held to be applicable because De Lema had not delivered his property to the hotel for safekeeping.

COMMENTARY:

As the court in its opinion noted, this case differs significantly from the Goncalves decision, where the evidence showed that there was some dispute as to whether the hotel had in fact provided adequate safe-deposit boxes for its guests' property. Where a hotel does fail to provide such a safekeeping facility, then the statute provides the hotel with no protection, and the hotel would then be liable for the full amount of the guests' loss.

PSCLR—MAR 85 (4)

Innkeeper Is Not Liable To Guest Assaulted On Premises Unless The Assault Was Or Should Have Been Anticipated By Innkeeper:

Meyers v. Ramada Inn of Columbus, *in the Court of Appeals of Ohio, Franklin County, decided March 15, 1984, reported at 471 N.E.2d 176.*

The plaintiff Delilah Meyers concedes and agrees that an innkeeper is not an insurer of the safety of its guests. In order to recover from an innkeeper as the result of being assaulted by a third person on the hotel premises, she must establish must that the hotel owner/operator knew or should have known about the danger propensities of the assailant and/or that the assault upon her was imminent.

Meyers was in fact assaulted and injured while she was a business invitee of the Ramada Inn motel in Columbus, Ohio. She sued the Ramada Inn, alleging that her injuries resulted from the hotel failing to provide adequate security for its guests. The innkeeper moved to dismiss the plaintiff's case, arguing that the assault on Meyers was a criminal and intentional act by a third person for which the defendant innkeeper had no legal responsibility. Jeff Polhemus, the food and beverage manager at the hotel, testified in his affidavit that the assailant Jeffrey Kelly had never been an agent or employee of the hotel, and that no employee of the hotel had any knowledge or forewarning of the attack itself or of Kelly's propensity or intentions to commit the attack. The plaintiff introduced evidence that Polhemus had told the assistant manager, prior to the assault, that Kelly had been loitering around the motel, had applied for a job several times, but had not been hired because Polhemus said he felt "very uncomfortable" about him and had said that there was something "scary" about him. The assistant manager had also commented that Kelly was "up to no good."

The court concluded that this evidence showed that the innkeeper had merely a vague concern or suspicion about Kelly's character. The hotel employees had never seen him do anything improper, and he had always obeyed their requests without animosity. He had never been observed carrying any weapons, nor had he been seen threatening or making improper advances upon the hotel guests. The court concluded:

> the mere fact that there may be some subjective concerns about an individual on the part of the manager of a motel, so that the manager would not hire the person as an employee, does not give rise to an additional duty to a patron of the motel to provide further security measures to protect against some speculative possibility of wrongful action by that person towards a guest of the motel.

Therefore, the court concluded that the Ramada Inn had no reason to know of any violent propensities on the part of the assailant Jeffrey Kelly nor that he might commit some attack on one of the motel's customers. The trial judge's grant of summary judgment effectively dismissing the case was affirmed.

COMMENTARY:

The present Ohio law requires that a motel/hotel conform to a standard of ordinary care in providing security for its invitees. Furthermore, if it can be shown that special circumstances exist, such as prior assaults on the property or where the motel/hotel is located in a high crime area, then this standard of ordinary care may require the owner/operator to provide additional security measures.

This Ohio decision, and earlier Ohio cases have established a more restrictive theory concerning innkeeper's liability to guests. As noted above, for an innkeeper to be held liable to an assaulted guest, the guest must prove that the innkeeper knew about the violent propensities of the particular attacker in question or that this attack on the plaintiff was imminent. This is a heavy burden for a plaintiff to carry in most instances. Other jurisdictions have not required the plaintiff to carry such a heavy burden of proof but rather have ruled that innkeepers can be held responsible and liable for assaults on their guests when the facts and circumstances of the entire situation have made the attack in question a foreseeable event, due to prior criminal activity in the area and other factors. Several earlier editions of **PSCLR** *have carried such cases from other jurisdictions.*

PSCLR – APR 85 (1)

Trial Court May Properly Limit Extent Of Opinion Testimony Given By Expert Witnesses:

Barbara Lou Otwell v. Motel 6, Inc., *No. 84-1540, in the United States Court of Appeals for the Eighth Circuit, on appeal from the United States District Court for the Eastern District of Arkansas, decided February 26, 1985.*

Barbara Otwell filed suit against the defendant Motel 6 alleging that negligence on the part of the defendant contributed to the death of her husband at the hands of assailants on the motel property. The jury returned a verdict in favor of the innkeeper, and Mrs. Otwell appealed. The incident in question occurred at approximately 1:00 AM when Mr. Otwell returned to his motel room at the Motel 6 in Little Rock, Arkansas, after having left the room to call a taxi cab. A short while later, while waiting in the room, someone knocked on the door; Otwell looked out the picture window beside the door and then opened the door. Two men then forced their way into the motel room, robbed Otwell and left. Otwell gave chase, and was fatally shot by the criminals.

The motel rooms at the Motel 6 lacked standard security devices, such as "peepholes" or chain locks on the doors. The rooms were not equipped with telephones.

During the trial, Mrs. Otwell called to the stand two expert witnesses in the field of hotel/motel security. The trial judge permitted these two witnesses to testify as to the security devices used and the security precautions taken by the defendant Motel 6 with those security precautions used and taken by other hotels and motels. However, the judge wouldn't permit the experts to testify as to the rate of crime at the defendant motel as compared with the rate of crime at other motels. Further, the judge would not permit the experts testify to the "standard of care in the security industry," or to their opinion of the cause of the incident. Mrs. Otwell claimed that the judge's exclusion of this testimony was error and requires a reversal and a new trial.

On appeal, however, the appellate court agreed with the trial judge, noting that the judge did permit evidence establishing that crimes had occurred previously at the defendant's motel, as well as evidence as to how the security devices used and security precautions taken by the defendant motel or lack thereof, compared with the security devices and precautions taken by other motels. The appellate court agreed that the cause of the incident was a matter which the jury, as laypersons, could themselves determine based upon the evidence. The jury could also determine whether the defendant motel's security practices fell below an acceptable standard of care without the assistance of expert witnesses's opinions.

The appellate court did not agree that the trial judge had committed reversible error when he instructed the jury that the innkeeper has a right to assume that "guests would recognize open and obvious dangers and approach such dangers in a reasonably manner."

Finding no reversible error in the trial record, the verdict and judgment of the lower court was affirmed by the Court of Appeals.

COMMENTARY:

It should be remembered that expert witnesses are not permitted to testify as to their opinions merely because they are experts. As in all cases where expert testimony is permitted, opinions may be given only if (1) the witness is a qualified expert due to special training and/or education in the particular field, and (2) the conclusion which the expert is able to reach is based upon his special training and/or education, and is outside the expected realm of knowledge of the average juror/layperson.

In other words, while a juror in a medical malpractice case may hear evidence of the fact that a physician mistakenly and improperly removed a patient's spleen, the juror would not have the specialized knowledge which would permit the juror to then draw the conclusion as to the significance of the removal of the spleen and the problems which such a removal may or may not cause the patient in the future. Only a witness with special expertise in the field of medicine would be expected to know the significance of a spleen removal.

On the other hand, if the medical malpractice case involved the removal of the wrong leg, it would not be necessary, or permissible to have an expert witness testify as to the significance of the loss of a limb, since the average juror could be expected understand the significance of such loss, without special training or education.

Likewise, in security cases such as above, some courts take the view that the significance of a property owner's failure to have certain security devices in place, such as "peepholes" or chain locks on doors, is a conclusion which the average juror can reach without the assistance of expert advice or opinions. This was the rationale for the court in the above case excluding the testimony of the security experts.

PSCLR – MAY 85 (1)

Innkeeper's Limitation of Liability Statute Will Not Shield Hotel From Guests' Loss Unless Hotel Strictly Complies With Statute:

Florida Sonesta Corporation d/b/a Sonesta Beach Hotel v. Sam Aniballi and Margaret Aniballi, his wife; Sam Aniballi and Margaret Aniballi, his wife v. Florida Sonesta Corporation d/b/a Sonesta Beach Hotel, *Nos. 84-2097, 84-2103, in the District Court of Appeal of Florida, Third District, decided February 5, 1985, rehearing denied March 6, 1985, reported at 463 So. 2d 1203.*

While visiting the Sonesta Beach Hotel in North Miami, Florida, Sam and Margaret Aniballi suffered the loss of $85,000 in jewelry when a hotel employee left for lunch with the contents of their hotel safe deposit box and never returned. Although Florida statutes limit a hotel's liability for its negligent loss of a guest's property to $1,000, the Aniballis sued the hotel for the total amount of their loss. They alleged that the hotel had breached a contract with them and that it had been negligent in hiring the employee who stole the jewels. The trial court ruled that the hotel was clearly negligent and therefore granted the Aniballis' motion for summary judgment, but limited the hotel's liability to the $1,000 statutory maximum. This judgment was affirmed by the appellate division and both parties appealed to the District Court of Appeal. The hotel argued that there was conflicting evidence as to whether it had been negligent at all, and the Aniballis argued that the statutory limitation on the hotel's liability should not be applied.

After the Aniballis had checked into the Sonesta Beach Hotel and while they were unpacking their bags, they noticed a sign in the room which read, "Be safe ... put valuables in the hotel's safe deposit box." Mr. Aniballi then went to the front desk and asked for a safe deposit box for his and his wife's valuables. The clerk gave him a card, "Safe Deposit Box - Statement of Value," and told him to print his name on it and to sign it. The card set forth

Florida's statute on the limitation of a hotel's liability for guests' losses, was dated, and indicated the safe deposit box number and the guest's room number. The clerk then led Aniballi into the safe deposit room, and explained that the box must be opened and closed with two keys - one held by the guest and the other by the hotel clerk. The safe deposit box was opened and Aniballi was left alone to place his belongings in the box. Afterward, the clerk returned and the two closed and locked the box. This procedure was repeated on three separate occasions before the jewels were discovered missing. An investigation revealed that a desk clerk who had been hired several weeks earlier had disappeared with the jewels and $5,500 from the hotel's cash drawer, that the employee's identification and references were fictitious, and that he had made an extra key for Aniballi's safe deposit box.

The applicable Florida statutes state in part:

> 509.101 Establishment rules; maintenance of guest register.—
>
> (1) Any operator of a public lodging establishment ... may establish reasonable rules and regulations for the management of the establishment and its guests and employees; and each guest or employee ... shall abide by such rules .... Such rules and regulations will be deemed to be a special contract between the operator and each guest or employee ... [and] shall control the liabilities, responsibilities, and obligations of all parties. Any rules or regulations established pursuant to this section shall be printed in the English language and posted, together with a copy of Section 509.111, ... in the office, hall, or lobby or another prominent place....
>
> 509.111. Liability for property of guests.—
>
> (1) The operator of a public lodging establishment is under no obligation to accept for safekeeping any moneys, securities, jewelry or precious stones of any kind belonging to any guest, and, if such are accepted for safekeeping, he shall not be liable for the loss thereof unless such loss was the proximate result of fault or negligence of the operator. <u>However, the liability of the operator shall be limited to $1,000 for such loss, if the [motel] gave a receipt for the property (stating the value) on a form which stated, in type large enough to be clearly noticeable, that the [motel] was not liable for any loss exceeding $1,000 and was only liable for that amount if the loss was the proximate result of the fault or negligence of the operator.</u> [Emphasis added.]

The Sonesta posted a notice to guests which contained the statutes verbatim on the interior of each room's bathroom door. However, the hotel was not in the practice of giving any documentation to a guest who used a safe deposit box; Aniballi was not given a receipt or any kind of document in connection with his use of the safe deposit box. Therefore, the District Court of Appeal of Florida ruled that the Sonesta Hotel had failed to strictly comply with the statute requirement of posting the "Notice to Guests" in a "conspicuous manner," "in the office, hall, or lobby or another prominent place" and of giving a receipt stating the value of the Aniballi's property. Because the statute sets forth provisions for a hotel to follow to limit its liability where it negligently handles guests' property in its possession, the court stated that a hotel is required to <u>strictly comply</u> with the statute's requirements in order to claim its benefits. The District Court of Appeal ruled that the Sonesta failed to strictly comply with the statute, and thereby relinquished the benefits of limitation of liability. The case was therefore remanded for further proceedings consistent with this ruling.

Although the hotel may subsequently be found negligent and be held liable for the actual amount of the Aniballis' property loss, the court held that there was no evidence that it acted "outrageously, or with fraud, malice or wantonness, or oppression so as to warrant a punitive damages award."

COMMENTARY:

The above ruling should come as no surprise to the well-informed innkeeper and hotel security director. In fact, what is surprising about this case is that the trial judge, as well as the three judges on the appellate division, failed to follow the clear precedent established by similar cases in the past, and that the case had to be taken by certiorari to the District Court of Appeals before the correct ruling was reached.

In any event, the case does illustrate the fact that it is not especially difficult to run afoul of a limitation of liability statute in particular state, and it certainly would be in the best interest of all hotel/motel owners and operators, and security directors, to periodically review the innkeeper's statutes in those jurisdictions in which their properties are located to ensure that all provisions of applicable limitation of liability statutes are being strictly complied with by their managers and desk clerks. These limitation of liability statutes serve a very useful purpose in the industry but the courts will not countenance any deviation from the clear provisions of the statutes.

Does the statute in your jurisdiction require the innkeeper to give a receipt, with the value stated thereon, when a guest deposits valuables for safekeeping? Is this being done, or is the established practice in your hotel the same as the Sonesta's? Also, does the statute require posting of the notice "in the office, hall, lobby, or other prominent place," and if so, do you have notices posted in the proper locations?

PSCLR – JUN-JUL 85 (3)

Contributory Negligence Of Hotel Guest In Theft Of Jewels From Room May Be Evidenced By Guest's Keeping Valuables In Room Without Informing Hotel Employees:

Herbert Schear, et al. v. Motel Management Corporation of America, et al., *No. 673, September Term, 1984, in the Court of Special Appeals of Maryland, decided February 14, 1985, reported 487 A.2d 1240.*

Mr. and Mrs. Herbert Schear of Dayton, Ohio, reserved the entire 12th floor of the Holiday Inn Hotel in Chevy Chase, Maryland, for several days for themselves, other family members, and guests. Visiting Chevy Chase to celebrate a grandson's bar mitzvah, they chose the hotel because it was located near their two daughters' homes. They had rented the 12th floor a few years before for another grandson's bar mitzvah and had stayed in room 1212 on about 10 or 12 other occasions. When placing this reservation, Mr. Schear spoke to the hotel's manager, Mr. Craig, and the two discussed security measures available at the hotel. Mr. Schear decided to hire an additional private security guard for two days of their stay to protect himself, his family and his other guests.

Mr. and Mrs. Schear arrived at the hotel early in the afternoon and were carrying what Mr. Schear described as "an extraordinary amount of jewelry" valued somewhere between $200,000 and $250,000, in addition to their clothing and furs. They brought the jewels with them to attempt to sell some of it, and had discussed this possibility with a Washington area jeweler by telephone. After settling into the room, Mr. Schear telephoned the jeweler and left a message on his answering machine which included his name, hotel address, telephone, and room number, stating that he had arrived with the "items." The Schears left their room later that afternoon; the jewelry remained in the suitcases in their room. Mr. Schear locked the room door and, as was his custom, tossed the room key on the hotel desk as he walked out through the lobby.

Upon their return, Mr. and Mrs. Schear found that their jewelry, a fur coat, three pieces of luggage and some of Mr. Schears' suits had been stolen. The Schears brought a negligence action against the hotel owner, the franchisor Holiday Inn, Inc., the management corporation, and the hotel manager, but the court granted the franchisor's and the hotel manager's motions for a directed verdict and entered judgment in favor of the management corporation and the hotel owner.

The Schears appealed on the grounds that the trial judge had refused to admit police computer printouts of criminal activity reports in and around the hotel; had improperly instructed the jury on the plaintiffs' contributory negligence and assumption of risk; had refused to admit the Holiday Inn corporation's loss prevention manual and consumer advertisements; and had instructed the jury on Maryland's "innkeeper's statute." The Court of Special Appeals of Maryland reviewed and affirmed the trial court's holdings.

The court reviewed the plaintiff/appellant's allegations of error concerning the judgments granted to the hotel owner and the management corporation. The plaintiffs argued that the computer printouts were relevant in proving that the hotel and its management knew or should have known of the high incidence of criminal activity in and around the hotel, but failed to take reasonable steps to protect its invitees. They acknowledged that the printouts constituted hearsay evidence, but contended that they should be admitted under the written business and public record exception to that rule. However, the court ruled that although police incident reports are admissible under this rule, the computer printouts are not because "they are merely lists of unverified complaints made by unidentified individuals." Each complaint listed on the printout had been investigated by a police officer and those police investigative reports would be admissible, but the computer printouts are not based on those reports and are not admissible.

The Schears contended that the judge's instructions to the jury on contributory negligence and assumption of risk by the plaintiff were improper because of insufficient evidence. The plaintiffs argued that nothing less than their failure to lock their hotel room door would warrant such instructions. The appellate court disagreed, citing the Schears' testimony that they had never bothered to deposit their valuables in that or any hotel's safe even though they traveled frequently and were aware that safety deposit boxes were available. The extra security guard which the Schears had hired for their stay was not scheduled to start until two days after the theft, and Mr. Schear did not mention to the hotel manager or to any hotel staff member that he and his wife were carrying an "extraordinary amount of jewelry." In addition, Mr. Schear stated in a pre-trial deposition that leaving a telephone message with complete details for a jeweler whom he had never met was not a "reasonably prudent" act. Finally, Mr. Schear did not take the simple precaution of handing his room key over to a staff member at the front desk, instead of simply tossing the key onto the front desk. A jury could reasonable derive from this evidence that the Shears had acted negligently.

As to the plaintiff's contention that the trial judge erred in refusing to admit the Holiday Inn franchisor's loss prevention manual and consumer advertisements, the court upheld the trial judge's ruling that these items were irrelevant. The loss prevention manual was not in effect when the Chevy Chase Holiday Inn was built, and in fact served only as suggestive, and the franchisor did not require that it be followed. The advertisements were also irrelevant; the Schears testified that they had chosen the hotel because it was located near their daughters' homes, and not because they had relied on advertisements of high security standards by Holiday Inn.

The court also reviewed the directed verdict granted to the Holiday Inn corporation. The plaintiffs argued that the Holiday Inns corporation was responsible for the negligent acts of the Chevy Chase property manager under the theory of agency. The court disagreed, explaining that the corporation had a franchise agreement with the owners of the Chevy Chase property and the nature of the agreement did not contain the elements integral to an agency relationship, which must include (1) the principal's right of control over the agent, (2) the agent's duty to act primarily for the benefit of the principal; and (3) the power of the agent to make substantive alterations in the legal relations of the principal. In the Chevy Chase property's franchise agreement with Holiday Inn, the element of control was most notably missing. The franchise agreement merely granted the franchisee the use (for a fee) of the "Holiday Inn System," which included the Holiday Inn name and identifying characteristics, trademarks, and service marks. Though the franchisor conducted regular inspections to ensure that the franchisee maintained certain prescribed standards, the Holiday Inn corporation did not control the day-to-day operations of its franchise hotels. Because it was not proven that the Schears had relied on the Holiday Inn name in choosing the hotel, the Holiday Inn corporation could not be held liable for the negligent acts of the franchisee.

The plaintiff/appellants challenged the judge's instruction to the jury on the the Maryland Innkeepers' Statute as improper because the statute was not applicable if, as claimed, the innkeeper was negligent. The appellate court reviewed the statute, which states:

> If any iron safe or other secure depository for the keeping of the money, jewelry, securities and plate belonging to guests is provided, then, unless such money, jewelry, securities and plate are deposited with such hotel or inn, or its agents, the hotel or inn shall not be responsible for the loss of such money, jewelry, securities and plate, or any part thereof, by robbery or otherwise, unless such loss occurred from collusion or negligence on the part of such hotel or inn or its agent; in the event that such money, jewelry, securities and plate are deposited with such hotel or inn, the liability of the hotel or inn for the loss of such money, jewelry, securities and plate shall not exceed the sum of $300, unless at the time of such deposit the money, jewelry, securities and plate so deposited shall be exhibited to the representative of the hotel or inn and the value thereof declared by the depositor to such representative. In no event shall any hotel or inn be under any obligation to accept for safekeeping any money, jewelry, securities and plate having a declared value in excess of $1,000, nor shall any hotel or inn be liable in excess of $1,000, for loss of or damage to any such property, whether or not tendered to it for safekeeping and whether or not accepted by it for safekeeping.

A condition of the limitation of liability required that a printed notice of the statute's provisions be posted in each of the guest rooms of the hotel or inn. Another provision limited the liability of an innkeeper for the loss of "proper baggage" of guests, items other than money, jewelry, securities and plate, to $300.

Since there was evidence that the hotel had strictly complied with the requirements of the statute by providing a safe and a notice in the Schears' room, the court approved the judge's instruction to the jury on this statute. The innkeeper's statute would not be rendered inapplicable by a finding that the innkeeper was negligent, but would rather trigger the exception to the innkeeper's limitation of liability.

Dismissing all other allegations of error by the plaintiff/appellants, the appellate court affirmed the judgment of the trial court.

PSCLR – JUN-JUL 85 (5)

Innkeeper's Limitation-Of-Liability Statute Protects Innkeepers Against Negligence Claims As Well As Contractual Claims:

Paulette Laubie, Wife of/and Andre Laubie v. Sonesta International Hotel Corporation, Archer Casbarian, C. Mark Stevenson, Sam F. Uchello and United States Fire Insurance Company, *No. 84-3518, Summary Calendar, in the U.S. Court of Appeals, Fifth Circuit, on appeal from the U.S. District Court for the Eastern District of Louisiana, decided February 7, 1985, reported at 752 F.2d 165.*

The facts of this case were previously reported in **PSCLR** (see the October 1984 issue at page three) and therefore are not repeated here. Suffice it to say at this point that the Louisiana Innkeeper's Statute, which has now been amended by the Louisiana Legislature to protect innkeepers against negligence actions as well as contract action, has been upheld on appeal. The decision as reported in the October, 1984, issue has been affirmed: No landlord or innkeeper, or his officers, clerks, agents, or employees, can be held liable for breach of contract or for negligence to any guests or party of guests occupying the same apartment or room for any loss of cash, jewelry, or other stated items, in any sum exceeding $500 (previously $100, until recently amended by the Louisiana legislature) unless provided for in writing by special agreement with the landlord or innkeeper and the guests.

PSCLR – JUN-JUL 85 (8)

To Recover Damages For Murder Of Hotel Guest, Plaintiff Must Prove Inadequate Security Was Proximate Cause Of Assault:

Barbara F. Mitchell, et al., v. Pearson Enterprises, a Utah corporation, et al., *No. 18848, in the Supreme Court of Utah, decided February 5, 1985, reported at 697 P.2d 240.*

The widow and children of a Delaware businessman who was murdered while staying at a Salt Lake City, Utah, Hilton Hotel brought suit to recover damages for his wrongful death which they alleged was due to the hotel's inadequate security measures and the hotel's breach of implied and expressed warranties of habitability and safe accommodation which it owed to its guests. The trial court dismissed the suit by granting summary judgment to the Hilton Hotel's owner. The plaintiffs appealed, and the Supreme Court of Utah affirmed the summary judgment, holding that there was no evidence linking negligence on the part of the hotel to the man's murder.

The victim, Donald P. Mitchell, was murdered in his room in the Hilton Hotel after having breakfast in the hotel coffee shop with a business associate. He was found by a hotel maid and the associate who, concerned when Mitchell did not show up for a meeting, had summoned the maid to open Mitchell's hotel room door. His body was lying face down on the floor, and he had two gunshot wounds in the back of his head. It was later determined that he had been shot where he was found lying. His wallet had been rifled and there was no cash inside it, but his Rolex watch and credit cards were still on his person. There was no evidence of forced entry or a scuffle, other than scratches on Mitchell's arm. Although the police who investigated came up with several theories – Mitchell surprised a burglar who entered with a passkey, Mitchell was forced to open his room to a robber who surprised him in the hallway or elevator, or Mitchell was executed gangland style – the identity of the murderer was never discovered.

The plaintiffs attempted to prove that the owners of the Hilton Hotel were liable for wrongful death damages because Mitchell's death had been caused by the hotel's negligence in providing security measures adequate for a hotel of its size, set-up, and location. The hotel building consisted of ten floors (in the high-rise section) and contained more than 340 guest rooms, three restaurants, three lounges, a lobby, a convention center, 11 entrances, numerous stairwells, elevators, and a parking lot. Nine of the 11 entrances were open 24 hours a day and all of the stairwells allowed direct access from the outside to the guestroom floors. At the time of the murder, there were no electronic cameras, alarms, or other devices at the entrances, stairwell doors, or the guestroom floors. A single security guard patrolled the entire premises at any given time. In addition, the hotel's Chief of Security had been a Salt Lake City police officer, but he had no previous supervisory experience. He was nonetheless solely responsible for setting up the hotel's security system, without any assistance from the Hilton franchisor. In doing so, he made no attempt to determine the level of criminal activity in the hotel's immediate vicinity. Hiring of security officers was based on the hotel's standard application form; no security experience was required and applicants' military and police records were not checked. There was no system to supervise and control access to the room keys and the master keys.

In addition to these aspects of the hotel's security program, there was evidence that there had been previous crimes in the hotel, including an attempted rape and thefts from guestrooms and in the lobby. There was proof that people sometimes wandered in off the street and onto the guestroom floors to sleep in the hallways. It was also shown that a maintenance employee of the hotel, with access to all master keys, was on probation from an arrest and conviction for attempted homicide-robbery. This employee was the husband of the maid who was responsible for the rooms which included Mitchell's. This maid was in fact in possession of the passkey to Mitchell's room at the time of the murder.

The appellate court agreed that a hotel owes its guests ordinary care to insure their safety, the extent of which care would depend on the circumstances and location of the hotel property. The court further acknowledged that a hotel's failure to provide adequate security could foreseeably result in allowing and even encouraging criminal activity on the hotel property. Quoting from a decision by the Wisconsin Supreme Court, the court enumerated the relevant facts to be considered in determining whether a hotel's security measures were in keeping with the ordinary care requirement:

> industry standards, the community's crime rate, the extent of assaultive or criminal activity in the area or in similar business enterprises, the presence of suspicious persons, and the peculiar problems posed by the hotel's design.
>
> A hotel's liability depends upon the danger to be apprehended and the presence or absence of security measures designed to meet the danger. The particular circumstances may require one or more of the following safety measures: a security force, closed circuit television surveillance, dead bolt and chain locks on the individual rooms, as well as security doors on hotel entrance ways removed from the lobby area.

However persuasive the plaintiff's evidence might be in demonstrating the hotel's negligence, there must also be evidence establishing a direct link between the hotel's alleged negligence and the injury. This is where the trial court, and the appellate court in its review, found the plaintiff's case to be insufficient. Without any direct evidence to show how or where the murderer encountered Mitchell and to show that the hotel's security measures or lack thereof were in any way related to the murder, then a jury would be forced to speculate on the circumstances leading up to the murder. Such speculation is not permissible. Therefore, the summary judgment granted to the defendant hotel was affirmed by the appellate court.

COMMENTARY:

Lest there be any misunderstanding, it should be emphasized that the Utah Supreme Court here did not find that the defendant hotel had provided adequate security. In fact, the Court specifically stated that the plaintiffs had presented sufficient evidence to raise material issues of fact as to the defendant's negligence in providing adequate hotel security, including but not limited to the following:

> *(1) hiring employees without regard to the safety of guests, including ex-convicts and inexperienced security personnel;*
>
> *(2) lack of guidance by the hotel/franchisor and the minimal capability of the franchisee to set up an adequate security system for the hotel as built and laid-out;*
>
> *(3) minimal supervision or control by anyone in the hotel management structure over room keys and pass keys;*
>
> *(4) 24-hour access to hotel and guest floors through multiple, unsecured and unsupervised doors, and stairwells;*
>
> *(5) no surveillance devices of any kind;*
>
> *(6) insufficient security personnel for structure size and complexity;*
>
> *(7) apparent lack of response by security personnel to prior instances of crime in the hotel;*
>
> *(8) failure to investigate and determine levels of criminal activity in the vicinity of the hotel so as to tailor security procedures to that level.*

The problem the plaintiffs here had was not in proving that the defendant hotel was negligent in providing security, but rather in proving that the negligence was the proximate cause of the hotel guest's murder in his room. For instance, if the plaintiffs could have shown that the assault on Mitchell was the result of one of the above factors, the case would have gone on to trial instead of being dismissed. If it could have been shown that the murderer had in fact used a passkey or had in fact gained access to the hotel through an unlocked door which reasonable care would require to be locked, then the plaintiffs would prevailed on the motion.

There is no question but that this decision will be cited by hotel owners and operators and other defendants in future cases where they are attempting to avoid liability when there is clear evidence of their negligence in failing to provide adequate security for guests, tenants, and other invitees. However, regular readers of this publication will recognize that this decision reflects a very conservative approach by the court, and it is doubtful that the same result would have been reached in a number of other jurisdictions around the country.

PSCLR–AUG 85 (1)

Innkeeper's Statute Provides No Limitation Of Liability For Hotel Owner If Guest's Property Is Lost Due To Theft Or Negligence By Hotel:

Asim Kutbi v. Thunderlion Enterprises, Inc., a Nevada corporation; Tod E. McClaskey, Edward Pietz, Roy Fordham, James A. McClaskey, an Oregon partnership, dba Red Lion Motor Inn/Portland Center, *A8305-02815, CA A30982, in the Court of Appeals of Oregon, decided May 1, 1985, reported at 698 P.2d 1044.*

While Asim Kutbi was a guest at the Red Lion Inn in Portland, Oregon, his room was entered in the middle of the night by someone who stole his clothing, jewelry, and luggage, which he valued in excess of $100,000. Kutbi sued the hotel's owner, Thunderlion Enterprises, alleging that the theft was the result of the negligence of the defendants, citing twelve specific acts or omissions by the defendants which contributed to the loss. The trial court granted the defendants' motion for summary judgment, and Kutbi appealed. The summary judgment was reversed by the Court of Appeals of Oregon and the case remanded for trial.

Oregon has an Innkeeper's Law which limits the liability of hotel and motel owners/operators in certain situations, for any loss of or damage to a guest's personal property. In order to enjoy the benefits of the statute's limitation of liability, a hotel must maintain a safe or vault for the storage of the guest's valuables, must provide locks or bolts on guest room doors, and must "conspicuously post" the text of the innkeeper's liability statutes. The limitation of liability would not apply if a guest had attempted to deliver valuables to the innkeeper for safekeeping and the innkeeper had refused or failed to accept the goods, but in any event the amount of the innkeeper's liability would be limited to $300. However, there is no limitation of liability if the loss is due to theft or negligence on the part of the hotel or its employees.

Kutbi's allegations of negligence included a claim that the hotel failed to change the locks on Kutbi's hotel room following a prior theft (where there had been no evidence of a forced entry) or following the loss of certain "master keys", or following "repeated and unexplained loss of keys" to Kutbi's room and the adjacent room. He also alleged that the hotel had permitted excessive duplication of the room keys designed for access to Kutbi's room and the adjacent room. He listed several other allegations concerning the hotel's inadequate video security system, the management's failure to warn guests and failure to maintain adequate security, etc. However, in the defendant's responses to the allegations, they provided no evidence to dispute Kutbi's claims that they had negligently duplicated the keys or failed to change the locks. Therefore, the appellate court ruled that issues of fact remain which cannot be dismissed by a summary judgment, and Kutbi's suit was remanded for trial on these remaining allegations.

COMMENTARY:

It is important to look at the exact wording of the last sentence of the Oregon Innkeeper Law to fully appreciate the above ruling:

> *...Every innkeeper or hotel keeper is liable for the loss of any property of a guest in his inn or hotel, whether or not the property has been accepted for safekeeping as provided in this section, if the loss is due to the theft or negligence of the innkeeper, hotelkeeper, or any of his servants.*

Therefore, unless the innkeeper can successfully negate the plaintiff's allegations of negligence, the plaintiff will be able to put his evidence before a jury and try to prove that such negligence existed, and that such negligence was a contributing cause of the theft of the plaintiff's property. If so, then there is <u>no limitation</u> on the amount of the innkeeper's liability, and the plaintiff may recover the full value of the loss.

PSCLR – AUG 85 (4)

Innkeeper's Statute Limits Hotel's Liability For Guest's Alleged Loss Of $10,000 Ring To $100:

Dr. and Mrs. James L. Philip v. Fairmont-Roosevelt Hotel, Inc., *No. CA-2711, in the Court of Appeal of Louisiana, Fourth Circuit, decided May 8, 1985, reported at 469 So. 2d 1140.*

While a guest at the Fairmont Hotel in New Orleans, Louisiana, Mrs. James Philip discovered that one of her diamond rings was missing. She immediately reported the loss to the hotel's security officers who investigated, but the ring was not recovered. She and her husband, Dr. James Philip, sued the hotel for the full amount of the loss.

A Louisiana Innkeeper's Statute limits a hotel's liability for the loss of a guest's property to $100, absent some proof that the hotel was negligent. The plaintiffs claimed that the ring had been stolen due to the hotel's negligence, and that the limitation of liability therefore would not apply. The trial judge, after hearing the evidence, concluded that the alleged theft was not proven, and that the limitation of the hotel's liability for the loss was limited to $100. The Philips appealed, and the trial court's decision was affirmed.

There was not much evidence to prove that the ring had been stolen. Mrs. Philip gave conflicting stories to the officers when she reported that her ring was missing, first telling the officers that the band of the ring was cracked and that she may have lost the ring while playing golf. It was only after her husband insisted she had left the ring in the hotel room, that she stated she had last seen the ring in her jewelry case. This statement was the only one which the security officers recorded in their report, though two officers recalled that Mrs. Philip changed her story. Casting more doubt on the alleged theft was the fact that the ring was the only item missing: various pieces of costume jewelry, a $5,500 bracelet, and a $650 gold ring all located in the same jewelry case, along with a large amount of cash and travelers checks in a suitcase, were all left intact. In fact, the only other evidence the plaintiffs offered to prove that a theft had taken place was Mrs. Philip's testimony that a security officer had told her that a suspect maid had "escaped." The officer testified a maid was being investigated for another theft in the hotel, and he had mentioned it to Mrs. Philip in an attempt to "console her."

The appellate court agreed that there was no evidence to support a finding of negligence on the part of the hotel, and that the statute limiting the hotel's liability was indeed applicable. Therefore, the court upheld the $100 award.

PSCLR – NOV-DEC 85 (1)

Hotel Is Not Liable For "Negligent Infliction Of Emotional Distress" For Wife Who Saw Live News Coverage Of Fire At Hotel Where Her Husband Was Staying:

Victor Scherr and Karen Scherr v. Las Vegas Hilton, Hilton Hotels, *Civ. B006810, in the Court of Appeal of California, Second District, Division 2, decided May 30, 1985, reported at 214 Cal. Rptr. 393.*

Karen Scherr was watching television in Los Angeles when news of a hotel fire was broadcast live from the fire's location, the Las Vegas Hilton Hotel. She became distressed because she knew that her husband was a guest at the hotel. She later learned that her husband was in the hotel at the time of the fire, and that he suffered physical injuries. She and her husband brought suit against the hotel's owner: Mr. Scherr's suit was to recover for his personal injuries and property damage, and Mrs. Scherr sued to recover for the hotel's "negligent infliction of emotional distress." Mrs. Scherr's suit was dismissed and her appeal is the subject of this case.

The theory of negligent infliction of emotional distress was created to allow certain bystanders to accidents to recover for mental injuries due to witnessing the accident. In the case of *Dillon v. Legg* (1968) the California Supreme Court set forth California's guidelines for determining whether a bystander's allegations of mental distress are actionable:

> (1) ... the bystander was located near the scene of the accident; (2) ... the bystander's shock resulted from a direct emotional impact upon him from the sensory and contemporaneous observance of the accident; and (3) ... the bystander and the accident victim were closely related.

The trial court determined that, though Mrs. Scherr certainly met the third requirement, the fact that she "witnessed the fire" on television was an intervening force which terminated any duty the defendant may have owed her On appeal, Mrs. Scherr pointed out that "intervening acts of third parties do not relieve a tortfeasor of liability if the third party's acts were reasonable foreseeable." She argued that, since the live television news broadcast of a major hotel fire is a foreseeable consequence of a fire, the issue should not be dismissed as a matter of law.

The appellate court, however, found this issue – whether or not the television broadcast was an "intervening" force – irrelevant:

> ...the decisive issue in this case is whether plaintiff, through whatever medium, received a sudden and severe shock by actually and contemporaneously witnessing not just the fire but the infliction of injuries upon her husband.

Although it could be argued that Mrs. Scherr's viewing of the hotel fire was actual and contemporaneous, she did not allege she witnessed the infliction of her husband's injuries via the television broadcast. Indeed, all she learned from the news broadcast was that her husband was probably in danger. The court also found inadequate her argument on appeal that her mental distress was due to the "high probability" that her husband was being injured. Mrs. Scherr could not have known, as she watched the television news broadcast, her husband was in the hotel or that he was suffering injuries at the time.

Stating that to hold otherwise under the circumstances would "do violence to reason and common sense," the Court of Appeal of California affirmed the dismissal of Mrs. Scherr's suit.

PSCLR – NOV-DEC 85 (2)

Hotel Is Not Held Liable For Accidental Drowning Death Of Adult Guest Who Could Not Swim But Who Voluntarily Walked Into Pool Until Water Was To Her Chin:

Mary Brown, duly appointed tutrix of Kary Daily and Myra Daily v. John Harlan, Fernando Bustamente and XYZ Insurance Company, *No. 84-CA-534, in the Court of Appeal of Louisiana, Fifth Circuit, decided April 15, 1985, motion to dismiss granted June 17, 1985, reported at 468 So. 2d 723.*

On July 1, 1981, Sandra Daily, her two minor children, her sister (Mary Brown), her daughter, her cousins (Charles Roach and Carolyn Jackson), and her boyfriend (Rogers Castellanos) all checked into the Airport Rodeway Inn Motel located in New Orleans, Louisiana, while the plumbing was being repaired at their home. They rented two rooms for the night, went to the rooms, and then decided to go to the pool. Castellanos asked them not to go into the water until he returned from getting some food, since he was the only member of the party who could swim. Nonetheless, the children, Daily, and Brown went into the pool that night while Castellanos was gone. The children stayed only a few minutes, but Daily and Brown walked into the pool, holding hands, until the water was up to their chins. Brown lost her footing, and lost her grip on her sister's hand. She managed to make it to the side of the pool and Jackson pulled her out. They then realized they couldn't see Daily, and began shouting for help. A hotel employee came out, pulled Daily from the pool, and attempted to revive her. Although she was taken by ambulance to the hospital, she died due to drowning.

Brown brought suit on behalf of her niece and nephew to recover from the Inn, alleging its negligence caused Susan Daily's death. She alleged that the motel was negligent because there was no rope separating the deep from the shallow end of the pool, the pool light was out, and the bottom of the pool was slippery due to the presence of algae. A jury heard the evidence and found that the hotel had not been negligent; Brown appealed. She argued that the trial court had erred in making jury instructions and in excluding certain evidence.

The appellate court reviewed all of the evidence, including that which was excluded from the jury, and found that the judge had not erred. The plaintiff had wanted an inspector from the Jefferson Parish Health Department to testify on inspections of the Inn's pool on July 13, 14, and 28. The inspector was not subpoenaed before the trial and, though the court gave the plaintiffs an extra day to locate him, he was not available to testify on the report. The report contained his subjective impressions concerning the condition of the pool and the court ruled that it was not admissible in evidence without his accompanying testimony.

The trial court did allow, but limited, the testimony of an expert witness, a medical microbiologist who specialized in the study and identification of fungi and algae relating to the human patient. The doctor was permitted only to testify about the properties and characteristics of algae in the New Orleans area in general, but was not permitted to make specific statements about the presence of algae in the Inn's pool on the evening Daily drowned. She stated algae could have a slippery effect and could be present but unobservable. The only evidence admitted concerning the state of the pool was the motel manager's testimony that as of July 13, there was a problem with the pool's chlorinator.

The court also reviewed the judge's instructions to the jury and found that the trial judge had incorrectly given the jury instructions on the standard of "ordinary care," and had incorrectly omitted instructions on the here-applicable theory of strict liability. The appellate court noted that previous Louisiana cases have held that the innkeeper owes his guests a "high degree of care and protection" and is required to to take reasonable precautions to "minimize the risk to his guests within his sphere of control." The appellate court specifically cited the Louisiana Supreme Court decision in *Kraaz v. La Quinta Motor Inns* (reported in the August 1982 **PSCLR**, case 2, page 2) and the U.S. Court of Appeals, Fifth Circuit decision in *Banks v. Hyatt Corp.* (reported in the April 1984 issue, case 1, page 1).

The jury was given instructions on ordinary negligence instead of the theory of strict liability. The burden of proof in a strict liability case is less than that that required in an ordinary negligence case. Under strict liability, the plaintiff must prove that the subject circumstances created an unreasonable risk of harm which resulted in injury to the plaintiff by the defendant. Under ordinary negligence, the plaintiff must additionally prove that the defendant knew or should have known of the risk of harm, but did nothing to prevent the resulting injury to the plaintiff.

However, the court also found that the decedent had voluntarily assumed the risk of harm. She was familiar with the pool: Castellanos had given her rudimentary swimming lessons in the Inn's pool on the day before she drowned in the same pool. She knew that neither she nor anyone present could swim, and discussed with Castellanos the dangers of entering the pool when he wasn't present. She could see that no lifeguard was on duty, that there was no rope in the pool, and that the pool was dark. As a 25-year-old woman without physical or mental impairments, she was fully capable of understanding the risks involved but entered the pool anyway.

Since it was clear that the decedent had voluntarily assumed the risk of harm, the defendant hotel could not be held liable for her death. The judgment dismissing the plaintiff's suit was therefore affirmed.

PSCLR – MAR 86 (3)

Hotel Is Granted Summary Judgment In Negligent Security Suit Brought By Husband After Wife's Suit On Same Incident Was Unsuccessful:

Ronald Hendrix Courtney v. Albert N. Remler, Devaul L. Henderson, Jr., Dwain E. Ball, Erling D. Speer, Bernie Hirsch, Bernie Slotin, Robert M. Crumley, James E. Wright, E.K. Van Winkle, Jr. d/b/a Hotel Investments, A Limited Partnership, *Civ. A. No. 2:83-2347-1, in the U.S. District Court for the District of South Carolina, Charleston Division, decided June 29, 1985, reported at 612 F. Supp. 967.*

This case was filed in the U.S. District Court for South Carolina after an incident in which Ronald Courtney and his wife Deborah Marie were assaulted in their hotel room during their honeymoon at the Islander Inn, in Hilton Head, South Carolina. Mrs. Courtney was sexually assaulted in the incident, and sued the owners and managers of the Islander Inn, alleging that the defendants were negligent and reckless in designing, constructing, maintaining, and operating the hotel. The trial of her suit ended in a verdict for the defendants; the court held that, although defendants should have foreseen the possibility of criminal assault on the Courtneys, defendants had acted reasonably to ensure the safety of guests at the Islander Inn. The verdict was upheld on appeal to the U.S. Court of Appeals.

Mr. Courtney now brings suit for loss of his wife's consortium, and for his personal injuries resulting from the same incident. The defendants argued that the resolution of Mrs. Courtney's suit in their favor required that this suit be dismissed as a matter of law.

The District Court examined South Carolina's law on "collateral estoppel" and found that a particular issue which has been "fully and fairly" litigated shall not be relitigated. The court therefore examined the issues which had been litigated and resolved in Mrs. Courtney's suit:

There was no evidence of prior assaults or other crimes against any person occurring on the Islander Inn premises.

The doorway to each guest room had a fluorescent light above it which illuminated the hallway and enabled the guest to see through the "observation port" located on the room side of the door. The port was configured with optics so as to allow magnification and wide vision of the hallway area around the room door. The door was a fire rated steel door with a steel frame, equipped with an automatic door lock and a deadbolt, and fitted with "security lips" to prevent the locks from being picked. In fact if the guest fully engaged the locks while inside the room, only the manager's "emergency key" could unlock the door from outside.

The hotel had implemented a security plan, which included both in-house and contract security personnel. In-house personnel provided security on a 24-hour basis through specific staff members charged with security responsibilities and a "security awareness" program for all employees. A private security guard company provided extra security during the night hours, and a uniformed guard was on duty the night that the Courtneys' room was broken into.

Based on these facts, the District Court found:

> Based on the lack of criminal activity on the premises in the past, one uniformed guard during the nighttime hours is reasonable security for a guest at the motel;
>
> The entire security program was reasonable under all the circumstances;
>
> [T]he design of the building is reasonable.
>
> The guest room doors, with all their locking mechanisms, provided reasonable protection for the safety of the guests.
>
> ***
>
> The Islander Inn, as a whole, was operated and supervised reasonably. The Inn obviously felt a need to protect its guests from criminal attacks by third persons, and measures were taken to ensure their safety. The measures were reasonable. The rape incident was a truly unfortunate and horrible experience, but the defendants cannot be held responsible. For this court to hold otherwise would be equivalent to making all hotels the insurer of their guest's safety.

The court concluded that Mr. Courtney's suit for loss of consortium and personal injuries resulting from the same incident raised no new issues of negligence against the defendants. As described above, the defendants were exonerated in Mrs. Courtney's suit for each of the allegations also raised in Mr. Courtney's suit. Therefore the District Court ruled that his suit was barred because of the prior judgment, and summary judgment was granted in favor of the defendants.

PSCLR–MAR 86 (4)

Insurance Policy Excluding Coverage For Jewelry Stolen From Hotel Room Unless Jewelry 'Attended By' Insured Requires Physical Presence Of Insured With Jewelry In Order For Insurance To Cover Loss:

John Michael Poland, Lead Underwriter for Those Certain Underwriters at Lloyds, London Subscribing to Policies Numbered HJ 27213 and HJ 27214 v. Dean Paul Martin and Dorothy Hamill Martin, *No. 84-5739, in the U.S. Court of Appeals, Ninth Circuit, on appeal from the U.S. District Court for the Central District of California, decided May 21, 1985, reported at 761 F.2d 546.*

Dean Paul Martin and his wife Dorothy Hamill Martin suffered the loss of jewelry valued at $365,994 when Mrs. Martin's suitcase was stolen from her room at the Pacific Plaza Hotel in San Francisco. The Martins filed a notice of loss and a reimbursement claim under their insurance policies with Lloyds, London. Lloyds refused to cover the loss, claiming that the policies in question contained a hotel and baggage restriction which excluded the Martins' claim. Lloyds filed for a declaratory judgment with the U.S. District Court, and the Martins counterclaimed for Lloyds' breach of statutory and contractual duties; Lloyds was granted summary judgment and the Martins appealed.

Hamill took the jewelry with her during a five day visit to San Francisco; she chose to keep the jewelry in a locked case in her hotel room, rather than to use the hotel's safe. Before leaving the hotel room, Hamill placed the locked jewelry case into her suitcase and hid the suitcase under a bureau. While she was out of the room, a maid opened the door for an unknown man who claimed that it was his room. When Hamill returned, she discovered that her suitcase along with all of her jewelry was gone.

The Martins' insurance policies with Lloyds contained this restrictive language:

> This insurance excludes the loss of or damage to any item of jewelry valued at $1000 or more while on the premises of hotels, motels, resorts or health facilities unless the said jewelry is attended by the assured or is contained in a locked safe or vault.

In arguing against the summary judgment motion of Lloyds, the Martins listed five unresolved issues of fact:

> (1) whether Hamill exercised due diligence in guarding her jewelry;
> (2) whether Hamill acted reasonably and prudently to guard her jewelry;
> (3) what was the proper construction of the word 'attended' in the hotel exclusion of the policy [The Martins asserted that it could mean 'to look after,' by taking precautions such as hiding the jewels]; (4) what was the proper construction of the baggage exclusion of the policy; and (5) whether the action of the hotel employee who let an unknown man into Hamill's room was a concurrent cause of the theft which took it out of the exclusions.

The U.S. Court of Appeals reviewed the alleged issues and ruled that the plain meaning of the policies' restriction required that the insured either place the jewelry in the hotel safe or be physically present with the jewelry in order to be eligible for recovery under the policies. Therefore, it was irrelevant whether or not Hamill exercised due diligence, or acted reasonably to protect her jewelry, since she did not comply with the policies' restriction.

The Martins' argument that the hotel maid's negligence in unlocking Hamill's hotel room for the thief along with Hamill's failure to remain with the jewelry constituted "concurrent causes" of the loss is also irrelevant. The appellate court therefore affirmed the judgment granted to Lloyds.

COMMENTARY:

Now that the Martins have failed in their effort to recover from their own insurance carriers, one might speculate that there will now be an effort made to recover from the Pacific Plaza Hotel. The Martins may claim that the maid's negligence in allowing the unidentified man into the Martins' room was a breach of the hotel's duty to safeguard the privacy and personal possessions of the hotel guests and constituted the proximate cause of the loss.

PSCLR – APR 86 (3)

Hotel Guest's Recovery For Theft Of Jewelry Is Limited By Statute Even Though Statute Was Passed After Theft Occurred:

Mr. and Mrs. George Peppard v. Hilton Hotels Corporation, *No. CA-3147, in the Court of Appeal of Louisiana, Fourth Circuit, decided January 15, 1986, rehearing denied February 26, 1986, reported at 482 So. 2d 639.*

During January of 1978 Mr. and Mrs. George Peppard were staying in the Hilton Hotel while visiting New Orleans. They chose to keep their valuables in the room in their suitcases instead of placing them in the hotel safe deposit box. Unfortunately, the room was burglarized and their valuables taken. The Peppards brought suit against the Hilton Hotel Corporation, alleging that the hotel was negligent in failing to provide them with adequate protection from theft. After a 1981 trial, the court ruled (in 1983) that the Hilton was negligent and the Peppards were awarded $500. The Peppards appealed the amount of the award.

The Louisiana civil code concerning innkeeper's liability was originally enacted to provide that

> an innkeeper, who provided a safe deposit box for use by his guests and otherwise notified them by posting the Louisiana Hotel Law, would only be responsible for the loss of personal property of guests 'by theft or otherwise' up to a $100.00 limit.

A 1982 Louisiana Supreme Court decision held that the Louisiana code would not apply if there was proof that the guest's loss occurred because of negligence on the part of the innkeeper, and that a guest could then recover full damages resulting from the negligence. (See *Laubie v. Sonesta International Hotel Corp.*, case two in the October 1984 **PSCLR**, and case five in the June-July 1985 issue.) The Legislature responded to this decision by amending the Louisiana code so that the innkeeper's liability for the loss of guests' valuables would be limited even if the loss was due to the innkeeper's own negligence. Effectively overruling the state Supreme Court's interpretation of the statute in *Laubie*, the amendment also increased the maximum recovery limit from $100 to $500.

The Louisiana Supreme Court ruled that, in spite of the defendant's negligence, the Hilton Hotel's liability was limited to the statutory maximum. The court expressed some doubt "as to the correctness of awarding the Peppards $500 in damages under the amended version of the article rather than the $100 maximum which was in effect at the time of the theft," but did not disturb the award.

COMMENTARY:

Generally speaking, a new statute can only govern conduct in the future, and can have no retrospective operation. However, there are three little-known exceptions to this general rule:

> *Laws which suppress or reduce penalties;*
>
> *Laws which are merely interpretive of previously-existing legislation; Laws which the legislature either expressly or impliedly declare to be retroactive.*

With regard to the interpretive law exception which is the exception on which the Peppard decision above was based, the existence of the exception is founded on the fact that the law does not establish a new rule as such, but merely determines the meaning of an already-existing law and may therefore be applied to fact situations which occurred prior to the enactment of the new interpretive law.

Therefore, the new law is not truly "retroactive" because it is the original law rather than the new interpretive law which establishes rights and duties between the parties. While all of that may sound well and good, it is little comfort to a plaintiff who finds himself in the position of the Peppards.

PSCLR – MAY 86 (2)

Innkeeper's Negligence Will Not Give Rise To Liability For Third Party Criminal Assault On Guest Without Evidence That Innkeeper Had Reasonable Grounds To Foresee Assault:

Burdine et al v. Linquist, *No. 71055, in the Court of Appeals of Georgia, decided January 7, 1986, rehearing denied January 23, 1986, reported at 340 S.E.2d 199.*

While removing luggage from their car to their motel rooms, Mr. and Mrs. James Burdine and Mr. and Mrs. Wallace Adolph were accosted by an armed man who forced them into the motel room, and robbed and assaulted them. The Burdines and Adolphs brought suit against the owner of the Journeys End Motel in Atlanta. The owner, Edwin Linquist, was granted summary judgment at trial, and the plaintiffs appealed.

The plaintiffs argued that the motel owner should have foreseen and prevented their assault. They alleged that poor lighting in the area immediately outside their motel room was a "defective condition" which subjected them to a unreasonable risk of criminal activity. The motel owner, they contended, knew that this dangerous condition existed because there had been several motel room burglaries in the two weeks prior to this assault.

The Court of Appeals of Georgia quoted a ruling from its decision in another case, *McCoy v. Gay* (reported in the August 1982 issue of **PSCLR**, case two, page two):

> Under the [applicable Georgia statute], the true ground of liability of the owner of property to an invitee who is injured thereon is the superior knowledge of the proprietor of the existence of a condition that may subject the invitee to an unreasonable risk of harm. ... It is the duty of a proprietor to protect an invitee from injury caused by the misconduct of employees, customers, and third persons if there is any reasonable apprehension of danger from the conduct of said persons or if injury could be prevented by the proprietor through the exercise of ordinary care and diligence.'

Generally, an innkeeper is not liable for guests' injuries due to an intervening criminal act, even in the event of the innkeeper's or his employee's negligence. Under Georgia law, however, if there is evidence that the innkeeper knew or should have known that there was a risk of criminal injury to his guests, then the guest would be able to recover for such injury from the innkeeper.

The defendant hotel owner argued that the prior burglaries were crimes against property and not against people, and therefore did not provide him with knowledge that there was a "defective condition" which put his guests at risk of an assault. The appellate court disagreed, and noted that at least one of the prior room burglaries occurred in a room which had dim outside lighting. Further, it is not necessary for the prior crimes to have occurred against people in order to be admissible to show that the subject assault was foreseeable. The court again quoted a previous Georgia case, *Bayshore v. Pruitt* (reported in the November-December 1985 **PSCLR**, case six, page five), in which it had ruled:

> In order for a party to be held liable for negligence, it is not necessary that he should have been able to anticipate the particular consequences which ensued. It is sufficient if in ordinary prudence he might have foreseen some injury would result from his act or omission and that consequences of a generally injurious nature might result.

Therefore, the Georgia Court of Appeals reversed the summary judgment granted to the defendant hotel owner Linquist, and remanded the case.

PSCLR—JUN-JUL 86 (5)

Hotel Guest Injured In Room Burglary Cannot Recover For Property Loss Due To Innkeeper's Statute And Cannot Recover For "Sentimental Value" Of Stolen Property As Part Of Personal Injuries:

Everett Royal and wife, Lois Royal v. Days Inns of America, Inc., *in the Court of Appeals of Tennessee, Western Section, at Knoxville, decided November 14, 1985, Application for Permission to Appeal denied by Supreme Court, March 24, 1986, reported at 708 S.W.2d 411.*

While Lois and Everett Royal were sleeping in their Days Inn motel room in Chattanooga, Tennessee, an intruder entered their room to burglarize it. When Mrs. Royal awoke to see a man taking her purse, she attempted to restrain him, but he fled through the room's open sliding glass door with her purse. Her purse contained cash for her and her husband's vacation and her "lifetime collection of jewelry, including her wedding band." The Royals sued Days Inns of America to recover damages for the hotel's negligence which led to their property loss and personal injuries. The claim for property injuries was dismissed under the Tennessee "hotel safe law," but the jury awarded Mrs. Royal $18,000.00 and Mr. Royal $2,000.00 for their personal injuries. The motel owner appealed, and the Tennessee Court of Appeals reversed the personal injuries verdict.

In its review of the defendant's appeal the Tennessee Court of Appeals examined the defendant's contention that there was no evidence to support the verdict that the Days Inn's negligence had caused the burglary. A plaintiff in a negligence suit must prove the existence of a duty on the part of the defendant to the plaintiff, and a breach of that duty by the defendant which caused the injury to the plaintiff. The appellate court affirmed that there was sufficient proof to support the jury's finding of negligence on the part of Days Inns: there was evidence that the sliding glass door through which the burglar fled was also his means of entry into the room, possibly due to a defective lock on the door. The evidence also showed that there had been break-ins at the motel in the past and that "many glass door locks had at one time been in need of repair."

The defendant also argued that the trial court erred in admitting evidence of other, unrelated criminal incidents on the motel premises during the 18 months prior to the plaintiffs' burglary, and that testimony by a local police lieutenant that "crime was greater at the Days Inn than at other nearby motels" unfairly influenced the jury. Fourteen prior incidents were admitted into evidence at the trial: four room robberies, one assault, three robberies outside rooms, three incidents of vandalism, and three incidents of hubcap theft. The appellate court stated that though such incidents did not have all the characteristics of the Royal's burglary,

> the severity of the prior incidents matches that of the events in this case. ... Such incidents could be relevant in establishing that general criminal activity on the premises was high, creating a climate conducive to other crimes specifically similar to the one in this case.

The appellate court stated that the decision to admit evidence of the prior crimes and the testimony by the police lieutenant was within the discretion of the trial court, and that the trial court had "wisely refused to admit evidence of some 103 additional criminal incidents" of lesser significance.

The defendant Days Inns also objected to the trial court's admission of testimony concerning the Royals' loss of property since that claim was dismissed under Tennessee's hotel safe law. In spite of this the judge permitted Mrs. Royal to testify about the "sentimental value" of her stolen jewelry to "establish the extent of mental anguish occasioned by its loss." However, the appellate court stated:

> Sentimental value of personal property is not recognized in Tennessee as an element of damages. ... Since the property damage claim was disallowed, and one cannot recover for the sentimental value of personal property, it was improper to allow this testimony as a basis for an award on mental anguish.

The testimony concerning Mrs. Royal's distress over the loss of the jewelry was heavily emphasized by her attorney in his presentation of the personal injury case. Therefore, the trial court's error in allowing this testimony was sufficiently prejudicial to the jury to require that the plaintiffs' verdict and damages award be reversed and the case remanded for a new trial.

COMMENTARY:

We have here a case in which there was no expert testimony presented by the plaintiffs to establish the standard of care owed by the hotel to the guests. However, as in the Roettger case in this issue, the appellate court agreed with the trial court on this point and found that the jury had sufficient information to enable the jury to make its own determination as to whether the defendant hotel had acted reasonably and prudently given all the facts and circumstances of the situation. Expert testimony is not absolutely necessary except in those situations in which the causal relationship between a cause and its arguable effect would not be known to the average juror but would only be known to an expert with training and experience in the particular field of knowledge under consideration.

PSCLR—AUG 86 (7)

Motel Owner Is Not Liable To Undercover Federal Agents Injured and Killed In Shoot-Out Where Owner Allegedly Knew Of Illegal Activity At Motel And Did Nothing To Stop It:

Esilda Rois etc., et al. v. Teresa Junco, Aldalberto A. Mastos, et al., Nos. 85-1336, 85-1565, *in the District Court of Appeal of Florida, Third District, decided April 1, 1986, rehearing denied May 19, 1986, reported at 487 So. 2d 331.*

A motel located in Miami, Florida, was the site of a shoot-out between narcotics offenders and undercover officers of the Federal Drug Enforcement Agency. One of the officers injured in that shoot-out and the estate of an officer killed in the fracas brought suit against the owner and operator of the motel. Their lawsuit alleged that the defendants had actual knowledge of several illegal drug transactions which had taken place on the motel premises, but had not taken any steps to prevent this criminal activity. They alleged that this negligence led to a failure in the DEA's undercover operation at the motel, and the subsequent shoot-out. Judgment was entered for the defendants, and the plaintiffs appealed.

The District Court of Appeal affirmed this judgment. Even if it was proven that the motel owner had known that illegal drug transactions were taking place on his premises, he could not be held liable to the offers, who were equally aware of the illegal activity and the accompanying danger. Moreover, the officers' lawsuit against the motel owner is barred under the "fireman's rule," which provides that a fireman, or police officer, who is injured as a result of being engaged in hazardous activities on private property in the performance of official duties cannot recover from the property owner.

The appellate court affirmed that no cause of action against the motel existed under the circumstances of this case, and the case was properly dismissed by the trial court.

PSCLR—NOV-DEC 86 (1)

Statute Limits Innkeeper's Liability To $750 For Guest's $50,000 Property Loss, Even If Innkeeper Guilty Of Gross Negligence:

Robert D. Kahn v. Hotel Ramada of Nevada d/b/a Tropicana Hotel and Country Club, *No. 85-1739, in the United States Court of Appeals for the Fifth Circuit, on appeal from the United States District Court for the Western District of Texas, decided September 10, 1986, reported at 799 F.2d 199.*

Robert Kahn, a jewelry broker, stayed at the Tropicana Hotel in Las Vegas while participating in a Blackjack tournament. He had a briefcase containing jewelry among his luggage because he planned to go next to Phoenix, Arizona, to meet with some customers. After Kahn vacated his room and checked out of the Tropicana, he left his luggage with a bellman, telling him that the contents of the briefcase were "important." About an hour later, Kahn returned for his luggage and learned that his bags were missing. Later, one valise and the now-empty briefcase were located on the golf course of the hotel.

Kahn sued the Tropicana to recover the value of his missing property, which he claimed to be $50,000, due to the hotel's ordinary negligence, gross negligence, breach of bailment, and "other violations of its duty as an innkeeper."

The suit was brought in the United States District Court because the "amount in controversy" was over $10,000, which is the minimum amount over which the federal district court can exercise jurisdiction in this type of claim. But, the district court ruled that the Nevada Innkeeper's Statute limited the hotel's total potential liability to a maximum of $750 thereby removing the action from the district court's jurisdiction. Kahn appealed to the U.S. Court of Appeals, which affirmed.

Nevada's Innkeeper Statute states:

> 1. No owner or keeper of any hotel, ..., in this state is civilly liable for the theft, loss, damage, or destruction of any property left in the room of any guest of such an establishment because of theft, burglary, fire or otherwise, in the absence of gross neglect by the owner or keeper.
>
> 2. If any owner or keeper of any hotel, ..., provides a fireproof safe or vault in which guests may deposit property for safekeeping, and notice of this service is personally given to a guest or posted in the office and the guest's room, the owner or keeper is not liable for the theft, loss, damage or destruction of any property which is not offered for deposit in the safe or vault by a guest unless the owner or keeper is grossly negligent. An owner or keeper is not obligated to receive property or deposit for safekeeping which exceeds $750 in value or is of a size which cannot easily fit within the safe or vault.
>
> 3. The liability of the owner or keeper under this section does not exceed the sum of $750 for any property of an individual guest, unless the owner or keeper receives the property for deposit or safekeeping and consents to assume a liability greater than $750 for its theft, loss, damage, or destruction in a written agreement in which the guest specifies the value of the property.

The second and third sections above were added in a 1979 amendment because the Nevada courts were granting large damage awards to hotel guests in spite of the then-existing statute. The plaintiff argued that section three referred "only to liability explicitly imposed by the [other section of the] statute." The Court of Appeals did not agree, stating that if the statute was read this way, then

> a grossly negligent innkeeper would be held liable for only $750 if property were taken from a guest's room (whether or not the innkeeper provides a safe) whereas an only slightly negligent innkeeper who provided no safe to his guests would be subject to unlimited liability for theft of property left in the hotel lobby.

The Court of Appeals noted that the third section of the statute was awkwardly worded, but that all three sections of the statute must be interpreted thusly: The first section shields the innkeeper from liability for the loss of a guest's property in the guest's room, in the absence of gross negligence. The second section shields the innkeeper from liability, if the innkeeper provided a safe and gave proper notice that it was available. The third section limits the innkeeper's liability for gross negligence with regard to property taken from the guest's room or elsewhere. The court stated that the proper interpretation of the statute will have the affect of limiting an innkeeper's liability to $750 "when a guest's property is lost as a result of the gross negligence of the innkeeper except when the innkeeper has waived its application."

The Court of Appeals acknowledged that its interpretation of Nevada's innkeeper statute limited the liability of even the grossly negligent innkeeper who did not provide a safe. However, the court determined that the Nevada legislature intended to limit the liability of innkeepers "in all instances of negligence except when the innkeeper consents to assume more responsibility." Therefore, Kahn cannot recover more than $750 even if he proved that the Tropicana Hotel was guilty of gross negligence. The action therefore cannot be brought in the federal district court because the amount in controversy is $750 and well below the $10,000 minimum jurisdictional limit.

PSCLR – NOV-DEC 86 (2)

Innkeeper Does Not Have A Duty To Protect Guest Sitting on Balcony Railing From Risk Of Falling Off Balcony:

Edward N. Eldridge v. The Downtowner Hotel a/k/a The Inn on Bourbon Street, Inc., and Liberty Mutual Insurance Company, *No. CA-4248, in the Court of Appeal of Louisiana, Fourth Circuit, decided June 18, 1986, reported at 492 So. 2d 64.*

During the Mardi Gras celebration in New Orleans, Edward Eldridge was a guest at the Downtowner Inn located in the French Quarter. While sitting on a second floor balcony of the hotel, Eldridge observed that people on other balconies were exposing themselves to the crowd below. Apparently swept away with the wild atmosphere of the Mardi Gras celebration, Eldridge climbed up on the balcony railing and exposed his buttocks. Unfortunately, he then fell off the balcony and sustained serious injuries.

Eldridge filed suit against the hotel, asking for 1.75 million dollars in damages, alleging that the hotel's negligence caused his injuries. He argued that the hotel had a duty to provide protective screens or uniformed guards to prevent patrons from falling off balconies, because such incidents were a foreseeable result of the wild atmosphere of Mardi Gras and "the fact that traditionally women would expose themselves from the balconies."

The trial jury returned a verdict for the defendant hotel and the plaintiff appealed.

The main issue on appeal was whether the trial court erred by not instructing the jury to apply the "duty-risk analysis of negligence liability," which inquires:

> I. Whether the conduct of which the plaintiff complains was a cause-in-fact of the harm;
>
> II. Whether there was a duty on the part of the defendant which was imposed to protect against the risk involved;
>
> III. Whether there was a breach of that duty; and
>
> IV. [Whether there were any resulting] Damages.

(The court here cited *Vicknair v. Hibernia Building Corp.*, reported in the March 1986 **PSCLR**, case seven, page seven and *Harris v. Pizza Hut of Louisiana*, reported in the January 1985 issue, case one, page one.)

Applying the above inquiry to the present case, the court noted that the evidence showed that Eldridge's own bad judgment caused his injuries. He chose to sit on the railing, lost his balance, and fell; no one pushed him and the railing did not collapse.

A hotel does not have a duty to protect guests from their own conduct; "a visitor assumes the obvious, normal or ordinary risks attendant on the use of the premises and owners are not liable for injuries to a visitor when those injuries result from a danger which he [the visitor] should have observed in the exercise of reasonable care." There was no evidence to support the plaintiff's argument that the defendant should have foreseen that a guest might fall from a balcony during Mardi Gras; there was no evidence that anyone had ever fallen from, or had even sat on, a balcony railing. The trial court, therefore, did not err by failing to apply the duty-risk analysis.

Eldridge also argued that the trial court had erred by not admitting evidence that other hotels had taken measures to prevent similar such incidents. Such measures were not required under Louisiana law, and the evidence was ruled to be irrelevant. The appellate court therefore affirmed the judgment of the trial court.

PSCLR—APR 87 (1)

Extent Of Prior Criminal Conduct On Premises Should Be Examined To Determine When And If Subsequent Criminal Conduct Was Foreseeable:

Daniel Moye, as administrator of the Estate of Cheryl Denise Moye v. A.G. Gaston Motels, Inc., and Otis Smith, *86-191, in the Supreme Court of Alabama, decided September 19, 1986, rehearing denied November 21, 1986, reported at 499 So. 2d 1368.*

This appeal was taken by the estate of Cheryl Denise Moye, deceased, after the trial judge granted summary judgment dismissing the estate's lawsuit against A.G. Gaston Motels, Inc. (the owner of a motel) and Otis Smith (the owner of a security service).

According to the appellate opinion, the innkeeper Gaston sponsored a dance for teenagers at his motel, in conjunction with a promotion by local radio station WENN. The dance was scheduled to last from 9:00 PM to 1:00 AM and was open only to teenagers up to the age of 19 (identification showing proof of age was required at the door). Gaston contracted with Otis Smith for the latter to provide four security guards at the dance. Certain employees from the radio station also assisted by collecting money at the door and checking identification. No alcohol, drugs, or weapons were permitted and everyone entering the dance was searched. Also, no one was permitted to leave and return later without again paying the admission price and undergoing another search.

The premises were checked periodically and the parking lot and the area outside the entrance to the motel, to keep those areas cleared for people who were coming into the dance. No alcoholic drinks were served at the dance, and there were no reports that anyone at the dance was in fact drinking. During the evening, one person was ejected around midnight due to some undisclosed disturbance. However there were no other disturbances during the evening.

Prior to the evening in question, the motel and radio station had sponsored at least eight dances, at which there had never been a single criminal incident.

Promptly at 1 AM, after the last dance, the lights were turned up. The security guards cleared and locked up the lounge where the dance was held. The last two security guards who were leaving walked a radio station employee who had collected the money to the motel office. Another radio station employee went with his wife to the radio station across the street from the motel. The area immediately in front of the motel was well lit, and approximately 15 teenagers were outside as the motel dance area doors were locked. According to one security guard, the teenagers seemed to be waiting for rides home, or heading to the convenience store next door.

Included among the group of teenagers were Cheryl and Daniel Moye, who were waiting for their ride home. As they waited, a group of young men who had been at the dance earlier returned to the motel and mingled with the teenagers waiting out front. One of the young men, Darron Burpo, had a pistol and Tara Foster saw it and ran over to the motel office to inform the security guard. She did not scream, shout, or indicate any distress or alarm, but simply motioned to the security guard in the office to "Come here." When she noticed that the young men had left, she returned to continue standing with her friends. Burpo, however, drove over to the convenience store next door where he got into an altercation with Terrance Oliver in the parking lot of the store. After Burpo and his friends got back into their car and began to drive away from the area, Oliver hit the car with a soft drink bottle as it drove by. Whereupon, Burpo fired one shot out of the car window with his pistol. The bullet hit and killed Cheryl Moye.

In reviewing the appeal filed by Cheryl Moye's estate from the dismissal of her suit, the Alabama Supreme Court noted, "It is difficult to impose liability on one person for the intentional criminal act committed by a third person." Noting that the reason for such difficulty is that most often the two essential elements of proof which must be present in a successful negligence action are absent: those being a duty owed by the defendant to the plaintiff and a showing of proximate cause. That is, the breach of the duty owed to the plaintiff caused his damages. However, the court went on to explain whether it is a lack of a legal duty owed or a lack of proof of proximate cause, "the key to either is foreseeability."

> It is the general rule in Alabama that absent special relationships or circumstances, a person has no duty to protect another from criminal acts of a third person.

The court then went on to review several cases which it had previously decided dealing with the issue of premises liability. In doing so, the court noted that it has never rendered an opinion imposing liability on a business owner/operator for injuries suffered as a result of a third party criminal act. In the case of *Law v. Omelette Shop, Inc.* (reviewed in the April 1986 **PSCLR**, case eight, page eight) the court dismissed the case of a restaurant patron who was assaulted in front of the restaurant. The court found that the restaurant had no duty to protect her against the criminal acts. In the case of *Henley v. Pizitz Realty Co.* (reviewed in the January 1985 **PSCLR**, case nine, page nine) the court affirmed summary judgment in favor of the owner of a parking deck where the plaintiff was attacked on the basis that the property owner had no duty to protect invitees against the criminal acts of third persons.

However, the court did recognize that such a duty could be imposed on the owner of a premises "in the exceptional case where the store owner possessed actual or constructive knowledge that criminal activity which could endanger an invitee was a probability," quoting from the earlier decision of *Ortell v. Spencer Companies* (reviewed in the February 1986 **PSCLR**, case nine, page six).

With regard to the second reason the Alabama court has generally refused to impose liability on premises owners for the unlawful conduct of third persons, that is the lack of "proximate cause," the court reviewed other previous decisions where it had been held that it was the intervening criminal act which was generally determined to be the proximate cause of the injuries suffered by the plaintiff, rather than any act or omission on the part of the premises owner. In the case of *Latham v. Aronov Realty Co.* (reviewed in the August 1984 **PSCLR**, case eight, page seven) this principle was explained as follows:

> Negligence alone does not afford a cause of action. Liability will be imposed only when negligence is the proximate cause of injury; injury must be a natural and probable consequence of the negligent act or omission which an ordinarily prudent person ought reasonably to foresee would result in injury. If, between the alleged negligent act or omission and the injury, there occurs an independent, intervening, foreseeable event, the causal connection between the alleged negligence and the injury is broken. [cite]

It therefore is clear that the key determinant is that of foreseeability. Someone who sets in motion a series of events by some act or omission on his own part cannot be held responsible for consequences which result from the intervention of another party. Unless, at the time of his acts or omission, the first party could or should reasonably foresee that the act of the intervening second party would occur, in which event the causal chain between the injury and the first party would not be broken. But if the intervening act of the second party cannot be reasonably foreseen, then the original negligent act or omission of the first party cannot be the proximate cause of the plaintiff's subsequent injury. It is for this reason, in the various fact situations presented to the court, the extent of prior criminal events occurring at or on the premises in question must be examined to determine whether the particular criminal act in question was known to the premises owner, or should have been known to the premises owner, and therefore was or was not foreseeable by the owner. For instance in the *Latham* case, the court's decision in favor of the premises owner was based upon the fact that the injured party was unable to present evidence proving that the premises owner had any cause to know from past experience that there was a probability of misconduct on the part of a third person which would harm or endanger the plaintiff. That is, the criminal conduct "was not reasonably foreseeable."

In the present case, the court concluded that

> The uncontroverted evidence in this case is that there had not been a single criminal incident at any prior teen dance at the Gaston Motel. This is an objective fact, and Moye submitted nothing in opposition to this fact, and Moye offers no other evidence to indicate that the defendants were on notice that this criminal act might occur.

The summary judgment for the defendants was affirmed.

COMMENTARY:

In an effort to convince the court that the criminal incident in question was in fact reasonably foreseeable, the defendants submitted the subjective opinions of two "expert" witnesses, Thomas Seals and Clifford Winkfield: In Seals's deposition he testified that he did not know of any specific, previous criminal misconduct at a teen dance at the motel in question, and stated that any opinion that he gave was not based on any prior criminal misconduct at the motel property.

Clifford Winkfield's affidavit included unverified and unauthenticated crime statistics for the "north precinct of Birmingham" for the prior calendar years. However, the plaintiff provided no proof as to where the north precinct is located, nor the size of the precinct, or where within its boundaries the particular crimes occurred. And, the court stated,

> *More importantly, neither the statistics, nor Winkfield's affidavit, nor Seals's deposition offers one shred of evidence that there had been any prior criminal incidents at the Gaston Motel or that Gaston or Otis Smith had notice or knowledge of any such incidents; ....*

In a supplement affidavit of Winkfield which was submitted by the plaintiff in support of a motion to reconsider the summary judgment, Winkfield testified that there had been a murder at the Gaston Motel lounge when a security guard was present, and a guest had been robbed in one of the motel rooms within the previous year-and-a-half preceding the death of Moye. However, the court dismissed this evidence, saying that such information would be insufficient as a matter of law to give rise to a duty to protect the deceased.

For other Alabama cases concerning the issue of premises liability see Simpson v. Wolf Ridge Corp., reviewed in the June-July 1986 **PSCLR,** *case nine, page 11 and Berdeaux v. City National Bank of Birmingham, reported in the May 1983 issue, case three, page three.*

PSCLR–AUG 87 (1)

Security Expert Witnesses' Testimony That Lack Of Security Precautions Was "Gross Negligence" Constitutes Inadmissible Legal Conclusion Which Expert Is Not Qualified To Make:

Mary Murrow v. Edith E. Daniels, Executrix of the Estate of Walter Clee Daniels, Edith E. Daniels, Individually, and Edith E. Daniels, d/b/a Henry Johnson's Motor Lodge & Restaurant, *No. 8511SC1208, in the Court of Appeals of North Carolina, decided May 5, 1987, reported at 355 S.E.2d 204.*

Mary Murrow, a 65-year-old retired school teacher from New York, was traveling alone to her home from a trip to visit her daughter. She had made this trip several times before, and on this occasion, she stopped at the Henry Johnson's Motor Lodge in Smithville, North Carolina, for the night. The hotel is adjacent to Interstate Highway 95, and was at that time owned and operated by the defendant Edith Daniels and her late husband Walter.

After checking in, Murrow went to her room. The room was located at ground level, and the front door to the room had neither a peephole nor a chain latch. There was only one window which faced the outer approaches to the room and that frosted-glass window was located in the bathroom over the bathtub.

At approximately 11 PM, Murrow heard a series of loud knocks on the door to her room. She asked who was there and what they wanted, but the only response was man's voice, loudly commanding that she open the door. She did not immediately do so, but more knocks followed and she against asked who was at the door and what they wanted. Again, the man commanded her to open the door. This happened several times, and she finally did open the door whereupon two intruders forced themselves into the room and assaulted her. The intruders threatened her with a knife, manhandled her, raped her several times, and otherwise abused her. Before leaving, the assailants took her money, tied her up, and urinated on her.

After Murrow freed herself, she called the registration desk of the hotel. Police officers had already been called to the hotel and were there, because the same assailants had made an earlier effort to break into another hotel room, but had been thwarted. The assailants were quickly caught and were subsequently convicted and sentenced for their assault on Murrow.

She filed this action against the innkeepers, claiming negligence, gross negligence, and breach of contract based upon her allegations that the defendants had failed to provide her with a safe place to stay, resulting in her injuries and damages. The hotel's defense was that Murrow had failed to take the precautions of looking out the bathroom window to see who was at her room door and calling the motel front desk to ask for help, but instead had opened the her room to the assailants, thus constituting contributory negligence on her part.

The trial judge dismissed the plaintiff's suit based on breach of contract before the trial began and dismissed the claim of gross negligence after the presentation of plaintiff's evidence at the trial. Therefore, only the plaintiff's claim of negligence was presented to the jury. The jury found that the innkeeper/defendants were in fact negligent, that Murrow had not contributed to her injuries and damages by her own negligence, and that she was entitled to a recovery in the amount of $50,000. The defendants moved the trial judge to grant a new trial, or in the alternative to grant a judgment in favor of the defendants notwithstanding the jury verdict. The judge denied these motions. The plaintiff, on the other hand, moved the trial judge to grant her a new trial on the issue of damages, claiming that $50,000 was inadequate. The trial judge granted her motion for a new trial on the damages issues only. The defendants appealed the denial of their motions and the granting of the plaintiff's motion.

In summary the evidence showed that the plaintiff incurred physical and emotional injuries. A psychologist, who qualified as an expert witness and had treated Murrow on more than 50 occasions, testified that the incident at the hotel caused her to have a permanent and totally disabling post-stress disorder. Several law enforcement officers and motel managers in the area testified that the Henry Johnson Motor Lodge was in a high crime area and that other crimes, similar to the one in question had occurred in the area previously. A security expert, qualified in motel/hotel security testified that the defendants' hotel lacked adequate security in several particulars. To the contrary, the innkeepers' evidence indicated that Murrow was not required to open the door to see who was at the door, but she could have looked out the bathroom window. Additionally, their evidence showed that the prior criminal activity occurred two miles away from the motel, that the security measures in effect at the hotel were in fact adequate, and that the injuries suffered by Murrow were temporary and not disabling.

The basis of the defendants' appeal is their claim that the evidence presented to the jury was insufficient to establish negligence on their part, but did in fact establish contributory negligence on the part of Murrow. The defendants further alleged that they were prejudiced by the trial judge's admission into evidence of illegal testimony, and that the trial judge had erred in deciding to grant a new trial on the damages alone. The appellate court decided that a new trial should be granted on all issues and not just on the damages alone, because the trial judge improperly charged the jury as to the applicable law and for other reasons, some of which are set out below.

The appellate court first discussed the innkeepers' argument that the evidence was insufficient to establish negligence on their part. They argued that the assault on Murrow was not reasonably foreseeable. The court cited the case of *Urbano v. Days Inn of America, Inc.*, (reviewed in the November-December 1982 **PSCLR**, case eight, page six) holding:

> An innkeeper owes a duty of reasonable care to his guests and that duty includes taking precautions to protect guests from the reasonably foreseeable acts of third persons.

Foreseeability can be shown by presenting evidence of prior criminal activity on the subject premises, and also in the area in which the hotel is located. Here, the plaintiff's evidence showed that the criminal activity rate was high at both the defendant's motel location and at the next interchange on the expressway, some two miles away where other hotel/motels were located. Over 300 crimes had been committed at the two interchanges during the four years preceding the attack on Murrow, and one-third of these were at the interchange where the Henry Johnson Motor Lodge is located. The defendants argue that the evidence of criminal activity was too remote to be relevant and therefore was inadmissible and should not have been presented to the jury. The appellate court disagreed, however, and ruled that such evidence does tend to show that the occurrence of similar crimes at the nearby interchange, two miles from the hotel, could or should have been foreseen by the defendants. The court therefore concluded that when the case is retried, this evidence may again be presented to the jury.

With regard to the issue of contributory negligence on the part of the plaintiff Murrow, the defendants argued that Murrow was negligent in three respects: (1) in failing to use the bathroom window to see the intruders before opening the door, (2) in failing to call the front desk upon seeing them outside her room (had she looked), and (3) in opening her door to persons who were loud and boisterous, and refused to identify themselves. The question of whether or not a plaintiff had been contributorily negligent is typically left to the jury to decide based upon the particular facts and circumstances of each situation. The court noted, to say the least, the plaintiff found herself in unusual circumstances, and whether her reaction to the circumstances amounted to contributory negligence is an issue on which reasonable people could differ.

The appellate court therefore decided that the trial judge had not committed error by refusing to instruct the jury that the plaintiff had in fact been guilty of contributory negligence in failing to protect herself. However, the court went on to say that the trial judge should have instructed the jury that they could consider whether or not the plaintiff's failure to look out the bathroom window to see who was at her door could have been a basis which they could consider in determining that she was contributorily negligent. Since the trial judge had failed to so instruct the jury, the appellate court decided that a new trial would be required on all issues. This was so regardless of the evidence that the bathroom window was located such that Murrow would have had to stand in the bathtub and slide aside an adjustable panel of frosted glass before she could see who was at her door. The court nevertheless felt that the jury should have considered this evidence as the defendants argued that it constituted contributory negligence.

The plaintiff had also called to the witness stand a Thomas Kindler, an expert witness on motel security. During his testimony, he opined that the defendants' hotel had inadequate security and that the lack of security amounted to "gross negligence" on the part of the defendants. The appellate court ruled that such testimony on the witness's part was illegal and the trial judge should not have allowed it to be heard by the jury. While a witness is permitted to express an opinion, particularly an expert witness, Kindler's opinion "amounted to a legal conclusion he was not qualified to make and the court is not authorized to receive." An expert witness, according to the rules of evidence, is not permitted to testify that a particular legal conclusion or legal standard has or has not been met by a party. The term "gross negligence" is a legal conclusion and should be left to a jury to weigh the facts and circumstances presented to them by the evidence, aided by the factual conclusions arrived at by the expert witnesses, and then to determine whether the acts and/or omissions by the defendant constitute "gross negligence" as that term is defined for them by the trial judge.

Upon the retrial, the appellate court made clear, the trial judge should not permit the security expert to testify as to his legal conclusion that the acts and/or omissions of the defendants amounted to gross negligence.

A new trial was ordered.

COMMENTARY:

There are basically two lessons to be learned from the foregoing opinion: The first is that, in conducting studies and gathering information in order to evaluate the efficacy of your own or a client's security program, statistics of similar crimes should be gathered from the authorities concerning criminal incidents not only on the premises in question, and not only on adjacent properties, but also statistics on other properties similarly situated which are not too remote in distance and crimes which are not too remote in time. As here, with businesses strung up and down an interstate highway, it is easy to conclude that criminal activity at the next interchange might be remote; however, such a conclusion, as seen above would be erroneous.

Second, for that increasing number of security professionals who are becoming expert witnesses, it is important to understand the distinction between giving one's opinion as to a factual conclusion, as opposed to a legal conclusion. While it is entirely proper and appropriate to state the former, the latter is inadmissible in a trial. Occasionally, and as apparently happened in the above situation, the use of a term, such as gross negligence can have both a factual meaning and a legal meaning and the expert may therefore inadvertently state what he intends to be a factual conclusion but which amounts to a legal conclusion. Therefore, it is most important that the expert review his testimony in detail with counsel prior to taking the witness stand before the judge and jury. Inadvertent error such as this can then be minimized.

PSCLR – AUG 87 (3)

Hotel Liable For Inadequate Security To Woman Raped At Hotel By Hotel Guest Who Invited Her To Room and Guest's Accomplices:

Elizabeth Daniel v. Days Inn of America, Inc., and Nadia, Inc., *No. 0933, in the Court of Appeals of South Carolina, decided April 13, 1987, reported at 356 S.E.2d 129.*

The plaintiff Elizabeth Daniel sued Days Inn of America, Inc., and its franchisee, Nadia, Inc., for money damages to compensate her for mental and physical suffering resulting from a sexual assault perpetrated by a hotel guest and his accomplices, allegedly as the result of inadequate security on the part of the hotel. At the trial court level, the trial judge sustained the defendant's motion for summary judgment and dismissed the plaintiff's action and she appealed. The appellate court, after consideration of the evidence presented on the motion for summary judgment, reversed the trial judge's dismissal of the action, and remanded the case for presentation to the jury. The appellate court's analysis follows:

Daniel went to the hotel of her own accord for the purpose of talking with Roscoe Brown, who had fathered Daniel's illegitimate child, about the possibility of a paternity suit and child support for the child. Brown was a paying guest of the Days Inn Hotel. Daniel stated in an affidavit which was submitted in opposition of the defendant's motion for summary judgment that she was not afraid to meet with Brown at the hotel since he had never before exhibited any "violent or weird propensities." She also said that she had agreed to meet him at the hotel because she was aware that Brown's wife knew nothing of their relationship, nor of the fact that Brown had fathered Daniel's child.

She also testified that the two of them talked about the paternity suit and the child support issue on a friendly basis for approximately two hours. At that point, Brown in what Daniel described as a joking tone mentioned that he would blackmail Daniel if she refused to dismiss her paternity suit against him. He then brought two other men, Braxton and Vaughn, into the room. The three males then assaulted Daniel, raping and torturing her for a period of approximately six hours; she lost almost half of her blood supply and her injuries required hospitalization.

Daniel was finally able to leave Brown's hotel room and went to the hotel front desk where she received some assistance from the "night auditor" who was on duty. The three assailants were apprehended and convicted of criminal sexual assault. Her suit against the hotel followed.

In her suit, Daniel claimed that the hotel had inadequate security and that it had failed to exercise ordinary and reasonable care to protect her. She claimed that the hotel had disregarded its duty of care toward her and had acted carelessly, grossly negligently, recklessly, willfully, and in a wantonly negligent manner. The hotel denied any negligence, denied that it owed any duty to Daniel, who was not a paying guest of the hotel, and denied that any negligence on the part of the hotel caused Daniel's injuries. The hotel also alleged that Daniel had contributed to her injuries by her own negligence and by assuming the risk of going to the hotel room with Brown in the first place and putting herself at risk in the first place.

The trial court, in granting the defendants' summary judgment motion, found that the hotel provided security which was "adequate and reasonable under all the circumstances" and he therefore concluded that the hotel had not breached any duty of care owed to Daniel. In this regard, the judge concluded that even if the hotel security had been inadequate, such failing on the part of the hotel could not be considered the cause of the assault on the plaintiff since the unexpected criminal acts of Brown, Vaughn, and Braxton were the actual cause of her injuries.

In her appeal, Daniel argued that whether the hotel owed her a duty, whether it had breached that duty, and whether that breach was the cause of her injuries, were all questions which should properly be addressed by a jury and should not be decided by a judge alone at the hearing on the defendants' summary judgment motion.

In order to win a negligence case, a plaintiff must show

(1) some duty or obligation to "conform to a particular standard of conduct" toward another;

(2) a breach of that duty of care;

(3) that such breach was the direct and proximate cause of the resulting injury;

(4) that there was in fact a resulting injury.

Considering first what duty, if any, the hotel owed to Daniel as a nonpaying guest of the hotel who had been invited there by a paying guest, the appellate court noted that the defendants had failed to take issue with the trial judge's ruling rejecting the defendants' arguments on this issue. Since they had taken no appeal of that ruling, the appellate court declined to interfere with or even to review the judge's finding in that regard. For purposes of the appeal therefore, Daniel was considered to be a guest of the hotel and the issue presented was whether the defendants had provided its guests with reasonable protection against assaults from criminals.

The next issue addressed by the appellate court was whether there was any factual issue concerning whether the hotel had breached its duty of care toward Daniel. The trial judge had found that the security at the hotel was reasonable under the circumstances and that no such duty had been breached. The appellate court found otherwise, noting that an innkeeper cannot be automatically exonerated merely because a criminal act was the direct cause of the plaintiff's injuries. Rather, the negligence on the part of the hotel can be established if it can be shown that the hotel either knew or should have known that its conduct put its guest at unreasonable risk of harm from the hands of a third person, even where the acts of the third persons are criminal in nature.

> Although a proprietor of a hotel is not an insurer of the safety of his guests against improper acts of other guests or third persons, he is bound to exercise reasonable care in this respect for their safety and may be held liable on grounds of negligence for failure to do so.

In an effort to prove negligence on the part of the hotel, the plaintiff introduced an affidavit from Robert Plummer, a "professional security agent" who both teaches and attends seminars in private security matters, including seminars involving hotel security systems. In his affidavit, Plummer averred he had reviewed the facts concerning the attack on the plaintiff, and concluded that while the security handbook of the hotel was a "very well written document," the hotel's personnel had failed to implement any of the policies and procedures set forth in the handbook to provide any security at the hotel. He specifically concluded that at a hotel the size of the one in question, it would have been reasonable that security patrols be made once or twice an hour. He also noted that the hotel had failed to hire security guards or to specifically direct employees concerning regular inspections of the hotel property, and he found this failure to be totally unreasonably, particularly since the security handbook pointed out that crime can be deterred by the presence of uniformed security guards.

Plummer also had reviewed the crime statistics for the county where the hotel was located, as well as for the entire state of North Carolina, had compared the statistics, and had concluded that the county had a "very high rate of serious crimes, such as murder and rape." He also found, according to his affidavit, that there had been an armed robbery, domestic disturbances, vandalisms, and a number of incidents of theft on the hotel property. Because of these incidents, it was Plummer's opinion that the hotel should have been placed on notice that they had security problems and that they should have taken steps to provide adequate and reasonable security for the protection of its premises.

Plummer went on to state that he believed that the attack on Daniel could have been prevented or aborted before she suffered through five or six hours of torture and rape, if the hotel had made regular patrols on an hourly or half-hourly basis. He stated that it was his professional opinion that the hotel had failed to use reasonable security procedures on the night in question to protect the hotel guests.

There was also evidence presented concerning other criminal statistics in the area, and this evidence showed that there were 50 incident reports regarding criminal activity at the hotel during the four year period immediately preceding the incident in question.

One of the Anderson County deputy sheriffs testified that sheriff's officers patrolled the hotel property, "between zero to three times a night, depending on how busy they were."

The hotel's night auditor did in fact patrol the hotel property at approximately 1:10 AM on the night of Daniel's attack, according to his affidavit, but he heard no disturbances. Daniel's affidavit showed that she had screamed "frequently" during the night, and a guest in the room adjacent to Brown's, Eleanor Hyde, confirmed that she heard intermittent "bloody" screaming over a period of several hours. She also heard raucous male voice and loud music coming from the room for several hours. The statements which Hyde heard convinced her that a male intended to have sexual relations with a female and the female did not wish to do so. Hyde feared for her own life due to the disturbance in Brown's room, and did not call the front desk or the police.

The court noted the similarities between this case and the case of *Urbano v. Days Inn of America, Inc.*, a 1982 North Carolina case (which was reported in the November-December 1982 **PSCLR,** case eight, page six; readers are referred to that previous issue for the details of that case). In that case, the facts presented to the court did raise questions concerning whether the hotel should have reasonably foreseen that the conditions of its premises were such that guests might be exposed to injuries by criminal acts of third parties, and whether reasonable care had been exercise to protect the guests from injury for such acts. Where such facts are presented by affidavits or otherwise in response to the defendant hotel's motion for summary judgment, then the questions should be resolved by a jury and not by the trial judge.

The fact that the attack on Daniel occurred over a five or six hour period also raises the question of negligence on the part of the hotel in failing to discover her during the attack and to prevent the continuation of the assault. The appellate court concluded that the record of affidavits, etc., presented to the trial judge included evidence which raised material issues of fact as to whether the defendants had breached the duty owed to Daniel to provide reasonable protection to her against injuries from criminal attacks, and because of the prolonged nature of the attack, the materials presented also raised an additional question as to whether the hotel breached its duty to care to the plaintiff to exercise reasonable care to discover her during the attack and to take measures to stop the assault before a period of five to six hours had elapsed.

The appellate court also concluded that it should be left to a jury to determine whether the criminal attack on Daniel was foreseeable by the defendants. Accordingly, the trial judge's grant of the motion for summary judgment was reversed and the case remanded for trial.

COMMENTARY:

This was a two-to-one decision by the South Carolina Court of Appeals, a special concurring opinion was written in which an additional finding of some interest was set forth: The concurring judge found that a jury could reasonably conclude from the evidence presented that the hotel's night auditor had in fact heard the "bloody screams" of the victim, but had chosen to ignore them.

> *After all, if the guest in the adjoining room could hear her screams through the door separating the two rooms, it would seem reasonable that the night auditor could hear the screams through the outside door to the room in which the rape was taking place.*

This judge felt that the evidence supported a conclusion that a jury, when confronted with the circumstantial evidence to the contrary, may not necessarily believe the auditor's testimony that he did not hear the screams.

The third judge, as indicated, dissented, finding that there had been no breach of any duty of care owed by the defendants to the plaintiff which amounted to being the proximate or direct cause of the plaintiff's injuries. As support for his dissent, this judge pointed out that the attack on the plaintiff took place in the motel room which she had voluntarily entered. The door had been closed and locked. The two accomplices of the paying guest had not forced themselves into the room, but had been invited by the hotel guest; also, the victim herself knew the guest.

The dissenting judge pointed out that the majority opinion simply assumes that the hotel owed a duty of care to protect an invitee from criminal acts which take place in the privacy of the guest's room. The dissenter said that there is no authority or legal precedent for imposing such a duty on the innkeeper in the privacy of the guest's room. He examined a number of other cases (including Virginia D. v. Madesco Investment Corp., reported in the September 1983 **PSCLR**, *case one, page one; and Kveragas v. Scottish Inns, Inc., reported in the October 1983* **PSCLR**, *case three, page three, and November 1984, case one, page one) which involved assaults on invitees either in the hotel's common areas, or where intruders had forcibly entered a motel room. His research found no case where such a duty to protect an invitee inside the privacy of the hotel room from other invitees legitimately inside the hotel room had ever been established.*

The dissenting judge also expressed his opinion that there was no evidence whatsoever which showed that the motel owners or operators could have or should have reasonably foreseen that Brown and his companions would attack Daniel in his room, nor that any hotel guest and his invited guests would attack anyone inside a hotel room. It was this judge's opinion that regardless of the defendant's duty to provide security at the hotel, such duty could not be extended so far as to protect such a guest in the privacy of a paying guest's hotel room.

It is our opinion that the dissenting judge's conclusions, given the current status of the law on a nationwide basis, more accurately reflect the law than the decision reached by the majority. Nevertheless, the majority rules and this is now the status of the law, at least in South Carolina, and this decision can be cited as persuasive authority in other jurisdictions. Since the hallmark of proximate cause is foreseeability, it would seem to us that the plaintiff should not be permitted to prevail unless the evidence establishes such foreseeability.

Nevertheless, the practical lesson to be learned from this decision is that all security personnel at hotel and motel properties must now understand that any and all suspicious disturbances, even within the private room of a paying guest must be reasonably investigated to insure that the duty owed to the guests inside the room is in fact being reasonably met. This decision clearly expands the duties and responsibilities of hotel owners and operators and their security directors, to require that proper training and retraining of their security officers, managers, and other hotel employees, includes an explanation of their duties in response to such disturbances and suspicious activities.

PSCLR—AUG 87 (5)

Statute Limiting Hotel's Liability To $300 For Loss Of Guests' Valuables Will Not Protect Hotel Which Fails To Provide Guests With Conspicuous Notice At Registration Desk Of Availability Of Safety Deposit Boxes:

Anthony Latini and Joan Marie Latini v. Lowes Corporation and LRI, Inc., *No. 85 Civ. 3275 (PNL), in the United States District Court for the Southern District of New York, decided April 8, 1987, reported at 657 F. Supp. 475.*

Anthony and Joan Marie Latini filed an action seeking compensation for the loss of their valuables while they were paying guests at the Regency Hotel in New York City. The hotel owners raised as a primary defense the New York statute which provides for limitation of liability to hotels for the loss of guests' unstored valuables. They filed a motion for summary judgment asking for dismissal of the plaintiffs' claim for all damages exceeding $300.

Although they had registered before arriving in New York, the Latinis checked in with the clerk at the registration desk when they arrived at the Regency Hotel on the night of March 8. While they were guests at the hotel, they stayed in their room all but four hours and checked out at 12:30 PM the following day. Before checking out, Mrs. Latini missed her purse, but she did not inform the hotel personnel, deciding instead that she would find it after she returned home. However, when she did check her luggage at home, and still could not locate the purse, she realized and concluded that it must have been removed from her luggage while she was staying at the hotel; whereupon she contacted the hotel management.

In the plaintiffs' suit, they alleged that jewelry valued at $24,345 was removed from their room at the Regency Hotel due to the defendant hotel owners' negligence in failing to properly secure the hotel premises. The hotel argued that there were safety deposit boxes available for such jewelry, and also alleged that notice of the availability of such boxes is provided to all guests in their preregistration information package, as well as by a series of some twenty-three notices posted throughout the hotel. There were two such notices, according to the defendants, which were located at the registration desk itself – one at the cashier's desk, and another on each of the hotel floors near the elevator. The plaintiffs, in response to the hotel's defenses, claimed that the two posters located at the registration desk were in fact obstructed and that the hotel staff did not inform them of the availability of the safety deposit boxes.

The General Business Law of the state of New York provides that an innkeeper who makes safety deposit boxes available for the use of their guests in storing jewelry and other valuables will not be liable to such guests for the loss of jewelry and valuables which they choose not to place in the hotel safe, in any amount exceeding $300. The Law also requires that the innkeeper notify the guests

> by posting a notice stating the fact that such a safe is provided ... in a public and conspicuous place and manner in the office and public rooms, and in the public parlors of such hotel....

The judge to whom the defendants' motion for summary judgment was presented reviewed certain precedent cases on this subject, including the case of *Goncalves v. Regent International Hotels Ltd.*, a 1983 decision from the New York Court of Appeals (reported in the June-July 1983 **PSCLR**, case four, page three) where it was held that such a limitation of liability which would otherwise attach to innkeepers should be applied only when and if the innkeeper "strictly adheres to the statutory provisions." In the present situation, the evidence shows that there were notices posted on the wall at each side of the registration desk at the Regency Hotel, but the hotel admitted that the notice posted at the cashier's desk was around the corner and therefore was not visible from the registration desk. In addition, the plaintiff Mr. Latini claimed that neither he nor his wife ever saw the notices due to the fact that they were obscured by a vase of flowers on one side and mail on the other side.

A security officer at the Regency, Owen Boyce denied that the vase of flowers obstructed the guests' ability to see the posted notice; however, Mr. Latini submitted a photograph showing the vase of flowers on the left side of the desk and the photograph itself suggested otherwise. The photograph did not support Mr. Latini's claim that mail obstructed the view of the other posted notice, but there was no evidence presented by the hotel to the effect that this second notice was "conspicuous" to those hotel guests who registered at the left side of the registration desk. In view of this evidence taken as a whole, the court concluded that there was a material, factual issue as to whether the posted notices required by the statute were in fact "conspicuous" as the statute specifically requires. The court therefore denied the hotel's motion for summary judgment and the case will proceed to trial so that the issue as to whether the notices were in fact "conspicuous" can be decided by a jury.

PSCLR—OCT 87 (2)

Where Plaintiffs Successfully Show Hotel's Breach Of Contract To Provide Security, Or Hotel's Gross Negligence For Failing To Provide Security, Guests Entitled To Damages Plus Pre-Judgment Interest:

MGM Grand Hotel, Inc. v. Herman Siegel, and Fredericka Siegel, *No. 84-1508, in the District Court of Appeal of Florida, Third District, decided April 14, 1987, on motion for rehearing May 26, 1987, reported at 506 So. 2d 451.*

Herman and Fredericka Siegel of Miami enjoyed traveling on gambling junkets to Las Vegas. They had gone to Las Vegas on a number of personal trips arranged through the MGM Grand Hotel, which had an office in Miami Beach. The MGM Grand provided air fare and free accommodations in return for which the Siegels agreed to gamble substantial amounts of money at the casino in the MGM Grand Hotel. On their 15th trip, the accommodations provided at the hotel were the same as on all the prior trips: They were housed in a penthouse suite located on the 26th floor of the hotel, which was protected by an iron gate and had around-the-clock armed security personnel controlling the entrance to the six penthouse suites on the floor.

On the evening before they were scheduled to return home, the Siegels went out for dinner and, upon returning to the 26th floor, found that the gate was open and that no security guard was in sight. Unknown to them in advance, the hotel had permitted a trade show to operate several hospitality suites on the 26th floor. Many people, all unknown to the Siegels, were roaming from suite to suite, enjoying free food and drink. Mr. and Mrs. Siegel went down the hallway to their suite, locked themselves in, and went to bed.

As they were packing to leave the next day, they realized that the money and jewelry (which had been left in their room while out the previous evening) was missing. They filed this suit to recover the value of their property loss, blaming the loss on the hotel's breach of contract in failing to provide appropriate security, and the hotel's alleged gross negligence in failing to provide appropriate security.

Unfortunately for the Siegels, the jury found damages in their favor in the amount of $100,000, but returned a verdict which was inconsistent: The jury's responses to certain written questions propounded to them by the trial judge were that (1) there was a contract between the Siegels and the MGM Grand which the hotel had breached, (2) this breach was the cause of the plaintiff's losses, and (3) the hotel was grossly negligent. But, the jury also found (4) that the hotel's gross negligence was not the legal cause of the plaintiff's loss!

Since the breach of contract and the gross negligence alleged were both based upon the hotel's alleged failure to provide appropriate security, the jury's answers are inherently inconsistent. Accordingly, the appellate court concluded that the case should be retried on the issue of liability alone. If the jury on the retrial determines that the hotel is liable, either for breach of contract, or for gross negligence, or both, for having failed to provide adequate security, then the damages are already established in the amount of $100,000 plus pre-judgment interest on that amount.

COMMENTARY:

The simple lesson to be learned from the above case is obvious: adequate security must be provided by an innkeeper for the protection of his guests and their property. The definition of "adequate" will, of course, vary depending upon the attendant circumstances.

In deciding the above case, the appellate court relied in part on the following case previously reported in **PSCLR**: *Honeywell Inc. v. Trend Coin Co., reported in the November-December 1984* **PSCLR**, *case twelve, page twelve.*

PSCLR – OCT 87 (6)

Hotel Guest Assumed Risk Of Injury By Running Into Dark, Unlit Area; Expert Witness Found To Be Unqualified:

Diane Ventura v. Roy E. Winegardner and John Q. Hammons, Trustees; Roy E. Winegardner and John Q. Hammons, Partners, doing business as Morgan Hills Operating Company, a partnership; Winegardner and Hammons, Inc., a corporation, doing business as the Holiday Inn, *No. 17186, in the Supreme Court of Appeals of West Virginia, decided May 15, 1987, reported at 357 S.E.2d 764.*

The plaintiff Diane Ventura was a college student at Rutgers University, where she was a senior and a member of the university tennis team. The team stayed at the Holiday Inn in Star City, West Virginia, while participating in a tennis tournament. On the night in question, Ventura's teammates surprised her by throwing water at her. In attempting to elude her friends, she ran some 30 feet beyond a paved sidewalk, into a dark and unlit area beyond the walkway, where she fell over a steep embankment. She injured her right knee, and has since continued to experience trouble climbing stairs, running, and jumping. The injury has also affected her ability to compete in tennis tournament play. She filed an action against Holiday Inn, the case went to trial, and the jury granted an award in her favor in the amount of $147,000.

Holiday Inn appealed, arguing that the plaintiff had failed to prove that the hazard in question, the steep embankment, was actually located on Holiday Inn property or that the Holiday Inn had knowledge of the dangerous condition caused by the hazard. The appellate court agreed with this argument, but concluded that it was clear from the evidence that the Holiday Inn personnel knew of the existence of the hazard since the area was inspected by the company's director of safety and since it was quite open and obvious: "Holiday Inn could hardly argue that it had never noticed the steep embankment a few feet from its property line."

The hotel also objected to the trial court judge's refusal to explain the legal theory of "assumption of risk" to the jury. The appellate court agreed that the trial judge should have instructed the jury on this legal defense since the evidence showed that the elements of this defense were present. These elements include: (1) a knowledge of the danger on the part of the plaintiff; (2) an appreciation of the danger; and (3) the plaintiff's voluntary choice to expose herself to the danger.

The evidence showed that the plaintiff Ventura was familiar with the surrounding area since she had been a guest at the hotel during the daylight hours and had passed the embankment. Furthermore, she had testified that she ran at full speed into the dark, unlit area while her eyes were unadjusted to the dark. She admitted that her field of vision "was only about five feet." The court concluded that she should have known that running into the dark is dangerous, that a reasonable person would appreciate that danger, and that she voluntarily chose to expose herself to that danger. Since the trial judge had failed to explain the existence and availability of such a defense to the hotel, the appellate court concluded that a new trial was required.

The appellate court also examined the qualifications of the various expert witnesses called to testify by the plaintiff to prove her injury. In the court's words, her experts "ran from the top to the bottom of the scale, superbly qualified, qualified, and unqualified." The court examined the qualifications of four witnesses and found that the most qualified was Ventura's physician, Dr. Kay Douglas Bowers, who testified to the extent of injury to the plaintiff's knee. He was an orthopedic surgeon, specialized in sports medicine, had been practicing 19 years, and had treated the plaintiff. In addition, Ralph Plummer, who testified that the steep bank was not property guarded, was determined to have "specialized knowledge that [would] assist the trier of fact." Although he had not done any specific research concerning unguarded embankments and slopes, the appellate court opinion found that his experience as a professor of industrial engineering at West Virginia University, as the director of the Occupational Health & Safety Engineering Program, and as consultant for safety evaluations of companies' facilities, made him qualified.

Another witness the court put in the "qualified" category was Frank Ferrante, who was Ventura's high school tennis coach. The court noted that, even though he had not observed her play tennis in college, he was nonetheless qualified since he had been involved in tennis since childhood, had played professional tennis, and had sponsored two players on the professional tennis circuit.

The "unqualified" expert witness according to the appellate court was Thomas Serpento, who had testified as to the plaintiff's "future earnings" as a tennis professional. The court noted that Serpento had a master's degree in vocational counselling and guidance and was employed by West Virginia University as Director of Human Resources, but he he admitted that he knew very little about tennis and had no expertise concerning the salaries of tennis professionals. He had arrived at his opinion concerning Ventura's future earnings by reviewing an article in *Tennis Week Magazine* which reported the earnings of the top 40 ranked tennis players in the country. He averaged those salary figures, projected a productive tennis life of approximately ten years for the plaintiff, and estimated that the present cash value of her future earnings was $737,317. The appellate court noted that

> experts may base their opinions on treatises and publications in their professions, but must first show the authoritative nature of the works.

While not being critical of *Tennis Week Magazine*, the appellate court concluded that its purpose was not the scientific enlightenment, but rather "the amusement of the readers, and [should] not to be relied on in any professional manner."

Summarizing, the court concluded that the trial judge had committed an error by allowing Serpento to testify as to his opinion of the plaintiff's future lost earnings.

A new trial was ordered.

COMMENTARY:

It is unfortunately a fact that some security professionals are under the mistaken impression that they are not responsible and cannot be held liable for any loss or injury which takes place beyond the property lines or boundaries of the property for which they are responsible. As is indicated in the above case, such is not the law. (See the case of Banks v. Hyatt Corporation, in the April 1984 **PSCLR***, case one, page one, as an example of a case in which an innkeeper was held liable for the death of a guest due to an assault which took place off the hotel property.) Such "technical defenses" as the existence of a property line do not establish nor preclude liability.*

The above opinion also points out the danger of an expert witness attempting to testify beyond the scope of his expertise. As more and more of our subscribers are entering the field of consulting in the area of private security, there is an increased need for each to ensure that he does not succumb to the temptation to testify and give opinions in areas in which he has not received the requisite training, or had extensive experience, or done research.

A comprehensive and accurate resume of one's educational, industrial, research, and other experiences, should not only be provided to the party retaining the expert, but should be reviewed thoroughly with that party in order to ensure that the witness has the requisite background to have an "expert" opinion on the question in issue, and more importantly, to ensure that the witness has a reasonable and reasoned basis for that opinion.

PSCLR – NOV-DEC 87 (3)

While Concealment Of Material Facts May Amount To Fraud When Direct Inquiry Is Made, Innkeepers Are Not Liable To Guests For Fraudulent Concealment Where No Direct Inquiry Was Made By Guest And Innkeeper Did Not Deceive Or Mislead Guests:

Hicks et al. v. Days Inns of America, Inc., *No. 74004, in the Court of Appeals of Georgia, decided May 20, 1987, reported at 357 S.E.2d 847.*

The plaintiffs Mr. and Mrs. Hicks were staying in the Days Inn Motel when they suffered the theft of several hundred dollars and a valuable antique pistol from their motel room. They brought suit against Days Inn, claiming they were entitled to recover for their injuries on two legal theories: (a) fraudulent concealment, and (b) negligence. Days Inn moved for summary judgment as to both counts; the trial judge granted its motion on the fraud and the negligence counts, and granted a partial summary judgment by limiting the innkeeper's liability to the plaintiffs to the amount of $100. The Hicks filed an appeal.

In their effort to set forth a cognizable claim of fraudulent concealment, the plaintiffs claimed that the Days Inn management were aware that the Hicks's motel room was not adequately secured to prevent the possible theft of their personal property in the room. They further alleged that the motel management had failed to properly warn them of the inadequate security and that such failure constituted "fraudulent concealment" of material facts, thereby giving rise to their cause of action against the motel.

Upon review of the evidence submitted to the trial court on the motions for summary judgment, the Court of Appeals found no evidence to indicate that the subject of security, adequate or otherwise, was ever discussed between the motel management and Mr. and Mrs. Hicks. The guests never asked about the motel's security measures to prevent theft from motel rooms, nor did they give any indication to the motel management that they had large sums of money or other valuables in their possession. Nothing in the record indicates any intentional act by the motel management to mislead or deceive the motel guests concerning security measures. The appellate court noted that,

> Concealment of material facts may amount to fraud when direct inquiry is made and the truth evaded, or where the concealment is of intrinsic qualities of the article which the other party by the exercise of ordinary prudence and caution could not discover.

The court also noted that the evidence presented did show that the motel had posted a notice, in accordance with Georgia statutes, on all motel room doors requiring guests to store any valuables in the motel safe. The sign notified guests that the motel's liability for any loss by the guest would be limited to $100 unless the guest gave a prior written notification to the hotel to the effect that his property was valued at more than $100.

The court concluded that the motel had posted the disclosures authorized by law for purposes of notifying its guests of the motel's limited liability for guests' losses. Furthermore, the motel management did not act fraudulently by failing to voluntarily provide additional information concerning security measures. Accordingly, the appellate court affirmed the trial judge's ruling which granted Days Inn's summary judgment on the fraudulent concealment claim.

With respect to the negligence claim, since the motel had posted the notice authorized by the Georgia statute, and since the Hicks had failed to provide the motel with a written notification that their property was valued in excess of $100, then any negligence on the part of the hotel which caused the loss of the guests' property would subject the motel to a maximum liability of only $100, pursuant to the innkeeper's statute.

Accordingly, the trial judge's grant of partial summary judgment on this count was likewise affirmed.

COMMENTARY:

In reviewing the various precedent cases cited by the appellate court in support of its decision affirming the trial judge's grant of summary judgment in the above case, it is interesting to note that one of the cases cited was decided in 1914, where it was held:

> *To hold that, without complying with the requirements of the notice authorized by ... statute, the guest could nevertheless recover as to articles covered by the notice, merely by showing want of [additional disclosures] on the part of the innkeeper ..., would practically make the statute nugatory.*

Jones v. Savannah Motel Co., 141 Ga. 530, 534.

These issues of motel security, lost property, posting of notices, limitation of innkeeper's liability, are not new. However, as is seen from time to time in this publication, innovative trial lawyers are constantly attempting to avoid the effect of limitation of liability statutes – such as by claiming "fraudulent concealment" – and they are occasionally successful.

PSCLR – JAN 88 (2)

While Generally A Property Owner Has No Duty To Control Behavior Of Third Parties, Innkeeper's Duty To Protect Guest Extends To Unreasonable Risks Arising From Acts Of Third Parties, Whether They Be Innocent, Negligent, Intentional, Or Criminal:

John Knodle, individually and as the administrator of the estate of Linda Kay Knodle v. Waikiki Gateway Hotel, Inc., Hyatt Corporation, Continental Air Lines, Inc., Northridge Industries, Inc., Continental-Kalakaua Hotel Corporation, and Waikiki Gateway Hotel Associates, and George Patrick Murphy, *No. 9392, in the Supreme Court of Hawaii, decided September 10, 1987, reported at 742 P.2d 377.*

John Knodle, father of Linda Kay Knodle, brought this lawsuit on his own behalf and on the behalf of the estate of his murdered daughter. The suit was originally filed against the murderer George Patrick Murphy, and Knodle's employer Continental Air Lines, and the owners and operators of the Waikiki Gateway Hotel in Honolulu. Murphy filed no defensive pleadings and a default judgment was entered against him. Continental Air Lines moved for, and was granted, a summary judgment and dismissed from the lawsuit and no appeal was taken. The negligence claims against the remaining defendants, the hotel owners and operators, were presented before a jury, resulting in a verdict on behalf of those defendants. The plaintiff appealed, arguing among other things that the trial judge erred (1) in refusing to allow the introduction into evidence of all reported criminal activity in and around the hotel, and (2) by failing to properly instruct the jury on the legal issues of proximate cause and foreseeability, and (3) by refusing to grant a judgment in favor of the plaintiffs in spite of the jury verdict for the defendants, or in the alternative a new trial.

After reviewing the record and transcript, the appellate court agreed with the trial court's ruling concerning the admissibility of evidence as to prior criminal activity at the hotel, but agreed with the plaintiff that the judge had improperly instructed the jury on the legal issues of proximate cause and foreseeability. (The appellate court therefore did not address the issue whether the judge had erroneously refused to grant the requested judgment notwithstanding the jury verdict or for a new trial.)

In summary, the facts of the case were that Knodle used one of her airline employee passes to take a Thanksgiving holiday trip and she was returning to her home in Chicago on a Continental Air Lines flight from Guam, with a scheduled stopover in Honolulu. The plane landed in Hawaii at approximately two AM and the Continental Air Lines crew who had worked the flight from Guam were expedited through customs and went on to the Waikiki Gateway Hotel. Knodle took longer to clear customs and later caught a ride with other Continental pilots, arriving at the hotel just before five AM

Upon arrival she went to the hotel front desk to register, where she spoke briefly with the assistant manager and the night auditor, with whom she had become acquainted by virtue of previous stays at the hotel. She placed a telephone call in the reception area which took approximately one-half hour, during which time the pilots who brought her to the hotel went on to their rooms. She then asked the assistant manager to give her a key to one of the rooms in which the two Continental flight attendants from the Guam flight were staying (where there was an extra bed for her). The assistant manager gave her a key to one of the rooms, which was located on the tenth floor. She put two of her suitcases on the hotel elevator, and left them there while she returned to the front desk to get her remaining luggage.

At exactly this time, George Patrick Murphy entered the lobby of the hotel, walked directly to the hotel elevator, and entered the elevator containing Knodle's luggage. Seeing this, the assistant manager called to him to "hold the elevator," which Murphy did, allowing Knodle to enter the elevator with the rest of her luggage. The door then closed, the elevator ascended, and the two hotel employees resumed their duties.

At approximately 6:30 AM, a hotel guest reported to the hotel night auditor that there was some luggage on the elevator. The auditor retrieved the bags and recognized them as belonging to Knodle. He and the assistant manager, however, decided not to call her at that time, because they believed that she was asleep in her room. At approximately 7:00 that morning, a hotel maintenance employee turned in a guest room key for the tenth floor, which he had found in a fourth floor hallway. And, approximately two hours after that Linda Knodle's body was discovered in a restroom near the fourth floor laundry room. She had been strangled to death sometime between the hours of 5:15 and 9:00 that morning.

The plaintiff's theories of the liability on the part of the hotel owners and operators were based upon his argument that the defendants were negligent in failing to provide "safe accommodations" and "adequate security to protect [Linda Knodle] from the unreasonable risk of physical harm" and that this constituted "wanton and reckless disregard for her safety."

The court noted that the law recognizes a special relationship between the innkeeper and his guest, such a relationship requires the innkeeper to take reasonable action to protect the guest against unreasonable risks which could harm the guest. This duty extends to include unreasonable risks which arise out of the innkeeper's own conduct, or out of the physical condition of the property itself. In addition, the innkeeper's duty includes unreasonable risks which arise from the conduct of third parties, whether such conduct is innocent, negligent, intentional, or criminal. This is so, in spite of the general proposition that one person is not responsible for (i.e., has no duty to control) the acts of a third party. However, because of the "special relationship" between the innkeeper and the guest, the guest has a right to rely on the protection by the innkeeper from those unreasonable risks which arise from the misconduct of third persons. (By way of illustration, a person who was assaulted on the hotel premises, but was not a guest of the hotel would have no right to rely on the hotel to protect him from the criminal acts of third persons, because no "special relationship" exists between the innkeeper and a non-guest.)

Therefore, whether the hotel had any duty to protect Knodle is a question of law, and therefore, a question to be decided by the judge. And what constitutes "reasonable" and "unreasonable" and the question as to whether the hotel's employees' acts were reasonable given the particular fact situation are questions for the jury to determine. In doing so, the jury should compare the acts of the defendants' employees with those of the mythical "reasonable person."

Another issue for the jury's determination is the presence or absence of a causal relationship between the conduct of the defendant and the damages to the plaintiff. This relationship has often been referred to by the courts as the issue of "proximate cause." However, the appellate court in this case found that the term "proximate cause" might have been an appropriate term to use to characterize a legally sufficient causal relationship, but opined that

> in the complex society of the late twentieth century, a negligent tort is as likely as not to have plural or concurring causes.

The Hawaii Supreme Court noted that some courts are adopting a "but for" rule, under which a defendant's conduct would be considered a legally sufficient cause of the plaintiff's damages if the damages would not have occurred "but for" that conduct; conversely, the conduct would not be a legally sufficient cause of the damages if the damages would have occurred in the absence of such conduct. The court went on to note, however, that the "but for" test was itself inadequate in that it may not cover situations in which there were two or more causes, none of which standing alone would have produced the damages.

Accordingly, harking back to an earlier Hawaii Supreme Court decision, the appellate court concluded that the most workable definition of legal or proximate cause was one in which the defendant's negligence was found to be a "substantial factor in causing plaintiffs' injuries." The court found that the trial judge had failed to properly instruct the jury as to this definition.

The supreme court also concluded that the trial judge should not have allowed the jury to determine whether or not the hotel owners and operators owed Linda Knodle a duty of reasonable care, since there was no reasonable inference that she was not a guest of the hotel. In other words, the trial judge should have instructed the jury that the duty in fact did exist owing to the fact that she was a guest, and the "special relationship" of innkeeper and guest was thereby established.

Due to these errors of law by the trial judge in instructing the jury on the applicable principles of law, the appellate court reversed the judgments entered on the jury verdicts for the defendant hotel owners and operators and granted the plaintiff a new trial.

PSCLR–FEB 88 (3)

Whether Hotel Should Have Foreseen Assault On Guest Can Be Shown By Evidence Of General Likelihood Of Harm To Guest, Degree of Criminal Activity In Vicinity, and Security Measures Employed By Land Owner:

Cathleen Meyers v. Ramada Hotel Operating Company, Inc., a corporation, Tolbert Enterprises, Inc., *No. 87-3224, in the United States Court of Appeals, Eleventh Circuit, on appeal from the United States District Court for the Northern District of Florida, decided December 15, 1987, reported at 833 F.2d 1521.*

The plaintiff Cathleen Meyers rented a room with a friend at the Ft. Walton Beach Ramada Inn, where she had previously stayed on several occasions. On this occasion, she was assigned a room on the fifth floor of a six story building known as The Tower, consisting of 194 rooms. On the night in question, at approximately 2 o'clock in the morning, Meyers went up to the fifth floor of the Tower by way of the hotel elevator, in order to get her driver's license; her friend waited for her in the parking garage. While in the hotel elevator, a male asked Meyers if he could use the restroom in her room, and she said, "No!" The elevator doors then opened on her floor, and she exited the elevator and walked toward her room. The unknown male also exited the elevator, walked in the opposite direction, but then just as Meyers was opening her door to her room, the man turned and forced his way into her room, where he sexually assaulted her. After he left, Meyers went to another room where those hotel guests notified security, and took Meyers back to her room.

The trial evidence record reflected that the management of the Ramada Inn had employed security services for approximately 10 years, mainly for the purpose of controlling noise, and for keeping uninvited guests from using the hotel swimming pool. At the time of the assault on Meyers, there were three security guards who were on duty, one of them patrolling the hotel Tower and the pool, while the other two patrolled the remainder of the hotel property. The guard patrolling the Tower, Lt. Ashmore, checked the building routinely every 30 to 40 minutes by walking down each hallway throughout the building. When he was notified of the assault on Meyers, Ashmore had just finished his walk-through of the building and had returned to the back of the building to the boardwalk area. He immediately went to Meyers room on the fifth floor, spoke with her about the assault, and positioned the other two security officers around the outside of the hotel Tower to look for any suspects who might run out of the building. Ashmore then again walked through the entire building, checking the stairwells as well. The assailant, however, was never found.

The hotel itself was owned by Tolbert Enterprises under a franchise agreement with Ramada Inns, Inc. Meyers brought her lawsuit against Ramada Hotel Operating Company, Inc., and others for negligently permitting her to be assaulted as she entered her hotel room. The trial judge dismissed the lawsuit, finding that that the hotel did not owe a duty to protect Meyers from the criminal assault because the assault on her was not foreseeable.

In an effort to show that the attack on her was or should have been foreseeable by the hotel, Meyers introduced evidence that 14 years before her assault the night clerk at the hotel had been robbed, and that within 17 months of her assault an individual had been charged with resisting arrest with force based on an incident at the hotel, five men had been arrested for trying to sell over two pounds of cocaine, and that there had been several arrests in the hotel itself for drug possession and trespassing. Also within 17 months of Meyers' assault, an assailant had kidnapped a woman at knife point from another hotel two miles away and had sexually assaulted her. Police records also showed a number of arrests for alcohol and drug related offenses at a park adjacent to the Ramada Inn, and within two miles of the hotel there had been some arrests for lewd conduct toward a minor child.

There were three separate bars in the hotel, each with a live band, which attracted non-hotel guests to the hotel, especially to the bar on the top floor of the Tower which required such non-guests to use the elevator or stairwells, all of which permitted access to the residence floors of the hotel, including the fifth floor where Meyers' room was located. In addition, Fred Tolbert, the owner of the Ramada Inn franchise, had been told that some of the cocktail waitresses were afraid of a man whom they believed was hanging around the bar at closing time.

Meyers appealed from the trial judge's order granting summary judgment to the hotel. The only issue presented to the appellate court was whether the evidence supported her claim that the attack on her was or should have been foreseeable by the hotel management. In deciding this appeal, the appellate court first noted that Meyers clearly was a "business invitee" to whom the Ramada Inn owed a duty of care, including the duty to guard her against dangers known to the hotel, or dangers which should have been known to the hotel. Furthermore, such a duty on the part of the hotel includes guarding her against reasonably foreseeable criminal acts by third parties. The court noted that, in order to determine whether such foreseeability existed, evidence as to the general likelihood of harm to the plaintiff, the level of criminal activity in the area, and the security measures employed by the owner, should all be considered. In addition, other evidence concerning the issue of foreseeability would include evidence showing compliance or noncompliance by the hotel with industry standards, the presence or absence of suspicious persons on the property and any peculiar security problems at the hotel (e.g., the number of bars, public access to the room floors, and stairwells being available to the public and not restricted to emergency use). Even the fact that the hotel owner had made some provision for security services may in and of itself be some evidence as to the hotel's foreseeability of harm to hotel guests. Accordingly, the appellate court reversed the trial judge and remanded the case for trial for presentation to a jury, finding that Meyers had presented sufficient evidence to create an issue of foreseeability by the hotel.

PSCLR – MAY 88 (3)

Innkeeper Not Liable For Loss Of Guest's Property Where Guest Failed To Provide List of Jewelry Samples In Bag Which Was Entrusted To Innkeeper:

Nova Stylings, Inc. v. Red Roof Inns, Inc., *No. 60849, in the Supreme Court of Kansas, decided December 11, 1987, reported at 747 P.2d 107.*

Darryl Kulwin, worked as a jewelry salesman for Nova Stylings. Kulwin made arrangements to meet with Panoria Ruston, a guest of the Red Roof Inn in Overland Park, Kansas, and to have dinner with her afterward.

On Kulwin's request, Ruston asked a desk clerk to keep "her bag" in the office of the inn while Ruston and Kulwin went out to dinner. Ruston informed the clerk that the contents were valuable and that the bag should only be returned to her. She did not, however, inform the clerk as to the specific contents of the bag. Neither Ruston nor Kulwin made any written agreement with the inn regarding the safekeeping of the bag.

While Kulwin and Ruston were having dinner, someone entered the manager's office with the clerk's knowledge and took the case. The bag, according to Kulwin and his employer, Nova Stylings, contained jewels owned by Nova valued at approximately $650,000. Nova Stylings subsequently sued Red Roof Inns.

Nova Stylings claimed that the clerk violated Kansas law by not asking for a list of contents of the bag and the value of those contents. Nova Stylings also claimed that the clerk acted negligently and unlawfully in failing to issue a receipt for the bag. Nova Stylings asserted that Red Roof was liable for its own negligence and for its clerk's negligence in failing to properly care for Kulwin's bag.

Red Roof filed for summary judgment. Red Roof claimed that Kulwin, not the inn, had failed to follow guidelines set out in the applicable Kansas statute as to liability of an innkeeper for negligence in caring for guests' property.

Red Roof also claimed that the statute eliminated liability of an innkeeper for any such negligence, and further limited any liability of an innkeeper to $250 for guests' property absent a specific agreement. The federal trial court requested the Kansas Supreme Court to settle these state law issues, and the Kansas Supreme Court agreed to do so.

The supreme court first cited the relevant Kansas statute. This statute (subsection B) states:

> No hotel or motel keeper in this state shall be liable for the loss of or damage to, merchandise for sale or samples belonging to a guest, lodger, or boarder unless the guest, lodger, or boarder upon entering the hotel or motel, shall give notice of having merchandise for sale or samples in his possession, together with an itemized list of such property, to the hotel or motel

> keeper, or his authorized agent or clerk in the registration office of the hotel or motel. No hotel or motel keeper shall be liable for any loss of such property designated in the subsection (B), after notice and itemized statement having been given and delivered as aforesaid, in an amount in excess of Two Hundred Fifty Dollars ($250), unless hotel or motel keeper, by specific agreement in writing, individually, or by an authorized agent or clerk in charge of the registration office of the hotel or motel, shall voluntarily assume liability for a larger amount with reference to such property. The hotel or motel keeper shall not be compelled to receive such guests, lodgers or boarders with merchandise for sale or samples.

Nova Stylings asserted that this statute simply limited the innkeepers common law strict liability for loss of or damage to a guests merchandise or samples.

The court, however, held that this statute unambiguously eliminated strict liability or negligence liability on the part of the innkeeper towards the guest for the guest's merchandise or samples except as specifically provided for in the statute. The supreme court cited a number of cases from other jurisdictions in support of its reading of the Kansas statute.

Kahn v. Hotel Ramada of Nevada (reported in the November-December 1986 **PSCLR**, case one, page one) held that a similar Nevada statute provided a limit of $750 on the innkeepers liability absent written consent by the innkeeper to greater liability for storage and protection of the guests property. The court stated that the Nevada statute limited the innkeepers liability "in all instances of negligence except when the innkeeper consents to assume more responsibility."

Missouri courts have similarly recognized legislative limitations on innkeepers' liability and rejected a plaintiffs argument that a similar Missouri statute simply limited the innkeeper's duty to act as insurer of the guests property. The court stated, "it appears that the legislature has completely covered the common law liability of the innkeeper—both as to strict liability and negligence liability." The court then cited Maine and Oklahoma cases interpreting similar statutes in a nearly identical manner.

The supreme court relied heavily upon these other cases in rejecting Nova Stylings's argument that the purpose of the law of negligence is to prod the parties into exercising due care, and that the interpretation of the Kansas statute adopted by the court would be a disincentive for innkeepers to protect guests' property. The supreme court quoted a Maine case, which stated,

> The hotel keeper is not a banker; and he is not in the business of operating a safe deposit vault except as an incident to operating a hotel. It is not, therefore, unreasonable to restrict his liability for such incidental services rendered to his guests within such limits as will meet their ordinary needs. Those who carry with them large amounts of money or jewelry must take other measures for their protection. The added cost for the hotel keeper providing for such protection even as against the willful act or negligence of an employee, is in the last analysis one of his costs of operation reflected in the rates charged to all. Why should those guests who do not need protection pay for the cost of those who do?

The supreme court next dealt with Nova Stylings's argument that stolen jewels fell under subsection (C) rather than subsection (B), of the Kansas statute. Nova argued that since subsection (C) applies to "money, jewelry, diamonds, or other valuable stones, articles of silver and gold, tickets, certificates of other valuables," it controlled in this case.

The court ruled that, while the stolen jewelry could technically be considered under subsection (C), the purpose of the statute dictated that this case should be viewed as falling under subsection (B). The court held that subsection (C), arguably more favorable to guests, did not apply to jewelry samples used by salesmen to show to customers and thereby generate sales.

Finally, the supreme court rejected the plaintiff's argument that an innkeeper has a duty to ask guests about the nature of their property. The court noted that absent a guest's disclosure of the value of his property, an innkeeper would not be aware of the need to make special arrangements for care of valuable property.

In other words, the guest is in a better position than the innkeeper to initiate action to protect particularly valuable property brought by the guest to the hotel or motel. The court concluded by stating,

> The failure of the guest in the present case to inform the innkeeper of the value of the property which he sought to leave in the motel manager's office, for failure to give notice of having merchandise for sale or samples, and for failure to give an itemized list of the property to the defendant precluded the defendant from determining whether it wished to accept the responsibility being placed on it.

PSCLR – JUN-JUL 88 (1)

Innkeeper's Right To Exclude Disorderly Customer Does Not Permit Innkeeper To Take Customer, Against His Will, To Private Place To Read Eviction Notice, Where Customer Was Willing To Leave Voluntarily:

Robert F. Simone v. Golden Nugget Hotel and Casino, trading as GNAC, Corp.; Robert F. Simone v. Golden Nugget Hotel and Casino, trading as GNAC, Corp., *Nos. 87-1410, 87-1443, in the United States Court of Appeals, Third Circuit, decided April 20, 1988, on appeal from the United States District Court for the Eastern District of Pennsylvania, reported at 844 F.2d 1031.*

Robert F. Simone was at the Golden Nugget Casino with two of his clients. The two clients were at a blackjack table in the casino, and Simone was in the Rangoon Saloon, a lounge near the casino floor. Golden Nugget Security Sergeant Nixon told the two clients that they would have to leave the casino, whereupon they began walking toward the exit, but instead of exiting, went into the Rangoon Saloon. Sergeant Nixon then notified Security Captain Long.

When Sergeant Nixon entered the Rangoon Saloon area, Simone approached him and inquired about the problem with his two clients. Nixon told Simone that the problem was "none of his business." Captain Long then arrived at the scene, and he decided to formally evict the two men, as well as Simone and another individual who had joined the group. Simone and the other three were ushered by seven or eight security guards from the saloon area into a hallway, and then into a service elevator.

Apparently, there was a good deal of shoving and pushing, although the evidence conflicted. Simone complained that he was thrown into the elevator with such force that he fell, whereas Sergeant Dixon testified that Simone was pushed only after he charged at the security officer. The elevator did not work, so the security guards took the four men to the valet parking area, near a stairway leading to the Security Command Office, where "formal evictions" took place.

As the group was en route to the Command Office, the police arrived in the valet parking area, having been summoned by Captain Long. The security guards requested that the four men be arrested. The four men and the police officers then struggled, whereupon the police handcuffed the four and took them to the police station.

Security Sergeant Dixon and another security guard, Officer Palmieri, were instructed by Security Captain Long to proceed to the police station and file complaints against the four individuals, which Dixon and Palmieri did. They gave signed, handwritten statements to the police, and they then signed blank complaint forms, on which the police later typed the charges. The final typed forms, however, said that Simone had struck the two guards, but this contradicted the written statements that the guards gave the police when they signed the complaints.

Simone was criminally prosecuted. Both Sergeant Dixon and Officer Palmieri testified at the criminal trial that Simone had not struck them as was stated on the formal charges, and their testimony complied with their own written statements.

Therefore, Simone filed this lawsuit against Golden Nugget for assault and battery, false imprisonment, abuse of process, and malicious prosecution. At the first trial, the jury found in his favor on the false imprisonment and abuse of process claims, awarding him $150,000 compensatory damages, plus $1 million punitive damages. Both parties filed post-trial motions, resulting in a new trial on the false imprisonment claim.

At the retrial, the jury ruled against Simone, and he then filed a motion for a new trial, which was denied. He then filed this appeal, claiming that the trial judge should not have ordered a retrial on the false imprisonment claim, and that the trial judge should not have granted judgment in favor of the Golden Nugget since the jury had found in favor of the plaintiff on his abuse of process claim. He also argued that the trial judge should have ruled in his favor on the malicious prosecution and the assault and battery claims, notwithstanding the jury's verdict against him on these issues.

With respect to the assault and battery claim, Simone argued that there could be no rational basis for the jury to find against him because, in his opinion, the evidence was clear that he had been physically grabbed and taken to the service elevator area. The appellate court noted, however, that his account of what happened was disputed by other evidence: Sergeant Dixon denied grabbing Simone and throwing him into the elevator, and testified that Simone had been loud and abusive in the saloon, and that Dixon and the other guards escorted Simone and the other individuals from the saloon but that he touched Simone by pushing him towards the elevator only in response to Simone's attack on him.

It was the jury's duty to determine the credibility of witnesses when their stories are in conflict, and the jury had found that the evidence presented by the Golden Nugget on this issue was more credible. Therefore, the court concluded that the jury was correct in finding that the touching of the plaintiff Simone by the security guards was justified under the circumstances. Accordingly, there was no error by the trial judge in refusing to set aside the jury's verdict on the assault and battery claim.

Simone also argued on appeal that the evidence clearly showed that he had been prosecuted unsuccessfully in the criminal court for an assault which he did not commit, and that this prosecution was instigated at the direction of the Golden Nugget security guards who had signed the blank formal complaints, and that therefore he was entitled to a judgment in his favor on the malicious prosecution claim at the first trial.

The appellate court, however, disagreed, noting that the evidence showed that Palmieri and Dixon went to the Atlantic City police station to file charges against Simone, that they did sign blank complaint forms, and that the charges which were later typed in by the police were incorrect. This evidence permitted the jury to conclude that Simone had failed to prove the requirement of malice on the part of the security guards.

The discrepancy between the written statements given by the security guards and the charges set forth in the formal complaint was the only evidence which could have demonstrated malice on the part of the guards in bringing the criminal prosecution against Simone. The evidence showed that the guards had given the police a handwritten statement which set forth a correct version of the facts, but, the jury could have found, that the police ignored or embellished on those statements in filling in the formal complaint form. Therefore, the mistake was due to the policemen's actions and not to the security guards'. Accordingly, the jury was justified in finding that there had been no malicious prosecution of Simone by the Golden Nugget employees.

With respect to the abuse of process claim on the first trial, the appellate court noted that the record did reveal certain "irregularities" in the procedures followed to effect Simone's eviction, arrest, and criminal prosecution. But the court also found that the record did not contain sufficient evidence to prove that the Golden Nugget Casino employees had used the arrest process for some purpose other than it was designed to accomplish. The appellate court therefore affirmed the trial judge's order which granted judgment in favor of the casino, in spite of the jury's verdict in favor of the plaintiff on that claim.

The court acknowledged the plaintiff's challenge to the Golden Nugget's formal eviction policy, emphasizing that the court, by ruling in favor of the defendant Golden Nugget, did not intend to approve or condone such a procedure. The court specifically noted that, while an innkeeper does have the right to evict a disorderly patron, the innkeeper does not have the right to forcibly take a patron who is willing to leave the establishment voluntarily, to a private place against his will merely in order to read the person an eviction notice.

At the same time, the court did not agree with the plaintiff Simone that the casino's effort to implement the formal eviction procedure constituted a sufficient basis for finding the casino was liable for an abuse of process. In order to prove an abuse of process, New Jersey law requires evidence that the defendant (here, the casino) had some ulterior motive and engaged in some improper act after process was issued against the plaintiff.

In summary, the irregularity involved in the security guards' signing of a blank complaint form, on which the police subsequently typed a misstatement, does not amount to an abuse of process by the Golden Nugget Casino.

Next, Simone argued that the trial judge should have granted him a new trial, after the second trial in which the jury ruled against him, finding that there had been no false imprisonment. It was clear that he had been detained by the security guards at the casino, leaving the question for the jury to decide whether the detention was lawful and carried out in a reasonable manner. But, according to the testimony of Security Captain Long, the appellate court noted the jury could have concluded that Simone accompanied the group from the saloon to the service elevator area willingly and voluntarily, and that he was detained at that point because he became disorderly and assaulted one of the guards. This testimony would provide a rational basis for a conclusion by the jury that there had been no false imprisonment of Simone by the security guards.

In conclusion, the appellate court found no reversible error on the part of the trial judge, with respect to his rulings made during either the first or the second trial, and therefore affirmed those rulings, resulting in judgments being entered for the Golden Nugget Hotel and Casino on all counts.

PSCLR—JUN-JUL 88 (4)

California's 90-day Statute Of Limitations On Actions For Personal Property Left At Hotel Does Not Prevent Plaintiffs' Action For Theft Of Car And Trailer From Motel Parking Lot, Filed After Expiration Of 90-day Period:

Brad M. Martin et al. v. Holiday Inns, Inc., *No. F007994, in the Court of Appeal of California, Fifth District, decided April 5, 1988, reported at 245 Cal. Rptr. 717.*

The plaintiffs registered at a Holiday Inn in Bakersfield, California, driving their personal vehicle, a 1979 Ford Bronco, and pulling a rental trailer which contained "all their personal and worldly possessions." The hotel management told them where to park the vehicle and trailer in the motel parking lot where, they said, security guards were on duty to safeguard guests' property. During the night, however, both the automobile and the trailer containing all of their possessions were stolen.

The Martins filed suit, claiming that the Holiday Inn was negligent in failing to secure the premises in the parking lot so as to prevent the theft of their goods. They argued that the hotel owed them a duty to inform them that there had been a rash of similar thefts, and the hotel had breached that duty by telling them to park in a designated space, without warning them of the prior thefts.

The defendant hotel argued that the plaintiffs had filed their lawsuit more than 90 days after the loss, and that their suit was barred by the provisions of a statute which reads:

> All civil actions for the recovery or conversion of personal property, wearing apparel, trunks, valises or baggage alleged to have been left at a hotel, hospital, rest home, sanitarium, boarding house, lodging house, furnished apartment house, or furnished bungalow court, shall be begun within 90 days from and after the date of the departure of the owner of said personal property, wearing apparel, trunks, valises or baggage from said hotel, hospital, rest home, sanitarium, boarding house, lodging house, furnished apartment house, or furnished bungalow court.

Holiday Inns argued that the Martins' auto and trailer came within the term "personal property," as provided for in the statute. However, the California appellate court construed the term "personal property" to be restricted to things like "wearing apparel, trunks, valises," etc., and concluded that the automobile and trailer stolen from the parking lot would not fall within the meaning of the statute.

Accordingly, the trial court's dismissal of the case as being barred by the California statute of limitations was reversed and was returned to the trial judge for further proceedings.

COMMENTARY:

Of significant import in the court's opinion in the above case was the court's notation in a footnote that the statute in question had been ruled on only once in its entire 67 years of existence, and that was in 1928. The court's point: Has this statute been overlooked in other cases?

Is there such a statute in your jurisdiction? Is it being overlooked? Or is it being used to defeat claims which are not brought within the 90-day period required by this statute, for example. Ninety days is a particularly short period and many plaintiffs and, it is safe to say, many lawyers for plaintiffs would probably be unaware of such a short statute of limitations.

It is, therefore, all the more important that defense attorneys defending any hotel, hospital, rest home, sanitarium, boarding or lodging house, or furnished apartment house or bungalow court, be aware of the 90-day limitation. The limitation would be a complete bar for any recovery for the conversion of the type of personal property described in the statute.

It would be equally important, of course, for defense attorneys in such cases to determine whether any such specialized statutes of limitations may be applicable to hotels, hospitals, rest homes, etc., in jurisdictions other than California.

PSCLR – JUN-JUL 88 (9)

Where Medical Evidence Provides Causal Connection Between Criminal Assault On Motel Guest And Jury Found Inadequate Motel Security Caused Assault, Motel Is Responsible For Guest's Damages Due To Subsequent Heart Attack And Psychological Problems:

Sarah B. Crinkley; James E. Crinkley v. Holiday Inns, Inc.; Traveler's Management Corporation, a/k/a Travco; American Health Home, Inc.; Frank Schilagi; C. Roger Harris, and Tola J. Litaker, individually; John A. Howard, Jr., individually and as partners in Summit Associates, *No. 87-1547, in the United States Court of Appeals, Fourth Circuit, decided April 7, 1988, on appeal from the United States District Court for the Middle District of North Carolina, reported at 844 F.2d 156.*

During the two weeks immediately preceding a visit by Mr. and Mrs. James Crinkley to Charlotte, North Carolina, to attend a weekend function at the Charlotte Civic Center, there had been a rash of assaults and robberies on guests at several motels in and around the Charlotte area. The criminals were referred to in media reports as the "motel bandits," who had victimized guests in various motels located throughout the metropolitan area and close to Interstate 85, which runs around the northern edge of the city.

The Crinkleys had tried to make reservations at the Holiday Inn located in Charlotte, close to the Civic Center itself, but it was booked. So they consulted a Holiday Inn directory and found the next nearest Holiday Inn, the Holiday Inn Concord, located approximately 20 miles north.

Brian McRorie, assistant manager of the Holiday Inn Concord, had become aware of the activities of the motel bandits through the media. In addition, he had been contacted by several members of the local sheriff's department concerning the motel bandits, and had been queried as to his plans to employ off police officers as security guards due to the increased criminal activity in the area.

McRorie contacted Jim Van Over about hiring additional help to patrol the Holiday Inn Concord. Van Over was the president of the Metrolina Innkeepers Association and also had some supervisory responsibility for the Holiday Inn Concord, as an employee of the defendant Travelers Management Corporation which managed the Holiday Inn Concord. In fact, Van Over had been interviewed for a newspaper article about the motel bandits. In that news article, Van Over had stated that his hotel, the Holiday Inn on Woodlawn Road in Charlotte, had employed additional security personnel for night patrols.

In response to McRorie's inquiries concerning additional security at the Concord motel, however, Van Over decided that extra precautions were not warranted at that location. Nevertheless, McRorie did advise his employees at the Concord property to be especially alert for suspicious activities and persons. Moreover, he had personally patrolled the premises between 8:00 and 8:30 PM, at about the time the Crinkleys checked into the motel and were assaulted.

When the Crinkleys had arrived and checked in, they parked their automobile in front of their assigned room and began unloading their luggage. Sarah Crinkley, while standing in the doorway to the room, saw a man come around the corner of the hotel and begin walking toward them just as her husband brought in the last of their belongings.

The man approached the Crinkleys' room and asked to speak with James Crinkley by naming him. He then began pushing his way into the room. Crinkley tried to resist, but the intruder succeeded in forcibly entering the guest room with the Crinkleys, whereupon the intruder brandished a firearm. The assailant began beating Crinkley and called for his accomplices. Two additional men entered the room who also beat Crinkley. They bound and gagged him, and placed a mattress over him.

The three intruders rifled the Crinkley's baggage, and threatened to kill Mrs. Crinkley if she did not remove her engagement ring and give it to them. They also bound and gagged her before taking flight.

After they were gone, she managed to untie her bonds, released her husband, and called the front desk for assistance. The police arrived in moments, and the Crinkleys were taken to a local hospital for emergency care. Crinkley's jaw was broken severely, requiring it to be wired together for approximately six weeks. He also sustained multiple injuries to his head and upper body due to the beating.

Mrs. Crinkley's condition was significantly more complicated, due in part to the fact that she had already been under a doctor's care for hypertension and obesity prior to the assault. Approximately 14 months after the incident, she suffered a heart attack and underwent first a balloon angioplasty, then bypass surgery.

Friends and family testified that she also underwent a drastic personality change, becoming anxious and withdrawn, and restricting her activities significantly. A psychiatrist diagnosed her as having post-traumatic stress disorder and major affective disorder.

This lawsuit was filed by the Crinkleys against several defendants with various connections to the Holiday Inn Concord. The basis of the lawsuit was that the various defendants were negligent in failing to provide adequate security to protect the Crinkleys from their assailants, and that such negligence on the part of the defendants was the cause of their resulting injuries and on-going medical problems.

At the trial, the Crinkleys relied heavily on Brian McRorie's testimony to establish that the assault by the third parties was reasonably foreseeable by the defendants. They also produced a security expert who testified that the security measures employed by the Holiday Inn Concord were not sufficient to deal with the then-existing threat.

Furthermore, doctors testified that Mrs. Crinkley's heart attack and psychological problems resulted from the stress she continued to experience as a result of the assault.

The jury returned verdicts against all defendants, finding that their negligence caused the injuries to the plaintiffs, and awarding $400,000 to Sarah Crinkley and $100,000 to James Crinkley in compensatory damages. The defendants filed this appeal.

On review, the appellate court first noted that, in general, a landowner or business operator is under no duty to insure the safety of his business guests on the property. There is only the general duty to exercise reasonable or ordinary care for their safety. And, ordinarily, such a duty does not include the protection of invitees from the criminal acts of others, although it may in certain circumstances.

For instance, such a duty exists on the part of owners of businesses which provide public accommodations for guests. (Here the court cited cases previously reported in **PSCLR:** *Murrow v. Daniels* in the August 1987 issue, case one, page one; *Brown v. North Carolina Wesleyan College, Inc.* in the March 1984 issue, case seven, page six; and *Urbano v. Days Inn of America, Inc.* in the November-December 1982 issue, case eight, page six.)

Further, it is the element of "foreseeability" which determines whether a duty to protect invitees against such criminal acts will or will not be imposed on a particular landowner or business operator in a particular situation. Foreseeability may be established by evidence showing prior criminal activity on the premises or in the area.

The defendants here argued that the activities of the "motel bandits" in the Charlotte area should not be considered as reasonably requiring them to foresee a threat to the Holiday Inn Concord, some 20 miles north. The appellate court disagreed, noting that all of the evidence, taken in its totality, was sufficient to allow the jury to conclude that there was a duty to take reasonable precautions.

There was evidence that seven motels in the area had been victimized in the two weeks immediately preceding the assault on the Crinkleys, and these assailants had been identified as the same persons in all seven incidents. On occasion, the group had struck more than one motel property in a single evening. Furthermore, many of the reported attacks were separated by several miles, and there are two highways providing convenient access to Concord from Charlotte. Finally, the Holiday Inn Concord was similar to several of the motels in which the Charlotte attacks had taken place. These facts and others made it reasonably foreseeable that the attacks would continue, according to the appellate court.

Of course, in addition to this circumstantial evidence, the plaintiffs were able to present evidence of actual notice to McRorie of the specific threat presented by the motel bandits. As stated above, he had even been contacted by local law enforcement officials, seeking to hire themselves out as security guards, specifically because of the threat posed by the "motel bandits." McRorie also went so far as to ask his supervisor Van Over about hiring additional security personnel, and he warned employees to take additional precautions. Therefore, the appellate court found that the jury was well within its authority under these facts to conclude that the attack on the Crinkleys was foreseeable by the defendants.

The appellate court then reviewed the testimony of Kenneth Prestia, the expert witness on hotel/motel security who had testified at trial on behalf of the plaintiffs. He had opined that security at the Holiday Inn Concord was insufficient for several reasons: The Holiday Inn Concord was accessible from several directions, and some of the points of access could not be seen by motel employees from the front desk. The fence around the property was of insufficient height to deter trespassers, and there were no "no trespassing" signs on the property. While signs alone would not stop intruders, Prestia had testified that, in his opinion, such measures do have a deterrent effect on criminals by giving the impression that the business operator maintains "heightened security posture."

Prestia had also pointed out that Holiday Inn had failed to institute any formal security plan specifically designed for the Concord property, and that it did not employ any security patrols whatever. Such patrols, he had stated, are especially important when there are numerous access points to a property, particularly where there is limited observation of the access points by employees. Furthermore, such patrols could be added sporadically, during periods of higher criminal activities or increased threats of such activities. The appellate court finally noted that no evidence was presented by defendants in opposition to Prestia's opinions of inadequate security at the Concord property.

The appellate court concluded that there was sufficient evidence for a jury to reasonably conclude that Holiday Inn and the other defendants had breached their duty to provide adequate security to protect the plaintiffs against the known and foreseeable risk that was created by the circumstances.

The appellate court then reviewed the defendants' argument that any negligence on their part was not the proximate cause of the assault suffered by the Crinkleys. The defendants pointed out that there is no reasonable probability that the installation of "no trespassing" signs, or additional fencing would have prevented the assault on the Crinkleys. They contend that the fact that McRorie had patrolled the premises himself at approximately the time of the assault proved that added security measures, such as patrol guards, would not have prevented the assault either. But, the appellate court found, it is not necessary for the plaintiff to prove that the particular protective measures which the innkeeper did not employ would in fact have prevented the particular injuries suffered.

Under North Carolina's law, according to the Circuit Court of Appeals, the jury could find proximate cause by accepting the expert Prestia's opinions to the effect that criminals will typically make a judgment as to the risk of apprehension presented by the level of security employed at a particular location, and that the particular measures employed by the defendants at the Holiday Inn Concord would not have acted as an effective deterrent. The jury then could reasonably conclude (though it was not required to do so) that the criminals who assaulted the Crinkleys "were indeed emboldened by the lax security measures to come on the premises, size up the situation, and plan their assault."

The defendants next contested the amount of damages awarded to the Crinkleys, claiming that there was not sufficient evidence to prove a necessary causal relationship between the attack and the subsequent heart attack suffered by Sarah Crinkley, as well as the subsequent psychological problems attested to by her doctors. The defendants argued that the expenses related to these problems should not have been submitted to the jury, and the jury should not have been entitled to include them in any award of damages.

The appellate court therefore reviewed the medical testimony of her psychiatrist, Dr. Silverman, and her cardiologist, Dr. Bennet, and her family doctor, Dr. Harris. These doctors had testified that her complaints were genuine and that the assault at the motel caused her mental illness, her heart attack, and the stress which led to heart attack. Expert testimony was also presented that it was medically and scientifically plausible that "significant stress could produce or accelerate atherosclerosis to the point of heart attack." The appellate court concluded that the damages related to the mental condition and the heart attack were properly submitted to the jury.

Finally, the appellate court rejected the defendant's argument that the awards were excessive.

The awards to the Crinkleys were affirmed in full.

COMMENTARY:

For those readers who may be interested in aspects of this decision, beyond the issue of inadequate security, and in reviewing the relationship and consequent legal liabilities among Holiday Inns as the franchisor to Holiday Inns-Concord as the franchisee, and Travco as the management company, a copy of the entire decision should be reviewed.

The franchisor, Holiday Inns, attempted to convince the court that they could not be found liable on the basis of apparent authority since the evidence would not support any finding of actual agency. They also argued that the evidence was insufficient in any event to support any finding of apparent agency.

The court, however, reviewed the franchise agreement and noted that Holiday Inns retained a significant degree of control over the method of operation of the Holiday Inn Concord, including the use of the Holiday Inns tradename appearing on numerous items in the motel. Holiday Inns also engaged in national advertising which promoted the national network system, and the advertising did not distinguish between company-owned and franchised properties. The Holiday Inns directory listing, which was consulted by the Crinkleys in an effort to select a motel, also does not distinguish between company owned and franchised properties.

Even though the franchise agreement may disclaim any agency relationship, such denial is not alone determinative of liability since "agency by estoppel" can apply to situations where no actual agency relationship exists.

The court concluded:

> *We think that a jury could reasonably conclude that the Holiday Inn Concord was operated in such a way to create the appearance that it was owned by Holiday Inns, Inc., and that this was one of the purposes of the franchise agreement.*

The court also noted that Sarah Crinkley had testified that she and her husband had relied on this appearance by using the Holiday Inns directory to select the Holiday Inn Concord.

There is also a strongly worded dissent by one of the circuit court judges on the appellate bench, based primarily on his opinion that the chain of liability and recovery, whereby the majority found Holiday Inns and the other defendants obligated to the defendants, to be an "exceedingly tenuous one consisting of a series of weak links."

The dissent noted:

> *Plaintiff contends that the motel's failure to post 'no trespassing' signs, extend its fences, and to keep a proper lookout, resulted in the assault on her by the 'motel bandits,' which in turn, resulted in a heart attack more than a year later. Then through a theory of apparent agency, Holiday Inns, Inc., is joined as a defendant in the action, despite the fact that the motel was run by an independent franchise and Holiday Inns, Inc., played no role in its management.*
>
> *The upshot of the majority's decision is, therefore, that an innkeeper not responsible for managing the premises is found, under a marginal theory of causation, to have proximately caused an assault by third party criminal ac-*

tors which in turn is found to have caused a heart attack occurring at a remote point in time. With all respect, speculation seems piled upon speculation until the label of tortfeasor can be pinned upon some deep pocket in the vicinage.

PSCLR–AUG 88 (1)

Hotel's Liability To Drugged Guest For Theft Of Certain Personal Property Is Limited To $500 By Statute:

Allen Chase v. Hilton Hotels Corporation, *Civ. A. No. 87-2118, in the United States District Court for the Eastern District of Louisiana, decided March 10, 1988, reported at 682 F. Supp. 316.*

The plaintiff, Allen Chase, sued the New Orleans Hilton Hotel, alleging that while a guest in the hotel, he was drugged by a female whom he had met in the Hilton and/or the bartender in Hilton's Rainforest Lounge. He contends that due to the Hilton's negligence, he passed out in his room and awoke the next day to find certain money and personal possessions stolen. In addition, he alleges that he suffered prostate gland and kidney damage from the drug, causing a loss of profits to two of his business enterprises and other damages totalling $575,000.

The Hilton Hotels Corporation brought a motion for partial summary judgment, requesting that the court rule, prior to trial, that Chase could not legally recover certain damages, whether or not the Hilton's negligence and ultimate liability could be established. Therefore, only the issue of the type and amount of damages Chase could recover, if he proved the liability of the Hilton Hotels Corporation, was before the court in this case.

Most significantly, the court granted the Hilton's request limiting Chase's potential recovery for the theft of his personal property to $500, based on a Louisiana statute, Article 2971, which provides:

> No landlord or innkeeper, or his officers, clerks, agents, or employees shall be liable contractually or delictually under the provisions of the foregoing six articles to any guests or party of guests occupying the same apartments for any loss of cash, jewelry, rare art items, furs, cameras, or negotiable instruments sustained by such guests or party of guests by theft or otherwise, in any sum exceeding five hundred dollars, unless by special agreement in writing with the proprietor, manager or lessee of the hotel or inn a greater liability has been contracted for.

However, the Louisiana statute does not apply if:

> (1) The landlord or innkeeper fails to provide a safe deposit facility for valuables.
>
> (2) The landlord or innkeeper fails to conspicuously post this article in the guest room and registration area.
>
> (3) The guest avails himself of the safe deposit facility provided by the landlord or innkeeper.

In this case, the Hilton did provide a safe deposit facility for its guests and conspicuously posted the provisions of Article 2971 in the guest rooms and registration area; Chase did not use the safe deposit facility. Therefore, the court held that the Hilton's liability for the particular types of personal property listed in the statute would be limited to $500. Specifically, Chase's loss of $1,000 in cash, a $100 traveler's check, and his Canon camera would be subject to the $500 limit.

However, citing another Louisiana statute, Article 2967, the appellate court held that Chase's Piaget wristwatch and gold money clip, although properly described as jewelry, are not covered by the limitation of liability act because they are ordinarily worn about the person. In addition, Chase's dictating machine, passport, credit cards, briefcase, lettercase, business papers, passport case, and billfold are not covered by the limitation.

The court further held that the 1981 case of *Laubie v. Sonesta International Hotel Corporation*, 398 So. 2d 1374 (La.1981), limiting the application of Louisiana's limitation of liability statute to suits based on contract, as opposed to suits (such as this one) based on tort, was no longer valid. That case was overruled by the Legislature in a 1982 amendment to Article 2971, specifically providing that the limitation applied to an Innkeeper's delictual (founded in tort) and contractual liability.

The court also analyzed whether Chase could recover damages for loss of income from Who's Who-Bachelor's, Inc. d/b/a Who's Who International, of which Chase owns 90 percent of the outstanding stock, and for damages due to the failure of a proposed joint venture between the People's Republic of China and a corporate entity in Australia in which he was a shareholder.

The court held that Chase, who sued as an individual, could recover for direct injury to him as a shareholder of Bachelor's Inc., but not for the injuries of the corporation. For example, he could recover for the reduced value of his corporate stock, but not for lost profits sustained by the corporation. The court stressed, however, that Chase would have to prove his loss with "reasonable certainty," and that the loss was caused by the alleged drug ingestion. The court expressed doubt that Chase could prove a causal connection between the drug ingestion, his kidney and prostrate damage, and the depreciated value of his stock in Bachelor's, Inc.

The court drew the line with respect to Chase's attempt to recover for the failure of the proposed joint venture with the People's Republic of China. The joint venture involved growing cashew nuts in Australia, with labor and technical assistance provided by Chinese citizens, and ultimate sale of the nuts in China. The court held that Chase could not recover for his personal loss of business opportunity or loss of profits related to the proposed cashew venture because any damages would be based on "conjecture or speculation." Recovery for such losses would violate the rule that damages must be proved with reasonable certainty, since a corporation was never formed to carry out the proposed joint venture, and since Chase admitted that the venture had, at best, a one-in-five chance of success. In addition, there was no "customary or foreseeable profit" for such a venture which the court could use as a guide. Furthermore, the court noted that Louisiana law does not allow damages for loss of profits from a crop which never commenced growing.

The court in this case significantly limited the Hilton's potential liability for losses sustained by Chase due to his alleged drugged state, without addressing the question of the Hilton's negligence. The question of the Hilton's liability for any of the other damages discussed in the case will have to be determined at the trial on the merits.

COMMENTARY:

The court noted that the opinion in Laubie v. Sonesta International Hotel Corporation, 398 So. 2d 1374 (La.1981) had since been overruled by an amendment to the applicable statutes. Later appeals of that case were reported in the October 1984 **PSCLR**, *case two, page three and in the June-July 1985* **PSCLR**, *case five, page six.*

The court also cited Peppard v. Hilton Hotels Corporation, which interpreted the amended statute and which was reported in the April 1986 **PSCLR**, *case three, page four.*

PSCLR – NOV-DEC 88 (1)

Hotel Operator Not Liable For Death Of Fireman Caused By Risks Normally Associated With Fires, Even Though Hotel Not In Compliance With City Fire Prevention Code:

Eileen Coglianese, as Special Administrator for the Estate of Edmond Coglianese, deceased, v. The Mark Twain Limited Partnership, beneficiary under Trust No. 23904, *No. 87-2227, in the Appellate Court of Illinois, First District, Third Division, decided May 18, 1988, reported at 524 N.E.2d 1031.*

Edmond Coglianese, a City of Chicago firefighter, died while fighting a fire at the Mark Twain Hotel in Chicago. His wife, Eileen Coglianese, sued the Mark Twain Limited Partnership, manager and operator of the hotel, for the wrongful death of her husband, alleging that the Mark Twain Hotel had caused his death by failing to comply with the building and fire prevention codes of the City of Chicago. Specifically, the interior walls of the hotel were not fire resistant and were highly combustible. When the fire broke out, the interior walls burned rapidly and produced large amounts of black smoke, soot, and other noxious gases. Although he was wearing a breathing apparatus, Coglianese suffocated to death from the smoke and gases released by the fire.

The trial court applied what has become known as "the fireman's rule" and dismissed the case against the hotel operator without a trial. Acknowledging that harm from fire is a reasonable risk of a firefighter's occupation, the courts have fashioned the "fireman's rule" as follows:

> While a landowner owes a duty of reasonable care to maintain his property so as to prevent injury occurring to a fireman from a cause independent of the fire, he is not liable for negligence in causing the fire itself.

Because the ordinary duty of a firefighter is to deal with fires, a firefighter assumes the risks normally associated with that function when accepting that employment. Thus, a landowner will only be liable when a firefighter is exposed to "undue risks beyond those inevitably associated with fighting a fire."

Ordinarily, the violation of the municipal code regulating fire safety precautions would be evidence of the hotel's negligence or failure to exercise reasonable care. However, in this case, the hotel owed no duty to exercise reasonable care with respect to a firefighter, except as to hazards unrelated to fire fighting. Since no duty was owed, no liability for negligence attached, although the municipal code was violated.

The court further held that even if the hotel operator did fall below the standard of reasonable care in failing to comply with the fire code, the fire code violations did not directly or "proximately" cause the death of Coglianese. His death was caused by the fire, not the construction or maintenance of the premises. Further, the court did not believe that the large amount of smoke and gas was an undue risk beyond that inevitably associated with fighting a fire. Coglianese was not subjected to a danger greater than that which a firefighter would reasonably anticipate in the performance of his duty.

The court, therefore, upheld the trial court's dismissal of Mrs. Coglianese's action against the hotel operator for the wrongful death of her husband.

COMMENTARY:

The "fireman's rule" is a long standing and well-respected principle of tort law, and has been a part of our legal heritage and tradition for many years. We have previously reported on other cases involving this same principle [see e.g., Rois v. Junco, in the August 1986 **PSCLR**, *case seven, page seven, and we do not anticipate any significant departure from this precedent in the foreseeable future.*

PSCLR – NOV-DEC 88 (3)

Innkeeper Not Liable For Loss of Guests' Property Left In Room When Guests Are Properly Notified Of Availability Of Safe And Innkeeper Disclaims Liability For Loss of Valuables Not Placed In The Safe:

Kates et al. v. Brunswick Motel Enterprises, Inc., *No. 76588, in the Court of Appeals of Georgia, decided July 15, 1988, reported at 371 S.E.2d 686.*

Harry B. Kates and Carolyn Kates lost certain items of personal property from the room that they occupied at the Holiday Inn I-95, owned by Brunswick Motel Enterprises, Inc. The Kates sued Brunswick to recover damages for the loss of their property.

The Holiday Inn provided a safe for the deposit of its guests' valuable articles during their stays. The hotel notified its guests of the availability of the safe and of the hotel's disclaimer of any liability for guests' valuables not placed in the safe. Three such notices were in the room occupied by the Kates. One notice was placed on a table, and two framed notices hung on the wall between the front door and the door to a connecting room. In addition, a framed notice hung on the wall behind the registration desk, and a notice was printed on the registration form signed by Harry Kates.

The trial court dismissed the case against the hotel owner based on the undisputed facts before it prior to trial. The appellate court upheld that decision, finding that the notice given by the hotel was adequate to insulate the hotel from liability for loss of valuables not placed in its safe.

Georgia law provides that

> [an] innkeeper may provide a safe or other place of deposit for valuable articles and, by posting a notice thereof, may require guests of the innkeeper to place such valuable articles therein or the innkeeper shall be relieved from responsibility for such articles.

This particular Georgia statute does not specify the location for placement of the notice of the safe's existence nor of the disclaimer of liability for articles not placed in the safe. The court held that this statute, applicable in this case, requires only that the innkeeper post a notice of the availability of the safe and of the requirement that guests place the valuables therein, in order to relieve the innkeeper of liability for the loss of items not placed in the safe.

The court noted that the notice was posted not only on the registration card, but also in the guest's room and behind the registration desk. These notices were sufficient, according to the appellate court, to comply with the "posting" requirement of the statute, and such posting of these notices relieved the hotel of any liability for the loss of valuables not placed in the safe.

The appellate court further held that two other Georgia statutes with stricter notice requirements, which were argued by the plaintiff, did not apply in this case. One, requiring that a notice containing the provisions of the statute be posted "in a conspicuous place in all rooms of the hotel occupied by guests," applies only to the limitation of liability of an innkeeper for valuables that are deposited in the safe. The second statute, requiring the innkeeper to post a copy of the statute itself "printed in distinct type, on the inside of the door of the guest's room," applies only to reasonable regulations other than those connected with the safe.

On the undisputed facts of this case, the court found that the notice given by Holiday Inn was sufficient, and the lower court ruled correctly in favor of the hotel owner.

PSCLR–FEB 89 (13)

Innkeeper Not Liable To Guest Injured In Criminal Attack When The Attack Was Not Reasonably Foreseeable:

T. Erwin Satchwell, III v. LaQuinta Motor Inns, Inc., *No. 87-472, in the District Court of Appeal of Florida, First District, decided November 2, 1988, reported at 532 So. 2d 1348.*

T. Erwin Satchwell, III, checked into a motel owned by LaQuinta Motor Inns. At around 11:15 PM, Satchwell left his door open behind him and went to pick up his bags. As he was carrying his bags into his room, he was attacked, beaten, and robbed.

Satchwell sued LaQuinta for negligence, alleging that it breached its duty to provide lodging which was reasonably safe from foreseeable criminal conduct of third persons by failing to provide adequate security on the premises and as a result of that negligence, he was attacked, beaten and robbed while in his room.

At trial, the judge refused to allow testimony concerning an incident within two months of the assault in which a car was stolen from the Inn's parking lot. Later testimony revealed that on the night of the assault, the night auditor was on duty assisting the guests and occasionally walking the area for security. The lighting outside of the building housing Satchwell's room consisted of 4" x 10" yellow lights placed between every two rooms. The Inn was adjacent to wooded and undeveloped property. There was testimony that the innkeeper often discovered vagrants sleeping in the trash can area on the perimeter of the premises. In the past, the Inn had employed two security guards during times of heavy business.

At the close of the evidence, the trial judge dismissed the case, finding that no reasonable person could find that it was foreseeable that Satchwell would be robbed and beaten on the night in question. Satchwell appealed.

The appellate court stated that "as a motel operator, [LaQuinta] was under a continuing legal duty to its business invitees to use ordinary care to keep the premises in a reasonably safe condition and [to] protect them from harm due to reasonably foreseeable risks of injury." LaQuinta's duty to provide security guards and other precautions to prevent criminal attacks would depend upon whether it was reasonably foreseeable that such attacks would occur in the absence of such precautions.

The appellate court initially agreed with Satchwell that the trial judge should have allowed the testimony regarding prior criminal activities that had taken place on the premises. The court held, however, that the error was "harmless" because, even considering that testimony, Satchwell did not present sufficient evidence to raise an issue as to the foreseeability of the criminal attack.

The court noted that Satchwell presented no evidence of any significant criminal activity against motel guests within five miles of the location of this motel. The security guards that were hired by LaQuinta were hired due to the need to police unruly guests rather than the need to protect against the criminal acts of trespassers. Satchwell presented no expert testimony and did not present evidence of reasonable precautions that a motel operator in LaQuinta's position should have taken under the circumstances. Satchwell did not demonstrate that LaQuinta had actual or constructive knowledge of any danger to motel guests from third party criminal assaults. There was no evidence that LaQuinta violated industry standards, that the design of LaQuinta's premises posed any peculiar security problems, or that suspicious persons had been present on the premises on prior occasions.

In sum, the court held that Satchwell failed to present sufficient evidence that LaQuinta knew or should have known that its guests were foreseeably subject to criminal attack by third parties. The appellate court, therefore, upheld the trial court's dismissal at the close of plaintiff's evidence.

COMMENTARY:

Factors to be considered in proving foreseeability include: (1) industry standards, (2) community crime rate, (3) extent of assaults or criminal activity in the area or similar business enterprises, and (4) the presence of suspicious persons and the peculiar security problems posed by the premises design.

Proof of similar incidents is pertinent but not essential to establishing foreseeability.

PSCLR – MAR 89 (6)

Motel's Limitation of Liability For Personal Property of Guests Does Not Limit Recovery of Damages for Personal Injury to Guests Assaulted in Motel Room:

Milt Millman v. Howard Johnson's Company, et al, *No. 87-2160, in the District Court of Appeal of Florida, Third District, decided November 15, 1988, reported at 533 So. 2d 901.*

Milt Millman, a jewelry salesman, was held at gunpoint, bound and gagged by an armed robber who broke into his locked room at a Key West Howard Johnson's Motel. Millman sued the Howard Johnson's Company for negligence, and a jury awarded him $50,000 in damages at trial. The jury then found Mr. Millman to have been negligent himself, and reduced his total award 50 percent, leaving him with $25,000.

After the trial, the court reduced the verdict and judgment to $500, pursuant to a Florida statute limiting a public lodging establishment's liability for property loss of its guests. The Florida statute provides:

> (1) The operator of a public lodging establishment is under no obligation to accept for safekeeping any moneys, securities, jewelry, or precious stones of any kind belonging to any guest, and, if such were accepted for safekeeping, he shall not be liable for the loss thereof unless such loss was a proximate result of fault or negligence of the operator. However, the liability of the operator shall be limited to $1,000 for such loss if the public lodging establishment gave a receipt for the property (stating the value) on a form which stated, in type large enough to be clearly noticeable, that the public lodging establishment was not liable for any loss exceeding $1,000 and was only liable for that amount if the loss was a proximate result of fault or negligence of the operator.

Millman appealed the trial court's reduction of his recovery.

The appellate court held that, even to the extent that the statute applies, it limits an innkeeper's liability only as to the value of personal property which is stolen from or lost by a hotel guest. The statute has no effect upon the damages recoverable for mental pain and suffering, anguish, humiliation, and other personal injury which are appropriate in a case such as this, involving assault and false imprisonment.

The appellate court found that Millman presented adequate evidence of his personal injury damages to justify the award of $49,000 above the $1,000 awardable as property loss. The appellate court, thus, reversed the trial court's decision and ordered the trial court to reinstate Millman's award of $25,000, in accordance with the jury verdict.

PSCLR – MAY 89 (8)

Hotel Owners Only Partially Liable For Injury To Guest Raped In Hotel Room, When Guest Negligently Opened Her Door To Stranger Late At Night:

Susan Wassell v. Wilbur L. Adams and Florena M. Adams, doing business as Ron-Ric Motel, *No. 88-1118, in the United States Court of Appeals, Seventh Circuit, on appeal from the United States District Court for the Northern District of Illinois, decided January 5, 1989, reported at 865 F.2d 849.*

Susan Marisconish travelled with her fiance's parents to Chicago to attend her fiance's graduation from naval basic training at the Great Lakes Naval Training Station. She and her future in-laws checked into the Ron-Ric Motel near the Training Station. The Ron-Ric, owned by Wilbur and Florena Adams, is a small, inexpensive motel that caters to the families of sailors at the Training Station.

The Ron-Ric is within four blocks of a high-crime area. The Adamses occasionally warned female guests not to walk alone in the neighborhood at night, but they did not warn Marisconish or her future in-laws.

One evening, upon returning to her room at the Ron-Ric, Marisconish locked the door, fastened the chain, and went to bed. She was awakened from a deep sleep by a knock at the door. She turned on a light and noticed that it was 1:00 AM She went to the door and looked through the peephole, but saw no one. She did not look through the pane of glass next to the door. Although the door was equipped with two locks and a chain, Marisconish unlocked the door and opened it all the way, thinking that her fiance had come from the base and, not wanting to wake her, was on his way to obtain a key from the innkeeper. Rather than her fiance, she was met by a neatly dressed man whom she had never seen before. He asked for a glass of water. Marisconish went to the other end of the room to obtain the glass of water. The man complained that the water was not cold and went into the bathroom himself to get some more water.

When the man went into the bathroom, Marisconish became nervous and hid her purse containing $800 in cash. There was no telephone in the room. The parking lot on which the motel rooms opened was brightly lit by flood lights.

The man asked Marisconish to join him in the bathroom. She refused, and he emerged from the bathroom naked. Marisconish fled from the room and beat on the door of the adjacent room. The man ran after her and grabbed her. Marisconish screamed, but no one responded. The motel had no security guard and the Adamses, who lived in a basement apartment at the other end of the motel, did not hear her screams.

The man covered Marisconish's mouth and dragged her back to her room. He gagged her with a wash cloth and raped her at least twice. After more than an hour, Marisconish persuaded the rapist to take a shower with her. To save herself from further harm, she tried to convince the rapist that she liked him. After the shower, she managed to slip away, get dressed, and flee in her car.

There had been a rape at the motel several years previously, and there had also been a robbery and an incident in which an intruder had kicked in the door to one of the rooms. These were the only serious crimes committed during the seven years that the Adams owned the hotel.

Marisconish subsequently married her fiance but suffered severe post-traumatic distress from the incident. A psychologist testified that she would need between $25,000 to $30,000 in psychological therapy as a result of the incident.

Marisconish sued the Adamses, charging them with negligence in failing to warn her or take other precautions to protect her against the assault. A jury composed of four women and three men found that the innkeeper had been negligent and such negligence had been a proximate cause of the assault. The jury assessed Marisconish's damages at $850,000. However, the jury also found that Marisconish had contributed to her injury through her own negligence. The jury found that Marisconish was 97 percent to blame for the attack and the Adamses only 3 percent. The jury, thus, awarded Marisconish only $25,500 in damages.

Under the Illinois comparative negligence law currently in place, a plaintiff may not recovery in cases in which the plaintiff's fault is more than 50 percent of the proximate cause of the injury or damage for which recover is sought. However, because Marisconish's suit was filed before the new law went into effect, Marisconish was allowed to recover for that portion of her injury that was not her fault, in this case, 3 percent. The trial judge entered judgment in accordance with the jury's verdicts, and Marisconish appealed.

The appellate court found that a jury could not reasonably find that the Adamses had consciously disregarded a high probability of serious physical harm to Marisconish. The court also found that the evidence supported the jury's apportionment of fault. The court found that the cost to the Adamses in hiring a security guard might be much greater than the monetary equivalent of greater vigilance on the part of Marisconish that would have actually averted the attack.

The court stated that, were it the trier of fact, it would have assessed the innkeeper's share of fault at more than 3 percent. However, the jury, and not the court, was charged with making this decision, and the reviewing court would not overturn the jury's verdict unless the trial judge abused his discretion in determining that the jury's verdict was not against the clear weight of the evidence. Because the jury's verdict was supported by the evidence, the court affirmed the trial judge's decision not to set it aside.

PSCLR – SEP 89 (2)

Motel Met Duty To Provide Reasonable Precautions To Protect Guest; Held Not Liable For Robbery And Shooting Of Guest:

Euel Franklin and Sue Franklin v. Paul Dupuis & Associates, a general partnership d/b/a Econolodge, and United States Fidelity & Guaranty Company, *in the Court of Appeal of Louisiana, Third Circuit, decided April 19, 1989, reported at 543 So. 2d 970.*

One morning at approximately 4:00 AM, Euel Franklin, a truck driver, stopped at the Econolodge motel in Lafayette, Louisiana. After registering, Franklin drove to the back of the motel and parked under a light. As he began unloading his personal gear, a gunman walked around the front of Franklin's truck and demanded Franklin's money. Franklin gave the robber all of his cash, and the robber proceeded to shoot Franklin in the abdomen.

Franklin and his wife filed suit against the motel claiming that the motel had breached a duty to take reasonable precautions to protect its guests from the crimes of other people. The jury decided in favor of the defendant, Econolodge. The Franklins appealed.

On appeal, the court acknowledged that

> innkeepers owe their guests a high degree of care and protection, and that reasonable precautions must be taken to protect patrons from the acts of criminals.

Econolodge had only one entrance to the property and there was an eight-foot high fenced topped with barbed wire surrounding the premises. The motel had been robbed only twice before, so there was apparently no need for extraordinary precautions. Additionally, there was evidence that the robber had "stalked" Franklin, following him by car onto the premises.

Under Louisiana law, a jury verdict should only be overturned only when the evidence is so overwhelming that reasonable men could not disagree as to its conclusion. However, in this case, the court allowed the jury verdict to stand. Basically, the court took the position that there was enough evidence to support the jury's conclusion that Econolodge had met its duty of care – in other words, Econolodge had taken reasonable precautions to protect its guests.

Parking Lots

PSCLR – JUN 82 (1)

Property Owner Not Responsible To Employee Injured In Unauthorized Area:

Ruth Nicoletti v. Westcor, Inc., an Arizona Corporation, d/b/a Metrocenter et al., *No. 15336, in the Supreme Court of Arizona, decided January 11, 1982, reported at 639 P.2d 330.*

This case explores the boundaries of liability a landowner owes a business invitee. Generally, landowners have a duty to invitees to maintain their property in a reasonably safe manner. But that duty may be reduced or even extinguished if the invitee goes beyond the particular area to which he/she is invited.

Ruth Nicoletti was employed by Rhodes Department Store, a major retail outlet at Metrocenter, a regional shopping mall in Phoenix. During the Christmas rush, employees were asked to park at a temporary parking facility on the far side of Metro Parkway, a public street which encircles the center and regular parking area.

On the night of November 30, Nicoletti and to co-employees left Rhodes, taking a direct route through the parking lot toward the special parking area instead of taking the sidewalks with pedestrian crosswalks which would have been a few steps further.

The shopping center's own parking area was landscaped with raised planters surrounded by a five-inch cement curb with ornamental shrubbery, a vine-like ground cover about one to one-and-one-half feet in height.

The shortcut Nicoletti took went through the shrubbery in the planter. As she later testified, she tangled her feet in the bushes, "made a headwade [sic] right through the center of the thing" and fell, injuring her hand and side.

Nicoletti sued Westcor, Inc., owner of the mall premises. Westcor obtained summary judgment, dismissing the case on the ground that Westcor had no duty as a matter of law to Nicoletti under these facts.

Nicoletti appealed, but the Arizona Supreme Court agreed with the trial court that a landowner's duty was reduced or extinguished if the invited person goes beyond the area to which she is invited. Further, while the landowner must also provide a reasonably safe means of entering and leaving the property, the owner does not have to provide the shortest route possible to any particular destination. An owner could be liable if he misleads an invitee into believing that a particular pathway may be used by invitees; in this case, Nicoletti failed to show that Westcor had done so. The court stated:

> A reasonable person could not have thought the planter was an appropriate means of egress from [Westcor's] property nor had [Westcor] misled her into a reasonable belief that it was a proper passageway.

Since Westcor had no duty to maintain the planter as a safe passageway, Nicoletti could not recover for any injuries she received while using the unauthorized passageway. The dismissal was affirmed.

COMMENTARY:

The result of this case would have been quite different if the plaintiff had been able to show that the landowner knew that people were using the route through the planter and did nothing to prevent it. Such would be construed as an implied invitation and would impose a duty on the landowner. (See MGA Theaters, Inc. v. Montgomery, 321 P.2d 1009, as an example of that situation.)

Therefore, every landowner and tenant should take steps (signs, barriers, etc.) to notify the public not to come onto or on those areas of the property where they are not invited. All employees and security personnel in particular should be instructed and reminded to warn people away from such areas, not to ignore or to acquiesce in the use of unauthorized areas or passageways, and to report the use of unauthorized areas.

And finally, all areas which the owner knows are being used by the public, whether they should be so used or not, should be inspected regularly for defects and dangerous conditions and maintained in a reasonably safe condition.

. Are all areas of your property which are off limits to the general public clearly marked with signs or barriers?

. Do your employees and security personnel enforce the restricted areas by warning people and requiring them to leave the areas?

. Have you instructed your employees and security personnel to report the use of unauthorized areas or passageways?

. Do your employees and security personnel report the use of unauthorized areas or passageways pursuant to your instructions?

. Do your employees and security personnel regularly inspect the premises for defective or dangerous conditions?

PSCLR – SEP 82 (10)

Shopping Mall Owner And Tenant Acquitted Of Liability For Homicide Of Mall Shopper; But Owner Liable To Tenant For Settlement Loss Because Of Lease:

Morris B. Phillips, et al. v. Equitable Life Assurance Company of the United States, et al., No. 12439, in the Court of Appeal of Louisiana, Fourth Circuit, decided April 15, 1982, reported at 431 So. 2d 696.

The husband of a woman killed in a mall parking lot sued a store and the mall owner, blaming his wife's murder on inadequate security precautions.

Morris Phillips' wife died of gunshot wounds several hours after being the victim of an attempted robbery in the Winn-Dixie parking lot in a New Orleans shopping center. After committing three other robberies within a two-day period, Mrs. Phillips' murderer was apprehended, subsequently convicted of her murder, and sentenced to life in prison.

Mr. Phillips claimed his wife's murder could have been prevented if Winn-Dixie and the shopping center had provided (1) adequate warning of the possible danger, i.e., criminal activity in the area; (2) sufficient lighting; and (3) adequate security.

At trial, the shopping center's manager testified that a security guard, an off-duty member of the New Orleans Police Department, patrolled the center daily from 10 AM to 9 PM and checked inside each store from time to time. The guard testified he was inside a store when Mrs. Phillips was shot at 7 PM. The guard's salary was paid by all of the mall store owners. In addition, many of the individual store owners employed private security guards who kept in tough with each other and the patrolling guard by telephone and a "beeper" device.

Phillips argued the high crime rate of the area required tighter security measures. Although 32 crimes were reported in the shopping center area during the year which preceded Mrs. Phillips' murder, an expert testified crime was still percent lower than average for similar areas.

Phillips also pointed to a lack of warning signs and to poor lighting as contributing to his wife's death. The only signs posted in some of the store windows were announcements that

New Orleans police patrolled the area. Additional evidence showed parking lot lighting to be poor, but testimony conflicted as to how much this contributed to Mrs. Phillips' murder.

The defendant store, Winn-Dixie, arrived at pre-trial settlement with Mr. Phillips for $15,000. The jury, however, found that neither Winn-Dixie nor the mall owner, Equitable Life Assurance Company ("Equitable"), were negligent. Nevertheless, Equitable was required by the court to repay Winn-Dixie the $15,000 under the provisions of its lease agreement. Phillips appealed the decision on the basis of inadequate security and Equitable appealed on the ground that it should not have to indemnify Winn-Dixie.

The Court of Appeal affirmed both decisions. First, the court explained, "It is well established that a store owner is under a duty to take reasonable care for the safety of its patrons, but it is not the insurer of the patron's safety." In this case, the record showed that Winn-Dixie and Equitable practiced adequate security precautions; the court found evidence in the record that the murder could have occurred even with a "vast increase in the security force."

COMMENTARY:

The duty of an owner or occupier of land to business invitees remains a much litigated subject. In this case, the major factors in the favorable decision to the landowner were the extensive precautions which had been taken in light of the area's crime rates, including a regular patrol the parking lot and stores, signs advising of the patrol, as well as the presence of various individual store guards.

. Do you systematically consult police records to determine the current crime trends in the areas for which your security people are responsible?

. Do you adjust your employee forces in light of those statistics?

. Are you aware of lease provisions which effectively shift the cost burden of adverse judgments or settlements from one party (a mall tenant) to another (mall owner)?

PSCLR – OCT 82 (3)

Restaurant Owner Who Provided Reasonable Care For Patrons' Safety Not Liable For Failing To Employ Security Guard To Prevent Homicide:

Toy Kelly, Janet Kelly, Denise Kelly Boswell, and Mark Kelly, by his Attorney and Next Friend, William Willard v. Retzer & Retzer, Inc. d/b/a McDonald's Restaurant, *No. 53106, Supreme Court of Mississippi, decided July 21, 1982, reported at 417 So. 2d 556.*

A McDonald's Restaurant was deemed to have provided adequate security where its assistant manager was responsible for checking and clearing the premises of loiterers every half hour. Because of this, the restaurant was able to avoid liability for a spontaneous shooting which resulted in a teenager's death in the restaurant parking lot.

On the night of September 25, 1979, the assistant manager of a McDonald's in Clarksdale, Mississippi, made his rounds and cleared the parking lot of some loitering teens. One of the teens, Lanny Byrd (who is white), left but returned about one hour and fifteen minutes later to retrieve his car. Some black teens walked by and one of the group, Roosevelt Grandberry, asked Byrd for a cigarette. Byrd's hostile refusal and racist remarks angered Grandberry and an argument ensued between the boys.

A few moments later, Rodney Kelly drove up and Byrd asked him if he had a weapon. When Kelly appeared to be getting a weapon for Byrd, Grandberry pulled out a gun which he had purchased a day earlier for $10. Showing Grandberry a scar which he claimed was a gunshot wound, Kelly told him that he was not afraid of guns. Grandberry then waved the .22 caliber pistol, fired a shot in the air, and then shot Kelly.

Only seconds before Kelly was fatally shot, Byrd alerted a manager in the restaurant of the trouble and police were immediately called. It was estimated that Kelly had been in the McDonald's parking lot for only two to four minutes before being shot.

Kelly's family brought a wrongful death suit against the restaurant and its owner alleging that prior crimes on the premises provided warning to McDonald's that security patrols by managers as part of their duties was inadequate to protect patrons. Twenty-eight police reports were introduced into evidence which described various crimes and attempted crimes occurring at the McDonald's Restaurant over the past three years: theft, burglary, assault, vandalism, armed robbery, and fraud.

The Kelly family wanted to have an expert criminologist testify on needed security precautions as indicated by prior crimes on McDonald's property, but the trial judge refused.

The restaurant owner testified that it was McDonald's normal policy for a manager to check the restrooms, lobby and dining areas, parking lot, and "surrounding areas" every half hour, ensuring that loiterers left the property. For the past six to eight months (out of the six years this McDonald's had been open) "off duty policemen or others" had been used to assist management in patrolling the premises on the busy weekend nights. A manager testified that the presence of such uniformed guards seemed to be no more effective than management teams patrolling the property.

The trial judge dismissed the case without allowing the jury to consider it, ruling that McDonald's had not breached its duty to exercise reasonable care for the safety of its patrons.

On appeal, the Supreme Court of Mississippi examined the question of McDonald's duty to protect its patrons from criminal attack. The court decided that given the "sudden and spontaneous" nature of the violence,

> ... coupled with McDonald's notice only seconds prior to the shooting and plaintiffs' decedent's voluntary intervention into the affray, we are of the opinion that McDonald's was not negligent in either failing to assist Kelly at the time of the encounter by not providing an armed security guard or by the assistant manager's failing to interject himself into the affray rather than call the police.

Noting that this was a first impression question for a Mississippi court, the court relied on similar decisions of the District of Columbia, Washington, Wisconsin, Louisiana, and Oregon courts.

The court did agree that the trial judge made an error in not allowing the criminologist to testify. This error did not change the appellate court's opinion that "the responsibility of enforcing the law is on the government chosen by the people of the area and does not necessarily rest upon the business involved" and judgment for the defendant McDonald's was affirmed.

PSCLR – NOV-DEC 82 (9)

Retailer Required to Provide Warnings or Adequate Protection of Patrons When Risk of Criminal Activity is Prevalent:

Helen Butler v. Acme Markets, Inc., *in the Supreme Court of New Jersey, decided May 11, 1982, reported at 445 A.2d 1141.*

Is a retail business required to protect its customers from criminal attack on its premises? The question of how much protection a business owes its customers, as invitees, is closely examined in this case.

While putting her grocery sacks into her car, Mrs. Helen Butler was knocked down by a purse snatcher in the Acme Market parking lot. In her fall, she suffered injuries to her head, neck, chest, and ankle.

Because she felt Acme had not taken sufficient steps to protect her, as a customer, from criminal attack, Mrs. Butler sued the store for personal injury damages and for wages lost during her recuperation. At trial, the store showed that it employed off-duty police officers as security guards and their duties included patrolling inside the store for shoplifters, preventing bad checks being passed, patrolling outside the store, and watching customers' packages while they got their cars. However, only one guard was on duty at a time.

Mrs. Butler's suit asserted that having only one guard on duty (who was inside the Acme store during the attack on her) was insufficient in light of the prior criminal activity at the Acme store. Furthermore, before she was mugged Mrs. Butler did not know there had been seven muggings in the past year at Acme (five taking place on nights in the past four months), although she shopped there regularly and lived in the area. Acme had posted no warnings to advise its patrons of such danger.

The trial jury found that while Acme "did not cause the crime," it neglected to "exercise reasonable care" to protect Mrs. Butler, and therefore awarded her $3600 in damages. The trial judge set aside the verdict, granting judgment to Acme, from which ruling Mrs. Butler filed her appeal. On appeal, the Superior Court, Appellate Division, reinstated the judgment in favor of Mrs. Butler.

Faced with these conflicting opinions between the trial judge and the appellate division, the New Jersey Supreme Court granted a discretionary appeal, and on review, affirmed the decision of the appellate division. The court directed the trial judge to reinstate the jury verdict and to enter judgment for Mrs. Butler.

The supreme court stated that although "an absolute obligation to prevent all crime" was not intended, a store owner:

> is subject to liability to members of the public while they are upon the land for [business purposes], for physical harm caused by the accidental, negligent, or intentionally harmful acts of third persons or animals, and by the failure of the possessor to exercise reasonable care to (a) discover that such acts are being done or are likely to be done, or (b) give a warning adequate to enable the visitors to avoid the harm, or otherwise to protect them against it.

The court continued:

> Here it was at least reasonable for the jury to determine that absent warnings, hiring one guard who primarily remained inside the store was an insufficient response in light of the known, repeated history of attacks on the premises.

COMMENTARY:

The opinion of the court goes into some detail to explain the rationale of imposing affirmative duties only on selected individuals based on their status (as landowners, common carriers, public utilities, etc.) and acknowledging the law's reluctance to force people to help one another. It is interesting to note that the court specifically pointed out that "trained private security services have now become widespread [and] they are regulated by state statute" and thereby declined to follow a 1962 precedent case which had concluded that "police work was specialized and there was no room for the private devices to operate in the field."

So it appears that the growth and very existence of the private security industry itself, and its emergence as a potent trained force against criminal activity, is now being cited by some courts as a reason for imposing greater duties on specific entities to protect the public.

PSCLR – AUG 83 (2)

Motel Will Not Be Held Liable for Attack on Patron If Attack Was Not Foreseeable:

McCoy v. Gay, et al., *No. 65117, in the Court of Appeals of Georgia, decided March 2, 1983, reported at 302 S.E.2d 130.*

After a patron of the cocktail lounge at the Downtown Motor Inn in Albany, Georgia, was robbed and beaten by an unknown assailant in a parking lot of the motel, he sued the motel owners because of their alleged negligence "in not controlling and maintaining their business premises in a safe and adequate manner for the protection of their invitees, customers and patrons."

Although not a guest of the motel on the night of the attack, the patron McCoy was drinking in the motel's cocktail lounge until early morning. He was robbed when he left the lounge and went to his car which was parked in one of the motel's lighted parking lots. His lawsuit asserted that the motel's lack of adequate security and "poor lighting" in the parking lot presented a defective condition which made the criminal attack possible.

Evidence was presented which indicated that two crimes had taken place on the premises in the three years before the parking lot robbery of McCoy: the first was a purse snatching at the door of the motel, one year earlier, and the second was an unarmed robbery on an outside stairway, three years earlier. Further, the motel employed only one security guard who made "periodic patrols about the premises." The guard was inside the motel when McCoy was robbed.

The trial judge directed a verdict for the motel owners and McCoy appealed. The Court of Appeals of Georgia held that "the evidence of these two prior robberies at the Inn did not meet the 'similarity' requirement so as to constitute a sufficient showing of [the motel owners'] knowledge of the 'litigated' dangerous condition." The only other evidence of a prior crime at the motel was a shooting in the parking lot ten years earlier, which the court considered "too remote and irrelevant."

Because the circumstances of the prior incidents of crime were not sufficiently similar in physical conditions or other factors, the court determined that the motel owners were not put on notice that the "periodically patrolled" parking lot presented a dangerous condition to their guests, invitees and patrons. Motel proprietors are obligated to practice ordinary care and diligence to protect their invitees and guests from injuries, particularly where there is "any reasonable apprehension of danger ... ." But, the motel proprietor who practices ordinary care will not be held liable to an invitee or guest who was the victim of an unforeseeable criminal attack on motel property.

COMMENTARY:

The true basis of liability of a property owner to a guest who is injured on the property is the superior knowledge of the proprietor of the existence of a condition that may subject the guest to an unreasonable risk of harm. And that is why the injured guest, to be successful, must prove that the owner had such superior knowledge, and was "put on notice" of the dangerous condition, by virtue of prior similar incidents.

(See, however, the case of Mullins v. Pine Manor College, reported in the August 1983 **PSCLR**, *case 10, where a plaintiff prevailed without proof of prior incidents because of the "special relationship" between a college and its students.)*

That also is why property owners, when they do become aware of the existence of dangerous conditions on their property, should take reasonable steps to correct the condition and/or to warn and protect their guests from same.

PSCLR – NOV-DEC 83 (10)

Where Shopping Center Should Have Known Of Likelihood Of Criminal Conduct By Third Persons In General Likely To Endanger Customers, Shopper May Recover For Criminal Assault:

Rosabel Brown v. J.C. Penney Company, Inc., a foreign corporation, and May Department Stores Company, a foreign corporation d/b/a Meier & Frank Company, and W.H. Shields, H.A. Anderson, Joseph Fought, and Allan Penney, d/b/a Valley River Center, and H.A. Ellsworth, *No. 78-8085; CA A22384, in the Court of Appeals of Oregon, decided August 10, 1983, reported at 667 P.2d 1047.*

Two days before Christmas in 1976, 76-year-old Rosabel Brown went shopping at the Valley River Shopping Center in Eugene, Oregon. At approximately 7 PM, she and her husband walked to their car in the shopping center parking lot. Upon reaching the car, Brown was knocked to the ground by a purse snatcher. She brought this action against the shopping center and others to recover for the injuries she suffered in the attack. The jury returned a verdict in her favor and she was awarded just over $20,000. The defendants filed this appeal.

The Oregon Court of Appeals noted that its Supreme Court had previously stated the law concerning a property owner's liability for the intentional acts of third persons:

> Since the possessor is not an insurer of the visitor's safety, he is ordinarily under no duty to exercise any care until he knows or has reason to know that the acts of the third person are occurring, or are about to occur. He may, however, know or have reason to know, from past experience, that there is a likelihood of conduct on the part of third persons in general which is likely to endanger the safety of the visitor, even though he has no reason to expect it on the part of any particular individual. If the place or character of his business, or his past experience, is such that he should reasonably anticipate careless or criminal conduct on the part of third persons, either generally or at some particular time, he may be under a duty to take precautions against it, and to provide a reasonably sufficient number of servants to afford a reasonable protection.

In order to prove that the shopping center should have reasonably anticipated that criminal conduct by assailants would be likely to endanger shoppers, Brown was permitted to introduce into evidence a computer print-out from the Eugene police department which listed crimes in the vicinity of the Valley River Shopping Center. The print-out listed all reports of criminal incidents which had occurred in that vicinity in the six months immediately preceding the incident. The print-out listed the investigating agency, the type of crime, the crime "name," the case number, the date and time, and the address at which the incident occurred. Over objection by the defendants, the court ruled that the computer print-out was properly admitted into evidence.

The plaintiff established that the defendants knew of the availability of the statistics from the Eugene police department. Her case further alleged that these statistics could have aided the shopping center in analyzing security risks in and around the center, but the evidence further showed that they had never requested such statistics. The amount of criminal activity increased during the Christmas season and the evidence supported the conclusion that this fact was known or should have been known by the shopping center's director of security who was a former police officer and who was aware of the existence of police records of reported criminal activity. Furthermore, extra security personnel had been hired to provide security during previous Christmas seasons, but no extra personnel were hired for the 1976 season.

The appellate court affirmed the verdict and judgment in favor of the plaintiff Brown. However, the case is currently before the Oregon Supreme Court on a second appeal; arguments were presented on December 5.

COMMENTARY:

The obvious lesson to be learned in this case is that statistics on criminal activity occurring in the vicinity of your business locations should be obtained and analyzed periodically in order to include them in the factors to be considered in deciding when and whether to increase the use of security measures at a particular location. According to this case, you will be charged with knowledge of such reports where available and the astute security manager will insure that this knowledge is used to his and his client's advantage rather than to his or his clients' disadvantage.

For further reading in this area, see Phillips v. Equitable Life Assurance Company of the United States, reported in the September, 1982, issue of **PSCLR**. *In this case, the shopping center owner was acquitted of liability for the murder of a customer by a third party which occurred in the center's parking lot.*

PSCLR—APR 84 (3)

Retailer Has Duty To Police Its Parking Area To Protect Business Guests From Foreseeable Criminal Assaults:

Susan Dailey et al. v. K-Mart Corporation et al., *No. CV 80-07-0644, in the Court of Common Pleas of Ohio, Butler County, decided September 2, 1981, reported at 458 N.E.2d 471.*

Susan Dailey was kidnapped at gunpoint from a K-Mart parking lot in Hamilton, Ohio, on July 19, 1979. She was taken to a deserted place, shot twice and left for dead.

She survived the attack and brought suit against the retailer K-Mart as the possessor of the parking area from which she was kidnapped, against Butler Improvement as the lessor of the K-Mart building, and against Marlou Realty as the lessor of the land to Butler Improvement, for failure to provide adequate security in the parking area. Ms. Dailey cited 49 separate incidents of serious crime occurring in or about the parking lot since March of 1968, claiming that such incidents were sufficiently numerous to put the defendants on notice that they should provide security guards in the parking lot.

All of the defendants filed motions for summary judgment. They claimed that the plaintiff's allegations, taken in the light most favorable to the plaintiff, could not sustain a verdict in favor of the plaintiff. The trial judge, noting that "This case raises novel questions which have never been addressed directly by the Ohio Supreme Court," first granted the summary judgment motion to Marlou Realty, holding that a landlord out of possession and control is not liable for injuries to its tenants, nor to business guests. The court found that Marlou Realty had no responsibility for the management of the property and no control over the use of the parking lot; the claim against it was therefore dismissed.

The trial judge then examined the question as to whether any duty was owed by the defendants K-Mart and Butler Improvement to protect Ms. Dailey from the criminal acts of third persons in the parking lot. Concluding that there is a duty on the possessor or occupier of land to police its business premises so as to protect business guests, the trial judge denied these defendants' motions for summary judgment. Whether the criminal acts perpetrated on the plaintiff were foreseeable by the defendants, and therefore placing a duty on the defendants to protect the plaintiff, was left for a jury to decide. Specifically, the judge ruled that "reasonable minds could differ as to the issue of foreseeability as to whether K-Mart should have provided security in its parking lot. ... As to defendant Butler Improvement, there is some evidence ... in the lease that they had responsibilities in regard to the parking lot." Also, it appeared from the evidence that Butler Improvement could be liable pursuant to an indemnification clause contained in its property lease with K-Mart which provided that:

> the landlord shall keep tenant insured against all statutory and common law liabilities for damages on account of injuries to property or person, including death, sustained by any person or persons while within said common areas.

The parking lot being on of the common areas listed in the lease, Butler Improvement's motion for summary judgment was likewise denied.

Following this denial of the motions for summary judgment by the defendants K-Mart and Butler Improvement, the case proceeded to trial and resulted in a verdict of two million dollars in favor of the plaintiff. An appeal was filed by the defendants; however, the appeal was dismissed after notification to the appellate court that the case had been settled by the parties.

COMMENTARY:

All of the cases on this question reported in previous issues of **PSCLR** *have emphasized the element of foreseeability. It is not therefore at all surprising that this Ohio trial judge ruled as he did. And perhaps of equal significance, it is not at all surprising that the case was settled prior to the issuance of an opinion by the appellate court. Sheer speculation may lead one to conclude that the defendants did not want to give the Ohio appellate courts an opportunity to affirm the above opinion by the Ohio trial judge, since this opinion significantly expanded the potential liability of occupiers of land for third party criminal conduct on their property.*

While the above decision is not the decision of an appellate court, it is reported in keeping with our commitment to our readers to report those significant trial court decisions which we believe will have major impact on the law of a particular jurisdiction. It is our opinion that the decision of the trial judge, as set out above on the motions for summary judgment by the defendants, would have been followed by the appellate court had the appeal been allowed to proceed and had the case not been terminated by a settlement.

The Dailey case also emphasizes the importance of closely scrutinizing lease provisions, particularly as they relate to indemnification. While it is not clear from the trial judge's written opinion, it may well be that the lessor Butler Improvement would not have been held liable but for the lease provision quoted in the opinion.

PSCLR—APR 84 (4)

Retailers Not Liable To Injured Customer Where Criminal Assault Was A Random, Isolated Event And Security Was Adequate:

Guy H. Miles, et ux v. Flor-Line Associates, et al., *No 83 CA 0132, in the Court of Appeal of Louisiana, First Circuit, decided November 22, 1983, reported at 442 So. 2d 584.*

On the evening of April 2, 1980, Barbara Miles and two of her children left the Cortana Mall in Baton Rouge, Louisiana, and walked to their car in the J.C. Penney parking lot. The children were several yards ahead when Miles was pushed from behind into the side of a moving car. She was struck unconscious by a metal object, and someone in the car grabbed her purse. It happened so quickly and with so little commotion that, when the children turned to see their mother lying on the ground, they assumed she had slipped in a puddle. Miles' injuries included fractured facial bones and a severely bruised arm.

Miles and her husband sued Sears, Roebuck & Co., J. C. Penney, Inc., and Flor-Line Associates (the company which provided Cortana Mall's security services) for her injuries and the loss of her purse. The trial court granted Sears' motion for summary judgment, and denied the Miles' claims against J.C. Penney and Flor-Line. The plaintiffs brought this appeal on the latter judgment, alleging that the trial court erred in not determining that Penney's and Flor-Line had a duty to warn her of and protect her from criminal attack in light of a history of criminal activities in and around the mall.

The Court of Appeal of Louisiana reviewed the duty of a store owner to its customers, and found:

> A store owner is under a duty to take reasonable care for the safety of his patrons, but <u>he is not an insurer of their safety</u>. ... a proprietor of a public place has a duty to protect his patrons from injuries at the hands of third parties <u>when it is in his power to do so</u>. [cites omitted, emphasis added]

The court also reviewed the criminal history of the area and the extent of security at the mall. A police computer printout of crimes occurring in and around the Cortana Mall in the years from 1977 to 1980 was entered in evidence. The printout showed that most crimes took place inside the mall and were against store owners. There were numerous examples of thefts from automobiles in the mall parking lot, but very few criminal assaults on individuals and only six purse snatchings.

Flor-Line Associates provided mall security for all of the stores in the mall. Depending on the amount of business at the mall, from two to six guards patrolled inside the building and in the parking lots. On busy weekends, a guard was posted on the roof. (Miles' purse snatching occurred on a Wednesday—two guards were working.) The security guards kept in touch by radio. In addition to security guards patrolling the parking lot by motor scooter, the parking lot was well lighted (at least according to a security expert testifying for the Miles).

Based on this evidence and the fact that the assault happened so quickly that even Miles' children did not realize she had been attacked, the Court of Appeal held that J.C. Penney and Flor-Line had exercised reasonable care, but simply did not have the opportunity to prevent the assault. The trial court's judgment denying Miles' claims was therefore affirmed.

PSCLR—APR 84 (12)

Trial Judge's Decision To Exclude Testimony By Purported Security Expert Upheld On Appeal:

Memory S. Tidmore v. Goudchaux's, Inc., et al., *No. 83 CA 0224, in the Court of Appeal of Louisiana, First Circuit, decided December 22, 1983, reported at 444 So. 2d 160.*

Memory Tidmore was the victim of an assault and purse snatching on February 5, 1981, while she was approaching the Goudchaux's store on Main Street in Baton Rouge, Louisiana. She had parked her car on the street immediately in front of the store entrance. As she walked toward the store on the sidewalk, she was thrown to the ground and her purse was taken by an unknown assailant. She received serious injuries to her leg which required surgery and resulted in a permanent partial impairment of her left leg. She sued Goudchaux's to recover for her injuries.

The trial evidence showed that Goudchaux's had a contract with an independent guard agency to provide outside security at its Main Street location. These guards were instructed to patrol the parking lots and the perimeter of the building at least once each hour. During the trial, Tidmore attempted to convince the trial judge to permit Russell Cook, Sr., to testify on her behalf as an expert in the field of private security. Upon inquiry by the judge, it was determined that Cook was employed as a sergeant with the Louisiana state police and that he was a co-owner of Roco Systems, Inc., a local security-safety agency. However, he had never been engaged in any security work for a retail establishment such as Goudchaux's, and the overwhelming majority of his security experience had been in designing and implementing security plans for state correctional institutions. The trial judge concluded that while Cook was well qualified in the field of security on governmental matters, he lacked expertise insofar as security in the private sector was concerned. The judge therefore refused to permit Cook to testify as an expert witness in the field of private security, and the plaintiff contended on appeal that this ruling was erroneous.

The appellate court found nothing erroneous in the trial judge's decision not to accept Cook as an expert. The court also held Goudchaux's did not owe a duty to protect Tidmore from the purse snatching, due to the fact the incident was not foreseeable.

> Unforeseeable or unanticipated criminal acts of independent third persons do not impose a duty to protect business patrons. It is only when the owner or manager of the public place has knowledge or can be imputed with knowledge of a third person's intended conduct that the duty to protect patrons is imposed.

The court concluded the instant case did not present a factual situation sufficient to impose a duty to protect or to warn based upon knowledge by Goudchaux's either actual or implied. The dismissal of the plaintiff's complaint was affirmed.

COMMENTARY:

Great care should be taken by security "experts" that they not attempt to hold themselves out as experts in fields in which they have not obtained sufficient education, experience, or training. As is seen in the above case, a security consultant who is retained by counsel to testify in a given case should thoroughly discuss with the attorney his own qualifications and background to testify as an expert witness in the particular case under consideration.

PSCLR—JUN-JUL 84 (10)

Shopping Center Liable To Customer For Purse Snatching And Injuries: Computer Printout of Area Crimes Admissible To Establish Foreseeability:

Rosabel Brown v. J.C. Penney Company, Inc., a foreign corporation, and May Department Stores Company, a foreign corporation dba Meier & Frank Company; W. H. Shields, H.A. Anderson, Joseph Fought, and Allan Penney, dba Valley River Center, and H.A. Ellsworth, *No. 78-8085; CA A22384, in the Court of Appeals of Oregon, decided August 10, 1983, reconsideration denied September 16, 1983, review allowed October 25, 1983, reported at 667 P.2d 1047.*

On December 23, 1976, Rosabel Brown and her husband did some Christmas shopping at Valley River Shopping Center in Eugene, Oregon. When they left the mall and walked to their car in the parking lot, an assailant grabbed Mrs. Brown's purse and knocked her to the ground. Mrs. Brown, a 76-year-old woman, suffered numerous injuries; she subsequently sued the shopping center owners, among others. A judgment was entered in her favor and the defendants here appealed.

The Court of Appeals of Oregon reviewed the case and held that Mrs. Brown was only required to show that the defendants

> should reasonably have anticipated that careless or criminal conduct on the part of third persons would be likely to endanger the safety of a visitor.

To establish this, Mrs. Brown's evidence included a computer printout from the Eugene Police Department which listed "all reports of all possible criminal incidents which occurred at 'Valley River' between July, 1976, and December, 1976." The Eugene Police Department's computer section supervisor testified that "every time a report of a crime is written" in the city or county, the information was stored in the computer, and that the reports were an accurate reflection of the investigating officers' reports. Each report included the investigating agency involved in the case, a number which indicated the type of criminal incident, the name of the crime, the case number, the date and time of the incident and the address where the incident actually occurred.

The defendants argued that the printout should not have been admitted because it was merely a "summary" report and could therefore be admitted into evidence if accompanied by the original underlying police reports. They also argued that the printout was irrelevant to the case. The defendants further assigned as error the admission of lay witnesses' testimony concerning certain personality changes in Mrs. Brown which occurred after the purse snatching. Finally, the defendants challenged the verdict because there was insufficient evidcne to prove that they knew or should have known there was a likelihood of danger to their visitors due to criminal activity by third persons.

The court examined these arguments, and ruled first that the computer printout was relevant and was properly admitted into evidence under Oregon statutory and case law:

> Transcripts of entries in official records stored within either a data processing device or computer are, if properly certified, admissible as primary evidence of the facts stated. * * *

The court also affirmed the admission of the lay testimony concerning their observations of Mrs. Brown's behavior before and after the incident.

As to the foreseeability question, the court noted the evidence presented by the plaintiffs included: (1) 268 reports of criminal incidents in or around the Valley River Shopping Center within 6 months before Brown's attack, (2) expert witnesses' testimony that criminal activity in the area increased the risk to Valley River shoppers, and (3) evidence that 70 criminal incidents occurred in December of 1976 alone – nearly twice as many as other months. This information was available to the Valley River director of security, who was himself a former police officer and was therefore fully aware of such police reports.

Moreover, although extra security personnel were hired during previous Christmas seasons, no additional security personnel were hired during the Christmas season of 1976. This evidence, according to the court, could have led a jury to conclude that the defendants knew or should have known that a higher level of security was needed during the Christmas season than other times. The judgment to Brown was therefore affirmed.

PSCLR – AUG 84 (2)

Restaurant Operator Will Be Held Liable For Fatal Beating Of Patron In Parking Lot If Operator Should Have Foreseen Assailant's Misconduct:

Kenneth Nappier and Pamela Nappier v. Emmet Kincade, Madge Kincade and White Castle System, Inc., *No. 45926, in the Missouri Court of Appeals for the Eastern District, Division Four, decided January 31, 1984, motion for rehearing and/or transfer to Supreme Court denied March 8, 1984, application to transfer denied April 16, 1984, reported at 666 S.W.2d 858.*

The parents of Darren Nappier brought this wrongful death suit against Emmett Kincade for inflicting injuries which led to their son's death, and against White Castle System, Inc., as the owner/operator of a White Castle restaurant in St. Louis, Missouri, where their son was fatally assaulted by Kincade on August 23, 1981.

According to the lawsuit, Darren Nappier had visited the restaurant on that date to purchase some food and, while he was waiting for his order, he was threatened by another restaurant patron, Emmet Kincade. They alleged that Nappier left the restaurant with his food, but stayed in the parking lot to eat; it was at this point that Kincade attacked him. Nappier died two days later as a "direct and proximate result" of injuries suffered in this attack.

The Nappiers alleged that the White Castle's owners/operators were negligent in failing to restrain, control, or remove Emmett Kincade when it became aware of his threats against Darren Nappier. Further, they alleged that the White Castle's owners/operators had failed to provide adequate security to protect its customers such as Darren Nappier when they knew that intoxicated and dangerous persons frequented the restaurant in the early morning hours, and that altercations had occurred between customers.

However, the plaintiffs did not allege that Kincade had, on previous occasions, acted in a hostile or threatening manner towards Darren Nappier nor towards anyone else. They also failed to alleged that Kincade was, at the time in question, intoxicated or that the White Castle employees knew that when intoxicated he became violent or dangerous. Furthermore, there was no allegation that any White Castle employee knew or had reason to know of Kincade's threats against Darren in sufficient time to remove Kincade from the premises, nor that any employee was in the parking lot in time to prevent the assault.

The suit was dismissed by the trial judge, and this appeal was brought before the Missouri Court of Appeals. The court first pointed out that the traditional rule has been that a "private person has no duty to protect another from deliberate criminal attack by a third person." However, the courts have made certain exceptions to this rule to require a business owner to exercise "reasonable and ordinary care to make his premises safe for his customers," and even to protect his customers from criminal attack under "special circumstances." Specifically, these circumstances would include a situation where the business owner knows or has reason to know that criminal acts by third parties are occurring or are about to occur on his property. The court quoted a California opinion which stated, "We are of the view that the law imposes a duty upon the owner of a restaurant to use reasonable care to protect his invitees against assaults by third persons upon the premises," and the Missouri court then added, "The proprietor's duty arises from his control over the premises and his power to restrain or expel offenders against the peace and safety of the premises."

In this case however, the court ruled that the Nappiers failed to allege that the White Castle's employees knew that Emmett Kincade had behaved threateningly toward Darren Nappier or anyone else on prior occasions and further, that they failed to allege that the White Castle's employees were on the parking lot at the time of the assault or knew of Kincade's threats against Nappier in time to prevent the assault. Therefore, the Court of Appeals ruled, the trial court had not erred in dismissing the lawsuit. The court did however reverse the judgment and remand the suit for retrial in order to allow the Nappiers to amend their suit in keeping with this ruling.

COMMENTARY:

The Missouri Court of Appeals was careful to point out that they were not holding, nor implying, that there was any duty owed by White Castle to the Nappiers' son as a result of any "special relationship" between a retailer and its customers. The appellate court specifically noted that the concept of a "special relationship" exists only when someone entrusts himself to the protection of another and then relies on that other person to provide a safe place. Such relationships which do in and of themselves impose a duty on one to protect another include the innkeeper-guest relationship, the common carrier-passenger relationship and the school-student relationship. Occasionally such a duty arises due to an employer-employee relationship.

Therefore, the court noted, no duty to protect Nappier existed on the part of White Castle merely because the retailer-customer relationship. The only way that White Castle could be shown to be liable would be if a "special circumstance" existed, such as knowledge on the part of the White Castle's employees that the alleged assailant Kincade was violent or dangerous.

The court also noted that there is a split of authority on whether the frequency of violent crimes on the premises in question constitutes "special circumstances" which would impose a duty on the property owner to protect customers against such criminal activity. The court cited New Jersey and North Carolina cases for the proposition that they do, and cited Tennessee and District of Columbia cases for the proposition that they do not. The Missouri court also cited the case of Virginia D. v. Madesco Investment Corporation, a Missouri Supreme Court opinion decided in 1983, and previously reported in the September 1983 edition of **PSCLR**. *That is a particularly interesting and important case in this area of law and points out the fact that there is still much disagreement among legal authorities as to the extent of liability of property owners in these cases involving either "special relationships" or "special circumstances," as can be seen where the Missouri Supreme Court was divided four to three in the Madesco Investment case.*

PSCLR—AUG 84 (8)

Shopping Center Owners And Contract Security Service Are Not Liable For Unforseeable Criminal Assault Upon Tenant's Employee In Parking Lot:

Woodrow Wilson Latham v. Aronov Realty Company, et al.; National Security of Alabama, Inc., *82-86, in the Supreme Court of Alabama, decided July 8, 1983, reported at 435 So. 2d 209.*

Woodrow Wilson Latham was employed by Giovanni's restaurant, a tenant of the University Mall located in Tuscaloosa, Alabama. After working until One AM one summer evening, he and three friends walked to their cars in the mall's parking lot. They were confronted by

Horace Hallman and Lonnie Stone, and an exchange of words led to an fight during which Stone hit Latham with a tire tool.

Latham brought a lawsuit to recover for the injuries he suffered in the attack, naming as defendants: Stone and Hallman; the owners of University Mall; the agent of University Mall, Aronov Realty; and the company contracted to provide security at the mall, National Security. The jury found in favor of Latham against Stone and Hallman and awarded him $35,000, but the judge granted motions for a directed verdict to the mall owners and Aronov Realty. Latham's appeal of the latter judgment is discussed in this decision.

The Supreme Court of Alabama first examined the duty of a storekeeper to invitees, stating that Alabama law "clearly imposes a duty upon storekeepers" to take reasonable precautions to maintain "reasonably safe premises for the use of their invitees." Because a tenants' employee is considered an invitee of the landlord under the law, the court held that Latham was an invitee, and it then examined the duty of a landlord to invitees for criminal assaults by third parties. In similar Alabama decisions (including, *Berdeaux v. City National Bank of Birmingham* which was reported in the May 1983 issue of **PSCLR**), the court has held that the key to liability is the foreseeability of danger to invitees:

> Negligence alone does not afford a cause of action. Liability will be imposed only when negligence is the proximate cause of injury; injury must be a natural and probable consequence of the negligent act or omission which an ordinarily prudent person ought reasonably to foresee would result in injury. ...

The court noted that there have been conflicting rulings in different jurisdictions on the duty of a landowner to protect invitees from criminal attack by third persons. (Among such cases cited by Latham in support of his allegations was a New Jersey case, *Butler v. Acme Markets, Inc.*, which was reported in the November-December 1982 issue of **PSCLR**.)

Latham argued that the fact that the mall had hired security personnel was sufficient to infer that the mall was aware of the possibility of criminal activity by third persons on mall property. He further claimed that the defendants knew of a prior theft of property and the abduction of another mall employee from the parking lot, and that the crowd which was gathered in the parking lot on the night of his assault should have alerted the guard to focus his attention on that area of the parking lot. He contended that a jury should determine whether the mall's one security guard was sufficient to fulfill the mall's duty to protect its invitees.

The Supreme Court of Alabama noted that the trial judge reviewed the development of Alabama law in this area. The court had not found an Alabama precedent and relied heavily on a decision by the Tennessee Supreme Court in a similar case, which set forth the following rule applicable to such situations:

> There is no duty upon the owners or operators of a shopping center, individually or collectively, or upon merchants and shopkeepers generally, whose mode of operation of their premises does not attract or provide a climate for crime, to guard against the criminal acts of a third party, unless they know or have reason to know that acts are occurring or about to occur on the premises that pose imminent probability of harm to an invitee; whereupon a duty of reasonable care to protect against such act arises.

The Alabama Supreme Court in the present action chose to uphold the trial judge in relying on this rule, and affirmed that Latham had failed to introduce evidence showing that the defendants knew, or had reason to know from past experience, that criminal activity on the part of a third person would endanger invitees. Concluding that "as a matter of law under the facts of this case, the criminal assault upon Latham in the parking lot was not reasonably foreseeable," the appellate court affirmed the trial court's decision to grant the mall owners and Aronov Realty its motions for a directed verdict.

PSCLR – SEP 84 (1)

Promoters Of Shopping Mall Events Not Liable For Parking Lot Injuries Absent Explicit Proof Of Their Duty To Protect Spectators In Parking Lot Area:

William Guthrie Kilpatrick III and Shirley Siler Kilpatrick, individually and as guardian ad litem for John Christopher Kilpatrick, a minor child, and Daniel G. Badgett and Barbara Tuck Badgett, individually and Daniel G. Badgett, as guardian ad litem for Lance Jeffrey Badgett, a minor child v. University Mall Shopping Center, a partnership, North Hills, Inc., a general partner, North Hills Properties, Inc., a general partner, Provident Life and Accident Company, a general partner; and University Mall Merchants Association, a corporation and Anheuser-Busch, Inc., a corporation, and Harris, Inc., a corporation, *No. 8315SC509, in the Court of Appeals of North Carolina, decided June 5, 1984, reported at 315 S.E.2d 786.*

Two young boys, John Kilpatrick and Lance Badgett, went to a performance of the Anheuser-Busch Clydesdale Horse Show which was held at University Mall Shopping Center in North Carolina. The show was sponsored, produced, and provided by the defendant shopping center, its owners, managers, merchant's association, and insurer. Mildred Cox also attended the performance along with a friend, and they watched the show from inside Cox's automobile. Cox and her passenger decided to leave before the performance was completed and Cox accidentally drove her automobile into the crowd of people watching the show and injured John Kilpatrick and Lance Badgett.

The boys' parents later settled with Cox and her insurance company, and subsequently filed the current lawsuit. The jury returned verdicts in the Kilpatrick case for a total of $66,000 for John Kilpatrick and his parents and in the Badgett case for a total of $45,000 for Lance Badgett and his parents. The trial judge reduced these jury verdicts to give credit for the settlements previously made with Cox.

All parties, including the plaintiffs, filed appeals. However before the appeals were decided, the shopping center, its owners, managers, merchant's association, and insurer, settled the cases for an undisclosed sum. The appellate court therefore had to consider only the appeals of Anheuser-Busch and Harris Inc., and the appeals by the plaintiffs.

In reviewing the evidence presented at the trial, the appellate court determined that neither Anheuser Busch nor Harris had any authority or power to control the parking lot itself. The evidence merely indicated that these defendants' duties were restricted meeting with representatives of the mall to select a site for the performance, making arrangements to keep the horses and crew, and controlling the Clydesdale horses during the show in order to prevent injury to spectators. There was no evidence that Anheuser-Busch or Harris breached any of these duties. Accordingly, the North Carolina appellate court determined that there was no "actionable negligence on the part of these defendants."

The plaintiffs argued that Anheuser-Busch and Harris should be held liable under a "joint enterprise" theory that the negligence of one party may be imputed to the others engaged in a joint enterprise." However, the appellate court rejected this argument, holding that such negligence of one may be imputed to another only where there exists both a "community of interest in the object and purpose of the undertaking and and equal right to govern the movements and conduct of each other" in carrying out the undertaking; the second element was not proved by the plaintiffs.

PSCLR—JAN 85 (2)

Plaintiff/Customer May Use Computer Print-Out Of Criminal Activity In Area To Prove Shopping Mall's Inadequate Security Protection In Parking Lot:

Rosabel Brown v. J.C. Penney Company, Inc., a foreign corporation; and May Department Stores Company, a foreign corporation dba Meir & Frank Company; W.H. Shields; H.A. Anderson; Joseph Fought and Allan Penney, dba Valley River Center; and H.A. Ellsworth, *No. 78-8085, CA A22384, SC 29951, in the Supreme Court of Oregon, In Banc, decided September 5, 1984, reported at 688 P.2d 811.*

The primary issue addressed by the Oregon Supreme Court in its decision on this case (a review of a decision by the Court of Appeals of Oregon which was reported in the November-December 1983 **PSCLR** at page 10) was whether a computer print-out prepared by the local police department could be properly admitted in evidence to show criminal activity in the area and to show that the defendant shopping mall owners and operators had failed to provide adequate security in view of the information provided in the print-out. The defendants were the owners and operators of the Valley River Center Shopping Mall in Eugene, Oregon, plus H.A. Ellsworth, who was the operations manager in charge of security at the mall. The plaintiff Rosabel Brown was a mall customer who was attacked by a purse snatcher in the parking lot of the mall, which attack resulted in severe physical injuries to her. She relied primarily on the following principle of law:

> Since the possessor is not an insurer of the visitor's safety, he is ordinarily under no duty to exercise any care until he knows or has reason to know that the acts of the third person are occurring, or are about to occur. He may, however, know or have reason to know, from past experience, that there is a likelihood of conduct on the part of any particular individual. If the place of character of his business, or his past experience, is such that he should reasonably anticipate careless or criminal conduct on the part of third persons, either generally or at some particular time, he may be under a duty to take precautions against it, and to provide a reasonably sufficient number of servants to afford a reasonable protection.

In this regard, Brown claimed that the shopping mall owners and operators had negligently failed to warn her of the danger so that she could protect herself and avoid the harm, and by failing to provide an adequate security force to protect her from danger.

In order to establish that the defendants were guilty of these failures, the plaintiff obtained and tendered a six page computer print-out prepared by the Eugene Police Department and containing the abstracts of 268 reported incidents of criminal activity in the immediate area of the Valley River Center occurring during the 6 months preceding her assault. The trial court had permitted the print-out to be introduced into evidence and the defendants complained on appeal that this was error and entitled them to a new trial. The appellate court disagreed, however, ruling that the material contained in the print-out was relevant and admissible.

The court also reviewed the evidence presented at the trial to determine whether there was sufficient evidence to obtain and support the verdict in favor of the plaintiff. The evidence showed that the defendant W.H. Shields had set up a security force the day the mall first opened and hired H.A. Ellsworth to head up the security force because of Ellsworth's extensive police experience. The security force's primary responsibility was to control the shopping mall and its parking lot and not the individual retail stores. Six or eight people were employed first, with one or two being assigned to the parking lot itself. The force was increased from time to time, and numbered from twelve to fourteen people at the time of Brown's assault. No signs or warning were posted to advise the mall's customers of the risk of assault, but Shields did instruct Ellsworth and the security force to keep undesirable characters off the premises.

Ellsworth had been the chief of police in Eugene previously, and he testified the police kept statistics compiled from reports by police officers to determine high crime areas and times. As police chief, he had utilized the statistics from these reports to provide increased protection for the public, and he testified that he had the same responsibility toward the mall's customers. He had not made any effort to obtain the police department's statistical data concerning the incidents of crime at the shopping mall. However, he believed that the records were available to him on request. He and his security officers at the mall were not aware of the assault on Brown until sometime later when they learned of the incident from the police department.

Where the defendants were aware of the need for security forces as here, the question is what steps would a reasonable person take to protect customers, if he knew or should have known that there were a large number of reported criminal incidents in the immediate area. Since there was evidence from which the jury could have found, and did find, that the defendants were negligent in failing to warn the plaintiff, and in failing to adequately protect her, the appellate court did not set aside the verdict in favor of the plaintiff Brown.

COMMENTARY:

This case suggests that security managers, as a part of the routine operation of their positions, should obtain and evaluate criminal incident reports compiled by local police or others concerning the geographical areas over which they have responsibility for promoting security services, in order to ascertain the type and extent of security protection to be provided in a given area. Also, updates should be reviewed periodically to ascertain whether changes in the extent and type of coverage provided should be recommended to management, or undertaken on their own initiative if permitted by existing manpower and other resources.

PSCLR – JAN 85 (9)

Parking Deck Owner Is Not Liable For Abduction, Assault, And Rape Of Customer Because Incident Not Foreseeable:

Lisa Diane Henley v. Pizitz Realty Company, *No. 83-362, in the Supreme Court of Alabama, decided July 27, 1984, reported at 456 So. 2d 272.*

Lisa Henley, kidnapped and taken from a private parking deck in her own car by an armed man, sued the parking deck owner to recover for injuries she suffered in the abduction and subsequent rape. Summary judgment was granted to the defendants, and was affirmed on appeal to the Supreme Court of Alabama.

Henley had parked her car in a parking deck owned by Pizitz Realty Company while she was at work in Birmingham, Alabama. She was given a parking ticket on which was written:

> This Contract Limits Our Liability. Please Read It. This Parking Deck rents space only. No bailment is created; and the Parking Deck is not responsible for loss or damage to car or contents due to fire, theft, vandalism, collision, or any other cause whatsoever. Please hold your children's hands. Do not leave children or pets in parked cars.

Upon returning to the parking deck at 5:40 PM, Henley proceeded through the lighted parking deck to her car, whereupon she was forced into her car by an armed man. He made her drive out of the deck, pay the attendant, and drive to an apartment where he raped her. Her assailant was subsequently arrested and convicted for the rape.

Henley alleged that Pizitz had knowledge that criminal attacks were likely to occur in the parking deck, but had negligently failed to prevent her attack. She supported this allegation with evidence that the deck was equipped with monitored closed-circuit television cameras, that the parking area was patrolled by a security guard, of the Limitation of Liability notice, and of the number of crimes which had occurred in the parking deck over a ten year period (ending with the month of Henley's attack). The crimes included a battery, six incidents of breaking and entering, two robberies, Henley's rape, six thefts and one incident in which a thief was shot by a customer. It was her contention that Pizitz expressed its knowledge that crime was likely through the Limitation of Liability notice and that it negligently failed to provide an adequate system of security, which would have prevented her injuries.

In considering Henley's arguments, the appellate court referred to cases decided in Alabama which set precedents in this area of the law, including *Latham v. Aronov Realty Company* (reported in the August 1984 **PSCLR** at page seven) and *Berdeaux v. City National Bank of Birmingham* (reported in the May 1983 issue at page three). The court did not agree that the parking ticket notice of limitation of liability imposed a contract on the parking deck owner to provide its patrons with protection against criminal attack. Likewise, the use of security cameras and guards or a history of sporadic criminal incidents did not provide the requisite basis of foreseeability of crime on the part of Pizitz. Since there was no proof that the defendants knew or should have known that the kidnapping and rape of a parking customer was likely, the Supreme Court of Alabama affirmed the summary judgment granted to Pizitz.

COMMENTARY:

The general rule in this area of the law is that there is no duty upon property owners or business operators either individually or collectively, or upon retailers generally, whose method of operation neither attracts nor provides a climate for crime, to protect against the criminal acts of third parties, unless they know or have reason to know that acts are occurring that pose imminent probability of harm to a customer or other invitee in which event a duty of reasonable care to protect against such act arises.

PSCLR – MAR 85 (6)

Security Company's Liability To Client's Customers Who Are Assaulted On Client's Premises May Be Determined By Contract Terms:

Pauline Brown and George Brown v. National Supermarkets, Inc., and Sentry Security Agency of St. Louis, Inc., and T.G. Watkins, *No. 48259, in the Missouri Court of Appeals, Eastern District, Division Three, decided August 8, 1984, motion for rehearing and/or transfer to the Supreme Court denied September 18, 1984, application to transfer denied November 20, 1984, reported at 679 S.W.2d 307.*

Pauline Brown was shot and seriously injured by an unknown third party in the parking lot of a National Super Market in St. Louis, Missouri. She sued National, its contract security company, Sentry Security Agency, and the security officer on the scene, contending that their negligence and failure to protect her, an invitee, while on the property was the direct cause of her injuries. Summary judgments were granted the defendants at trial; on this appeal to the Missouri Court of Appeals, the judgments were reversed and the case was remanded to the trial court for trial by jury.

The plaintiffs offered evidence showing that there were sixteen reports of robberies involving firearms, seven reports of strong arm robberies, and 136 reports of other crimes on National's property in the two years prior to Mrs. Brown's assault. This evidence was used by the plaintiffs to show the foreseeability of criminal activity in an effort to establish "special facts and circumstances giving rise to a duty on behalf of the defendants to protect store patrons against assault." The defendants argued that reports of previous crime in the area did not require National to protect Mrs. Brown from being injured by the criminal conduct of an unknown third person. The court, however, citing two previous decisions (*Meadows v. Friedman Railroad Salvage Warehouse* , reported in the November-December 1983 issue of **PSCLR** at page 5, and *Nappier v. Kincade*, reported in the August 1984 issue at page 2), stated:

> ... there is no general duty for the owner of a business to protect a plaintiff against the intentional criminal conduct of unknown persons, the duty can arise where special relationships or special circumstances exist. ... [including] frequent and recent occurrence of violent crimes on the business premises.

The appellate court here also noted that it is necessary for the plaintiff to specify the crimes which had preceded the attack in question, and Mrs. Brown had included a detailed list of prior violent crimes on the property. Since a question of fact was therefore presented as to the extent of National's duty to Mrs. Brown, the appellate court ruled that the summary judgment to National was error.

As to the suit against Sentry and its guard, the appellate court stated that case law did not "clearly establish whether a security company or security guard working for a business owner under contract have a duty to protect patrons from crime." Although the general rule is that a private person does not have a duty to protect another from the deliberate, criminal acts of a third person, a security company may assume such a duty by agreeing to provide security for a business. The extent of that duty, however, depends on the terms of the contract and the circumstances of the injury. The Browns claim that the contract between National and Sentry establishes its duty to protect patrons. Although the contract itself was not entered into evidence, the National store manager testified that, although he had not actually read the contract, he understood the contract to require security services both outside and inside the store. However, the security guard (T.G. Watkins) testified that he was never instructed to patrol the parking lot. This contradictory testimony clearly indicates the presence of factual questions which should not be dismissed on summary judgment. Therefore, the court also reversed this judgment.

Both cases were remanded to the trial court for further proceedings.

COMMENTARY:

This case emphasizes the necessity to review carefully the substantive portions of contracts when purchasing or selling security services. As is noted in the above case, the specific provisions of the contract will be relevant in helping to determine the full extent, and the limitations, of the security company's obligations to protect property and/or patrons from various types of risks.

For instance, if Sentry Security Agency had been hired (and contracted with) only to provide National Supermarkets with protection from shoplifters, the court may well decide that Sentry would not be negligent in failing to protect a patron who was assaulted outside in the parking lot. (Although, of course, National Supermarkets may still be liable to the plaintiff for such an assault.)

If your contract merely reads that the security company will provide security guard services for specifically named times, places, and dates, but fails to also specify the risks which the security company is being hired to protect against, then it can be readily seen that there may be future problems concerning whether or not the services contracted for are intended to include the particular loss in question.

PSCLR – MAY 85 (3)

Hospital May Be Held Liable For Inadequate Security Resulting In Parking Lot Assault Of Doctor: Foreseeability May Be Established Without Proof of Prior Similar Incidents:

Mervyn Isaacs, et al. v. Huntington Memorial Hospital, et al., *L.A. 31983, in the Supreme Court of California, decided February 28, 1985, reported at 695 P.2d 653.*

Dr. Mervyn Isaacs, an anesthesiologist affiliated with Huntington Memorial Hospital, was shot and severely injured by an unknown assailant in one of the hospital's parking lots. He sued the hospital and its insurance company, alleging that security measures were inadequate to protect invitees from criminal attack on hospital premises, and that the insurance company was liable because it had participated in decisions which weakened the hospital's security measures. The insurance company was granted a motion for summary judgment before trial; the hospital was granted judgment after trial because the court held the plaintiff's evidence to be insufficient. Isaacs appealed.

A property owner who invites the public onto his property for business purposes is subject to liability if those invitees are harmed by the accidental, negligent, or intentional acts of third persons. Liability will arise in the event that the risk of harm was reasonably foreseeable, but the property owner failed to take reasonable steps to discover the potential for harm, to warn visitors of potential dangers, or to take other steps so that the harm could be avoided. In reviewing this case, the California Supreme Court explored the question of whether foreseeability of harm could be established only by evidence of prior acts of an identical or similar nature, or whether foreseeability could be otherwise proved.

The plaintiff Isaacs and his wife went to the Huntington Memorial Hospital located in Pasadena, California, at about 8:30 in the evening so that he could check some patients, while his wife visited a friend. He parked his car across the street from the hospital's emergency room and the physician's entrance in the research parking lot (one of the hospital's five parking lots). When he and his wife returned to the car at about 10:00 PM, Dr. Isaacs was grabbed from behind by a man who held a gun to his chest. Isaacs moved slightly and was shot in the chest; the assailant fled and was not apprehended.

There was evidence presented by the plaintiff to show that the hospital was located in a "high crime" area; according to police reports, numerous assaults, threatened assaults, and thefts had occurred on hospital property in the three years preceding the assault on Isaacs. The incidents included several threatened assaults with deadly weapons in the emergency room area. A Huntington security guard confirmed that the emergency room area was the site of many incidents of "harassment," and a security director from another hospital testified that the emergency room, surrounding areas, and nearby parking lots have the "highest potential for violent acts ... [and] ... a lot of criminal elements." One doctor testified that it was not unusual for the emergency room area to be "scary" and physically threatening.

On the night that Dr. Isaacs was shot, there were three unarmed security guards on duty at the hospital: one inside the emergency room, another in the employee parking lot (a considerable distance from the emergency room area), and a third patrolling the second floor of one of the hospital buildings. The guards were uniformed and carried only flashlights. Television cameras – one in the employee parking lot and others inside the hospital – were also used to monitor activity on the hospital premises. An escort service was available for hospital staff members, but the evidence was not clear whether the doctors were aware of or had ever used the service.

Lighting in the research parking lot on the night of Dr. Isaacs' shooting was variously described by witnesses as poor and dimly lit and as fair to good. However, it was shown that a few lights which were normally illuminated were not operating that night.

Two security experts testified that the hospital's security measures on the night of the shooting were "totally inadequate." They said that there was an insufficient number of guards for the size of the hospital and for the scope of their responsibilities. Other problems cited by the experts included the hospital's failure to adequately administer the security force, to arm the guards, to monitor the parking lots with television cameras, to provide a way to notify the police in an emergency, and to warn the public via signs that the area was guarded. One expert concluded that these weaknesses in the hospital's security left the research parking lot in particular "totally devoid of any deterrents or security" on the night Dr. Isaacs was shot.

The trial court dismissed the suit because it held that the plaintiff did not introduce evidence to prove the hospital could reasonably have foreseen the attack on Dr. Isaacs or that it had had any notice of prior crimes of the same or similar nature in the hospital parking areas. On review, the Supreme Court of California took a different view, and held that the extent of a property owner's duty to protect invitees must be determined on a case-by-case basis. The appellate court commented that foreseeability is "a somewhat flexible concept," and is only one factor to be considered. Before this case, many California courts had applied a more rigid concept of foreseeability which held:

> In the absence of prior similar incidents, an owner of land is not bound to anticipate the criminal activities of third persons, particularly where the wrongdoer was a complete stranger to both the landowner and the victim and where the criminal activity leading to the injury came about precipitously.

According to the California Supreme Court, however, this rule is "fatally flawed," and would lead to results that are contrary to public policy: "Surely, a landowner should not get one free assault before he can be held liable for criminal acts which occur on his property," as well as to inconsistent results—how similar must a prior incident be to give rise to reasonable foreseeability? The court felt that it was erroneous to conclude that a property owner could not foresee an incident unless one just like it had already occurred. The court instead concluded that foreseeability must be analyzed in light of all the facts, and that prior incidents are but one element in such an analysis.

The appellate court reviewed all of the facts of this case and ruled that when considered as a whole, whether or not Dr. Isaacs's attack in the research parking lot was foreseeable by the hospital should have been submitted to a jury for determination. The jury should also have had the opportunity to pass judgment on whether the hospital's security measures were adequate or inadequate under the circumstances. Therefore, the judgment in favor of the defendant hospital was reversed and the case remanded for trial.

The court affirmed the summary judgment granted the hospital's insurer, however, because there was no evidence that the insurer had any substantive role in the hospital's operations or security policy decisions.

COMMENTARY:

Owners/operators/security personnel who maintain parking lots as a part of their property operations, whether they be hotels, shopping malls, apartment complexes, retail establishments, factories, professional buildings, etc., should take particular note of the following comment included by the California Supreme Court in the above decision:

"Parking lots, by their very nature, create an 'especial temptation and opportunity for criminal misconduct ....' " Accordingly, those owners/operators/security personnel concur an "especial" duty and obligation to warn their patrons/customers/clients of the dangers associated with the use of a particular parking lot and to take reasonable precautions to protect visitors from such dangers.

As has been predicted in many of our commentaries in the past, we see again with the above decision, a broadening of the owner/operator's potential and actual liabilities to the public, the courts' reasoning being that the owner/operator is better able to know of the dangers on a particular property and also is better able to take the necessary steps to protect them from those dangers. We again predict that there will be even further expansion of the legal doctrines providing for invitees to hold the owners/operators liable in damages for their injuries.

Readers may also be interested in knowing that the California court also held that the trial judge had erred in ruling that the plaintiff could not present evidence of incidents which had occurred more than three years prior to his own injury; the Supreme Court held that this limitation was too arbitrary and the plaintiff should have been permitted to introduce evidence of incidents which had occurred four years prior to the shooting of Dr. Isaacs. The court stated that "the only test is one of reasonableness" and that there is no rule requiring that prior incidents must occur within any particular given period of time in order to be relevant and therefore admissible.

PSCLR – AUG 85 (3)

Business Owes No Duty To Protect Invitees Once They Leave Premises, Even If Business's Limited Parking Requires Off-Premises Parking:

Louis C. Steinmetz et al., Minors, etc. v. Stockton City Chamber of Commerce et al., *Civ. 24496, in the Court of Appeal of California, Third District, decided May 31, 1985, reported at 214 Cal. Rptr. 406.*

A woman, identified only in the court's opinion as decedent, was stabbed and killed after she left a "business mixer" sponsored by the Stockton, California, City Chamber of Commerce and hosted by the California Human Development Corporation (CHDC). The mixer was held at the CHDC business office which was located in a rural area industrial park containing manufacturing and wholesale businesses. Because the CHDC office had 25 parking spaces and several hundred people attended the mixer, the decedent and others parked off the CHDC premises inside the industrial park. When the decedent left the mixer and returned to her car parked one block from CHDC's office, an unknown assailant stabbed her. Her family sued the Chamber of Commerce, and CHDC, for her wrongful death. The family also sued Charles Whiteside, the owner of the industrial park from whom CHDC leased its office space. Summary judgment, entered in favor of CHDC and the Chamber, was affirmed on this appeal.

The plaintiffs alleged in the lawsuit that the Chamber, CHDC, and Whiteside had a duty to the decedent as a business invitee, and had breached the duty to protect her by failing to provide a safe place to park. The plaintiffs contended that the absence of security personnel, adequate lighting, or other forms of security precautions in the parking areas had contributed to the decedent's death. The defendants all claimed that they owed no duty to the decedent, because the industrial park had not been the site of violent criminal activity. In addition, the Chamber and CHDC argued that they could not be held liable for decedent's death because they did not own and had no control over the premises where the murder took place. It was on this issue that the Chamber and CHDC were granted summary judgment.

In affirming the summary judgments, the California Court of Appeal quoted from a recent California Supreme Court decision:

> a defendant cannot be held liable for the defective or dangerous condition of property which it [does] not own, possess, or control. Where the absence of ownership, possession, or control has been unequivocally established, summary judgment is proper. (*Isaacs v. Huntington Memorial Hospital*, reported in the May 1985 **PSCLR** at page three.)

The court found that the evidence established that the murder took place off the property leased by CHDC, and on property which was not under the possession or control of either CHDC or the Chamber. Further, there is no precedent in the law to extend a landowner's liability for an invitee's injuries resulting from a criminal attack off the landowner's premises. Indeed, a landowner would not have the right to enter another's property to install lighting or to station security guards. The plaintiffs argue that invitees, including the decedent, were forced to park off the premises due to the extremely limited amount of CHDC parking spaces. However, the court rejected the notion that a landowner is required to insure safe ingress and egress for invitees to his property because it would be unreasonable and impossible to define the scope of such a duty.

Since the plaintiffs CHDC and the Chamber of Commerce owed no duty to the decedent to provide protection on property which was not within their possession or control, the summary judgments granted those defendants were affirmed. This case will proceed against the one defendant who is the actual owner of the industrial park, namely Whiteside.

PSCLR – JAN 86 (1)

Designer Of Hospital's Parking Lot Lighting System Cannot Be Held Liable For Unexpected Criminal Attack On Hospital Visitor In Lot:

Herbert J. Crochet, individually and as tutor of the minor, Donald Ray Derouen, Jr. v. Hospital Service District No. 1 of Terrebonne Parish, Louisiana, known as Terrebonne General Hospital, Carolinas Hospital & Health Services, Inc., Eugene L. Rizzo and Thomas N. Whitney, a Professional Corporation and Don Chesson & Associates, Inc., *No. 84 CA 0811, in the Court of Appeal of Louisiana, First Circuit, decided October 8, 1985, reported at 476 So. 2d 516.*

Linda Derouen, abducted from the parking lot at Terrebonne General Hospital and later killed by her kidnapper, was survived by her minor child. The child's guardian, Herbert Crochet, brought a wrongful death action against a number of defendants including the parking lot owner, a consultant to the parking lot owner, the designer of the parking lot, and the designer of the parking lot lighting system. The trial court granted summary judgment to the lighting system designer, Don Chesson & Associates, and the plaintiffs appealed.

The Court of Appeal of Louisiana reviewed the trial court's findings that Don Chesson was not liable to the plaintiffs for any negligence leading to Derouen's death. The court stated that in Louisiana a duty-risk analysis is applied to determine a defendant's negligence liability:

> Did defendant owe a duty to plaintiff?
>
> Was this duty breached?
>
> Was the breach of duty a substantial factor in bringing about harm to plaintiff, i.e., was it a cause-in-fact of the harm which occurred?
>
> Do the risk and harm encountered by plaintiff fall within the scope of the protection afforded by the duty breached?

To hold a defendant liable, the answer to each of the above questions must be affirmative.

The court then reviewed the concept of duty to invitees under Louisiana law, noting that a store owner has a duty to take reasonable steps to provide for the safety of his invitees. However, a store owner is not the insurer of the invitee's safety and will not be held accountable for unanticipated criminal acts upon an invitee. Only if the store owner is or should be aware of the potential for criminal conduct on his property by a third person, will the store owner then have a duty to protect patrons from that criminal conduct.

In this case, Chesson & Associates, a contractor of the parking lot owner, likewise does not owe a duty to protect parking lot patrons from unanticipated and unforeseeable criminal injury. The kidnapping and subsequent murder of a parking lot patron is not a foreseeable risk of a poorly designed lighting system, and could occur even with the best possible lighting system in place. The risk incurred by Derouen does not fall within a duty owed her by Chesson & Associates, and the appellate court affirmed the trial court's decision to grant that defendant's motion for summary judgment.

PSCLR—AUG 86 (5)

Owner of Open Air Parking Lot Will Not Be Held Liable For Unforeseeable Criminal Assault Of Tenant: Insufficient Evidence Of Prior Crime To Put Owner On Notice Of Danger:

Nancy J. Ronk v. Parking Concepts of Texas, Inc., *No. 2-85-007-CV, in the Court of Appeals of Texas, Fort Worth, decided June 4, 1986, reported at 711 S.W.2d 409.*

Nancy Ronk was assaulted and nearly kidnapped by an unknown assailant in an open air parking lot near her office building in Fort Worth, Texas. She sued the owner of the parking lot, Parking Concepts of Texas, Inc., alleging the defendant liable for her criminal assault. She contended that the defendant had breached its duty to protect her, a monthly tenant, from the foreseeable danger of criminal activity at the lot. Parking Concepts was granted summary judgment, and Ronk appealed.

Ronk was accosted by her assailant in the middle of the day just after she parked and got out of her car. The man "suddenly and without warning or provocation" struck Ronk, hit her repeatedly, and dragged her back into her own car. In the struggle Ronk lost consciousness; her assailant was unable to drive out of the lot because the car stalled and hit a curb. Ronk awoke and fled the car.

Ronk alleged that the parking lot was located in a "high crime" area and that a number of crimes had occurred on or near the parking lot. In spite of this, she contended, there were no security features, security personnel, or adequate warnings about the danger to tenants in the parking lot.

Parking Concepts did maintain a manned parking booth in a parking area on the other side of Ronk's office building, but it was not equipped with a telephone. Ronk argued that this proved that the defendant breached its duty and that this breach constituted negligence.

Reviewing the evidence presented by Ronk on her allegations, the appellate court noted that to sustain a suit for negligence under Texas law, Ronk had to show that (1) Parking Concepts had a duty to protect her as an invitee, (2) it failed to fulfill that duty, and (3) she suffered damages as a result.

The court ruled that Parking Concepts did owe Ronk a duty to "exercise ordinary care to keep the premises in a reasonably safe condition" to prevent reasonably foreseeable harm. The appellate court came to this conclusion after reviewing recent decisions by the Supreme Court of Texas which defined foreseeability, including *Nixon v. Mr. Property Management* (which was reported in the November-December 1984 **PSCLR**, case seven, page five):

> Foreseeability means that the actor, as a person of ordinary intelligence,

> should have anticipated the dangers that his negligent act created for others.... Usually, the criminal conduct of a third party is a superseding cause relieving the negligent actor from liability. However, the tortfeasor's negligence will not be excused where the criminal conduct is a foreseeable result of such negligence.
>
> *** Evidence of specific previous crimes on or near the premises raises a fact issue on the foreseeability of criminal activity.

The appellate court also noted that in *Castillo v. Sears Roebuck & Co.* (reported in the April 1984 **PSCLR**, case ten, page eight) the Texas Supreme Court ruled:

> There is no duty upon the owners of operators of a shopping center, individually or collectively, or upon merchants and shopkeepers generally, whose mode of operation of their premises does not attract or provide a climate for crime, to guard against the criminal acts of a third party, unless they know or have reason to know that acts are occurring or are about to occur on the premises that pose imminent probability of harm to an invitee; whereupon a duty of reasonable care to protect against such an act arises.

However, the court ruled the evidence insufficient to support Ronk's contention that Parking Concepts had breached this duty to her. The attendant on duty at the parking booth located in the other lot was completely unaware that the assault on Ronk occurred, until a man came up to the booth and asked where the nearest telephone was located. Moreover, the attendant stated in her affidavit that she had been working at the parking lot booth for two years and had not heard of any criminal incidents on or near the parking lot.

Ronk did submit affidavits from six of her co-employees stating their familiarity with the area as potentially dangerous, and introduced photostatic copies of police reports. The reports described 15 separate criminal incidents reported as occurring on and near the Parking Concepts lot in the two years prior to Ronk's assault. The reported incidents included two thefts of cars and three thefts of property from the parking lot; and five thefts of property, two burglaries, one assault, one criminal trespass, and one weapons violation at the adjacent office building.

Parking Concepts argued that none of the prior criminal incidents listed established a pattern of criminal violence of the nature of the crime committed against Ronk. Ronk, however, contended that the "prior, similar incidents rule" was not applicable here; the correct standard should be the "totality of circumstances" rule set forth in the California case of *Isaacs v. Huntington Hospital* (reported in the May 1985 **PSCLR**, case three, page three). In *Isaacs*, the California Supreme Court ruled that "While prior similar incidents are helpful to determine foreseeability, they are not required to establish it. Other circumstances may also place the landowner on notice of a dangerous condition." The Texas Court of Appeals agreed, and noted that this position was affirmed in the Texas Supreme Court's decision in *Nixon*:

> It is not required that the particular accident complained of should have been foreseen. All that is required is that the injury be of such a general character as might reasonably have been anticipated...

In *Isaacs*, the circumstances other than prior similar incidents were extensive and were firmly established by the evidence (see the report in the May 1985 **PSCLR**). In this case, however, no such evidence was presented to show that Parking Concepts should have foreseen or could have prevented the assault on Ronk. The Texas Court of Appeals therefore affirmed the summary judgment granted the defendant.

COMMENTARY:

In reviewing the applicable precedents to assist in reaching its decision in the Ronk case, the

Texas Court of Appeals also reviewed the case of Allright, Inc. v. Pearson decided by the First District Division of the Court of Appeals of Texas, sitting in Houston, on April 24, 1986 (reported at 711 S.W. 2d 686).

In that case, a monthly tenant of a parking garage had been robbed at gunpoint and her car had been stolen by an assailant who had lain in wait for her inside the multi-story parking structure. The jury had found in favor of the tenant, concluding that the parking lot owner/lessor had been negligent in failing to provide adequate security in the parking lot, failing to provide a safe place for the tenant to park, and failing to warn the tenant that the owner did not provide security after a certain time of day. The appellate court affirmed the award in favor of the tenant and rejected the parking lot owner/lessor's contention that the attack on the tenant had been unforeseeable because, so the owner alleged and argued, it had no knowledge of any dangerous condition before the attack and had been aware of no previous similar attacks on the premises, and could not have foreseen that there was any imminent harm or danger to the tenant or other patrons in the parking garage.

The appellate court noted, however, in the Allright case the tenant had presented testimony from a certified security consultant who had testified that "it was reasonably foreseeable that a person bent on robbery could enter, hide on upper levels, and lay in wait for persons coming in, whether or not such incidents had occurred in the past." The appellate court in the Allright case found therefore that the tenant had presented evidence from which a jury could conclude that the attack on the tenant was, or at least should have been, foreseen by the defendant parking lot owner/lessor.

And, the appellate court in the Ronk case above found that the factual situation presented by the Allright was different because Ronk had failed to present any expert witness testimony regarding the issue of foreseeability! While we have seen many other cases where the court has held that the testimony of an expert witness is unnecessary, we see here that such testimony made the crucial difference to the Texas appellate court.

PSCLR – AUG 86 (8)

Restaurant Owes No Duty To Protect Patrons In Unattended Off-Premises Parking Lot, Even Though Restaurant Provided Valets For Another, Closer Parking Lot:

David Warren and Marie E. Warren v. The Lombardo's Enterprises, Inc., *No. 49967, in the Missouri Court of Appeals, Eastern District, Division One, decided February 4, 1986, motion for rehearing and/or transfer denied March 11, 1986, application to transfer denied April 15, 1986, reported at 706 S.W.2d 286.*

The Lombardo's Enterprises operates a restaurant in St. Louis, Missouri. There are two parking lots: one is located next to the restaurant and is attended by valets; the other is located a block away from the restaurant and is for self-parking by customers. David Warren and his wife visited Lombardo's one evening and found that the valet parking was full: a valet directed Warren to the other lot. After parking his car, Warren was robbed and shot by an unknown assailant. He and his wife sued Lombardo's, alleging the restaurant was negligent for failing to provide parking attendants or some type of security surveillance of the distant parking lot and for failing to warn him of the danger in using the distant parking lot. The trial judge granted summary judgment in favor of Lombardo's.

The suit contended that Warren's assault was foreseeable because a number of violent crimes had occurred in the general vicinity of the parking lot in the three years prior to his assault. Warren offered newspaper reports as evidence of six such incidents; one was a robbery of an individual and the others were armed bank robberies. None of the incidents occurred on property owned by Lombardo's.

The court, relying on two previous decisions by the Missouri Court of Appeals, reviewed the status of the law in Missouri concerning a business owner's liability for third party criminal conduct:

> In Missouri, the owner of a business property is not liable for misconduct or negligent acts of third persons unless the third persons are acting under direction or control of the owner, or where their actions could have been reasonably anticipated and guarded against... There exists no general duty to protect a plaintiff against the intentional criminal conduct of unknown third persons. [*Meadows v. Friedman Railroad Salvage Warehouse*, reported in the November-December 1983 **PSCLR**, case five, page five. See also *Nappier v. Kincade*, reported in the August 1984 **PSCLR**, case two, page two.]

The plaintiff relied on the case of *Brown v. National Supermarkets* (reported in the March 1985 **PSCLR**, case six, page five), in which a negligence cause of action was upheld based on evidence of prior crimes on the same premises. Although the court noted that special facts may give rise to a business owner's liability for third party crime, it upheld the ruling in *Meadows* which states, "Such allegations [of prior crimes] are insufficient 'special facts' to create a duty on the part of defendants, even in those jurisdictions upholding a duty where incidents of actual crime had occurred on the premises or nearby." The appellate court stated that the determination of whether a business owner has a duty to protect invitees from criminal activity is "one of fairness and policy." The court continued:

> The responsibility for criminal activity is that of the criminal. Exceptions to the general rule are simply methods to place liability for the results of such irresponsibility upon one who bears no responsibility for the illegal conduct itself. Imposing tort liability on the landowner or business is basically a method to spread the risk of injuries from crime to the public as a whole, rather than to impose its burdens on the victims alone...Whether it is wise to utilize tort recovery to implement such policy, thereby creating a duty on private citizens and businesses to undertake the governmental duty to protect citizens and possibly creating a large private body of armed security guards, is...a determination for elected representatives. ... Courts are ill-equipped to make these kinds of public policy determinations.

The plaintiffs also argued that the restaurant had voluntarily assumed the duty to protect its patrons in the parking areas by supplying valets at the adjacent parking lot; the restaurants' failure to do likewise at the distant parking lot was a breach of that assumed duty. The court rejected this argument, noting that "people hired to park cars are not security personnel, and their function is not protection of customers."

The Missouri Court of Appeals affirmed that Lombardo's did not have a duty to provide protection to patrons in its parking lots and was therefore not liable for Warren's injuries.

PSCLR – SEP 86 (3)

Patron Cannot Recover For Criminal Assault From Restaurant Or Parking Lot Owner Without Showing Existence of "Special Circumstances":

Terry G. Vorbeck v. Carnegie's at Soulard, Inc., Anthony's Produce, Inc., Joseph Sanfillipo, and Delores Sanfillipo, *No. 49901, in the Missouri Court of Appeals for the Eastern District, Division One, decided February 4, 1986, reported at 704 S.W.2d 297.*

In this case, Terry Vorbeck was shot by a robber in a parking lot located behind and adjacent to the restaurant, Carnegie's at Soulard, to which she was proceeding. She sued the restaurant owner and the parking lot owner for negligence. She claimed that the defendants had failed in their duty to protect her from assault and to warn her of any danger. Arguing that the defendants should have foreseen the danger, she offered evidence that 18 incidents of crime had been reported to have occurred on the same premises in the three years prior to her assault. The incidents included 12 reported larcenies and six reported burglaries or forcible entries.

The trial court dismissed the lawsuit, and the Missouri Court of Appeals affirmed the dismissal.

The court cited three cases which set out the applicable law in Missouri: *Meadows v. Friedman Railroad Salvage Warehouse* (reported in the November-December 1983 **PSCLR**, case five, page five), *Nappier v. Kincade* (reported in the August 1984 **PSCLR**, case two, page two), and *Brown v. National Supermarkets, Inc.* (reported in the March 1985 **PSCLR**, case six, page five). The general rule is that a property owner has <u>no duty</u> to protect visitors from third-party criminal conduct <u>unless</u> there are "special circumstances" or "special facts" which impose a duty. Each of the three cited cases described the requisite special circumstances giving rise to the exception as situations where prior <u>violent</u> crimes on the property put the property owner on notice of the potential for third party crime on the property. The Missouri Court of Appeals emphasized that recovery was permitted only where there was evidence that the prior crimes were of a violent nature.

In this case, the prior crimes which occurred on the property were non-violent – larceny and burglary"where the victim was not present at the time. Therefore, these prior incidents did not give rise to the "special circumstances" exception to the general rule.

PSCLR – OCT 86 (6)

Owner Of Parking Garage Has Duty To Provide Security To Customers Or To Warn Customers Of Lack Of Security:

Allright, Inc. v. Carolyn Pearson, *No. 01-84-0525-CV, in the Court of Appeals of Texas, Houston (1st District), decided April 24, 1986, rehearing denied May 22, 1986, reported at 711 S.W.2d 686.*

This case is an appeal by Allright, Inc., owner of a parking garage, from a jury verdict and judgment of damages awarded to Carolyn Pearson, a customer of the Rice Rittenhouse parking garage in downtown Houston, Texas. Pearson had been robbed in the garage, and her automobile stolen. (Note: this case was briefly discussed in the Commentary section of the

Ronk v. Parking Concepts of Texas, Inc. case which was reported in the August 1986 **PSCLR**, case five.)

She sued Allright for negligence; at the trial the jury determined that Allright did not provide adequate security for its customers on the premises, nor did it provide a safe place for Pearson to park her car, nor did it warn her that there would be no security provided after a certain time of day. The jury concluded that these omissions on the part of Allright constituted negligence and contributed to the cause of the robbery. The jury went further and determined that these omissions by Allright constituted a "heedless and reckless disregard for the rights of others ...." And, finally, the jury concluded that Pearson had signed a written contract with Allright which she had not read prior to the robbery, and that Allright knew or should have known of a dangerous condition existing on the premises prior to the robbery.

Having made the above determinations, the jury then awarded damages to Pearson: $1,578.10, property damages for the loss of her automobile; $5,000, for past physical pain; $30,000 for past mental anguish; $1,000 for probable future mental anguish; and $50,000 exemplary damages. Allright appealed the jury's finding, their verdict, and judgment; in the appeal, it argued 62 reasons that the verdict and judgment should be set aside.

Among its many arguments, Allright asserted that it did not have a duty to provide security in the garage, nor to provide a safe and secure place to park, nor to warn customers that it provided no security after a certain time of day. In response to this argument, the Texas Court of Appeals held:

> The operator of a premises has a duty to an invitee to exercise ordinary care to keep the premises in a reasonably safe condition so that the invitee will not be injured. This includes the duty to inspect the premises to discover dangerous conditions. [Cites.] The operator of a premises is charged with knowledge of any dangerous condition that a reasonable inspection would have revealed, if a reasonably prudent person should have foreseen a probability that the condition would result in injury to another. [Cites]
>
> A business operator has the same duty of reasonable care to foresee and prevent a business invitee from being injured by the criminal acts of third persons. [Cites.]

Allright attempted to convince the Court of Appeals that it had no duty to Pearson, according to earlier decisions by the Texas appellate courts in the cases of *Nixon v. Mr. Property Management* (reported in the November-December 1984 **PSCLR**, case seven, page five), *Castillo v. Sears, Roebuck & Co.* (reported in the April 1984 issue, case 10, page eight), and *Campos v. South Texas Beverage Co.* (reported in the March 1985 issue, case one, page one). The Texas Court of Appeals rejected Allright's argument, finding that each earlier case was distinguishable on the facts. The appellate court therefore soundly rejected the appellant Allright's contentions, and concluded that the parking garage owner did in fact have a duty to provide security, or a safe and secure place to park or to warn customers that it provided no security during certain times of the day.

Allright also argued that there was no evidence to support the jury's conclusions that it had in fact failed to provide security, or a safe and secure place to park, or to warn customers that it provided no security during certain hours. To determine this issue, therefore, the court reviewed the evidence in some detail to see whether the jury had had sufficient evidence presented before it to arrive at these conclusions of negligence on the part of the garage owner.

The evidence showed that Pearson worked at a law firm in downtown Houston, and that she had parked her car at a different parking garage for some 15 months prior to the incident in question. The other garage had a full time attendant on duty, but the garage closed for renovation; she then went to the Rice Rittenhouse building, which was operated by Allright. She had seen an advertisement indicating that parking was available in that garage. She specifically noted that there was an attendant's booth at the garage which was centrally located, close to the one ramp which led to the upper levels of the building so that an attendant in the booth could see anyone entering or exiting.

She also spoke with the garage manager about monthly parking, and she was advised that she would have to walk up and down the ramp as the elevator did not operate. She inquired as to the hours of the garage's operation and was told that it was open until 8 PM. She was not told, however, that the garage manager left each day at 5:30 PM, nor were there any signs announcing that the attendant left at any particular time. Pearson concluded from this conversation that the garage was open and attended until 8 PM and she testified that she would not have contracted for a parking space in the building if she had known that no attendant was present after 5:30 each evening.

Pearson therefore contracted for a parking space, paid for same, signed as requested, and affixed the required sticker to the rear view mirror as directed by the garage manager.

The first day she parked at the garage, Pearson arrived at 8:15 AM and parked in the first available space, which happened to be on the seventh level of the building. Upon leaving work, she returned to her car at 5:15 PM and saw that the attendant was on duty in the booth.

The next day, she arrived at approximately 8:30 AM and once again found a parking space on the seventh level. On that day, she had to work late and did not return to her car until 7:15 PM. The attendant was not present, but the entrances to the garage were open and the vehicles were parked on each level she passed as she walked up to her car. When she arrived at the seventh level, her own car plus two others were still there. As she approached her car, a young man ran toward her. She tried to get into her car, but the man drew a handgun, made her walk to her car, open the door, and give him her keys and purse. He then got into her car and drove away. Pearson testified that she was terrified and emotionally drained, but finally summoned the strength to walk down the ramp. She found a businessman on the third level who drove her to the ground level of the building where she telephoned the police and a friend. She went home with her friend because she feared that the robber would come to her home because he had her keys and identification from her purse.

She returned to the garage the next day to request a refund and to report the robbery. She asked why there was no attendant on duty the previous evening, and was informed that Allright did not have a guard after 5:30 in the evening.

Allright admitted that there was no attendant or guard on duty at the parking garage at the time of the robbery, nor was there one on duty during all hours that the garage was open for business. Allright claimed that it had no knowledge that a robber was on its premises on the date Pearson was assaulted.

Dale Hudler was Allright's supervisor of garages throughout the city of Houston, and he testified that the company operated six garage buildings in Houston at the time of the incident, and that four of the buildings had attendants on duty after 5:30 PM, that the company knew that the Rice Rittenhouse garage stayed full most of the time, and that a number of its customers were female. He also stated that it was impossible to see from the outside whether people were on the upper levels of the garage building, but that the attendant's booth provided a clear view of anyone going up or down the ramp. He confirmed that the attendant at the garage in question left at 5:30 each evening and that no attendant or security guard was present until the 8:00 PM closing time when an employee from another garage location would lock the gates.

Next, Robert Waldhuber testified on behalf of the plaintiff as an expert witness. The court specifically noted that Waldhuber was a security consultant and "accredited as a Certified Protection Professional by the American Association (sic) for Industrial Security." Waldhuber testified concerning the standard of care which should be implemented for a multi-level parking garage, and stated that the location in question was considered to be a "high crime" area, and that the Rice Rittenhouse facility did pose certain security problems due to uncontrolled access to the building. He also stated that the garage lighting by itself was insufficient to meet a minimum acceptable standard of security, because the garage should have a security guard present at all hours the garage was open for operations. He further stated that it was his opinion that the management of such a parking facility should at least post some notice to its customers advising them when and if the garage was unattended, and should include a warning to that effect in their contracts with customers. Finally, Waldhuber concluded that, in his opinion, "it was reasonably foreseeable that a person bent on robbery could enter, hide on upper levels, and lay in waiting for persons coming in, whether or not such incidents had occurred in the past." (Emphasis added.)

The appellate court disagreed with Allright's argument that it had no knowledge of any dangerous condition existing prior to the time of the incident, it had no knowledge of prior similar incidents on the premises, and the incident therefore was "not foreseeable." In support of its holding to the contrary, the court specifically cited the expert Waldhuber's testimony to the effect that a reasonably prudent person should and could have anticipated that such an incident could occur. The court noted that Waldhuber's testimony was more than enough to support the jury's findings that the garage operator, Allright, failed to provide adequate security, and failed to provide a safe and secure place for the plaintiff to park, or had failed to warn her that no attendant would be present after 5:30 PM

In its next allegations of error, Allright argued that there was no evidence by which the jury could conclude that Allright had been "heedless and reckless" in disregarding the rights of others, and that therefore the jury's award of $50,000 punitive damages should be set aside. In support of this argument, Allright claimed to have conducted safety studies, but no evidence was presented concerning such studies except that one of the supervisors claimed that he had "checked with a few" other parking garages to find out what they did to protect their patrons during the evening hours. Allright also argued that the garage in question was "regularly patrolled," however, the supervisor of the Rice Rittenhouse garage admitted that the only protection in the evening hours came from the garage lights. The appellate court also noted that D.M. Carothers, the founder, chairman emeritus, and consultant of Allright, admitted that no one was on duty in the garage after 5:30 PM due to the fact that there was no need to have anyone present to collect parking fees after that time. The court concluded that:

> Allright's overriding concern with revenue, as distinguished from customer safety, is sufficient basis for the jury to have concluded that it acted with heedless and reckless disregard of the rights of others. [Cites.] We find the evidence is sufficient to support the jury's award of exemplary damages.

Allright then attempted to escape liability by arguing that Pearson had signed a written contract whereby the garage owner's liability was limited to $100. After reviewing the evidence on this issue, the appellate court concluded that whether the contract signed by Pearson contained a limitation of liability was waived because the contents of the signed contract were disputed. The issue of whether Pearson had signed a contract which contained a limitation clause was never presented to the jury.

And, finally, the appellate court granted the plaintiff Pearson's motion to add interest on the verdict awarded, beginning at the time of her injury to the date of judgment. Applying the Texas statute regarding pre-judgment interest, the appellate court computed that she was entitled to interest in the amount of $14,641.17.

Judgment affirmed, with interest.

COMMENTARY:

There was one dissenting opinion from the above holdings, the dissent would have granted Allright's motion for rehearing because he believed that the damages awarded were excessive and improper as a matter of law.

Perhaps the most noteworthy aspect of this case, insofar as our subscribers are concerned is the court's fairly heavy emphasis on the testimony given by the expert security consultant, Waldhuber, who testified on behalf of the plaintiff. The appellate court specifically pointed out that Waldhuber had been "accredited as a Certified Protection Professional by the American Association (sic) for Industrial Security." While it is unfortunate that the court mis-identified the correct name of ASIS (the American Society for Industrial Security), it nevertheless points out that courts throughout the country are becoming more cognizant of the professionalism and standards established by ASIS.

It is becoming more and more important, therefore, that security consultants who hold themselves out as "experts" in the private security industry should have a recognized professional certificate, such as the CPP issued by ASIS, in order to bolster and confirm their credentials as in their fields of expertise. For any of our subscribers who are interested in obtaining information about the organization, their headquarters are located at 1655 North Fort Meyer Road, Arlington, VA, 22209, (800) 522-5800.

PSCLR– NOV-DEC 86 (3)

No Recovery For Family Of Victim Murdered In Mall Parking Lot In Suit Against Mall Owners And Merchants' Association For Negligent Security Practices:

Albert F. Martinko, Administrator of the Estate of Michelle Martinko, deceased v. H-N-W Associates, A Limited Partnership, d/b/a Westdale Mall; Westdale Associates, d/b/a Westdale Mall Merchants' Association; Ernest Hahn, Inc.; Harry Newman, Jr.; Leroy H. Brettin; Robert P. Lawrence; and Stephen E. Gordon, *No. 85-1303, in the Supreme Court of Iowa, decided September 17, 1986, reported at 393 N.W.2d 320.*

Michelle Martinko's body was discovered in her car in the parking lot of the Westdale Mall Shopping Center in Cedar Rapids, Iowa. Her car apparently had been broken into, and she had been killed in the car at about 10 p.m. of the previous night. The mall closed at 10 p.m., and Michelle had been seen shopping in the mall only 30 minutes prior to closing. The murder remains unsolved.

Michelle Martinko's father sued the owners and operators of the mall, and the mall merchants' association, alleging that the security measures at the mall were not sufficient to provide reasonable security for Michelle's safety. The trial court granted the defendants' motion for summary judgment, ruling that the defendants had no duty to protect the victim from the criminal conduct of third persons and therefore could not be held liable for Michelle's murder. Martinko appealed.

The appellate court agreed with the plaintiff's contention that the following rule should be applied to this case:

> A possessor of land who holds it open to the public for entry for his business purposes is subject to liability to members of the public while they are upon the land for such a purpose for physical harm caused by the ... intentional harmful acts of third persons ... and by the failure of the possessor to exercise reasonable care to
>
> (a) discover that such acts are being done or are likely to be done, or
>
> (b) give a warning adequate to enable the visitors to avoid the harm or otherwise to protect them against it.

The key question is therefore whether the defendants should have foreseen the criminal assault on Michelle Martinko at the mall, based on prior incidents. Because the mall had only been open for two months, the plaintiff argued that the owners of Westdale mall should have looked at the nature of crime occurring at other malls it owned around the country. On appeal, the plaintiff argued that the trial court had erred in refusing to admit evidence of crime occurring at those other malls, because it would show that the mall should have foreseen and prevented Michelle's murder. The appellate court ruled that malls are not generally recognized as a likely site of third-party crime, so crime occurring at other malls would not prove that crime was foreseeable at the Cedar Rapids mall.

The plaintiff argued that this ruling permitted the defendants " 'one free' crime before they can be held liable," and that the defendants had foreseen the potential for danger to patrons in the parking lot because they had a security team patrol the parking areas after the mall closed each night. The plaintiff contended that a jury should be called upon to decide whether the defendants failure to employ a security force during mall hours was unreasonable and negligent. The appellate court, however, disagreed because the evidence showed that the mall employed a security force after the mall closed to protect the mall's and the mall tenant's property from vandals and burglars and not to protect patrons who would have no reason to be on the property after the mall closed.

The appellate court affirmed that summary judgment was properly granted to the defendants because they had no duty to protect the plaintiff's daughter.

COMMENTARY:

What do you think about the plaintiff's argument that the owners of every new shopping mall can allow at least one crime to occur before they have to begin worrying about liability to their patrons? Do the shopping mall owners actually get "one free crime" before they have to begin taking precautions to protect patrons, or to warn them of dangers?

Of course not. The absence of criminal activity does not mean that there is no duty to protect customers if the owners have reason to know that the criminal conduct is about to occur. And, if it can be shown that factors exist which would reasonably cause the owners to have such knowledge, then they can be held liable even for the first crime on the property. Such other factors which could be used to show that a reasonable person would have cause to anticipate such criminal misconduct on the premises would include the general location of the business, as well as the character of the business itself. In the above case, however, the plaintiff was unable to present any evidence which would indicate that the mall owners should have foreseen a criminal attack on a patron. And, the fact that the mall owners and the retailers had decided to employ security officers to protect their property during the hours the businesses were closed does not reasonably lead to the conclusion that the owners anticipated that violent assaults on patrons would occur.

There was, however, a dissenting opinion by three justices, who believed that the evidence of violent assaults at the malls owned by the defendants in other cities was sufficient to allow a jury to conclude that a reasonable person would have foreseen a violent assault at the Cedar Rapids Mall where the plaintiff's daughter was murdered. In other words, the dissenters felt that there was sufficient evidence to allow the issue of foreseeability to be decided by a jury, rather than simply have the case dismissed by the court without ever being considered by a jury.

PSCLR – NOV-DEC 86 (7)

Patron Shot In Parking Lot Of Business After Closing Time Is Licensee And Cannot Recover Without Showing Owner Willfully Or Wantonly Caused Injury To Patron:

Armstrong v. Sundance Entertainment, Inc., *No. 71762, in the Court of Appeals of Georgia, decided June 20, 1986, rehearing denied July 7, 1986, reported 347 S.E.2d 292.*

The plaintiff Mr. Armstrong was shot in the parking lot of the Sundance Entertainment nightclub in Albany, Georgia, after the club had closed for the night. The evidence showed that Armstrong had left the club and had sat for some time in his car in the club parking lot before the events occurred which led to his being shot and his resulting injuries. There was no evidence whatever that the Sundance Entertainment as owner of the nightclub had acted willfully or wantonly in causing the injury to Mr. Armstrong.

The owner or operator of a business property owes his business guests, or invitees, the duty to exercise ordinary care in keeping the business premises safe and to warn such invitees of any danger on the premises. If a business guest does not go beyond that part of the property to which the business invitation extends, then he remains an invitee and the business owner owes him a duty of ordinary care. If, however, the invitee does go beyond that part of the property to which he is invited he then loses his status as an invitee and becomes a "licensee." The duty owed by the owner or operator of a business is different for a licensee than for an invitee: the standard is to refrain from willfully or wantonly causing injury to the licensee.

The court concluded that the requirement that an invitee not go past the physical limits of the invitation should also extend to the time requirements, in that an invitee who uses the business property after the time to which the invitation applies becomes a mere licensee. Therefore, since the nightclub which Armstrong had visited was closed and had been closed at the time of the assault on him, he was at that time a mere licensee and not an invitee of the business. Therefore the defendant business did not owe him a duty to exercise ordinary care in keeping the premises safe, but merely had the duty to refrain from willfully or wantonly causing injury to him.

Therefore the summary judgment for the owner was affirmed.

COMMENTARY:

We do not believe that the above case should be read to mean that all invitees become licensees immediately upon the closing of the doors of the establishment in question. Remember that all the facts and circumstances of a given situation must be taken into consideration in determining a proprietor's duty of care and whether that duty has been properly exercised. Obviously, the length of time the business has been closed would be a relevant factor, as well as prior incidents in the parking lot immediately following closing time.

It would be our opinion that an invitee retains his status as an invitee for a reasonable period of time to permit him to go to his car in the parking lot and to leave the premises. The fact that the business's doors are locked immediately upon his exiting them would not, standing alone, deprive him of his status and rights as an invitee while going to his car in the establishment's parking lot.

PSCLR—JAN 87 (10)

Business Owner May Be Liable Where Customer's Violent Assault Of Other Customers In Parking Lot Is Foreseeable:

Stephen L. Bishop, et al., v. Fair Lanes Georgia Bowling, Inc., *No. 86-8086, in the United States Court of Appeals for the Eleventh Circuit, on appeal from the United States District Court for the Northern District of Georgia, decided November 12, 1986, reported at 803 F.2d 1548.*

Stephen Bishop and two friends attended a "Midnight Madness" Bowlathon at a Norcross, Georgia, Fair Lanes Bowling Alley from approximately one AM to approximately 2:30 AM on a Saturday night. During this time, Bishop and his friends (the Bishop group) noticed that a pitcher of beer was missing from their table; one of the group questioned the three bowlers using the adjoining lane, Steven Burke and his two friends (the Burke group), as to the whereabouts of the Bishop group's beer. The Burke group became angry at the inquiry and a loud and hostile conversation ensued, during which no bowling on either lane took place, for a period of several minutes to half an hour. No blows were struck at that time, although a member of the Burke group did suggest that the parties "go outside to resolve the matter." Two members of the Bishop group complained to the bowling alley manager, specifically telling him that the Burke group was drunk, loud, unruly, making a lot of commotion, and that they had acted in an aggressive manner. They also asked that the manager keep an eye on the members of the Burke group for the rest of the evening.

The Burke group continued to direct verbal harassment toward the Bishop group throughout the remainder of the session. Also, as the bowling alley was closing, another caustic exchange ensued between Steven Burke and Stephen Bishop. The Bishop group decided to wait a few minutes after the Burke group left the building in an effort to avoid any additional confrontation or unpleasantness in the parking lot. However, a violent melee took place in the parking lot when the Burke group accosted Bishop and Deborah Smith at their car, and Bishop was beaten unconscious. Smith fled and had an employee of the bowling alley call the police.

The District Court noted that there was evidence that the Burke group members "were visibly intoxicated when they purchased beer by the pitcher at Fair Lanes," and the bowling alley continued to sell beer to the Burke group. The bowling alley management did not separate the two groups by placing them on different lanes, nor did they speak to the Burke group to control their behavior.

In reviewing the status of the law governing commercial premises liability cases, the court concluded that an owner or occupier of land "may be held liable if he fails to exercise ordinary care in keeping his premises safe."

> Furthermore, it is the duty of a proprietor to protect an invitee or guest from injury by a third person if the host is reasonably aware of the probability or likely possibility of such an act by a third party and such injury could be avoided through the exercise of ordinary care and diligence. ... In order to affix liability in the case of an illegal act, the host must have had reasonable grounds to believe that a particular criminal act is likely to occur.

(The court here quoted the Court of Appeals of Georgia in the case of *Donaldson v. Olympic Health Spa*, reported in the November-December 1985 **PSCLR**, case five, page four.)

The business owner's duty of case does not attach until the danger to the customer is apparent or foreseeable rather than the result of a sudden unexpected action. In conclusion, if the potentially violent acts of the Burke group were reasonably foreseeable based upon the information about which the bowling alley management was or should have been aware, then the bowling alley owner is liable if the management failed to take any action to avoid the potential physical clash between the two groups. And, based upon the conflicting evidence, it is for a jury to determine whether the assault was a sudden act of which the business management had no warning, or was the culmination of confrontations, harassment, and taunting of which the management was made aware well before the physical violence took place.

Therefore, the U.S. Court of Appeals reversed the District Court trial judge's decision granting summary judgment to the defendant Fairlanes Georgia Bowling, Inc. The case was remanded to the trial court for presentation to a jury.

PSCLR—MAR 87 (8)

Although Mall Provides Security Patrols In Mall Parking Lot, Plaintiff Cannot Recover For Vehicle Stolen From Mall Parking Lot Absent Proof That Theft Was Caused By Inadequate Security:

Garrett Ciba v. Woodbridge Center, Inc., *in the Superior Court of New Jersey, Appellate Division, decided February 24, 1987.*

Mr. and Mrs. Garrett Ciba went to the Woodbridge Shopping Center in Woodbridge to pick up some jewelry which they had left to be repaired in a store in the mall. They went to the mall in their 1979 B200 Dodge customized van, which they parked only a few spaces away from the front entrance of the store. They locked the vehicle before going into the store, and when they came back outside they discovered that the van had been stolen. They immediately called the police who referred them to the shopping center security force who in turn drove the Cibas around the mall parking lots in an unsuccessful attempt to locate the vehicle. The Cibas' van was never recovered. They filed suit against the owners of the Woodbridge Center Shopping Mall, alleging that the mall owners were liable to them in damages for the loss of their van, claiming that their van was stolen because of inadequate security in the parking lot.

The parking lot itself has four separate quadrants located around the main building, and it contains some 9,000 parking spaces. The mall itself has from 15,000 to 25,000 people shop there daily. The mall has over 1,800,000 square feet with an outer perimeter road of 1.3 miles in length.

At the time of the theft, the owners of the mall employed a security force consisting of 15 people. Also at the time in question, six of the security people were actually on duty and two of them were assigned specifically to patrol the parking areas. These two security officers, Jeff DeSena and Barbara Breck patrolled the parking areas in a small truck with the shopping mall's emblem on the side and red lights on the front hood. Their primary duty was to provide safety in the parking lot, and to assist people who had lost their cars or had difficulty in starting their cars, and to look for any stolen vehicles or damage to vehicles. The Woodbridge police force also patrolled the shopping center parking lot as a part of their regular duties.

The record of prior incidents at the mall indicated that a substantial number of cars had been stolen from the parking lot: over 325 thefts in 1982 and 249 in 1983; 144 vehicles were reported stolen in the first six months of 1984.

Testimony was presented by the defendant shopping mall from a public safety and private security management and training consultant, Anthony N. Potter, Jr., who was qualified by the court as an expert witness in the field of security. He testified for the jury that professional car thieves operate with "speed and nonchalance" and that it takes a thief "approximately three to five seconds to get into a locked car." He explained that professional thieves frequently work in pairs with one actually taking the car while the other is the "spotter." He also explained that the ignition can be bypassed in mere seconds.

The defendant shopping mall moved to dismiss the case at the end of the presentation of evidence, but the trial judge reserved ruling on the motion and permitted the case to be decided by the jury. The jury returned a verdict in favor of the plaintiff. Thereafter, the trial judge ruled on the defendant's earlier motion to dismiss, and granted the motion, thereby effectively setting aside the jury's verdict for the plaintiff. The plaintiff appealed.

The appellate court agreed with the trial judge's conclusion that the Cibas had failed to prove that the loss of their van had been caused by any breach of duty on the part of the shopping mall or its security force. The appellate court also agreed with the trial judge's conclusion that the shopping mall did indeed owe some duty to the Cibas with regard to the protection of their vehicle while it was in the shopping mall's parking lot; but the court also agreed with the trial judge's conclusion that they had failed to prove that "even if there was a breach of that duty, the breach was the proximate cause of the theft."

The appellate court affirmed the trial judge's dismissal of the plaintiff's case, thereby setting aside the jury's verdict.

COMMENTARY:

This is about as strong an opinion in favor of shopping mall owners and operators as they could hope for. This decision in effect says that no matter how small, or large, the mall security force, the theft of vehicles from the mall parking lot cannot be due to the size of the security force employed. In fact, at one point in its opinion, the appellate court stated, "... we see no reason to find a basis from which the jury could have concluded that multiplying the size of the security force would have eliminated or reduced car thefts."

In other words, the court is saying even if the security force had been multiplied several times over, the nature of car theft being what it is, that is, a crime which can occur in only seconds and cannot be detected by a reasonably careful and observant security officer, no liability can attach to a shopping mall because it cannot be proven that the lack of security was the cause of the theft. Put another way, the presence of more security officers would not have prevented the theft.

For an additional decision concerning the theft of a vehicles from a parking lot, see Villella v. Sabine, Inc., reported in the January 1983 issue of **PSCLR**, *case eleven, page eight.*

PSCLR – MAR 87 (10)

Retailer's Duty To Keep Premises Safe Does Not Include Duty To Provide Crossing Guards, Warning Lights, Or Other Traffic Control Devices At Pedestrian Crosswalk Across City Street:

Josianne L. Laumann v. Debra Lee Plakakis and Adams-Millis Corporation, *No. 8618SC738, in the Court of Appeals of North Carolina, decided January 12, 1987, reported at 351 S.E.2d 765.*

Josianne L. Lauman was attempting to cross a busy city street in High Point, North Carolina, when she was struck by an automobile driven by the defendant Debra Lee Plakakis. Lauman had been visiting the defendant Adams-Millis Corporation retail outlet store, where she had just made a purchase. She was attempting to cross the aforesaid city street to return to a parking lot, which was also owned by the defendant corporation and which was provided for customer parking. There were, however, no crosswalks, warning signs, signaling devices, or crossing guards at the location where plaintiff attempted to cross the road, nor were there any such safety measures present at any location for customers to cross the road.

Lauman therefore filed this lawsuit against both the driver of the car which struck her, Plakakis, and against the Adams-Millis Corporation. In addition, Plakakis filed a separate claim against the corporation, claiming that the corporation should indemnify her as to any liability she may have to the plaintiff, or at least should contribute in damages to any recovery the plaintiff may obtain due to the negligence on the part of the corporation in failing to provide proper safeguards for pedestrians crossing the road.

On motion by counsel for Adams-Millis Corporation, the trial judge dismissed both the plaintiff's allegations against the corporation, as well as the defendant Plakakis's cross-claim against the corporation. This appeal followed.

In her appeal, the plaintiff argued that the corporation should be held liable for the injuries she received since the corporation was well aware that its customers crossed the busy city street at the very same place where she was injured when she attempted to cross. In spite of such knowledge, the plaintiff alleged, the corporation took no action to provide any safeguards for such customer-pedestrians. The plaintiff showed that the corporation did not install warning signs, lights, or any sort of cross walk for the use of its customers.

The plaintiff also argued that the corporation had in effect encouraged its customers to cross at the place where she was injured by having built a fence around the parking lot on the other side of the street from its retail outlet and had built the lot in such a way that the only opening in the fence through which both vehicles and customers entered and exited the parking lot was onto the busy city street, a point some sixty feet distant from the nearest traffic signal.

The appellate court, however, agreed with the trial judge and rejected plaintiff's arguments. The owner of a retail establishment does have a duty to use ordinary care to keep the business premises reasonably safe for customers, as well as to warn customers of any hidden dangers known to the retailer. Moreover, such duty includes the parking lot provided by the retailer for the use of his customers. Nevertheless the court concluded that the injury occurred on the city street over which the retailer had no control. And, the "duty to provide traffic control on public streets in a municipality is charged by statute to the city." The court found that the retailer's duty to keep his premises reasonably safe applies only when the customer is on the business premises proper, or where the retailer through some affirmative action created the dangerous condition which caused the injury to the customer.

Accordingly, the dismissal of the plaintiff's case against the corporate business owner was affirmed.

PSCLR—APR 87 (8)

Manager Of Parking Garage Does Not, By Virtue Of Management Capacity, Stand In Place Of Owner, And Thus Is Not Entitled To Limited Liability Afforded Owner:

Claudia Berry v. The Habitat Company and Darryl K. Matthews, and General Parking Corporation, *No. 85-0887, in the Appellate Court of Illinois, First District, Fifth Division, decided January 23, 1987, reported at 504 N.E.2d 153.*

Claudia Berry lived with her husband in the River Plaza apartment complex in Chicago, where they leased their residential apartment unit and a parking space for their car in the underground parking garage in the same building. The garage had four levels, the top level being available for public parking, the second level for employees of a newspaper located immediately next to the apartment complex, and the lower two levels for tenant parking.

During their third year as tenants, Berry was assaulted while in the tenant parking area of the garage. As a result of the injuries she suffered in the assault and attempted rape, she filed suit against Habitat as the designated agent of the apartment complex and the lessor of the parking spaces, against General Parking Corporation as the manager of the garage, and against her assailant Darryl K. Matthews. Insofar as her suit against General Parking is concerned, she claimed that that defendant "operated, managed, controlled, and provided security" for the garage and that it had failed to keep a door which led to the two tenant levels of the garage locked and that it had failed to properly maintain a security system which had been installed but which was not operational. General Parking filed a motion with the trial judge for summary judgment, asking that it be dismissed as party defendant in the case. After considering the plaintiff's response to this motion, the court granted this defendant's motion. Berry appealed, and the only question presented for decision by the appellate court was whether the trial judge was correct in granting summary judgment to General Parking Corporation.

Berry argued on appeal that General Parking had viewed its responsibilities in terms of the law of premises liability, which law says that the duty of a premises owner or occupier to an invitee is limited since it only arises with "the knowledge of danger and the likelihood of injury." General Parking, in its motion for summary judgment, relied on such landlord/tenant decisions to support its position that it had only limited liability to the garage tenants. The appellate court said that General Parking's effort to place itself in the shoes of the premises owner and to apply the premises liability laws relating only to ownership, however, was "patently misplaced." As the manager only of the parking facility, General Parking was only there to control the premises for the specified purpose of securing and managing it. It was not the landlord and should not have been permitted to claim the benefits of laws applicable only to landlords.

The court also noted that the record in the case showed that there had been frequent criminal activity in the parking facilities over a period of years preceding the assault on Berry and these occurrences were known to management. "Given the frequency of prior criminal activity in the facility, defendant does not appear to have undertaken reasonable security measures."

The appellate court therefore ruled that the trial judge was incorrect in granting summary judgment to General Parking Corporation, reversed the decision, and remanded the case to the trial court for further proceedings.

PSCLR—JUN-JUL 87 (1)

Employee Of Shopping Mall Tenant May Recover Damages For Injuries Sustained In Abduction And Rape Attributable To Inadequate Security In Shopping Mall Parking Lot:

Jardel Co., Inc., and John A. Robbins Co., Inc. v. Kathleen Hughes, *in the Supreme Court of Delaware, decided March 23, 1987, rehearing denied April 16, 1987, reported at 523 A.2d 518.*

Kathleen Hughes worked at the Woolco Store in the Blue Hen Mall in Dover, Delaware. On the day in question, she left work at approximately nine PM, closing time. She exited through the mall arcade, intending to go out through the back entrance of the shopping mall, next to the parking lot where she had left her car. At the entrance to the movie theatre, she was stopped by two men, later identified as Mark Reed and Harry Turner, who had just been denied entrance to the theater because the box office personnel thought that they were intoxicated. The two men asked Hughes for a cigarette; she spoke with them briefly and gave them a cigarette and proceeded to her car. They followed her to her car, which was parked outside the mall exit, approximately 50 feet from the doorway. When she tried to enter her car, the two men forced themselves into the vehicle, and drove away with her. They took the victim to a remote site near the mall where they assaulted and raped her. The assailants tried to run over her with the car as she lay on the ground unconscious; then they put her in the car and set fire to it. She regained consciousness and made her way to the nearby street where a motorist took her to the hospital.

She incurred serious personal injuries, and was hospitalized for six days. She received severe lacerations to her face and scalp, a cerebral concussion, and a fractured skull resulting in a permanent displacement of her eye and double vision. She also continued to suffer psychological symptoms of post-traumatic stress disorders for the four years between the incident and this suit.

There were two separate security firms which provided security at the Blue Hen Mall at the time of her assault. While the mall was open for business, the interior was patrolled by security guards employed by Bennett Security Service, in a contractual relationship with the Mall Merchant's Association on behalf of the retail tenants. After business hours, at approximately nine PM, Globe Security Systems, Inc., took the primary responsibilty for security, and its security guard/employee's primary concern was the exterior of the shopping mall, including particularly the shopping mall parking lot area.

Hughes filed her suit against Globe Security Systems, Inc., Jardel Company, Inc., the owner of the mall, as well as its parent corporation, John A. Robbins Co. The jury returned a verdict in favor of Globe Security Systems, absolving it of liability, but against Jardel Co., Inc., and against John A. Robbins Co. in the amount of $530,000 compensatory damages, plus $250,000 punitive damages, finding that the defendants had failed to provide adequate security in the mall parking lot.

Rita Rodriguez was the vice president of both Jardel and Robbins, and she had the ultimate authority for security expenditures by the companies. John Green was the mall manager for the company and he had discussed the security measures at the mall with Rodriguez. About a year prior to the assault on Hughes, Green and Rodriguez had discussed the problem of increased littering, loitering, and automobile racing in the parking lot of the mall, and they were concerned about minor theft from and damage to automobiles. Because of their conversations, Rodriguez instructed Green to contact security companies about providing guard services. Before obtaining bids from any companies, Green spoke with the Dover police chief to determine his impression of the level of criminal activity at the mall and to obtain whatever statistics were available concerning such criminal activity. However, he was told that since the mall was not a separate statistical "grid" or "unit" in the police's statistics, no such information could be obtained.

Jardel Company subsequently contracted with Globe Security Systems for security services at the mall. The contract did not particularize the number of security officers which Jardel Company would use, but Jardel decided to use one guard for each shift during the night beginning at closing time. This decision for the company was made by Rodriguez in conference with Green, who had discussed the requirements with the Globe personnel prior to the signing of the contract. At the trial of Hughes' claims, Globe representatives testified that they had told Green that the mall parking area was too large for only one guard, but Green testified that he never received a specific recommendation to that effect from Globe. He did admit that Globe had requested, after the contract had taken effect, and on more than one occasion, to increase the number of guards. When Globe made that request to Green, he asked them for a justification for the additional expense and the main reason that they gave was the lack of radio contact with the police. The police had declined to allow access to their radio band by anyone other than sworn police officers. Globe suggested that this communication gap could be remedied by putting an additional guard inside the mall who would then be in radio contact with the guard outside in the parking area, and the inside guard would then be able to request police assistance by telephone. On considering Globe's request, Green discounted it, concluding that the same radio contact could be carried out by maintenance personnel who were already employed and who remained inside the mall during the night.

Green received his knowledge of security conditions at the mall from Globe's supervisory personnel, plus he reviewed the security reports of each day for the previous night prepared by the Globe security guard. He also had conversations with employees of Bennett Security Service (who patrolled during the day) and with various tenants inside the mall.

At the time that Hughes was being forced into her vehicle by the assailants in the rear area of the parking lot, the Bennett security guard was still on duty inside the mall and the Globe security guard was in his patrol vehicle in the front parking lot area outside the Woolco Store.

In its defense, the Jardel Company claimed that it owed no duty to its tenant's employee to provide security for her safety. Jardel argued that its obligation was merely to provide a rental structure which was free of physical defects, but not necessarily free of criminal conduct by third parties. The appellate court noted that whether the mall owner was required to provide security in the mall by its tenant Woolco or otherwise, it had voluntarily undertaken to provide security. Having done so, even voluntarily, Jardel had the obligation to perform the task in a reasonable manner commensurate with the dangers to which the security program was directed. (The court specifically noted the Pennsylvania case of *Feld v. Merriam*, reported in the October 1983 **PSCLR**, case one, page one, and June-July 1985, case nine, page ten.) "The crucial inquiry is the extent of danger which a landlord must reasonably foresee."

Jardel argued that the Delaware court should follow the decisions in other jurisdictions, specifically Tennessee, South Carolina, and the District of Columbia, asserting that without prior indication of a <u>specific</u> assault of the type which harmed the plaintiff, a property owner has <u>no liability</u> for injuries caused by the intervening criminal misconduct of third parties. Thus, Jardel argues, evidence of general criminal activity in the mall parking lot created no duty on the mall owner to anticipate a violent assault and rape. Jardel argued that private companies should not be forced to establish their own police forces in order to ensure the safety of their customers.

The Delaware Supreme Court disagreed, holding:

> Jardel's proposed foreseeability standard – limited to specific crimes – is unrealistic. Criminal activity is not easily compartmentalized. So-called 'property crimes' such as shoplifting, may turn violent if a chase ensues and ... family quarrels may become violent with the risk that deadly weapons may be used. Moreover, the repetition of criminal activity, regardless of its mix, may be sufficient to place the property owners on notice of the likelihood that personal injury, not merely property injury, will result. [citing cases from North Carolina, Pennsylvania, and New York]

The court therefore concluded that although a property owner is not an insurer or guarantor of the safety of the public, any more so than are police agencies, nevertheless there is some obligation to use reasonable care to protect customers from the criminal conduct of third persons.

The Supreme Court of Delaware therefore adopted the legal principle that prior incidents of criminal activity give rise to a duty to foresee specific criminal misconduct. It is then left to a jury to determine whether the conduct of a particular owner is reasonable under all of the facts and circumstances presented. The result in the instant case shows that the jury concluded that, under the facts and circumstances presented, the level of protection provided by Jardel Company for its customers and tenant employees was insufficient based upon the known history of criminal activity in the area. (The court also cited with approval the New

Jersey case of *Butler v. Acme Markets, Inc.*, reported in the November-December 1982 issue of **PSCLR**, case nine, page six.)

Jardel also argued on appeal that certain evidence that the plaintiff tendered, which she obtained from the Dover Police Department should not have been admitted. During the trial, the plaintiff called a Dover police officer to testify. The officer presented documentation showing that police had responded to 394 incident reports at the Blue Hen Mall in the two-and-one-half years immediately preceding the assault on Hughes, and that over 90 percent of the incidents were property and non-personal crimes. Significantly, the remaining incidents included a kidnapping at gunpoint, an armed robbery, sexual molestation, indecent exposures, and purse snatchings. Approximately half of all of these incidents occurred in the parking area. Jardel maintained that the evidence should have been excluded because its mall manager Green had not been able to obtain such information when he made inquiries of the police chief while assessing the need for security at the mall. Of course, some of the incidents were known to Jardel because the police practice was to respond when summoned by mall security personnel and such reports were recorded in the daily security logs which were regularly reviewed by Green. The court rejected Jardel's argument and ruled that the crime statistics presented by the police were relevant and the evidence was properly admitted for the jury's consideration.

Another issue raised by the defendants on appeal was the question of joint responsibility for security and the joint verdict against both Jardel Company and its parent company Robbins. The appellate court nonetheless affirmed the joint judgment, based upon its finding that the evidence at trial showed that the parent company exercised direct operational control over security decisions, as well as other management areas, while day-to-day management of the mall was left to the Jardel Company as the designated landlord. The two companies had common officers and directors, and Rodriguez was vice president of both companies and in that capacity was responsible for financial decisions which affected the Blue Hen Mall. Furthermore, it was her decision to employ only one security guard in the parking lot area, and her approval was needed before Green could designate the number of personnel which Globe Security could employ.

Jardel was successful in convincing the court that the award of punitive damages was not warranted and was not supported by the evidence. The court noted that the imposition of punitive damages is permissible only in situations where a defendant's conduct, although unintentional, was particularly reprehensible, i.e., "reckless or motivated by malice or fraud." Most states acknowledge that punitive damages serve two purposes: to punish wrongdoers and to deter others from similar misconduct. Inadvertent conduct, mistakes, errors of judgment are not sufficient to constitute justification for an award of punitive damages. "It is not enough that a decision be wrong, it must result from a conscious indifference to the decision's foreseeable effect." Hughes did not contend that Jardel's conduct was intentional or malicious, and the court concluded that the evidence did not show that Jardel's actions indicated "a conscious indifference to the rights of others."

> It may have been an error in judgment and thus negligence, for Jardel not to opt for a larger security force, but it can hardly be said that Jardel turned its back on a known risk.

Accordingly, the appellate court set aside the $250,000 punitive damages award.

In summary, the Supreme Court of Delaware affirmed the judgment of the trial court with regard to the award of compensatory damages against both Jardel and Robbins and reversed the award of punitive damages.

PSCLR – SEP 87 (3)

Inadequate Lighting In Parking Lot Is Sufficient Basis For Imposition Of Liability On Property Owner For Invitee's Injuries Incurred In Fall:

Laverne Wallace v. Slidell Memorial Hospital, *No. 86 CA 0479, in the Court of Appeal of Louisiana, First Circuit, decided May 27, 1987, reported 509 So. 2d 69.*

While visiting her sister, a terminal cancer patient staying in Slidell Memorial Hospital, Laverne Wallace slipped and fell in the hospital parking lot, sustaining a fractured right wrist and contusions to her right eyebrow. For several months thereafter, she had to take pain medication and continued to experience intermittent, but chronic pain. Doctors indicated that she was likely to develop traumatic arthritis of the wrist in the future.

She sued Slidell Memorial Hospital and Gibbs Construction Company, Inc., the general contractor performing renovations at the hospital at the time of the plaintiff's fall. She claimed that both the defendants had been negligent in failing to maintain the parking lot in a safe condition, in failing to light the parking lot properly, in failing to warn visitors of the dangerous condition of the lot in the area of her fall, and in failing to barricade the area to prevent her and other visitors from entering. Gibbs' insurance carrier brought a claim against the hospital, claiming that the maintenance of the parking and the lighting for the lot were the hospital's responsibility. In turn, the hospital filed a cross-claim against Gibbs and its insurer, claiming that they should indemnify the hospital for any damages it may have to pay the plaintiff due to the provisions of the construction contract.

According to the court's summary of the record, Wallace resided in California and had come to Louisiana to visit her ailing sister. At the time of her visit, the hospital was being renovated by Gibbs, and the parking situation was "helter-skelter" and the lot itself was not in good condition. During the 11 days immediately preceding her fall, Wallace had visited her sister daily and had parked in various locations, "wherever space was available." On the day of her fall, she had difficulty locating a parking space so she put her car in a spot perpendicular to the front of the hospital. Wallace indicated that she left the hospital at 8:15 p.m., looked out the exit and turned left toward her car. She followed the sidewalk until it began to curve. She noticed that the parking lot was very dark, but that straight ahead of her she could see light reflecting off the top of her white car. She further said that she then left the sidewalk to walk directly to her car, stepping over mud or stones until she was only a few steps from her automobile. She then stepped into a hole, striking her head on the ground and knocking herself unconscious. When she awoke, she walked back to the hospital emergency room and received treatment for the injuries to her head and wrist.

The trial judge concluded that the hospital premises were defective in that the parking area was not properly lighted and, in that the hole which Wallace stepped into was not barricaded and there were no signs warning of the hole. The court concluded that both the hospital and the general contractor were negligent and awarded a verdict in favor of Wallace in the amount of $103,936.23, which was reduced by 10 percent due to her own contributory negligence. The court also found that the contractor and its insurance carrier should reimburse the hospital under the indemnification provisions of the construction contract. All defendants appealed.

On appeal, the defendants first argued that the trial judge had erred in deciding that the parking area lighting was inadequate. They pointed to the fact that a hospital maintenance worker had testified that there was a light standard in the parking lot, as well as two lights on the side of the building, although the record indicated that the worker did not know whether the lights were lit on the night in question. He did testify that the hospital security personnel routinely reported any lighting problems to the maintenance department, which was responsible for replacing burned out lights. The appellate court noted that plaintiff's testimony to the effect that, other than the light reflecting off the top of her car, it was "black." Also, a claims adjuster who was at the scene approximately two weeks later testified that there were no overhead lights in the parking lot which would shine down to the parking line.

The appellate court therefore declined to find that the trial judge was in error in concluding that the parking lot lighting was inadequate.

The general contractor said that the area where the plaintiff fell was under the control of the hospital, and the hospital said that the area was under the control of the contractor, thereby each attempting to place the fault on the other defendant. The court noted that although the construction work had not yet been completed and was still on-going at the time of Wallace's fall, and the hospital had submitted a list of corrections to the lot to Gibbs, the hospital had begun using the parking area notwithstanding. Hospital maintenance workers "striped" the lot and removed the barricades around the parking area, and people began parking their cars in the lot. The appellate court concluded that the trial judge had not erred in deciding that the contractor was negligent in failing to barricade the area, nor had the trial judge erred in deciding that the hospital was negligent in either removing or allowing the removal of the barricades.

And finally, the appellate court reviewed the provisions of the contract between Slidell Hospital and Gibbs Construction, and found that the indemnity provision in the contract clearly stated that the contractor would indemnify the hospital for any injuries caused "in whole or in part" by the contractor's negligence. Accordingly, the appellate court also found that the trial court was correct in granting judgment in favor of the hospital against the contract on the indemnity clause.

All of the trial judge's rulings were therefore affirmed, and the judgment in favor of the plaintiff Wallace was allowed to stand.

PSCLR – SEP 87 (5)

Retailer And Security Agency's Alleged Negligence Causing Assault On Plaintiff Can Not Be Proved By Evidence Of Prior Nonviolent Crimes In The Area:

Pauline Brown v. National Super Markets, Inc., Sentry Security Agency of St. Louis, Inc., and T.G. Watkins, *No. 51634, in the Missouri Court of Appeals, Eastern District, Division Two, decided April 7, 1987, motion for rehearing and/or transfer denied May 20, 1987, application to transfer denied July 14, 1987, reported at 731 S.W.2d 291.*

Pauline Brown filed this appeal after the defendants she sued won a jury verdict in their favor and against her claim that they had been negligent in failing to protect her from an attack by an unknown assailant in the parking lot of a National Super Markets store. The trial court rejected her appeal for the reasons below stated. (See the March 1985 **PSCLR** for a full report of the prior appeal in this case.)

In her appeal, Brown claims that her assault while in the parking lot owned by National Super Markets resulted from the defendants' negligence. She also claimed that during the two year period immediately preceding the attack on her, there were 16 robberies involving firearms on National's property, plus seven strong arm robberies reported, plus 136 other crimes reported. Brown argued that this criminal activity which was known to the defendants created special facts and circumstances, giving rise to a duty by the defendants to protect the customers including the plaintiff against attack in the parking area. The defendants denied that the plaintiff's assertions of prior crimes on the premises constituted any such "special circumstances" which could give rise to a duty on the part of the defendants to protect plaintiff from criminal assaults by unknown third parties.

In that previous appeal, the appellate court ruled that the trial judge had committed error by granting a summary judgment for the defendants, and ruled that the plaintiff had in fact properly alleged a cause of action against the defendants. The appellate court there held that while a retailer generally has no duty to protect customers against intentional criminal misconduct by unknown third parties, nevertheless such a duty may exist when and if there is some "special relationship" which is shown to exist, or when special facts and circumstances are shown to exist. Such special facts and circumstances can be shown by proof of frequent and recent violent crimes having occurred on the premises.

The case therefore proceeded to trial by jury, which as previously indicated resulted in a verdict in favor of the defendants and against the plaintiff. The evidence presented at the trial showed that the plaintiff had parked her automobile in the middle of one of the parking lots at National's retail outlet in St. Louis, Missouri, and another car pulled beside her when she returned to her vehicle from the store itself. A man emerged from the passenger side of the second car and ordered the plaintiff to "get back in that car, lady." When Brown tried to run back into the store, the assailant grabbed her and shot her in the side and fled. Brown was seriously wounded by the gunshot.

Evidence was presented to the jury during the trial that 22 violent crimes had occurred on National's property during the 22-1/2 months immediately before the incident in question. Furthermore, the evidence revealed that beginning two years before the assault on Brown, an uniformed armed guard from Sentry Security Agency was at work and on-duty whenever the National Store was open. The retailer introduced records indicating that the security guards were on duty at the time that the plaintiff was shot in its parking lot.

Two security guards who were listed as having been on duty at the time in question gave different stories regarding the instructions which each of them received concerning the security procedures on National's parking lot. One (Watkins) testified that he had been expressly told <u>not</u> to go onto the parking lot, while the other testified that other guards as well as himself were on the parking lot regularly. The plaintiff said she had never seen a security guard in the parking lot area during her many previous visits to the store.

During the trial, the plaintiff had attempted to present evidence to the jury to the effect that there had been 14 purse snatchings on the property, but the trial judge excluded this evidence, concluding that the purse snatchings were not similar to Brown's assault because they had not involved any physical contact between the perpetrator and the victim. The trial judge decided to admit into evidence only police reports of prior violent crimes which had taken place on the premises. The judge did, however, permit police reports of the 22 violent crimes to be introduced into evidence and he allowed testimony concerning some of the details of these other incidents.

The plaintiff also relied on the prior Missouri case of *Virginia D. v. Madesco Investment Corp.*, reported in the September 1983 **PSCLR**, case one, page one.

As another basis of appeal, the plaintiff argued that the trial judge had committed error when he excluded certain offered evidence to the effect that a bank parking lot located directly across the side street from the National Supermarket parking lot had had no crimes on its premises for a two year period of time preceding the assault on the plaintiff. A security officer had patrolled the bank parking lot during that time. The plaintiff was prepared to present evidence to the jury that a St. Louis city detective would testify that no crime, violent or otherwise, had been reported on that bank's parking lot. However, National Supermarkets objected to the introduction of this evidence on the grounds that the plaintiff had not shown that there was a similarity of conditions between the bank's parking lot and the supermarket's parking lot, and the court sustained this objection and refused to permit the plaintiff to introduce this evidence. The plaintiff, of course, was anxious to show that the evidence concerning the security guard on the bank's parking lot was relevant to prove that the assault on the plaintiff would not have occurred had the supermarket taken similar precautions as did the bank. According to the plaintiff,

> The bank's experience tends to prove the relative deterrent effect of the presence of the uniformed security guard on a parking lot in the same neighborhood. ...

The appellate court, however, rejected the plaintiff's argument and ruled that the trial judge had not abused his discretion in refusing to permit the plaintiff's evidence regarding the bank parking lot and its use of a security guard on its premises.

In conclusion, the appellate court affirmed the rulings of the trial judge, and this litigation was resolved in favor of the defendants.

COMMENTARY:

It is somewhat surprising to this writer that the trial judge in the above case refused to permit the plaintiff to present evidence concerning the security precautions taken in the parking directly across the street from the supermarket, and that the appellate court affirmed this exclusion by the trial judge. As has been indicated in a number of decisions previously reported in **PSCLR**, *security precautions taken by owners and/or operators of business and/or premises in the same neighborhood are frequently admitted into evidence to show the standard of care utilized by other persons who are similarly situated as the defendant in the subject case. More often than not it is this writer's opinion that such evidence will be admissible, and that the refusal to permit such evidence will be reversed on appeal.*

PSCLR—OCT 87 (5)

Where Criminal Act Of Third Party Constitutes Unreasonable Risk Of Harm To Customer, Retailer May Be Liable For Failing To Summon Police Promptly For Failing To Anticipate The Danger:

John Hardin, et al. v. Munchies Food Store, *et al., No. 18742-CA, in the Court of Appeal of Louisiana, Second Circuit, decided June 10, 1987, reported at 510 So. 2d 33.*

John Hardin and his girlfriend, Judy Tabor, were awarded verdicts of \$167,066.52 and \$55,141.05 respectively for injuries received when they stopped at defendant's convenience store and were assaulted by a third party. The defendants paid the verdict to Hardin, but moved the trial judge to grant to the defendants judgment notwithstanding the jury's verdict (judgment n.o.v.) in Tabor's case. The trial judge granted the motion, setting aside the verdict in her favor, and she filed this appeal.

The appellate court first noted the heavy legal burden which must be carried for the movant (here, the defendant Munchies Food Store) to be successful in obtaining a judgment n.o.v. Such a judgment should be granted by a trial court "only when the evidence points so strongly and overwhelmingly in favor of the moving party that reasonable men could not reach different conclusions, ..."

In reviewing the evidence presented at the trial to determine whether the trial judge erred in granting the motion, the appellate court also stated that every reasonable and legitimate inference that can be drawn from the evidence by the jury must be read in the light most favorable to the respondent (the party against whom the motion for judgment n.o.v. is made; here, the plaintiff Tabor).

John Hardin and Judy Tabor were driving home in his truck late at night after spending an evening out. They stopped at defendant Munchies Food Store for cigarettes, which Hardin went into buy while Tabor waited in the truck. While Hardin was inside, a large man who had been standing near the front door of the store came up to Tabor and began talking to her through the partly opened window of the truck. Because his remarks to her were "sexual, crude, and offensive," she attempted to ignore him. Hardin came out and saw the man, Oliver King, standing by the driver's side of his truck,and told him to leave Tabor alone.

There were several witnesses to the subsequent events. The Munchies' cashier on duty, Jerry Green, another Munchies' employee, Herbert Coleman, and a customer, Tommy Johnson, were all present inside the store. An argument ensued between King and Hardin which lasted approximately two to three minutes, at the end of which King said to Hardin, "There isn't going to be any trouble," whereupon King pulled a heavy pipe from his back pocket and hit John in the head. Hardin fell to the ground with blood streaming from his mouth and nose.

Tabor immediately jumped out of the truck and ran to help Hardin, but King pulled a knife and threatened to "cut her open." She jumped into the truck and King began slashing at her with the knife through the partly opened window. She got on the floor to avoid him and, during this initial phase of the incident, she struck her cheek against the rear view mirror. She was screamed throughout the entire event.

The assailant King then returned to Hardin, who was unconscious and savagely beat him. Tabor opened the passenger door and screamed for help to two men standing in front of the Munchies Store. Neither of them responded, but merely watched the beating. (In fact, one man in a Munchies' t-shirt actually threw up his hands and shouted an obscenity in response to Tabor's cries for help.) Tabor attempted to run over the assailant King with the truck, but King threatened to stab Hardin if she moved the truck. He continued beating John, pounding his head against the pavement. The bystander in the Munchies' t-shirt shouted, "You gonna kill the man." King again returned to Hardin's truck, yelled at Judy, and beat on the windshield. One of the bystanders then told King that they had called the police and "You gonna have to book." Before fleeing, King went back to John and stole his gold chain, cigarettes, and wallet. Tabor estimated that the entire incident lasted approximately 20 to 25 minutes, but Hardin suffered from amnesia and could not testify about the length nor the events of the incident.

The Munchies' customer, Johnson, testified that he first noticed the assailant when he heard the loud arguing outside and he told the cashier Green to call the police, but that Green said that it would "be over before the police could arrive." Johnson said that the Munchies' cashier could see the entire incident very easily through the front door of the store. Johnson said that he was too scared to intervene and that, 12 or 13 minutes into the assault, he again told Green and the off-duty Munchies' employee Coleman to call the police, but they stated that the fight was "almost over." Johnson also testified the employees did go outside and tell King that they were going to call the police, but King did not stop the beating until several minutes later when the employees told him that they had in fact called the police. King immediately took Hardin's property and left the scene. Johnson estimated that the incident lasted 18 to 20 minutes.

The off-duty Munchies employee, Coleman, said that he had been at the store for approximately 15 to 20 minutes when he first heard the assault in progress. He testified that he instantly went to the front of the store and, as soon as he knew what was going on, called the police.

The Munchies' cashier, Green, knew the assailant King and knew that he was an ex-convict, but testified that King had never caused any problems at the store before. He denied seeing or hearing anything until the fight was "well in progress." He did admit, however, he did not immediately call the police, believing that Coleman had already done so. Green also testified that he had received no special training from Munchies as to any procedures for calling the police, but that there was a procedure for doing so in certain cases. Green also admitted that he did not tell the police the assailant's identity. Green testified that he had quit his job at Munchies three days after the incident, because he was afraid of King.

Testimony from the police officers showed that the first officer was on the scene three minutes and 18 seconds after Coleman placed the call to police. Officer Jones, who arrived on the scene first, had been stationed only a block and a half from the store and could have responded instantly if called earlier. Other officers were also nearby.

The assailant Oliver King was arrested later that night and was subsequently prosecuted for attempted second degree murder and armed robbery, convicted and sentenced to 75 years in the state penitentiary. He did not testify at trial.

As stated above, the trial resulted in verdicts for both Tabor and Hardin against the convenience store's insurance carrier, Safeco Insurance Company, which paid Hardin's verdict.

The trial judge, as noted earlier, granted the motion for the judgment n.o.v. against Tabor according to his written opinion for the following reasons:

(1) there can be no award of damages to a plaintiff merely for witnessing an injury being inflicted on a third party; and

(2) there can be no recovery to a plaintiff merely for being subjected to obscene and abusive language such as the assailant used toward Tabor; and

(3) the evidence conclusively proved that Tabor's jaw injury had occurred at the very beginning of the incident as she jumped back into the truck after her first unsuccessful effort to assist Hardin; the judge felt that this injury had happened so early in the course of the event itself that

> even if Munchies' employees had responded by calling the police within a reasonable period of time of receiving knowledge of events going on outside the store, Tabor would have received these injuries.

In addition, the judge accepted evidence that her jaw problem pre-dated the incident itself and found that the medical treatment she required was due to this prior condition. Therefore, he ruled that her injury was not the result of any fault on the part of the defendant Munchies' employees.

In its review, the appellate court looked to its own 1986 decision in *Ballew v. Southland Corporation* (reviewed in the May 1986 **PSCLR**, case three, page three), concerning the duty of a retailer to its customers. While a retailer does have the responsibility of taking reasonable care to protect its customers from injury, such responsibility does not include protecting them from unforeseeable and unanticipated criminal misconduct by third parties. The duty to protect from such activities arises only when the retailer has knowledge of the assailant's intention to harm the customer, and the retailer has the power to do something about it. Furthermore, the retailer's duty toward the customer can be fulfilled by calling the police upon learning that the customer is in danger from attack by a third party. Therefore the question presented on this appeal was whether the Munchies' employee knew, or should have known, of the potential misconduct by the assailant and whether he had the ability to protect Tabor by calling the police. In reviewing the evidence to answer that question, the appellate court concluded that there were four salient facts which supported the jury's verdict of liability:

(1) "The cashier knew of the potentially dangerous situation well in advance of the first blow." The witness Tommy Johnson unequivocally told the cashier to call the police at the beginning of the argument. Furthermore, Green knew King's background and reputation, and based upon his statement that "it would be over" before the police arrived, the jury could infer that the cashier obviously anticipated trouble.

(2) "The jury was certainly entitled to conclude that a prompt call [to the police] would have prevented the injury [to Judy Tabor]." The police were already in the area and could have been on the scene in a remarkably short period of time—just over three minutes. The record shows from the time the witness Johnson sensed the danger and told the cashier to call the police, the argument continued for the next two to three minutes. Immediately after which, Judy rushed to assist Hardin, was chased by King into the truck, and injured her jaw. It is a close question as to whether there was a three minute interval during which the police, if they had been immediately called, could have arrived, stopped the event in progress, and prevented Tabor's injury. However, the appellate court held that the jury could reasonably conclude that such was the case from the evidence presented.

(3) "Oliver King fled as soon as Green specifically told him that the police had been called." Even after King's first blow on Hardin, the injury to Tabor could have been avoided if Green had immediately told King that the police had been called. King's would have then responded by instantly fleeing, as he did when he was finally told that the police had been called. The jury, therefore, was entitled to infer from the evidence that the employee Green

had negligently failed to fulfill his responsibility to provide reasonable care and protection to his customers by telling King earlier that the police had been called.

(4) "[T]here is the testimony of Dr. Gamble who said that [part of the injury to Tabor's jaw] could have resulted from screaming or yelling as well as from striking the mirror." According to the court, based upon the medical testimony, the jury could have reasonably inferred that Tabor's 20 minutes of screaming could have aggravated her injury, and that most of her screaming would have been abated had the police been promptly called.

In summary, the appellate court for the above enumerated reasons concluded that the trial judge had committed error by setting aside the jury verdict in favor of Judy Tabor. (The case was also remanded to the trial court for decision on Safeco's motion for a new trial, which the trial judge had not ruled on. The appellate court did note that the trial judge had, in stating his reasons for granting the judgment n.o.v., indicated that he would have reduced her award to $15,000 if he had not granted the judgment n.o.v. Presumably, he may well now do so.)

PSCLR—OCT 87 (8)

Security Guard, Under Contract To Restaurant And Ordered By Restaurant Not To Leave Building, Is Not Liable To Restaurant Patron Wounded In Assault In Parking Lot:

Jonathan R. Jones v. Samuel Williams, individually and d/b/a Williams Private Patrol Services, Inc., and White Castle System, Inc., *a Delaware corporation, No. 90273, in the Court of Appeals of Michigan, decided April 14, 1987, released for publication July 17, 1987, reported at 408 N.W.2d 426.*

The plaintiff was assaulted by a gunman on Halloween night in a White Castle restaurant parking lot in Detroit. He brought this action against the White Castle and against the security company which had contracted with the restaurant to provide security services.

The plaintiff Jonathan Jones entered the restaurant and placed an order for food to be taken out. After receiving his order, he left the store and returned to his car, as did another customer, Ibrahim Ameen Barakat. As they approached their respective automobiles, a Thunderbird car containing three or four men backed toward them at a high rate of speed and stopped near them. One of the men in the car got out of the car and removed a shotgun from the trunk. Barakat and the plaintiff continued to run toward their cars, trying to escape the gunman who pointed the gun directly at the two. Barakat found himself in the line of fire between the gunman and Jones, so he fell to the ground and attempted to roll underneath his car. The gunman then shot at Jones, returned the shotgun to the trunk of the car, got back into the automobile, and the car sped away. Barakat got up and ran back into the restaurant and notified the White Castle security officer to call the police, which he did. The police arrived quickly, but the assailants were not apprehended.

Both the defendant White Castle and the defendant Williams Private Patrol Services, Inc., moved the trial court judge for summary judgment in their favor. The trial judge granted the motions, concluding that neither the restaurant nor the security service had any legal duty to prevent the injuries which were sustained by the plaintiff, and also that no act or omission on the part of the defendants could be considered the cause of the injuries sustained by the plaintiff. The plaintiff filed his appeal from these rulings.

The plaintiff argued on appeal that the defendants had a duty to protect him from the criminal assault and that their failure to meet that duty resulted in his injuries. Plaintiff also argued that if there had been a security guard in the parking lot itself, or if the defendants had taken other steps, such as installing a drive-in window, the assault would have been prevented. For these reasons, he claimed that the trial judge should have allowed the questions of responsibility and causation to be presented to the jury, rather than dismissing the case prior to trial.

The appellate court disagreed, and affirmed the trial judge's ruling, finding that White Castle had no legal duty to make the parking lot secure. The appellate court further held that White Castle had no duty to provide security guards in the parking lot. According to the appellate court, it could be inferred from the evidence that the criminal was specifically after the plaintiff Jones, because the assailant did not shoot Barakat while he was in the line of fire, but waited until Barakat fell to the ground before firing at Jones. The appellate court concluded that the restaurant had no duty to protect Jones from an unforeseeable "assassination attempt."

With regard to the defendant security guard, the appellate court reviewed the contract between the White Castle and the security company, which provided for

> uniformed police and guard service at the White Castle restaurants located in the greater Detroit metropolitan area, as instructed by the White Castle Detroit management.

The contract also provided that the security service would "maintain order" on the property, both inside the restaurant proper as well as inside the adjacent parking lot, and required that the security officer assigned to the restaurant would direct traffic on the parking lot, "at such times as shall be specified by White Castle."

The appellate court therefore held that the security guard had no duty to the plaintiff since Jones was not a party to the contract between the restaurant and the security company. Moreover, the guard had merely complied with the contract by following the restaurant management's instructions not to leave the restaurant itself.

In conclusion the court stated:

> It is one thing to require apartment building owners to equip their buildings with adequate locks and light. It is another thing to require fast food restaurants to provide armed guards for adjacent parking lots so that assassination attempts against patrons can be thwarted. Even the presence of guards could not prevent assassination attempts. If there is no duty to provide guards, and guards could not have prevented the attack here in question, other methods, such as a drive-up window, would not necessarily prevent such attempts either.

Accordingly, the trial judge's decision was affirmed and the plaintiff's case was dismissed.

COMMENTARY:

In reaching the above decisions, the Court of Appeals of Michigan also relied upon the case of Williams v. Cunningham Drug Stores, reported in the April 1986 **PSCLR**, *case six, page seven. The Jones court also relied upon Fillare v. Union Oil Company, reported in the November-December 1985* **PSCLR**, *case three, page three.*

PSCLR – OCT 87 (10)

Where Retailer Know Or Should Know Of Existence Of Pot Hole In Parking Lot, He Is Liable To Customer Who Trips And Sustains Injury, Even Though Lease Requires Lessor To Maintain Parking Lot:

Kenneth W. Miller, et al. v. Great Atlantic & Pacific Tea Co., et al., *No. 86 CA 0542, in the Court of Appeal of Louisiana, First Circuit, decided May 27, 1987, rehearing denied July 20, 1987, reported at 510 So. 2d 695.*

Commercial Properties Development Corporation owned the grocery store in question, and leased it to A&P. The lease required Commercial Properties to maintain the parking lot. The manager of the A&P store would periodically instruct his bag boys to report to him any pot holes in the parking lot which might be dangerous to customers. The manager would then contact the A&P home office, which would in turn report the problem to Commercial Properties; the latter would then, according to the terms of the lease, be responsible for repairing the pot holes.

Mrs. Miller went to shop at the A&P store. After completing her shopping she pushed a cart full of groceries back to her car, taking a different path than she had taken entering the store. Accordingly, she did not have any opportunity to see a pot hole in that path. While pushing the cart in front of her and checking for traffic, she tripped in the pot hole and injured her left ankle. She sued A&P and Commercial Properties.

The trial judge determined that the particular pot hole in which the plaintiff tripped was dangerous because of its close location to the A&P grocery store. The trial judge further concluded that the size, depth, and location of the pot hole was such that the bag boys should have seen it and reported it to the A&P manager, who should have then alerted the home office to report it to Commercial Properties; such did not occur, however.

The jury assessed damages in her favor as follows: general damages – \$50,000; loss of past and future wages – \$50,462; loss of consortium – \$10,000; medical expenses – \$9,287.37; for a total of \$119,438.37.

On appeal, the appellate court first concluded that the owner of the property, Commercial Properties, was "clearly liable" because it was the owner of the property where the incident occurred. But the lease did not relieve A&P of its duty to protect its customers from falling into pot holes which they knew, or should have known, existed. (The court said, "the fact that plaintiff frequently shopped at the A&P does not impose a duty on her to cross the parking lot 'as though it were mine field'.")

The defendants also argued that the verdict in favor of the plaintiff was excessive and should be reduced. They claimed, for instance, that the medical expenses award should have been reduced because the award included expenses for surgery to Miller's right ankle, which was not injured when she fell in the A&P Parking lot. The appellate court noted, however, that the medical testimony established that the surgery on her right ankle had been required because the right ankle had been placed under extra stress and tension following the injury to the left ankle. The appellate court rejected the defendants' argument that Miller had weak ankles which may have "predisposed her to an injury of this sort." The appellate court concluded that the award was proper because Miller's injuries were extremely painful, she had limited use of her legs for nearly a year, the surgery left visible scarring on her ankle. In addition, the appellate court concluded that the evidence showed that her injuries played a part in causing Miller's separation from her husband, and therefore concluded that the \$10,000 awarded as loss of consortium was also authorized.

The defendants appealed additionally for a reduction in the award for lost income, because the evidence showed that Miller was not employed at the time she was injured. The appellate court also rejected this argument, noting that she had previously been employed and was in fact seeking employment at the time of the accident. Moreover, she had returned to work after the injury, but was required to quit because of the pain caused by the injury.

The verdict and judgment in favor of the plaintiffs was affirmed.

COMMENTARY:

Once again we see that the job of protection of assets includes not only protection of loss from burglary, embezzlement, shoplifting, and the like, but also from physical hazards on the property which can cause injury to invitees.

Also, with regard to the defendants' argument that the damages should be reduced because Miller had weak ankles prior to the fall, one is reminded of the analogy that the tortfeasor cannot justifiably claim that he should not be held responsible for having negligently destroyed a truck full of eggs, merely because he thought they were golf balls! "The tortfeasor takes his victim as he finds him, and is responsible for the consequences of his tort"—even if the victim turns out to be a vulnerable egg instead of an invulnerable golf ball.

PSCLR—JAN 88 (1)

Inference Attaches From Store's Hiring Of Security Company That Store And Security Company Have Notice Of Potential Danger:

Shaun Marshall v. David's Food Store and A&R Security, Inc., *No. 84-2494, in the Appellate Court of Illinois, First District, Fifth Division, decided May 29, 1987, rehearing denied November 10, 1987, reported 515 N.E.2d 134.*

Shaun Marshall parked her automobile in the customers' parking lot at the defendant David's Food Store in Chicago, entered the store and made her purchases, and returned to her car, at which point she was attacked by three assailants at gun point. She was abducted and taken to a remote location, where she was sexually assaulted by the men. She sued David's Food Store and its contract security company, A&R Security, which had been hired by the retailer to protect customers in the store and the adjacent parking lot.

In her lawsuit she complained that the defendant David's owed a duty of reasonable care to protect her and other patrons in its parking lot, and she alleged that it was or should have been foreseeable to David's that a criminal assault might occur in the parking lot. She also set forth an exhibit to her lawsuit of a list of 14 prior incidents of criminal activity which had taken place within five months before her assault, and within two blocks of David's Food Store.

David's and the security company both moved to dismiss the lawsuit, arguing that the plaintiff had failed to aver that the defendants were aware that the parking lot area was dangerous or were aware that individuals had been victimized in the area. The trial judge granted the defendants' motion and dismissed the plaintiff's complaint on the ground that the plaintiff had failed to allege enough factual basis to support a conclusion that the defendants had or should have foreseen a criminal attack occurring on a patron in the parking area. The trial judge had ruled:

> The mere fact that there has been criminal activity in the surrounding area is not enough to place the individual proprietor on notice.

However, on appeal the appellate court ruled that the trial court was in error when it said that Marshall should have alleged detailed evidence in her lawsuit to support the conclusion that the criminal attack on her was foreseeable by the defendants. The court noted that the law in Illinois provides that,

> A defendant who has actual or constructive notice of a potentially dangerous condition has a duty to take reasonable precaution to protect its invitees,

and in the present case, Marshall had alleged that either or both David's and A&R Security had such knowledge and that the attack was foreseeable by them.

The appellate court also pointed out that the defendants were in the best position to know the extent of prior criminal activity in the area. Furthermore, since the defendant store had contracted with the defendant security service to protect its customers, the court stated that it can therefore be inferred that they must have had some notice of the presence of potential danger in their parking lot. Marshall's suit established the fact that the security company assumed the duty of exercising care to protect the customers of the retailer, and the issue was then presented whether the security company breached that obligation; and such issue must be resolved by the trier of fact, the jury.

Accordingly, the trial court's dismissal of the plaintiff's complaint was reversed and the case was returned to the trial judge for further proceedings.

COMMENTARY:

This case is one of the first instances in which an appellate court has concluded that the merely because a business owner and operator has hired a security service, he can therefore be deemed to have knowledge of potential danger. This is a somewhat incongruous position, for it puts the business owner/operator in the quandary of either failing to hire a security service and being sued for failing to provide security, or hiring such a service and then being charged with knowledge of danger in the area. It is decisions like these which plague security directors as well as business owners and operators.

PSCLR – JAN 88 (10)

Retailer Has No Duty To Protect Invitees From Criminal Acts Of Third Parties, Absent Knowledge That Criminal Act Was A Probability:

Suzanne Smith Childers v. Winn-Dixie Stores, Inc., et al., *85-1365, in the Supreme Court of Alabama, decided September 4, 1987, reported at 514 So. 2d 879.*

This lawsuit was brought by Suzanne Smith Childers, her husband David L. Childers, and her daughter Suzannah Lane Childers, against Winn-Dixie Stores, Inc., Birmingham Realty Company, and Russell Cunningham. The suit arose out of the following fact situation:

Childers and her daughter went shopping at a Winn-Dixie store in Birmingham, Alabama, and after completing their shopping, they returned to their automobile in the parking lot. While they were placing their purchases in the car, two assailants approached them and snatched Childers' handbag. She chased the young thieves, but could not catch them.

In their lawsuit, the mother and daughter alleged that the defendants were negligent in failing to provide security guards or adequate lighting in the parking area, and that the defendants had failed to do so "willfully and intentionally with full knowledge of the very danger that caused plaintiffs' injuries." The husband David Childers filed his action for loss of his wife's consortium.

The defendants filed motions for summary judgment which, after hearing and consideration of the evidence, the trial judge granted, thereby dismissing the plaintiff's claims. They appealed.

In their appellate brief, the plaintiffs asked the appellate court whether "this court ever intends to require of owners and occupants of retail establishments in this state" any responsibility to invitees to protect them from criminal activity and attack. In response, the Alabama Supreme Court noted that they had in fact ruled in previous cases that a store owner may have a duty to employ reasonable precautions to protect customers from criminal assaults in the exceptional situation where the retailer has either actual or constructive knowledge that criminal misconduct endangering the customer was a probability. At the same time, the supreme court noted that it is difficult to impose such liability on a defendant for the intentional criminal misconduct on the part of an unknown third party, and in fact the supreme court noted that it had not yet found a fact situation egregious enough to impose such a duty. In reviewing the facts of the Childers' suit against Winn-Dixie, the court particularly noted that there had been only one purse snatching in the two years preceding the attack on Childers and concluded that there was a complete lack of evidence showing that Winn-Dixie had any knowledge, either constructive or actual, that there was any probability of criminal misconduct of the same type as harmed Childers.

Accordingly, the Alabama Supreme Court affirmed the trial judge's grant of summary judgment dismissing the plaintiffs' claims.

COMMENTARY:

The cases cited by the Alabama Supreme Court include *Law v. Omelette Shop, Inc.,* *reported in the April 1986* **PSCLR**, *case eight, page eight;* *Henley v. Pizitz Realty Co.,* *reported in the January 1985* **PSCLR**, *case nine, page nine; and* *Ortell v. Spencer Companies, Inc.,* *reported in the February 1986* **PSCLR**, *case nine, page six.*

As we have noted in previous Commentaries regarding the Alabama courts' position on liability for third party misconduct, the Alabama courts have taken, and continue to take, a conservative position on this issue. At the same time, even a court in a "liberal jurisdiction" would probably have affirmed the grant of summary judgment in the above case, because of the lack of foreseeability due to the fact that there had been insufficient proof of knowledge of similar incidents on the defendants' part. Nevertheless, the Alabama court has noted its willingness to revisit the issue and to impose such liability if and when the "exceptional" case is presented.

PSCLR—APR 88 (9)

Neither Shopping Center Nor Owner Of Bar Located Therein Have Duty To Protect Intoxicated Customer From Unavoidable Accident In Parking Lot, Even Though Bar Owner Had Security Guards Positioned At Door:

Michael Winters v. Protective Casualty Insurance Company, Inc., *et al., Civ. A. No. 86-540, in the United States District Court for the Southern District of Louisiana, decided February 3, 1988, reported at 678 F. Supp. 144.*

The various parties to this action include Michael Winters, the plaintiff in this action, who claimed to have been seriously injured when he was dragged beneath an automobile driven by the defendant Lawson Wilder. In his complaint, the plaintiff Winters averred that Wilder had become intoxicated due to drinking alcoholic beverages which he had obtained from the employees of the defendant H.B.M., Inc., d/b/a Murphy's Bar, and that Wilder's intoxication was such that he failed to see the plaintiff Winters lying on the ground in the parking lot outside the bar in front of Wilder's automobile.

Winters also claims that his injuries resulted from the negligence of the Murphy's Bar employees and their serving alcohol to Wilder in sufficient quantity for him to become intoxicated, as well as for failing to provide adequate security in the parking lot itself outside the bar. Winters also claims that his injuries resulted from The University Shopping Center employees' negligence in failing to provide adequate security personnel in the parking lot. (The bar was located in the shopping center.)

Other parties in the case included the Protective Casualty Insurance Company, which insured Wilder, and State Farm Mutual Automobile Insurance Company, Employers Mutual Casualty Company, Tulane University Chapter of Sigma Nu Fraternity, National Sigma Nu Fraternity, Beta Phi House Corporation, and the United States Fire Insurance Company. In addition, the Constitution State Insurance Company, Mount Vernon Fire Insurance Company, and Blue Cross and Blue Shield United of Wisconsin were also involved in this case.

The matters before the court at this stage are the following: a motion for summary judgment filed by the defendant H.B.M., Inc., d/b/a Murphy's Bar and its insurance carriers Constitution State and Mount Vernon Fire; also, a summary judgment motion filed by Theodore Cangelosi d/b/a The University Shopping Center. The first motion for summary judgment (by the Murphy's group) was opposed by plaintiff, as well as by Blue Cross and Blue Shield United of Wisconsin (as intervenor), and by the Sigma Nu Fraternity and its insurer. The second motion for summary judgment is opposed by the plaintiff Winters and the intervenor Blue Cross and Blue Shield.

With respect to Winters' claim against Murphy's for selling alcoholic beverages to Wilder, the defendant Wilder pointed out that, at the time of the accident in question, Louisiana law provided no liability for retailers of alcohol for injuries caused by people to whom the alcohol had been served (i.e., "dram shop" liability), unless the retailer had committed some affirmative act thereby increasing the danger caused by the customer's intoxicated condition. And, Wilder pointed out that the plaintiff Winters had failed to allege any evidence of any such affirmative acts.

But the plaintiff Wilder also claimed that Murphy's is liable upon yet another theory: its failure to provide adequate security guards in the parking lot outside the bar. And the defendants responded to this assertion by arguing that Louisiana law placed no burden upon the bar owner to provide security in the parking lot in order to protect customers from unforeseeable occurrences resulting from the acts of third parties.

Citing the case of *Harris v. Pizza Hut* (reported in the January 1985 **PSCLR**, case one, page one), the judge in this case noted that the Louisiana Supreme Court had previously held that a retailer who hires security personnel in order to protect itself and its customers can be held liable for injuries which are due to negligence on the part of the security personnel.

Therefore, where a retailer assumes the duty of providing a security guard, as did Murphy's Bar in this case, then the court must decide whether the guard acted reasonably in preventing the injury which occurred. The courts of Louisiana determine negligence liability by first finding whether the risk of injury which befell the plaintiff came within the scope of the duty of care owed to him by the defendant; that is:

> Did the defendant have a duty to protect the plaintiff from accidents of the type which occurred in this case?

Whether or not the risk which caused harm to the plaintiff was the type of risk which the defendant had a duty to prevent or protect the plaintiff against, must be determined on a case by case basis. In the present case, the facts are clear that the incident occurred in The University Center parking lot, and that there are several businesses located in the mall, including Murphy's Bar. One parking lot services all of the retailers in the mall, and there are no specific parking spaces allotted to any particular business establishment.

The incident which injured the plaintiff occurred in darkness, probably after the midnight hour. Furthermore, the plaintiff Winters has not alleged that he was so obviously intoxicated or so visibly unable to take care of himself when he left the bar, that the security personnel should have seen the degree of his inebriation and taken steps to ensure his safety. He merely alleged that the bar had failed to provide adequate security in the parking lot area.

Taking all of these factors into consideration, the judge found that the risk of the harm which caused injury to Winters was beyond the scope of any duty to protect him which was undertaken by the defendant Murphy's Bar. Simply stated:

> The duty imposed on the business which hires security guards is to take reasonable care to protect the premises and patrons; it does not encompass the risk that a patron will walk out of the bar, lie down near or under a parked vehicle, go unnoticed by the driver of the vehicle and his passengers, and be dragged beneath the car because they did not notice him.

Moving on to the summary judgment motion filed by the shopping mall itself, Winters argued that under the particular facts of this situation, the shopping mall owners owed him a duty to prevent him from being injured not only by the negligence of the defendant Wilder, but also from his own negligence. He further argued that the shopping center itself should have employed security personnel and was negligent for failing to do so, since it was "foreseeable that an intoxicated person would get into a car and cause injury to others in the parking lot." This argument was rejected by the court, with the judge stating that a retailer has a duty to take reasonable care to protect his customers, but he "is not the insurer of their safety."

The court concluded that the plaintiff could point to no evidence which would have put the shopping center owner on notice that the particular injury which occurred to the plaintiff was likely to occur or was otherwise foreseeable. That being the case, the defendant shopping mall owner also could not be found liable to Winters.

Accordingly, the trial court granted both the summary judgment in favor of the owner of Murphy's bar, and the owner of the shopping center.

PSCLR—SEP 88 (6)

Jury's Award Against Shopping Mall For Injuries Incurred In Parking Lot Assault Held Not So Grossly Inadequate As To Require New Trial:

General Syndicators of America, a Florida Corporation v. Ruth Green and Donald Green, *No. 87-898, in the District Court of Appeal of Florida, Fifth District, decided on April 14, 1988, reported at 522 So. 2d 1081.*

Ruth Green, an employee of a store located in Palatka Mall, in Palatka, Florida, was attacked in the parking lot behind the store one evening after work. As she approached her car, a young man accosted her and attempted to grab her purse. In the ensuing struggle, she was thrown to the ground and broke her hip. Her injuries, including the broken hip, were substantial.

Green and her husband sued General Syndicators of America (GSA), contending that, as owner and manager of Palatka Mall, it breached its duty to Green to provide adequate lighting and security at the mall and in the adjacent parking areas. In its pre-trial papers, GSA admitted that, at the time of the incident, GSA owned and managed the Palatka Mall. In spite of this admission, the trial court allowed testimony by the mall manager, Eileen Gallion, that Sofran Corporation managed the mall at the time of Green's injuries. Gallion testified on cross examination that Sofran, not GSA, made all the decisions regarding mall security, and that Sofran's employees were "running the show."

The trial court submitted the case to the jury with an instruction allowing GSA to escape liability if it proved that Sofran was an independent contractor responsible for managing the mall. In spite of the instruction, the jury apparently found that Sofran was not an independent contractor charged with managing the mall, and returned a verdict in favor of the plaintiffs. Finding GSA liable for 60 percent of the injuries and Ruth Green responsible for 40 percent, the jury awarded Ruth Green $54,000 in damages and Donald Green $5,000 for his loss of consortium.

The plaintiffs moved the court to grant them a new trial on the grounds that the judge had confused the jury by instructing them on the issue of independent contractors, and that the awards themselves were inadequate for the injuries Green had suffered. The trial court decided that the plaintiffs were entitled to a new trial because the jury may have been confused by the instruction regarding independent contractors. GSA appealed the grant of a new trial.

The reviewing court held that it was not necessarily improper for the trial judge to instruct the jury that, if an independent contractor relationship existed, GSA would be insulated against liability for negligence by the independent contractor. But, even if the instructions to the jury had been improper, the plaintiffs were not harmed by the error since the jury clearly found that GSA, not Sofran as an independent contractor, was liable for the negligence causing Ruth Green's injuries.

In addition, the appellate court felt that the damages awarded to the plaintiffs, "although they were on the low side," were "not so inadequate as to compel the conclusion that the jury was swayed by passion, prejudice, or that it erroneously applied a principle of law or fact." In light of the medical testimony presented, the total award was not so grossly inadequate as to warrant a new trial.

Accordingly, the granting of the new trial was held to be erroneous, and the jury verdicts stand.

COMMENTARY:

Although it turned out to be only a collateral issue in the above case, the issue of independent contractors insulating property owners from liability for inadequate security continues to be a "hot topic" of continual litigation in the field of private security. Regular readers will recognize this and remember that hiring a contract security agency will not necessarily protect your company from liability if the contract agency fails to provide legally adequate protection on your premises to your customers and employees.

PSCLR—OCT 88 (1)

Convenience Store Owner May Be Liable To Customer For Criminal Assault By Third Party On Vacant Lot Adjacent To Store Premises:

The Southland Corporation and Jan Lee, Inc., v. The Superior Court of The State of California for the County of Los Angeles, Victor Spencer, real party in interest, *No. BO31418, in the Court of Appeal of California, Second District, Division 3, decided August 5, 1988, reported at 250 Cal. Rptr. 57.*

One evening between 7:00 and 9:00 PM, plaintiff Victor Spencer and his friend, Neil McNeeley, parked their car in a vacant lot adjacent to a 7-Eleven store. As Spencer returned to his car after purchasing some items from the 7-Eleven store, three unknown individuals attacked him on the vacant, unpaved lot approximately ten feet from the store's property.

The three assailants, described as "a punk rock trio," came from the direction of the 7-Eleven and attacked Spencer from behind. Although Spencer and McNeeley told police they had noticed a group of kids leaning against the wall of the 7-Eleven store when they arrived, they could not describe them or identify them as the assailants.

Spencer sued the owner and operator of the 7-Eleven store, franchisor Southland Corporation and franchisee Jan Lee, Inc., to recover for his injuries. Spencer alleged that their negligence led to his assault in the vacant lot adjacent to the store.

Neither Southland Corporation, nor Jan Lee, Inc., owned or leased the vacant lot where Spencer had parked and was assaulted. However, the store's lease did provide for "nonexclusive use" of the vacant lot for passage of 7-Eleven customers to and from the store and for extra parking. Moreover, the defendants acknowledged that many 7-Eleven customers parked in the vacant lot because the marked and paved parking spaces on the store premises itself often proved insufficient to provide parking for all of the customers.

Spencer, a frequent 7-Eleven customer, never used the store's paved parking spaces, but always parked in the adjacent vacant lot for his own convenience. He believed the lot "to be under the control of 7-Eleven and the adjacent business, Phillips'." The court noted that neither Southland nor Jan Lee, Inc., did anything to discourage their customers from using the vacant lot for parking, such as posting signs or erecting a fence. Further, the store very likely realized increased sales and benefited commercially from its customers' use of the lot, according to the court.

There was evidence that the store and the adjoining lot served as a hangout for local juveniles and that fights frequently broke out among them. The store manager testified that he often asked loitering juveniles to leave and sometimes called the police to enforce his request.

Southland and Jan Lee, Inc., attempted to have the case dismissed without trial because Spencer was not injured on property "owned, operated or controlled by Southland" and because Spencer's injuries were not foreseeable to them, because there had been no similar incidents at the 7-Eleven in the seven previous years.

Spencer countered that there exists "a duty to patrons 'off the business premises' when circumstances causing injury are within the range of defendant's reasonable supervision and control." The trial court agreed with Spencer, holding that there were sufficient facts to raise an issue for a jury to decide regarding Southland and Jan Lee, Inc.'s control of the vacant lot.

On review of the trial court's decision not to dismiss the case, the appellate court framed the issue as whether a business proprietor owes a

> duty of care with respect to patrons injured by the criminal acts of third persons which occur on property, adjacent to the business premises, which is neither owned nor possessed by the proprietor and, if such duty exists...whether such criminal acts were reasonably foreseeable.

The court concluded that the existence of such a duty depends upon the "proprietor's actual or apparent control of the adjacent property." The court held that there were sufficient facts in the record to raise issues of fact as to whether Southland and Jan Lee, Inc., had such apparent control of the adjacent property, as well as whether the criminal conduct leading to Spencer's injury was foreseeable.

The court reasoned that a business proprietor owes a

> clear duty of care to his customers...to protect them, when they come upon his land, even from the wrongful acts of third persons which the proprietor has reasonable cause to anticipate.

Further, the business proprietor's duty to his customers may extend to property over which he exercises actual or apparent control even if he does not actually own or possess the property. The court stressed that the primary issue, in determining whether to extend liability to injury occurring off a business premises, is whether or not the proprietor exerted control over the area in question. The court found that, in this case, the question of the extent of the use of the adjacent lot by Southland and Jan Lee, Inc., and the consequent control over it which they apparently exercised, was an issue for the jury to determine.

Facts leading to the conclusion that Southland and Jan Lee, Inc., may have exercised sufficient control over the vacant lot so as to create a duty to customers using the lot, included:

> 1. only eight marked parking spaces were provided on the store's premises, and these often proved inadequate;
>
> 2. customers, including Spencer, regularly used the adjacent lot to park while shopping at the store;
>
> 3. [the store's] lease apparently authorized the non-exclusive use of the adjacent lot for customer parking;
>
> 4. [Southland and Jan Lee, Inc.,] were aware that their customers regularly used the lot and took no action to limit or discourage such use;
>
> 5. [Southland and Jan Lee, Inc., may have] realized a significant commercial benefit from their customers' use of the lot;
>
> 6. the [7-Eleven] premises and the adjacent lot had become a hangout for local juveniles, among whom fist fights sometimes broke out; and
>
> 7. the store employees had, on a number of occasions, taken action, including the request of police assistance, to remove juvenile loiterers from both the store premises and the adjacent lot.

A second issue of fact to be determined by a jury was whether or not the harm to Spencer was foreseeable. A business proprietor has a duty to protect customers from foreseeable injury, even to the extent of controlling the conduct of third parties. The duty to control the misconduct of third parties only applies to conduct that the proprietor has a reason to anticipate and a reasonable opportunity or means to prevent. The court, therefore, ordered the case to proceed to trial on the issues of Southland's and Jan Lee, Inc.'s apparent control of the vacant lot and the foreseeability of the criminal attack on the plaintiff Victor Spencer.

COMMENTARY:

The court cited, in support of its holdings, the following cases previously reported in **PSCLR**: *Peterson v. San Francisco Community College District, reported in the February 1985* **PSCLR**, *case seven, page seven; Isaacs v. Huntington Memorial Hospital, in the May 1985* **PSCLR**, *case 3, page 3; Noble v. Los Angeles Dodgers, Inc., in the August 1985* **PSCLR**, *case eight, page seven; Gregorian v. National Convenience Stores, Inc., in the February 1986* **PSCLR**, *case five, page four; and Cohen v. Southland Corporation, in the September 1984* **PSCLR**, *case four, page four.*

One justice filed a rather impassioned dissent which expressed an opposing view to the majority holding:

> *The umbilicus of the [majority's] reasoning is that inadequate parking existed and Spencer, who has never parked in the spaces provided, had a subjective and reasonable belief that Southland controlled the unpaved lot. [The majority] speculate as to increased sales from this off-premises parking and a realization of commercial benefit.*
>
> *Unrealistically, [the majority] comment that [Southland] did not erect signs or fences to discourage customer parking in an adjacent area open to all citizens. In any event, the record reflects that in a seven-year period prior to this event no claim for a similar incident, no other crimes or injuries had been*

reported as concerns Southland's business operations. So much for triable issue of foreseeability on the question of a reasonably perceived risk.

Most sadly, the fact that the store manager would ask loitering juveniles to clear the area and, on occasion, would summon law enforcement to preserve the calm and safety of the community, is now used against [Southland] on the question of control. 'Danger invites rescue. The cry of distress is the summons to relief.'

A caring society is ennobled and enriched when citizen observers sound the alarm which seeks a protective response. The majority view has the practical effect of discouraging activities which benefit the victims amongst us. Do they seek to sanction the deafening silence which responded to the cry of Kitty Genovese?

... this elastic extension of exposure under the guise of triability represents a costly and unwarranted imposition of legal burden.

PSCLR – FEB 89 (4)

Property Owner Not Liable To Individual Criminally Assaulted On Premises, When Individual Injured Was Permitted, Rather Than Invited, To Enter The Property:

Suzanne Figueroa and Luis Figueroa v. Evangelical Covenant Church, d/b/a North Park College, *No. 85 C 1916, in the United States District Court for the Northern District of Illinois, Eastern District, decided October 7, 1988, reported at 698 F. Supp. 1408.*

Suzanne Figueroa was assaulted and raped by an unknown assailant in the parking lot of North Park College after dropping off her child at an adjacent day care center. Figueroa and her husband sued the Evangelical Covenant Church, owner of North Park College, for the injuries she sustained in one of North Park's parking lots.

On the morning of the assault, Suzanne Figueroa parked her car in the North Park parking lot and dropped off her child at the nearby Northeastern Illinois University Childcare Center in Chicago, Illinois. Upon returning to the parking lot, Figueroa was abducted at gunpoint by two individuals.

The assailants drove Figueroa in her car to a different location, where they sexually assaulted her for several hours. The assailants also attacked her with a knife, slicing through her right hand and across her throat.

North Park permitted parents to use the parking lot when they dropped off or picked up their children from the day care center. There was, however, no written agreement or lease regarding the use of the parking lot between North Park or either Northeastern Illinois University, which leased the day care center, or the Yugoslavian Seventh Day Adventist Church, which owned the building.

North Park employed a security force of off-duty Chicago police officers to patrol its approximately 25-acre campus. Approximately 800-900 students attend North Park College, and about 300-500 students live on campus. There is generally no more than one officer on duty during a week other than the Director of Security, Harry Tannehill. The officers are required to patrol North Park campus, including the parking lot where Figueroa was attacked.

The Figueroas sued North Park for negligence in allowing the parking lot abduction and in failing to exercise reasonable care in the voluntary undertaking of providing security for persons using the parking lot. North Park moved to have the case dismissed before trial, claiming that it did not owe a duty to protect Figueroa from the criminal acts of third persons.

Under Illinois law, as a general rule, a person has no duty to protect another against criminal acts by third persons, unless there is a "special relationship" between the parties. The four special relationships giving rise to a duty to protect another from harm include: (1) carrier-passenger, (2) innkeeper-guest, (3) business inviter-invitee, and (4) one who voluntarily takes custody of another under circumstances so as to deprive the other of his normal opportunities for protection. The Figueroas contend that Suzanne was a business invitee on North Park's premises when she was abducted.

The court stated that a person is an invitee if he enters the premises of another by express or implied invitation; his entry is connected with the owner's business or with an activity that the owner conducts or permits to be conducted; and there is a mutuality of benefits or a benefit to the owner. In addition, a property owner owes a duty to protect a business invitee against criminal acts of third persons on his premises only when the harm is reasonably foreseeable.

North Park argued that Figueroa was a "licensee" of North Park rather than a business "invitee." A licensee is one who goes on the premises of another with the owner's express or implied consent, to satisfy his own purpose rather than for the mutual benefit of himself and the owner or for a business purpose.

In this case, the court pointed out that North Park had permitted patrons of the day care center to use the parking lot but did not invite them. Further, Figueroa did not enter the North Park parking lot for any reason connected to North Park's business. Figueroa merely used the parking lot as a convenient place to park while she dropped off her child at the day care center.

The court concluded that since Figueroa was a licensee, rather than an invitee, North Park did not owe a duty to protect her from criminal attack by a third party on its premises.

The Figueroas also alleged that North Park voluntarily undertook to provide security services and negligently performed the undertaking.

The court held that North Park's security department undertook to provide protection for those persons lawfully on the premises owned by North Park. So, although North Park had no general duty to protect against criminal attacks, it gratuitously undertook to secure protective services.

However, there was no evidence that any actions of North Park increased the risk of harm over that which existed without North Park's security measures. Further, there was no evidence that Figueroa's reliance upon the provision of security by North Park had harmed her.

In addition, the court noted that North Park's undertaking to provide security did not make it an insurer of the absolute safety of persons entering its property. The court held that the injury that occurred was not "foreseeable as likely to happen by a reasonably prudent person."

In sum, the court held that North Park did not owe a duty to protect Suzanne Figueroa from the criminal acts of third parties and, therefore, dismissed the case without a trial.

COMMENTARY:

The court noted that the distinction between invitees and licensees as affecting the duty owed by a property owner to entrants has been abolished by Illinois statutory law, not yet in effect at the time of the events of this case.

The plaintiff above also argued that she should recover pursuant to the "Good Samaritan Doctrine," which has been stated as follows:

One who undertakes, gratuitously or for consideration, to render services to another which he should recognize as necessary for the protection of a third person or his things, is subject to liability to the third person for physical harm resulting from his failure to exercise reasonable care to protect his undertaking, if (1) his failure to exercise reasonable care increases the risk of such harm, or (2) he has undertaken to perform a duty owed by the other to the third person, or (3) the harm is suffered because of reliance of the other or the third person upon the undertaking.

PSCLR – FEB 89 (5)

Convenience Store May Be Liable To Injured Customer For Negligent Design Of Its Parking Lot When Many Similar Accidents Have Occurred:

David Cohen and Lynn Cohen, his wife, v. Inez Schrider, Ariel Service Station, Inc., d/b/a Las Olas Chevron and The Southland Corp., d/b/a 7-Eleven, a foreign corporation licensed to do business in Florida, *No. 87-2207, in the District Court of Appeal of Florida, 4th District, decided October 26, 1988, rehearing denied November 21, 1988, reported at 533 So. 2d 859.*

While plaintiff David Cohen was using the telephone located outside of a 7-Eleven store, a car "lost its brakes" and struck and injured Cohen. Cohen sued Southland Corp., the owner of the 7-Eleven store, alleging: (1) that the parking lot was negligently designed; (2) that the defendant knew or reasonably should have known that the vehicles were able to move onto the sidewalk and endanger the lives of the customers; (3) that the defendant failed to install blocks, which are a common design feature of commercial parking lots; and (4) the defendant failed to warn store patrons of the dangerous condition of the parking lot.

The trial court determined that Cohen's injury was not foreseeable and dismissed the case without a trial. Cohen and his wife appealed.

To recover, Cohen had to prove that the injury that he suffered was likely to occur, given the design and condition of the premises. The court phrased the issue: "Was it reasonably foreseeable to the defendant-owner of the premises that a person in the plaintiff-invitee's position may be injured on account of the faulty design or condition of the premises?"

The appellate court stated that "unless it can be determined that no reasonable persons could differ in concluding that the accident was unforeseeable, then a jury should decide the question; not the judge."

The appellate court noted that the Loss Prevention Manager for 7-Eleven stated in pretrial testimony that approximately ten accidents similar to this one had occurred in the three preceding years. In addition, ten similar accidents occurred in the two years afterwards. The Loss Prevention Manager testified that 7-Eleven is looking for ways to prevent such accidents and has had discussions to determine whether parking bumpers would help.

Based upon this evidence, the appellate court held that a jury question had been raised regarding whether it was reasonably foreseeable to Southland that a car would "lose its brakes" and roll through the parking lot, onto the sidewalk in front of the store, and strike a person.

The appellate court declined to follow a twenty-seven-year-old Florida case, *Schatz v. 7-Eleven, Inc.*, 128 So. 2d (Fla. Dist. Ct. App. 1961) in which a convenience store was held not liable on similar facts. The court held that the passage of the twenty-seven years, and the knowledge and experience of 7-Eleven as testified to by its Loss Prevention Manager, rendered the prior decision obsolete.

Therefore, the appellate court reversed the trial court's decision dismissing the case against Southland and remanded the case for trial.

PSCLR—FEB 89 (12)

Parking Lot Owner Not Liable For Criminal Acts Of Third Party Without Knowledge Or Imputed Knowledge Of The Intended Criminal Conduct And The Power To Protect Against It:

Christopher E. Coblentz v. North Peters Parking, Inc., R. Guerico & Son, Inc., Sentry Insurance Company, John Doe and United States Fidelity and Guaranty Company, *No. 88-CA-0433, in the Court of Appeal of Louisiana, Fourth Circuit, decided October 11, 1988, reported at 533 So. 2d 98.*

Plaintiff Christopher E. Coblentz was shot in a parking lot operated by North Peters Parking, Inc. Coblentz sued the operator and owner of the park, among various other entities, for failing to protect him from criminal activity in the parking lot or to warn him of potential danger from persons attempting to make unauthorized collections in the lot. The case was tried before a jury, and the jury returned a verdict for the defendants. Coblentz appealed.

After work one evening, shortly before Mardi Gras, Christopher Coblentz and three friends, Brad Leppo, James Ray, and Marvin Lawhorn, had several drinks before driving to the French Quarter in New Orleans, Louisiana around 8:00 or 9:00 PM. They parked in a parking lot at 200 North Peters Street, entering the lot off Iberville Street. As they walked across the lot, a man approached and asked them for a $5 parking fee. Leppo testified that he thought the man was a hustler. The four men told the man that they would pay when they returned. Coblentz, who was walking behind the other three men, got into an argument with the man. The man shot Coblentz in the head, jumped into a truck and drove off. The assailant was never apprehended.

Donald Ware, who was cleaning a window in Shananagan's Restaurant across the street, witnessed the shooting. Ware testified that on more than ten occasions he had seen several different people collecting money at night for parking. He had seen people remove a cable blocking the driveway to open the lot at night. He testified that he had told the lot's daytime attendant, Alvin Turner, about the activity, as well as Roy Guerico Jr., vice president of North Peters Parking, Inc. He did not remember if he had spoken to Guerico before or after the shooting, however.

Alvin Turner, the daytime operator, testified that the lot closed when he left work at about 5:30 or 6:00 PM, or whenever it got dark, but that the signs listing the rates of the lot did not indicate the operating hours. The lot was enclosed by a cable running between posts and had two entrances which were closed by placing an additional post in the middle of each driveway. Turner stated that he had left the Iberville Street entrance open in the evening to allow the cars remaining in the lot to exit. The parking lot had no lights the night of the shooting.

Turner testified that the lock on the North Peters Street entrance had been broken once and the cable was down when he arrived for work. Turner acquired a larger lock and informed Guerico, but did not call the police. He testified that he was aware that some unauthorized person was in the lot at night collecting money, and that he had informed Guerico once or twice about the problem prior to the shooting. Roy Guerico, Jr., disputed this, and testified that he was not aware that someone had been illegally collecting for night parking until two months after the shooting. He stated that this was the only time Turner had mentioned the problem.

Raymond Pendleton, a security expert, testified that crime in the area of the lot increased at night and during Mardi Gras. During Mardi Gras that year, there were 26 calls to the police relating to the entire 200-block of North Peters Street. He testified that the parking lot's lighting was poor and the security inadequate. He stated that the parking lot should have been secured at night, with signs placed at each entrance.

A business establishment is under a duty to exercise reasonable care for the safety of its patrons. This duty does not extend to unforeseeable or unanticipated criminal acts by third parties. Only when the owner, management, or employees of a business have or should have knowledge of a third person's intended injurious conduct that is about to occur and which is within the power of the owner, management, or employees to protect against, does the duty arise. The duty can be discharged by calling the police at the time the proprietor knows or should reasonably anticipate that the third person poses a probable danger.

The court expressed doubt that Coblentz and his friends should be considered patrons of the parking lot. The court noted Leppo's testimony that he thought the man who asked for money was a hustler and that they did not have to pay to park. Even if Coblentz and his friends could be considered patrons of the lot, the duty to patrons "arises only when the owner has knowledge, or knowledge can be imputed, of a third person's intended criminal conduct and it is within the owner's power to protect" patrons from such conduct.

The court held that the information presented at trial was not sufficient to put the lot owner on notice that someone would commit a criminal act against the plaintiffs. Therefore, the court upheld the jury's verdict in favor of defendants as not "manifestly erroneous."

PSCLR – MAR 89 (9)

Business Owners Must Take Reasonable Steps To Protect Patrons From Foreseeable Criminal Conduct By Third Parties:

Opal Madden v. C&K Barbecue Carryout, Inc., and Ozzie Carr; James Decker, et al. v. Gramex Corporation, et al., *Nos. 70263, 70248, in the Supreme Court of Missouri, En Banc, decided September 13, 1988, rehearing denied October 18, 1988, reported at 758 S.W.2d 59.*

In this case, the Missouri Supreme Court addressed two separate cases involving a business owner's duty to protect patrons from criminal conduct by third parties. The first case involved Opal Madden, who was accosted by an unknown male while leaving C&K Barbecue in St. Louis, Missouri. The unknown male approached Madden and requested a jump start for his car or a ride home. Madden tried to get rid of the man, but he drew a gun, forced his way into Madden's car and kidnapped her. He drove Madden to another location where he physically and sexually assaulted her.

Madden sued C&K Barbecue for negligence seeking damages for personal injuries sustained as a result of the assault and kidnapping. Madden contended that C&K Barbecue failed to provide adequate security to protect its patrons and failed to warn its patrons of the danger present on the premises. Madden alleged that C&K Barbecue was the scene of numerous violent crimes over the three-year period preceding her assault. These crimes included six armed robberies, six strong-arm robberies, one assault, and one purse-snatching. The trial court dismissed Madden's claim without a trial.

The second case involved Gary and Donna Decker, who were assaulted after Christmas shopping at a shopping center comprised of a Schnucks Market and a Grandpa Pigeon's Department Store. When the Deckers returned to their car, they were abducted by two unknown assailants in the shopping center parking lot. The assailants murdered the Deckers after leaving the shopping center and dumped the Deckers' bodies in a vacant field in East St. Louis.

The Deckers' parents and son sued Schnucks Twenty-Five, Inc. and Gramex Corporation, the owners and operators of Schnucks and Grandpa Pigeon's, for the wrongful deaths. They alleged that the store owners were negligent for failing to provide adequate security in the parking lot area, for failing to protect Gary and Donna Decker from criminal assault, abduction, and murder, and for failing to warn the Deckers of the danger of criminal attacks in the parking lot.

These plaintiffs alleged a history of prior crime in the vicinity of the shopping center, including one armed robbery, one purse-snatching, and 45 assorted thefts. The plaintiffs contended that the store owners owed its patrons a duty to take reasonable precautions to ensure their safety. As in the *Madden* case, the court dismissed the action without a trial.

In order to proceed to trial on a negligence case, a plaintiff must allege facts which, if proven, demonstrate: (1) the existence of a duty on the part of a defendant to protect the plaintiff from injury, (2) breach of that duty, (3) injury to the plaintiff, and (4) a causal connection between the breach of duty and injury to the plaintiff. Generally, there is no duty under Missouri law to protect business patrons from the criminal acts of unknown third persons. However, a duty may be imposed under certain circumstances:

> The touchstone for the creation of a duty is <u>foreseeability</u>. A duty of care arises out of circumstances in which there is a foreseeable likelihood that particular acts or omissions will cause harm or injury."

The court noted that in the Madden case, Madden had alleged that C&K Barbecue was the scene of numerous violent crimes against persons during the three-year period immediately preceding Madden's assault. The court felt that these facts, if established at trial, would be sufficient to put C&K on notice of the possibility that patrons might be exposed to the danger of criminal acts of unknown third persons, including the danger of kidnapping and assault. The court held that the trial court should not have dismissed the case without a trial at which Madden could attempt to establish the alleged facts.

In the *Decker* case, the court noted that the plaintiffs had alleged a history of prior crimes <u>on the defendant's property</u>, including one armed robbery, one purse-snatching, and multiple thefts. The court felt that, even if established at trial, these facts might not be sufficient to establish a duty on the part of the store owners to protect their patrons from criminal conduct. However, the plaintiffs also submitted sworn testimony setting forth a lengthy list of prior crimes occurring at the <u>shopping center</u> over a three-year period. The list included four armed robberies, assault, assault with a deadly weapon, and flourishing a deadly weapon. The court felt that these facts, if established at trial, would be sufficient to establish a duty of care on the part of the store owners to take reasonable measures to protect their patrons from third party criminal attacks. A jury should have to decide if the property owners satisfied this duty of care.

COMMENTARY:

Two judges dissented from the court's decision in this case, stating that property owners should not be liable for the criminal acts of unknown third persons. One judge stated, "Today, the court opts for crime prevention in business localities through utilization of private police forces financed by consumers through higher prices." The second judge agreed, stating, "Every person in Missouri is going to find the cost of guards, security systems, and higher liability insurance premiums added to the cost of the groceries and products sold by the business establishments."

PSCLR–APR 89 (11)

Transit System Liable For Rape of Patron In Parking Lot When Prior Criminal Acts On The Premises Made The Criminal Activity Foreseeable:

Metropolitan Atlanta Rapid Transit Authority v. Allen, *No. 77201, in the Court of Appeals of Georgia, decided Oct. 4, 1988, rehearing denied, October 24, 1988, certiorari denied, December 1, 1988, reported at 374 S.E.2d 761.*

At 9:00 p.m. in late December, Allen was raped at knife-point in the parking lot of the Metropolitan Atlanta Rapid Transit Authority's ("MARTA") Hightower Rail Station in Atlanta, Georgia. Allen had parked her car in the MARTA parking lot that morning in order to take the train to work. She was raped when she returned to her car from the train late that evening. Allen sued MARTA to recover for the injuries sustained, alleging that MARTA had failed to provide adequate security in the parking lot. At trial, the jury awarded $250,000, and MARTA appealed.

On appeal, MARTA claimed that the trial court should not have allowed the jury to hear evidence that 10 armed robberies had occurred in the same parking lot prior to the year of Allen's rape, six of them in the same year. MARTA stated that the prior offenses did not occur under conditions substantially similar to those surrounding plaintiff's rape, since none of them involved rape and all of them occurred during daylight hours.

The court noted a proprietor must protect patrons from injury caused by the misconduct of third persons "if there is any reasonable apprehension of danger from the conduct of said persons or if injury could be prevented by the proprietor through the exercise of ordinary care and diligence." A proprietor will not be responsible for an act causing injury unless the proprietor had "reasonable grounds for apprehending that such criminal act would be committed." Such reasonable grounds can be shown by evidence that criminal acts have occurred on the premises in the past if there is a "substantial similarity" between the prior criminal acts and the criminal act giving rise to the plaintiff's claim. It is not required that the offenses be identical. "All that is required is that the prior [incident] be sufficient to attract the owner's attention to the dangerous condition which resulted in the litigated [incident]."

The appellate court concluded that the prior offenses were admissible on the issue of whether MARTA knew or should have known that there was an unreasonable risk of danger from criminals at the Hightower Station. The court noted that the occurrence of criminal activity in broad daylight would clearly suggest the likelihood that patrons using the parking lot would be exposed to danger from criminal assault in the darkness. The court stated that the prior offenses established a pattern of increasing criminal activity at the Hightower Station.

The court also held that the incident reports concerning the prior offenses were admissible to show that the offenses had been reported to MARTA, placing MARTA on notice of the criminal activity.

The appellate court rejected MARTA's further contention that the trial court should have admitted evidence of crime statistics throughout the MARTA system and within the general vicinity of the Hightower Station. MARTA asserted that the evidence should have been admitted to show the adequacy of MARTA's security system as evidenced by the extremely low incidence of crime throughout the system and at Hightower, even though the Hightower Station is situated in a very high crime area. The court stated, however, that "MARTA's diligence on the occasion in question could no more be established by comparing the crime rate on its rail system and with the crime rate in the surrounding community than could a driver's exercise of diligence on a particular occasion be established by proof that he had a better-than-average driving record."

The court also rejected MARTA's argument that Allen "was well aware of the danger of being out alone in the City of Atlanta at night and yet voluntarily chose to undertake those risks." Therefore, the jury should have considered a reduction of Allen's recovery in proportion to her own negligence. The appellate court noted that the trial court did instruct the jury that "if, by the exercise of ordinary care, [Allen] could have avoided the consequences of [MARTA's] negligence if any," Allen could not recover for her injuries.

The court noted that Allen certainly had no indication that a rapist was lurking near her car and that her mere use of the MARTA rail system at night could not be considered presumptively negligent. The court would not accept the proposition that Allen's duty to exercise ordinary care for her own safety required her either to abandon her car in the lot overnight or to make other arrangements to retrieve it and return home.

The appellate court, thus, upheld the trial court's decisions in the conduct of the trial and affirmed the judgment in favor of Allen for $250,000 in damages.

COMMENTARY:

The Metropolitan Atlanta Rapid Transit Authority has its own full-time security force which has authority for patrolling the subway stations proper, and also the adjacent parking lots which are owned and managed by MARTA. The assault in the above case took place on property under the control of the defendant.

Once again, the issue of foreseeability was found to be determinative of the finding of liability against the defendant. It is imperative that security managers conduct continuous and ongoing examinations of the adequacy of the security measures employed on the properties under their control in view of the status of conditions then existing and the increasing or decreasing levels of criminal activity in the area, in order to convince a judge and jury that all reasonable security measures have been undertaken under all of the controlling facts and circumstances of the situation.

PSCLR – JUN-JUL 89 (5)

Restaurant Not Liable For Criminal Assault Of Patron In Parking Lot, Notwithstanding Evidence Of Prior Violent Crime On Premises; Expert Testimony Regarding "Premises Security" Excluded As Unnecessary:

Jerral Pickle v. Denny's Restaurant, Inc., *No. WD 40480, in the Missouri Court of Appeals, Western District, decided December 20, 1988, motion for rehearing and/or transfer to Supreme Court denied January 31, 1989, reported at 763 S.W.2d 678.*

Early one morning, plaintiff Jerral Pickle was shot in the chest while in the parking lot of the Denny's Restaurant located at 1600 Broadway, Kansas City, Missouri. Pickle had driven to the restaurant with his wife in order to apply for a job. The parking lot where Pickle parked his truck was reserved for restaurant patrons and was controlled and maintained by Denny's Restaurant, Inc. Before exiting his vehicle, Pickle was accosted and robbed by an assailant armed with a shotgun. The assailant demanded that Pickle turn over his truck keys. When Pickle refused, he was shot in the chest at point-blank range, causing severe, permanent, and disabling injuries, requiring surgery on two occasions. Pickle suffered a shattered sternum and the loss of a lung. He incurred medical expenses of approximately $50,000.

Pickle sued Denny's Restaurant, Inc., for negligence in failing to make the parking lot reasonably safe for business visitors. The case was tried before a jury, and the jury found in favor of Denny's Restaurant, Inc. Pickle appealed, alleging several errors by the trial court.

In the two years and eleven months prior to Pickle's shooting, there had been thirteen violent crimes against persons on the Denny's Restaurant premises, including the parking lot. Pickle was allowed to present evidence at trial of nine of the thirteen crimes which occurred within two years and three months of his assault. These nine crimes included one armed robbery and two physical assaults that were perpetrated against employees of Denny's and of which Denny's had actual knowledge.

One week prior to Pickle's attack, two young men sitting in a tow truck in the parking lot were robbed at gunpoint, kidnapped from the lot, and sodomized. One or more Denny's employees observed the robbery in progress.

Pickle alleged that, despite the history of violent crimes against persons at Denny's, Denny's had not hired an armed security guard or a security service to patrol and secure the parking lot. Pickle alleged that the only safety measures taken at Denny's were lighting in the parking lot area, a fence with a gate that could be locked to close off the back alley access to the parking lot, and an off-duty police officer employed to work occasional Friday and Saturday nights to maintain order inside the restaurant.

On the night before Pickle was shot, Denny's employees had failed to lock the back gate, and the weekend security guard was not on duty. Pickle's assailant entered and exited through the back gate to the parking lot.

Under Missouri law, whether business owners have a responsibility to protect their patrons from criminal attack by unknown third parties depends upon the facts and circumstances of a given case. "The touchstone for the creation of a duty is foreseeability. A duty of care arises out of circumstances in which there is a foreseeable likelihood that particular acts or omissions will cause harm or injury." Evidence that numerous violent crimes have occurred on or about the premises is sufficient to establish a duty on the part of the business owner to take reasonable measures to protect patrons from third party criminal attacks. Whether the business owner has satisfied that duty is a question for determination by a jury.

In this case, Pickle appealed from the jury's verdict against him, claiming that the trial court should have allowed him to present evidence of all thirteen, rather than merely nine, crimes committed on the Denny's premises prior to his assault. The four crimes about which the trial court did not allow testimony, however, were dissimilar from the crime in this case, in that they involved robbery of the cash register inside Denny's or the attack of a person who was walking through the alley behind Denny's and was not a Denny's patron. The appellate court held that the four excluded incidents were merely cumulative and had little, if any, probative value in the case. The awareness of Denny's employees of previous crimes perpetrated in the parking lot was established in this case by an abundance of evidence. The failure to admit testimony on four more crimes did not prejudice Pickle's case.

Pickle also asserted on appeal that the trial court should have instructed the jury that knowledge gained by an employee while the employee was performing his job duties, is imputed as knowledge of the employer when the knowledge pertains to a matter which might be expected to affect the employer's interest. At trial, Pickle elicited testimony from former Denny's managers that the Denny's management was aware of crimes occurring in its parking lot. The court found that Denny's responded by taking security measures to prevent such crimes, including lighting and fencing the parking lot and hiring the security guard on Friday and Saturday nights. The court assumed that the jury had the intelligence of "reasonable men" and was capable of transferring knowledge to Denny's management because the evidence showed that management was, in fact, aware. Therefore, the court held that there was no prejudice in failing to instruct the jury concerning imputed knowledge.

On appeal, Pickle also complained that the trial court had erred in refusing to permit expert testimony on the subject of premises security. The trial court had concluded that the subject of premises security was within the common knowledge of the average lay juror and did not require expert testimony. The appellate court noted that the methods necessary to secure the Denny's parking lot need not be sophisticated. Denny's was not required to provide "the best security plan known to mankind; instead [Denny's] was held to a standard of 'ordinary care'." Denny's was required to use the care in securing its parking lot that "an ordinarily careful and prudent person would use under the same or similar circumstances." Therefore, the appellate court agreed that the testimony of a security expert was not necessary at trial.

Finally, Pickle argued on appeal that the trial court should have allowed a witness who was robbed in the Denny's parking lot to testify that, after he was robbed, he was kidnapped and sodomized. Pickle claimed that this testimony would counter Denny's assertion that had Pickle turned over the keys to his truck, he would not have been shot. This witness had turned over his keys and was kidnapped and sodomized. Without this evidence, Denny's was able to argue that Pickle caused his own injury by failing to turn over his own keys.

The appellate court, however, held that other than the "potential" to counter Denny's argument about Pickle's failure to cooperate with his attacker, the evidence was not probative. While the appellate court stated that it disagreed with the trial court's decision to exclude the testimony regarding the sodomy, the exclusion of the evidence did not justify reversal of the jury's verdict. The appellate court, thus, affirmed the judgment in favor of Denny's Restaurant.

COMMENTARY:

An expert will be allowed to testify as to his opinion on a particular matter relevant to the issues before the court when and if the expert has the experience and expertise which gives him some special knowledge above and beyond the knowledge the average layperson would have on the subject.

In addition, the expert's opinion may be relevant and necessary in the issue of causation which may not be readily apparent to the lay person. However, in the above case, the court determined that the expert testimony which the plaintiff intended to present to the jury was not "sophisticated" nor otherwise beyond the realm of knowledge of the average lay person juror, and therefore would not in any way assist the jury in determining whether the defendant Denny's Restaurant was or was not negligent in fulfilling its obligation to its customers of providing reasonable security precautions.

The Missouri court here took the conservative view of the use of expert witnesses – most courts have recognized the particular expertise and experience of qualified security professionals which goes well beyond the normal range of knowledge on the subject which the average juror would be expected to know. Nevertheless, the case does emphasize the need for continued training on the part of the security professional who holds himself or herself out as an expert witness. He or she must be prepared to convince the judge that his knowledge, education, and experience goes beyond that of the average layperson, and encompasses sophisticated facts, techniques, theories, etc., not commonly known to the average layperson. Only if this test is met will the trial court permit an expert to testify.

PSCLR – AUG 89 (2)

Store Owner Not Required To Provide Security Guards To Protect Patrons From Criminal Assault In Parking Lot, Notwithstanding That Such Security Was Previously Provided:

Evelyn Theis v. Malik Abduloor, individually and Malik Abduloor, d/b/a Serv-Well Vintage Shop, jointly and severally, *Docket No. 101539, in the Court of Appeals of Michigan, Submitted August 16, 1988, decided October 6, 1988, released for publication February 24, 1989, reported at 435 N.W.2d 440.*

After plaintiff Evelyn Theis left the Serv-Well Vintage Shop, owned by Malik Abduloor, she was attacked and mugged by two people who held her, hit her over the head, and dragged her across the parking lot. Theis sued Abduloor, contending that Abduloor had reason to know or should have known of the high incidence of criminal activity in the area and on the premises of his shop. She alleged that, knowing it was a high crime area, Abduloor breached his duty to her to keep the premises safe, provide adequate lighting, to make sure that competent security was present, to maintain constant control and surveillance over the premises, and to come to her aid when she was mugged.

Theis alleged that, after the incident, she was told by the owner or manager of the shop that the shop was located in a high-crime area and that it was their custom to escort women from the shop to their cars. Theis also stated that she had seen store personnel near the door watching over the parking lot in the past, but that she had not seen anyone watching on the night she was mugged.

The trial court found that Abduloor did not owe a duty to protect Theis from criminal attack in the parking lot. Theis appealed.

The appellate court held that because of the special "invitor-invitee" relationship, a business owner owes a duty to its patrons to exercise reasonable care to protect them from an unreasonable risk of harm. This duty, however, is not absolute. Whether the owner's conduct constituted reasonable care is one the court should determine as a matter of law. As a matter of law, the duty of reasonable care a merchant owes his patrons does not extend to providing armed, visible security guards to protect customers from criminal attack. "The merchant is not an insurer of the safety of his [patrons], and for reasons of public policy he does not have the responsibility for providing police protection on his premises."

Applying the above law, the appellate court held that Abduloor did not owe Theis a duty to provide security guards to protect her from criminal assault by unknown third parties, <u>even if security had been provided in the past</u>. The court, thus, upheld the trial court's dismissal of the case against Abduloor.

COMMENTARY:

The ruling in this case should give some comfort to those security directors, retailers, business owners and operators, etc., who have expressed concern that discontinuing certain security measures (e.g., cancelling a security guard service) would automatically make them liable if a subsequent loss occurs. Such is not necessarily the case. Just as an increasing crime rate in and around a particular property may indicate a need for the addition of security measures, likewise the decrease of risk can warrant reduction.

PSCLR – SEP 89 (17)

Business Owner Held Liable For Injuries Suffered By A Non-Customer With Whom There Was A "Business Relationship":

Jennifer Smith v. Walgreens Louisiana Company, Inc., ABC Insurance Company, Mutual Savings Life Insurance Company, XYZ Insurance Company and John Doe, *No. 88-CA-1161, in the Court of Appeal of Louisiana, Fourth Circuit, decided April 13, 1989, reported at 542 So. 2d 766.*

Jennifer Smith was an employee of WNOL-TV in New Orleans. WNOL had arranged a deal with Walgreens Drug Store, which was located near the TV station, so that WNOL employees could park in the Walgreens parking lot in exchange for free publicity for the drug store. One evening at around 6:00, Smith was returning to her car, which was parked in Walgreens' lot. At the time, she was carrying several bags and was walking under an umbrella because it was raining. When she opened the door to her car, she claimed that she was shoved into the car by a man she had passed in the parking lot. According to Smith, the man forced her to drive to another location where he robbed and raped her.

Smith filed this case against Walgreens. In the suit, she claimed that Walgreens had a duty to provide her with a reasonably safe place to park under the agreement between Walgreens and WNOL. She claimed that Walgreens had breached this duty and was therefore responsible for the assault she had suffered.

The trial court, although there was some dispute as to the facts, assumed that there was in fact an arrangement between Walgreens and WNOL regarding parking and that Smith had permission to park in the lot. Even after assuming these facts to be true, the trial judge rejected Smith's case on the legal grounds that Walgreens had no duty to provide her a safe place to park because she was not a customer of Walgreens and had not parked in the lot for the purpose of doing business with the drug store. Basically, the judge ruled that there was no duty because she was not a "business invitee."

The appeal court reversed the ruling and held that Smith's case should go to trial on the merits. The appeal court stated that under Louisiana law, a business has two duties to its invitees. First, the business must take reasonable steps to protect those it invites onto its premises from the predictable risk of assaults. Second, if a business owner has voluntarily undertaken the duty of protecting invitees, it must perform that duty with due care.

The first major issue in this case is whether the attack on Smith was "foreseeable." Under Louisiana law, the duty to protect invitees does not extend to a requirement that a business protect its invitees from unforeseeable attacks by third parties. The appeal court explained that one way of proving that a specific crime was foreseeable, and therefore the responsibility of the business owner, is to consider evidence of past crimes in the same vicinity or on the same premises. In this case, the court reasoned that there was not enough evidence one way or the other so as to prove that the attack on Smith was either foreseeable or unforeseeable. Based on that, the court ruled that the case should go back to the trial court so that such matters could be resolved.

The second issue that the appeal court dealt with was whether Smith was a business invitee. Unlike the lower court, the appeal court stated that:

> [I]f the plaintiff in the instant case had permission to park in the lot because her employer was extending consideration to the defendant, the relationship between the parties is very similar to the relationship between merchants and patrons since the merchant is gaining a benefit.

The court concluded that if Smith had permission to park in the lot, then Walgreens had the same duty to protect her from attack as it had for its customers. Again, the appeal court held that there was insufficient evidence to determine whether in fact Smith had permission to use the lot. The court therefore held that the case should return to the lower court for a trial on those issues.

Recreation/ Entertainment

PSCLR–JUL 82 (7)

Stadium Owner's Duty To Warn A Baseball Game Spectator Of Potential Hazards Questioned By Appellate Courts:

Kathy Lynn Falkner and David Falkner v. John E. Fetzer, Inc., A Michigan Corporation, *Docket No. 47033, Court of Appeals of Michigan, decided February 19, 1982, reported at 317 N.W.2d 337.*

Can a stadium owner be held liable for damages when a spectator is injured at a baseball game by a batted ball? The trial jury in this case said yes, and found the stadium owner's negligence to be the proximate cause of the spectator's injuries, and awarded $258,850 in damages to the plaintiff.

Kathy Falkner was struck by a batted ball while attending a baseball game at Tiger Stadium, and she filed suit against the stadium owner for its negligence in failing to make the stadium reasonably safe for the spectators. Her specific allegations of negligence included:

> (a) Failing to provide protective screening in the area in which Plaintiff KATHY FALKNER was a spectator.
>
> (b) Failing to inform Plaintiff KATHY FALKNER of the availability of seats in areas protected by screening.
>
> (c) Failing to offer Plaintiff KATHY FALKNER a choice between protected and unprotected seats.
>
> (d) Failing to provide a sufficient number of protected seats in areas of particularly high hazard.
>
> (e) Failing to warn Plaintiff KATHY FALKNER by signs (or other reasonable means) of the high hazard to her from batted balls.

The trial court dismissed the first four allegations. The fifth issue was submitted to the jury, and the jury found the defendant stadium owner to be liable for its negligence in failing to warn patrons of danger.

The Michigan Court of Appeals recognized that there is generally no duty to warn of a risk as obvious as being hit by a batted ball at a baseball game. But the evidence presented by the plaintiff showed the magnitude of the risk to be far greater than commonly believed, and the question concerning duty to warn was properly submitted to a jury.

The Court of Appeals also held that a disclaimer printed on the ticket and an announcement at the beginning of each game would not necessarily constitute an adequate warning. The court ruled that the adequacy of a warning is a jury question.

In the final analysis, however, the Court of Appeals reversed the judgment in favor of the plaintiffs and held the trial court should have granted a directed verdict on all counts based on the Falkners' failure to show that Mrs. Falkner would have taken precautions to prevent her injury if in fact she had received proper warning of the risk. The Court of Appeals summarized the elements necessary to prove negligence as being: "(1) that the defendant owed a duty to the plaintiff; (2) that the defendant violated that duty; (3) that the defendant's breach of duty was a proximate cause of the damages suffered by the plaintiff; and (4) that the plaintiff suffered damages."

Since the element of proximate cause was missing, according to the majority opinion, the plaintiffs were not entitled to judgment based on negligence, and their award was set aside.

COMMENTARY:

While the stadium owner has won this particular round in the Court of Appeals, this case warrants further observation. It should be noted that this was a 2-to-1 decision by one division of the Court of Appeals and an application for leave to appeal further has been filed in the Michigan Supreme Court.

PSCLR – AUG 82 (10)

Duty Of Property Owner To Warn Of Hidden Dangers:

Mary Grant, as Trustee for the Heirs of Gregory Grant, Deceased v. The City of Duluth, a municipal corporation, and Jeff Vollman, *Docket No. 81-1117, in the United States Court of Appeals for the Eighth Circuit, on appeal from the United States District Court for the District of Minnesota, decided March 11, 1982, reported at 672 F.2d 677.*

A property owner's duty of care to prevent injury to trespassers was explored in this appeal of a judgment for damages of $224,531 awarded a widow for the wrongful death of her husband who was killed while trespassing on amusement park property after hours.

Twenty-two-year-old Gregory Grant and three friends were celebrating graduation at a party on grounds adjacent to an amusement park which had been closed and locked for the evening. The park was leased by Jeff Vollman from the city of Duluth, Minnesota. After drinking "quarts" of beer, Grant and a friend ignored the no-trespassing signs and climbed over the 6'10" solid chain link and barbed wire fence to go down a thirty-foot slide. The men raced to climb the scaffolding of the slide instead of using the stairs. Grant went first, colliding with a chain midway down the slide. His resulting injuries proved fatal three days later.

At the trial, the evidence showed Vollman was concerned about trespassers and had taken "about all the measures he was able to take" to prevent unauthorized entry and use of the park. In addition to the fence and at least ten "no-trespassing" signs on the property, two chains were placed on the slide after the park closed – one at the top and one in the middle. A bag was hung from one of the chains to enhance visibility.

Vollman further testified there was a sign in front of the slide which stated, "Danger – this slide is chained and locked nightly for your protection." Although there was a dispute as to how well-lighted the park was on the night in question, Vollman testified that he had previously warned a group of young people, which included Grant, to stay out of the park and not to use the slide because it was chained and locked.

The jury determined the proper amount of damages to be $349,330 and then applied the Minnesota law as to comparative negligence. They found 30 percent of the negligence attributable to Grant and 70 percent to Vollman, and therefore awarded $224,531 to Grant's widow.

The Court of Appeals reversed the award; it was satisfied that Vollman had taken sufficient steps to keep people out of the park after hours and that Grant had deliberately ignored the warnings. The only remaining question was whether sufficient steps were taken by the possessor of land to warn trespassers of the particular dangers and risks which were involved, including the chain on the slide.

The Minnesota Jury Instruction Guide Rules concerning "Duty of Possessor to Trespassor – Injury Caused by Condition of the Premises" state:

> A possessor of land is under no duty to a trespasser to maintain his premises in a reasonably safe condition. If, however, a possessor of land knows or from facts known to him has reason to know that trespassers regularly use certain portions of his premises and if he creates or maintains an artificial

> condition on his premises which he knows is likely to cause death or seriously body harm, and which is such that he has reason to believe that trespassers would not discover it, then he has a duty to use reasonable care to warn a trespasser of the danger or risk involved, unless the trespasser is already aware, or from facts known to him should have been aware, of the condition of the premises and the risk involved.

The Court of Appeals believed the jury could have been misled as to the proper duty of the land possessor.

> The jury could have reasonably found, from the description of the chain across the slide, that Vollman created or maintained an artificial condition on his premises which he knew was likely to cause death or serious bodily harm...[or] that Vollman had reason to believe that trespassers would not discover the dangerous or artificial condition. The proper question for the jury then to ask...was whether Vollman used 'reasonable care to warn' of the danger or risk involved. Even if Vollman knew that trespassers regularly entered the amusement park and used the slide, he did not have a duty to maintain the premises in a reasonably safe condition.

Minnesota utilizes a "comparative fault" standard in negligence actions. "Recovery is permitted only from persons whose negligence is greater than that of the person seeking recovery." The Court of Appeals concluded that "Grant's negligence and reckless conduct far exceeds the claimed negligence of defendant Vollman." The court was convinced that the jury misunderstood the standards applicable to Vollman's duty of care owed Grant, reversed the judgment and remanded the case for a new trial.

COMMENTARY:

The Grant case shows that a property owner cannot turn his back on dangers which he knows about, or should know about, without using reasonable care to warn even a trespasser of the risk. Simply placing a "no trespassing" sign would not be sufficient to warn of the specific risk of a hostile animal or other particular risk.

. Do your routine procedures include a regular property inspection so that "hidden dangers" would be discovered?

. Have your personnel been alerted to look for and report potentially dangerous conditions?

. Do you maintain an effective follow-up system to insure that reports do not go unheeded and problems which have been reported are corrected or conspicuous warnings of the particular risk posted?

PSCLR – NOV-DEC 82 (16)

The State, As Owner/Operator Of Stadium, Has Duty To Provide Reasonably Safe Spectator Seating Area:

Richard Benjamin, as parent and natural guardian of Thomas Benjamin v. The State of New York, *Claim No. 64784, in the Court of Claims of New York, decided July 16, 1982, reported at 453 N.Y.S.2d 329.*

On November 16, 1979, Thomas Benjamin, an 11-year-old youngster, was struck and injured by a puck while watching a hockey game at the State University arena in Oswego. He had been seated in the sideline seats near the players' bench. The boy sustained a depressed skull fracture requiring surgery, followed by a period of convalescence, and partial disability lasting about one year. This suit followed.

The court ruled that the State of New York, like any other owner or occupier of land, is under a duty to exercise "reasonable care under the circumstances" to prevent injury to those who come to watch games played at its facilities. Since the State provided spectators with sideline seating near the players' bench, it had "a duty to make certain that such an area was reasonably safe for its intended purpose."

An expert testified that in similar facilities it was customary to protect the area near the players' bench, or to eliminate spectator seating in that danger zone. The State here had done neither, and it was the expert's opinion that the State failed to provide adequate protection.

The court found that it was no defense that no accident of the same kind had ever happened before, reasoning that such an accident was reasonably to be anticipated.

Young Mr. Benjamin recovered $24,000 as compensatory damages for his injuries.

COMMENTARY:

The question of contributory negligence, or assumption of the risk, was also addressed by the court. However, the judge ruled that a young boy of Thomas' age, intelligence, and development would not necessarily have fully appreciated the danger and thus could not be said to have assumed the risk.

PSCLR – FEB 83 (5)

Movie Producers And Theater Owners Not Held Responsible For Violent Acts Of Third Persons Allegedly Caused By Content Of Film:

Tony Bill et al. v. The Superior Court for the City and County of San Francisco. Jocelyn Vargas, etc., et al., Real Parties in Interest, *A017025, in the Court of Appeal, First District, Division 2, decided December 9, 1982, reported at 187 Cal. Rptr. 625.*

Jocelyn Vargas, a teenage girl, was shot as she and her friends walked down a San Francisco street only minutes after seeing "Boulevard Nights," a movie containing scenes of graphic violence, at the Alhambra Theater. Should the theater and the movie's producer be held liable for failing either to warn the girl and other patrons that such a movie might attract violent people, or to provide security in and around the theater? Jocelyn Vargas felt that she was victimized by an unknown assailant who "was a member of the general public prone to violence ... who had been attracted to said Alhambra Theater by the showing of said violent movie ... . "

The degree of responsibility owed to the public by movie producers, theater owners, and even television networks has been a source of some controversy in recent years. In this case, "Boulevard Nights" producers Tony Bill and Bill Benenson, their production company Eastside Productions, Inc., director Michael Pressman, and the distributor Warner Brothers, Inc. ("the defendants") were sued by Jocelyn Vargas and her mother for wounds suffered by Jocelyn in the shooting.

The defendants requested that the court grant them summary judgment because there were no questions of fact, only questions of law. Their request was denied and they appealed to the California Court of Appeal which ruled that summary judgment should be granted in favor of the defendants.

Jocelyn Vargas' suit alleged that the defendants knew that "Boulevard Nights" was extremely violent, and that it would attract persons "who were prone to violence and who carried weapons ... [and] would, or were likely to cause grave bodily injury upon other members of the general public at or near the showing of [the] movie." The Vargas' suit further alleged negligence on the part of the the defendants for failing to warn her and to protect her from the dangers which she contended were caused by showing the movie.

Evidence was presented that, before the film was released, Pressman and Benenson had discussed the possible need for security at some of the theaters showing "Boulevard Nights." Benenson, in fact, talked to Warner Brothers about placing a guard or guards outside the Los Angeles theater where the movie was to open. Apparently, these discussions were brought about by some violent incidents at showings of another film, "The Warriors." The two movies had been linked in some newspaper stories, because of similarities in subject matter.

Vargas contended that these conversations proved that the defendants were aware that there might be violent incidents at showings of "Boulevard Nights," even though advertisements for the movie implied that it could be safely attended. This implication of safety supposedly arose out of advertisements which invited the public to attend the movie.

The Court of Appeal felt that placing a responsibility on movie producers to consider the possible effect of a movie's subject matter on certain members of society came dangerously close to censorship. That is, if a movie producer who was planning to make a movie about violence would be held liable for the acts of the viewers, then the producer would have to consider

> ... the potential for liability to patrons for acts of violence on the part of persons over whom the producers would have no control. ... [Likewise, if] they were held to have a duty to warn potential patrons of the risk of attending their movie, they would have to anticipate that the warning would deter substantial portions of the public from attending it – as Jocelyn Vargas says she would have been deterred if a warning had been given. And if, under such circumstances, they were held to be responsible for providing security protection at and in the vicinity of every theater at which the movie is shown, including public streets, the attendant costs might be substantial indeed.

Exposure to such potential liability would have, according to the court, "a chilling effect upon the selection of subject matters for movies," and would permit persons who were offended by certain subjects "to dictate, in effect, what is shown in the theaters of our land." The court reiterated:

> ... the general First Amendment principle that when speech is of such a nature as to arouse violent reaction on the part of the lawless, the first obligation of the government is to maintain the peace and enforce the law, ... and not to silence the speaker.

The court conceded that there might be certain situations where the courts would impose the responsibility to warn or to protect persons without violating First Amendment principles. In this case however, the court found that the defendants did not have a duty to warn Vargas and other potential patrons that it might be dangerous to attend the film: "It seems unlikely that a volume of such warnings [in advertisements or on theater marquees] would contribute measurably to the effective protection of the public."

Therefore, the California Court of Appeal granted summary judgment to the producers, the director, and the distributor of "Boulevard Nights."

COMMENTARY:

We are all familiar with those cases where a retailer has been held liable for injuries to a patron caused by the violent acts of a third party. Generally, the retailer is held responsible for the resulting injuries to the patron if it can be shown that the violence was "foreseeable" by the retailer, and that the retailer failed to warn the patron of the known danger or to take reasonable precautions to protect the patron from the danger.

The primary problem with the above case seems to be that the plaintiff blamed her attack on the content of the movie itself. What if she had not done so? What if, to speculate a bit, an entirely innocuous movie were being shown, such as "Heidi" or "Bambi" or a similar "non-violent" film? And, to take the hypothetical one step further, what if the theater owner was aware of ten violent attacks on patrons immediately outside the theater within the previous month? There is little doubt that a property owner would be responsible for any attack caused by his failure to warn patrons of the known and foreseeable danger and to take reasonable steps to protect them. This is so regardless of the particular product or service being sold by the owner.

However, the interesting issue presented in the above case required the court to retreat somewhat from the traditional rulings rendered in cases such as mentioned in the hypothetical. The court analyzed the competing social interests and benefits and decided that the higher good was to allow movie producers and theaters to market movies regardless of content and regardless of any tendency to incite violence.

PSCLR – FEB 83 (11)

Patron On Amusement Park Ride Was Owed Duty Of Ordinary Care Only, Not Extraordinary Care, As Ride Was Not A "Public Conveyance":

Timothy R. Harlan, b/n/f Aaron A. Harlan v. Six Flags Over Georgia, Inc., *No. 38875, in the Supreme court of Georgia, decided November 30, 1982, reported at 297 S.E.2d 468.*

The standard of care owed passengers of amusement park rides was explored in this case.

On April 30, 1977, Timothy Harlan was a patron of "Six Flags Over Georgia" amusement park in Atlanta, Georgia. After paying the admissions price which allowed patrons to ride any of the park's "amusement devices," Harlan rode "The Wheelie." Because he was allegedly injured during the ride, Harlan brought a lawsuit against Six Flags Over Georgia, Inc.

Harlan contended that amusement park rides, such as "The Wheelie" are "public conveyances" making Six Flags "a carrier owing its passengers an extraordinary duty of care under [Georgia law]." In support of his contention that Six Flags was liable for his alleged injuries, Harlan cited Georgia statutes which (1) defined passengers as "those persons who travel in some public conveyance by virtue of a contract, express or implied, with the carrier as to the payment of the fare," and (2) required that public conveyances "exercise extraordinary diligence to protect the lives and persons of its passengers." However, the Georgia

Supreme court found the standard of care owed to passengers of amusement park rides to be ordinary, not extraordinary, care.

In its decision, the court pointed out that passengers of trains, buses, railroads, and even elevators are traveling from one place to another with the expectation of being carried "safely, securely and without incident to their destination." Passengers of amusement park rides, on the other hand, "intend to be conveyed thrillingly to a place at, or near to, the point they originally boarded, so that carriage is incidental. There is no transport involved ... ." Therefore, a ride such as "The Wheelie" would not fall within the definition of a public conveyance under applicable Georgia statutes.

PSCLR – MAY 83 (10)

Amusement Park With Fence Too Short To Prevent Access To Water By Small Children May Be Liable For Child's Death In Spite Of Parent's Foreseeable Negligent Supervision Of Child:

Marietta Goode, as personal representative of the Estate of Joel Goode, on behalf of Marietta Goode, individually, and Harry Goode, individually, v. Walt Disney World Co., a Delaware corporation; Columbia Casualty Company, and Lloyds of London, *No. 81-1431, in the District Court of Appeal of Florida, Fifth District, decided December 22, 1982, rehearing denied January 31, 1983, reported at 425 So. 2d 1151.*

Four-year-old Joel Goode was visiting Walt Disney World near Orlando, Florida, on August 11, 1977. At about 11:20 PM he was playing in a fenced grassy area near where his mother and several other relatives sat in an ice cream shop. Joel had climbed over the fence to get inside and was joined by his 13-year-old cousin and two of his brothers. Eventually the other children left, and the cousin joined Joel's mother. They looked over a little later and Joel was gone. When they could not find him, Disney personnel were notified that the small boy was missing.

It was almost 3 hours before his body was discovered in one of the park's waterways about 150 feet from where he was last seen playing. The 5-foot deep waterway was used for a Swan-boat ride and ran around the Disney castle. The castle's moat was separated from the grassy area where Joel had been playing by a walkway, another fence and an area of grass and shrubbery. The ice cream shop where Joel's family sat was on the far side of this walkway.

Mrs. Goode sued Disney for negligence in maintaining its premises by failing to build a fence which young children could not climb and by creating a hazard for patrons by negligently designing the moat. Disney denied liability and alleged in turn that Mrs. Goode had not properly supervised her child.

Disney's motion to dismiss the suit was granted by the trial court which ruled that the only evidence as to the cause of the child's death was that his mother failed in her duty to control the child. Mrs. Goode appealed.

The Florida Court of Appeals reversed the judgment which had been granted to Disney World. It noted that the fence around the moat was in fact shorter than the one designed for that purpose. If indeed there was negligence in using a shorter fence, the court determined that a jury could find that injury to patrons was a foreseeable result, particularly in a place such as Disney World where millions of children visit each year. Furthermore, the court ruled that any negligent supervision by the mother could not serve as an intervening cause to shield Disney from liability, since her negligent supervision, "similar to that of thousands of other parents which occurs annually at the Magic Kingdom," was foreseen by Disney World. The court continued, "Indeed, the evidence was that at the Kingdom's Lost Children Center there were 11,420 'lost children' reported in 1977." If an intervening cause is foreseeable, it can not insulate a defendant from all liability.

The case was reinstated and will proceed to trial.

PSCLR – JAN 84 (9)

Operators Of Public Entertainment Places Owe Duty Of Reasonable Care To Protect Patrons From Third Persons:

James Bearman and Christenna Bearman v. University of Notre Dame, *No. 3-882A223, in the Court of Appeals of Indiana, Third District, decided September 22, 1983, reported at 453 N.E.2d 1196.*

The plaintiffs Christenna Bearman and her husband attended a Notre Dame football game on October 27, 1979, leaving shortly before its conclusion. As they were crossing through a parking lot and approaching their car, they saw two men who were fighting and were apparently drunk. After one of the men fell, the men split up and walked away from each other, one of them walking past the Bearmans. Minutes later, the man fell into Mrs. Bearman from the rear. She was knocked to the ground, and suffered a broken leg. There were no security personnel in the area at the time of the incident.

The Bearmans sued the University of Notre Dame, contending that it had not taken proper precautions to protect Mrs. Bearman as an invitee. The University argued that it had no notice or knowledge of any particular danger to Mrs. Bearman, and that the University could not be held liable for the unanticipated acts of third persons. The trial judge agreed and granted the University's motion for judgment, and the Bearmans appealed.

The appellate court noted that

> ... the operator of a place of public entertainment owes a duty to keep the premises safe for its invitees [and] this duty includes a duty to provide a safe and suitable of ingress and egress ... and a duty to exercise ordinary and reasonable care to protect a patron from injury caused by third persons. ... However, the inviter is not the insurer of the invitee's safety. ... Before liability may be imposed on the inviter, it must have actual or constructive knowledge of the danger.

The University personnel were well aware that alcoholic beverages were customarily consumed in and around the stadium both before and during football games. In addition, the University was also aware that "tailgate" parties were held in the parking areas around the stadium. The appellate court concluded that while there was no proof that the University knew of any particular danger concerning the drunken man who had injured Mrs. Bearman, the University nevertheless had reason to know that some people in the area do become intoxicated, which could pose a threat to the general safety of other patrons.

Therefore, the appellate court reasoned that the University had a duty to take reasonable precautions to protect its patrons from injuries caused by the acts of third persons at the football games and in the immediately adjacent parking areas. Accordingly, the court remanded the case to the trial judge with directions that the case proceed to trial by jury to determine whether the protective measures employed by Notre Dame were or were not adequate, and if not whether such inadequacy caused Mrs. Bearman's injuries.

COMMENTARY:

The general rule of law applicable to the above situation is set forth as follows:

> *A possessor of land who holds it open to the public for entry for his business purposes is subject to liability to members of the public while they are upon the land for such a purpose for physical harm caused by the accidental, negligent, or intentionally harmful acts of third persons or animals and by the failure of the possessor to exercise reasonable care to*
>
> *(a) discover that such acts are being done or are likely to be done, or*
>
> *(b) give a warning adequate to enable the visitors to avoid the harm, or otherwise to protect them against it.*

PSCLR – SEP 84 (10)

Amusement Center Liable For Injuries Caused By Rowdy Patrons Whom Center Operators Failed To Control Or Eject:

Cassandra Dobard v. Skate Country, Inc. and Empire Fire & Marine Insurance Company, *No. CA-1659, in the Court of Appeal of Louisiana, Fourth Circuit, decided June 6, 1984, reported at 451 So. 2d 1231.*

According to plaintiff's evidence, Cassandra Dobard and her friends went skating at the defendant's skating rink. There were two boys who were "skating real fast," "cutting in front," and "clowning" for a 30 to 45 minute period before Dobard was injured. Dobard's companions were almost hit by the boys and one of them had complained about them to the rink "referee." One of them also asked a rink employee to move the offending skaters from the floor, and when the rink employee failed to do so, the plaintiff's friend stopped skating out of fear for her own safety.

The defendant's evidence, on the other hand, showed that the floor guard at the rink had removed the "wild" skaters from the floor for a period of approximately 20 minutes prior to the accident. The floor guard, Jose Santana, testified that he originally warned them about their conduct, and when they failed to heed his warning, he ordered them off the floor for 20 minutes. They later promised to skate properly and he allowed them back on the rink. He testified that this was his normal procedure for handling unruly skaters. He also testified that the two boys had been on the floor for only about 30 seconds when the accident happened.

In any event, the boys caused Dobard to fall and fracture her left arm. She filed suit against the rink owner and operator, and against its insurer. The evidence showed that she incurred medical bills of almost $2,000, and lost wages of approximately $600, and had to undergo surgery to have a plate and screws implanted to realign the fractured bone. She also had a lengthy convalescence period and treatment. Her doctor testified that the surgery was successful and that Dobard had no significant permanent impairment of her arm, and also that she could resume all regular sports and work activities. The jury returned a verdict in her favor in the amount of $37,600, and the defendants appealed.

It is well known that any operator of an amusement park or center, while not an insurer of a customer's safety, is nevertheless liable for its own negligence which causes a patron injury. It is also well known, and has been recognized in previous court decisions that some amount of bumping and jostling is an ordinary risk of participation in roller skating. Nevertheless, a skating rink patron assumes only the ordinary and usual risks of rink skating, and should be able to rely on the rink operator to protect the patron from the danger of wild or boisterous fellow patrons.

The Court of Appeals noted that the jury had chosen to believe the plaintiff and her witnesses's evidence as opposed to the evidence presented by the amusement center owner and that the jury was able under the evidence to conclude that the rink operator had knowledge of the danger to the plaintiff and others, and had a duty to protect the plaintiff, and was negligent in exercising that duty by failing to control or eject the rowdy skaters. The appellate court therefore allowed the verdict for the plaintiff to stand.

COMMENTARY:

There was another skating rink case decided by the Louisiana appellate court only last year in which the defendant amusement center operator was successful. The rink was found not liable to a patron injured by an unidentified couple who had dashed onto the floor from a nearby snack bar area.

The evidence in that case indicated that the boisterous conduct of the unidentified couple which had occurred some 30 minutes prior to the accident was too remote in time or place to be connected with the injury itself, and also that the couple had not engaged in any misconduct immediately before the accident which would place any duty on the rink operator to take any action against the couple.

In other words, the couple's "sudden and precipitous conduct" was not foreseeable by the rink operator, unlike the ongoing and observable misconduct by the two skaters who injured Ms. Dobard.

PSCLR – FEB 85 (11)

Spectators Injured In Foreseeable Stampede Of Patrons May Recover Against Business Operator For Failing To Provide Adequate Security Protection:

Mary J. Ollison v. Weinberg Racing Association, Inc., an Oregon corporation dba Portland Meadows Race Course, and Oregon Concession Services, Inc., an Oregon Corporation; and

Tommy Lee Nichols v. Weinberg Racing Association, Inc., an Oregon corporation dba Portland Meadows Race Course, and Oregon Concession Services, Inc., an Oregon corporation, *A8201-00435 & A8201-00436; CA A26274, in the Court of Appeals of Oregon, decided September 12, 1984, reported at 688 P.2d 847.*

The plaintiffs Ollison and Nichols were spectators at the Portland Meadows Race Course's "Fan Appreciation Night." The defendant Oregon Concession Services, Inc., operated the concession services at the race track, and sold beer there on this particular night for less than half the usual price. During the course of the evening, another patron by the name of Thurman fired a weapon, thereby causing a "stampede" of other spectators during which the plaintiffs Ollison and Nichols were trampled and received personal injuries.

In addition to suing the concessionaire, plaintiffs brought an action against the defendant Weinberg Racing Association, Inc., as the operator of the race track, alleging that the Racing Association had failed to protect them from the dangerous acts of others when the Association knew or should have known of the likelihood of such dangerous acts occurring based on past experience. They also alleged that the harmful acts of others should have been anticipated based upon the fact that beer was being sold for less than half the usual price, and they further alleged that the defendants were liable for failing to provide adequate security personnel to police the racetrack premises to prevent the foreseeable harmful acts by third persons.

The Racing Association agreed that it owed a duty to the spectators to exercise reasonable care to protect them from harmful acts by other patrons; however, the Association contends that such duty does not require them to provide security personnel, nor does it extend to protecting spectators from injury under the "highly unusual and unforeseeable circumstances alleged here."

The general rule of law is that an individual has no duty to protect another from the harmful acts of third persons. However, the court noted that previous decisions in Oregon have held:

> A possessor of land that holds it open to the public for entry for his business purposes is subject to liability to members of the public while they are upon the land for such a purpose, for physical harm caused by the accidental, negligent, or intentionally harmful acts of third persons or animals, and by the failure of the possessor to exercise reasonable care to '(a) discover that such acts are being done or are likely to be done, or (b) give a warning adequate to enable visitors to avoid the harm or otherwise to protect them against it.'

(Cases cited in support of this holding include *Brown v. J.C. Penney Co., Inc.*, reported in the November-December 1983 **PSCLR** at page 10, and *Torres v. U.S. National Bank of Oregon*, reported in the January 1984 issue at page three.)

The court went on to note that the possessor of land must have some sort of notice before the duty to provide security personnel arises, since the possessor is not an insurer of the patron's safety. But, if past experience shows that there is a likelihood of misconduct by others which is likely to endanger invitees, or if the place or character of the business operations should cause the business owner or operator to reasonably anticipate negligent or criminal activity on the part of others, then the duty may arise to take reasonable precautions to prevent such misconduct, including the obligation to provide a reasonably sufficient number of security officers to provide reasonable protection to invitees.

The Racing Association argues that it should not have been expected to "anticipate a patron would shoot a gun and thus cause a stampede." However, the exact manner of injury to a patron need not be foreseen; rather, the proper question to ask is whether the injury is of the general type to be anticipated from the misconduct. The court noted that while our imagination may not be able to dream of <u>every</u> possible drunken act that could possibly cause harm in a crowd, nevertheless, the plaintiff may be able to present evidence to convince a jury that the sale of alcohol at a greatly-reduced price at a crowded racetrack should make the racetrack owner/operator foresee that some spectator's misconduct could cause a stampede resulting in injury to bystanders.

The court therefore ruled that the case should not have been dismissed and that the plaintiffs should at least have the opportunity to attempt to prove their case to the jury. Furthermore, the court noted that, "although it is possible that evidence would not establish that additional security personnel could have prevented the plaintiffs' injuries, we cannot say that plaintiff could not adduce evidence that lack of adequate security was a substantial factor."

The court then considered the plaintiff's allegations against the concessionaire, Oregon Concession Services, Inc. The plaintiffs claim that this defendant was negligent for providing the patron Thurman with alcoholic beverages when he was already visibly intoxicated. Concession Services argues however that the peculiar circumstances of this particular incident were not foreseeable to their employee who served the alcohol to Thurman; that is, the employee could not have reasonably foreseen that Thurman would fire a pistol, causing a stampede, and that the plaintiffs would be injured thereby. The appellate court noted, in response to this argument, that the determination of foreseeability depends on how the harm is described, and concluded that the plaintiffs here may be able to produce sufficient evidence to convince a jury that the concessionaire should have been able to foresee that an intoxicated racetrack spectator who is served more alcoholic beverages will use a weapon or otherwise cause some disturbance in the crowd resulting in injury to bystanders.

Therefore the case will now proceed to trial against both the Weinberg Racing Association and the Concession Services Corporation.

COMMENTARY:

In an effort to help us understand the concept of "foreseeability," it may be helpful to review the Oregon court's definition, as follows:

> *The specific question before us is, then, whether plaintiff's injury and the manner of its occurrence was so highly unusual that we can say as a matter of law that a reasonable man, making an inventory of the possibilities of harm which his conduct might produce would not have reasonably expected the injury to occur. Stated in another way, the question is whether the circumstances are out of the range within which a jury could determine the injury was reasonably foreseeable.*

Foreseeability is a rather subjective concept, especially when it comes to considering results of actions retrospectively. That is, now that we know that certain acts resulted in injuries, it is perhaps easier to conclude that the results should have been foreseen from the act. However, if as the court says, the result is so "highly unusual as to be out of the range within which a jury could determine that the injury was reasonably foreseeable," then the trial judge should dismiss the action. As an example of this, see the case of Santiago v. New York Housing Authority, February 1985 **PSCLR**, *case twelve.*

PSCLR – AUG 85 (8)

Injured Party Must Establish Causal Relationship Between Negligence By Landowner And Injuries Sustained In Order To Recover Damages:

Phillip M. Noble and Marline Noble v. Los Angeles Dodgers, Inc., *Civ. B009809, in the Court of Appeal of California, Second District, Division 2, decided May 30, 1985, reported at 214 Cal. Rptr. 396.*

Philip and Marline Noble and their friend David McIntosh were returning to their car in the Dodgers Stadium parking lot after attending a night baseball game. They found two drunks standing by the car, one vomiting and the other urinating on the car. Phillip Noble said something to them, whereupon they shouted obscenities at him. When McIntosh approached them, one of the drunks struck him. Phillip Noble was then knocked down and injured when he tried to defend McIntosh. Mr. and Mrs. Noble subsequently sued the Los Angeles Dodgers, claiming that the Dodgers Corporation had failed to provide adequate security in the Stadium parking lot. The jury returned an award of damages in favor of the plaintiffs. The Dodgers appealed.

An owner of property is not an insurer of the safety of invitees on the property. He has only a duty to take reasonable measures to protect them from foreseeable injury, including the control of the conduct of third parties if necessary. The fact that anyone can foresee that a crime will be committed at any time, at any place is not enough, standing alone, to hold a property owner liable when a crime does take place on his property. The law requires that, in order to recover, a plaintiff must prove more than that negligence existed on the part of the defendant: where the specific conduct of third parties is brought to the attention of the property owner far enough in advance of the injury to give the owner opportunity to try to prevent the injury, then the causal connection between the owner's failure to act and the injury is obvious. For example, where one has been hit by a thrown bottle at a wrestling match, and the management had known that patrons were fighting and throwing objects for some six to seven minutes prior to the injury, but took no action to prevent or stop such conduct, then the causal connection is established.

The court noted that in the present case, the direct cause of the injuries to the Nobles was the conduct of the drunk who struck Philip Noble. The plaintiffs attempted to show that the Dodgers were negligent in failing to deter anyone from injuring them by producing an expert witness who testified that the Dodgers security measures were not adequate. He testified that in his opinion there should have been seven more people employed by the Dodgers for security purposes, and that they should have been deployed differently than they actually were. He, of course, could not and did not say that the addition of seven security personnel, or a different deployment, would have prevented the injury to Mr. Noble. In substance, he merely testified that he believed his method of policing the Stadium parking lot was preferable to the one used by the Dodgers organization.

The court therefore concluded that the present action is a "classic example" of a plaintiff establishing merely "negligence in the abstract" by showing that the Dodgers' security system did not comport with the security expert's notion of adequate security, but the plaintiff nevertheless failed to prove any causal connection between any such negligence and the resulting injury to him.

The appellate court therefore reversed the judgment in favor of the Nobles, finding that there was simply insufficient evidence to support it.

COMMENTARY:

The appellate court in the above case had some comments worth noting concerning the use of security experts as witnesses:

> *It appears that a growth industry is developing consisting of experts who will advise and testify as to what, in their opinion, constitutes 'adequate security.' The $64 question, of course, is 'adequate for what?' As noted, in each case where such testimony would be relevant, the security in existence has already proven to be inadequate to prevent the injury which did occur.*
>
> *The question then to be determined by the jury is what reasonable steps could have been taken to prevent the plaintiff's injury? The purpose of a trial in this type of case is not simply to critique defendant's security measures and to compare them to some abstract standards espoused by a so-called 'security expert.' The objective is to determine whether a particular defendant should, under the circumstances, be held liable for a plaintiff's injury because of a failure to prevent the criminal actions of third party. We submit that causation is a critical question*

These are thoughts worth studying for those of who, from time to time, are called upon to render our opinions on security matters.

PSCLR – JAN 87 (9)

Business Owner Is Not Liable For Injuries To Invitee Resulting From Unexpected Shooting Incident Even Though The Assailant Created A Disturbance On The Premises Two Weeks Before:

Linda Haddox and Gary Haddox v. Suburban Lanes, Inc., *No. 16223, in the Supreme Court of Appeals of West Virginia, decided October 28, 1986, reported at 349 S.E.2d 910.*

Richard and Linda Stump were going through an apparently messy divorce, when Stump entered the Suburban Lanes bowling alley in Morganton, West Virginia. Stump began firing shots at his wife, striking another bowler, Linda Haddox, and a police officer who was visiting his own wife at the alley. Haddox and her husband filed suit against Suburban Lanes, Inc., alleging that its negligence had caused her injuries. The trial court held that Suburban Lanes could not have foreseen Stump's behavior as a matter of law, and dismissed the action. The plaintiffs appealed.

Stump had visited the bowling alley earlier during the same evening as the shooting, and attempted to reserve a lane but could not. He was not armed at that time, and he left without any problem. Shortly after, when the Haddoxes were entering the bowling alley, they noticed Stump sitting in his car in the parking lot; they informed a Suburban Lanes employee that he was outside. Haddox was aware of the situation between the Stumps and and wanted to leave, but "everyone convinced [her] that [she] was overreacting to the situation." Even though Haddox felt uncomfortable, neither she nor any other bowler or employee at Suburban Lanes testified that they anticipated Stump would shoot anyone.

Stump had attempted to speak with his wife before she came to the bowling alley, and had phoned her at the bowling alley. On each occasion, she refused to speak with him, advised him to contact her attorney, and asked him not to attempt to speak with her again. Mrs. Stump discussed this with a friend, Judith Mackey, who called her husband, police officer Benjamin Mackey. He was on police duty that night, but stopped by the bowling alley. Mackey asked him to try and come back later when she and Mrs. Stump would be leaving to ensure that Stump wouldn't bother them in the bowling alley parking lot. He told them he would try, but that he didn't expect Stump to enter the building that night. A few minutes later Stump did enter and began shooting, striking Haddox and Officer Mackey.

Stump had created a minor disturbance at the Suburban Lanes bowling alley two weeks before the shooting took place. He had had an argument with his wife and a few other people. The manager of the bowling center had called the police and had him removed, because Stump was loud and obnoxious. He left without resistance, and the police arrested him for public intoxication.

The plaintiffs contended that Suburban Lanes was negligent in permitting Richard Stump to enter the premises, and in failing to remove him from the premises before he caused Haddox's injuries. They asserted that the defendants knew or should have known that Stump was capable of violent and dangerous conduct, and that he would attempt to injure Mrs. Stump and those around her.

The appellate court noted that the defendant did owe Haddox, a business invitee, a "duty to exercise ordinary care to keep and maintain the premises in a reasonably safe condition." The defendant, therefore, could be held liable if it negligently failed to perform this duty and this negligence proximately caused the plaintiff's injuries to the plaintiff. However, "a person is not liable for damages which result from an event which was not expected and could not reasonably have been anticipated by an ordinarily prudent person."

The focus of the appellate court's review was then whether Suburban Lanes could have or should have reasonably foreseen that Stump would enter the bowling alley, and begin shooting at and injure patrons. However, the record showed that everyone was surprised by Stump's violent conduct, including Haddox, Mrs. Stump, and Officer Mackey. The shooting was completely unexpected and probably could not have been prevented. The appellate court therefore affirmed the trial court's determination that plaintiffs could not establish the required foreseeability and that Suburban Lanes was therefore entitled to summary judgment as a matter of law.

PSCLR – NOV-DEC 88 (7)

City And Private Concert Promoters May Be Liable For Injuries Incurred As A Result of Breakdown In Crowd Control During Free Concert In City Park:

David Rotz v. The City of New York and Paramount Pictures Corporation and Anaid Film Productions, Inc., *in the Supreme Court of New York, Appellate Division, First Department, decided September 1, 1988, reported at 532 N.Y.S.2d 245.*

Twenty-five-year-old David Rotz attended a free concert given by Diana Ross in New York City's Central Park. A "tremendous crowd" gathered at the concert. According to Rotz, the crowd was "jammed in like sardines." As Rotz was standing, listening to the performance, a commotion erupted, and "everybody started running and they just ran on top of everybody." In the course of the stampede, Rotz was knocked down and trampled, incurring a serious fracture to his left leg. As the crowd ran, people shouted "get out of the way, there's a lion."

Rotz sued the City of New York, as the owner of the park, Anaid Film Productions, Inc., the producer of the concert, and Paramount Pictures, Inc., the holder of a license to create and broadcast a videotape of the concert.

Rotz alleged that the city, Anaid, and Paramount operated, supervised, arranged, managed, and controlled the concert at which Rotz was injured and that they were negligent in failing to maintain the concert site in a "reasonably safe, suitable, and well-controlled condition" and in failing to provide for the safety and well-being of those in attendance in the concert by failing to properly and adequately supervise the large crowd.

The "Permit Agreement" between the city and Anaid made Anaid responsible for all arrangements with respect to the concert, including security, and provided that Anaid would release the city from and indemnify the city against any liability from claims arising as a result of the concert.

The "Deal Memorandum" between Paramount and Anaid gave Paramount the right to televise and videotape the concert in return for substantial financial obligations assumed by Paramount. These obligations included setting aside certain funds to be applied against police overtime charges incurred in connection with the concert, for which Anaid and/or Paramount would be responsible up to the sum of $10,000.

Paramount claimed that it was merely the purchaser of television rights to the concert and that it had no role in the management or control of the event. Paramount further argued that the precipitating cause of the stampeding – shouts that a lion was coming – was an unforeseeable event that precluded any liability for the injuries that Rotz incurred.

The city claimed that the "Permit Agreement" with Anaid placed all responsibility for security arrangements on Anaid and that, as to any claim based on inadequate police protection, the city owed no "special duty" to Rotz. Further, even if it did owe a duty to Rotz, due to its ownership of the park, the city argued that the cause of the stampede was unforeseeable and precluded any liability for negligence.

The trial court dismissed Rotz's complaint against both the city and Paramount, based upon its conclusion that the events leading to the stampede were unforeseeable and, therefore, cut off any liability for either party's negligence in failing to take adequate crowd control measures.

The appellate court agreed with the trial court's determination that the city, in its capacity as the owner and operator of the public park, owed a duty to those who had been invited to enter the park. This duty included the obligation to provide "an adequate degree of general supervision of the crowd invited by exercising reasonable care against foreseeable dangers under the circumstances prevailing." The appellate court disagreed with the trial court's decision, however, that the danger in this case was unforeseeable. The issue of foreseeability, the court noted, is generally left to the jury, even when the underlying facts are essentially undisputed.

The tort of negligence consists of a duty of care owed to another and a breach of that duty. The duty owed to Rotz, as one invited to attend a concert in the city's public park, was the exercise by the city of reasonable care against foreseeable dangers, including the provision of an adequate degree of general supervision of activities that might endanger those within the park. The degree of care that is reasonably necessary depends upon the circumstances of each case and is proportional to the dangers to be avoided.

The question as to what risks the city should be expected to have anticipated can best be determined by a jury in light of "common contemporary experience." A jury could find that, in the absence of adequate supervision and crowd control, it was reasonably foreseeable that "disorder, unruliness, a melee, or a riot could erupt from some cause ignited by the vagaries of myriad individuals 'jammed together' in a heightened atmosphere."

The court noted that the city had anticipated this risk in limiting its liability in the "Permit Agreement" with Anaid. While that agreement held the city harmless as against Anaid and provided for the city's indemnification by Anaid, the agreement's provisions did not relieve the city from the duty owed to Rotz.

The court also felt that a jury could reasonably find that adequate crowd control measures could have prevented, or at least contained, a riot or stampede at the concert and that the city failed to exercise the reasonable care necessary under the circumstances to avoid the foreseeable risk of such events.

The court distinguished the foreseeability of the exact manner in which the harmful event was precipitated and the foreseeability of the general risk involved. The court felt that the general risk reasonably to be anticipated from the dynamics of a large, closely packed crowd was the outbreak of disorder, of necessity precipitated by the act of some third person, resulting in injury to those in attendance at the concert.

While the intervening act of a third person may often relieve a party from liability for negligence, there is no relief from liability if the intervening act, itself, is a normal or foreseeable consequence of the situation created by the negligence. The court concluded that a jury should determine whether the city failed to provide adequate crowd control measures to avert the risk of injury to persons in the crowd and, if so, whether the intervening conduct of third persons in initiating the stampede was a normal or foreseeable consequence of the city's failure to adequately supervise the crowd.

The court further held that Paramount's role in the control, management, and supervision of the concert, given the monetary obligation undertaken by Paramount and the fact that it was to be physically present to videotape the concert, could only be determined after further investigation of the facts. Thus, Rotz's complaint against both Paramount and the city was reinstated for further investigation and trial.

PSCLR – APR 89 (5)

Sports Facility Has No Duty To Prevent Intoxicated Invitee From Climbing Reasonably Adequate Fence Designed To Prevent Invitees From Falling Into Adjacent Tunnel:

Thomas Edwards v. California Sports, Inc., *No. B028127, in the California Court of Appeal, Second District, Division 2, decided December 28, 1988, reported at 254 Cal. Rptr. 170.*

A jury awarded Thomas Edwards $2,000,000 in damages, reduced by 73 percent for negligence the jury attributed to him for his own injuries, thereby assessing a $540,000 verdict against the defendant, Sports, Inc. The award was set aside on this appeal to the California Court of Appeal.

The defendant operates the stadium known as the Forum in Englewood, which hosts athletic and entertainment events, including the Los Angeles Lakers basketball games. There is a driveway from the parking lot into a tunnel under the arena, which is flanked by retaining walls. Above the walls are metal railings designed to prevent patrons in the parking lot from falling onto the driveway tunnel. It is customary for fans to gather at the railings to watch players leaving the building, but there had never been an injury in 12 years of the Forum's operation. Edwards, while intoxicated, climbed onto the guard rail and fell onto the driveway below. He incurred significant head injuries. There was no evidence that he had been pushed, jostled, or thrown over the railing.

At trial, a safety engineer testified that, in his opinion, a higher fence would have prevented Edwards' fall. The engineer speculated that there were several ways that the accident may have occurred, including the possibility that Edward had climbed over the fence and had fallen. Another witness, referred to as an expert on "crowd psychology and control," offered his opinion that alteration of the fence or deployment of security personnel could have prevented Edwards' fall. However, as the appellate court noted, there was not even a hint of evidence that those matters had any relevance to the case, nor was there any evidence that this witness had any expertise in building design or construction.

The appellate court stated the issue succinctly,

> The fundamental inquiry is whether the duty of a landowner to exercise reasonable care in preventing any injury to persons on the premises required such ... extreme measures which might have prevented this particular injury.

Evidence showed that the fence was 50 inches in height and that it would be about chest high for an adult of average height. Such person standing or walking near the fence, would be prevented by the fence from being pushed or falling into the driveway. The court concluded that the fence was clearly adequate for the purpose for which it was designed. As the jury found, the cause of plaintiff's injury was primarily due to his own conduct. The court went even further, concluding that the jury's apportionment of 27 percent of the fault to the arena was based on the presumption of a duty which, according to the court, never existed.

The appellate court therefore set aside the verdict in favor of Edwards and remanded the case to the trial judge with direction to enter judgment in favor of the defendant.

COMMENTARY:

In a sidebar to the opinion, the appellate court noted that there must be some limitation on the distance that society should go in indirectly supervising such activity through use of the tort liability merely to protect individuals from their own negligence, daring, stupidity, or self-destructive impulses. Society should not require that property owners become the insurers of invitees' safety, and at least so far, society and the courts have declined to do so.

The law still is that the property owner must use reasonable care to keep his property in a reasonably safe condition, and that he must warn others of latent or concealed dangers. But, a property owner is not liable for injuries which result from perils which are obvious or which should be observed by invitees in protecting their own safety.

PSCLR – MAY 89 (5)

Professional Rescuer May Not Recover For Injuries Incurred In Responding To Summons By Proprietor Absent Gross Negligence By The Proprietor:

Raymond Spicuzza v. Christopher C. Fonseca, et al., *No. CA 9206, in the Court of Appeal of Louisiana, Fourth Circuit, decided December 13, 1988, reported at 537 So. 2d 272.*

Last Warning Promotions contacted the St. Bernard Parish Police Jury to lease the St. Bernard Civic Center for a concert. The lease agreement required that Police Jury furnish twenty uniformed security personnel at Promotions' expense. Promotions assumed "full responsibility for the character, acts and conduct of all persons admitted to [the] premises." Promotions also agreed "to have on hand at all times and at its own expense, a sufficient number of policemen for the regulation of traffic, maintenance of order, and protection of persons and property."

During the concert, a small riot broke out amongst the spectators, prompting Civic Center Security to call the St. Bernard Parish Sheriff's Department for assistance. Raymond Spicuzza, a deputy with the Sheriff's Department, was dispatched to the Center along with other deputies in response to the call. In the course of attempting to disperse the crowd, Spicuzza was kicked by one of the concert attendees and injured his left leg.

Spicuzza sued Promotions, and the trial court dismissed the case because Promotions owed no duty to protect professional rescuers such as Spicuzza from injury.

Generally, a proprietor owes a duty to those invited onto his premises, that duty being to exercise reasonable care to protect such "invitees" from injury. This duty includes calling the police when necessary. However, as the court noted, "A proprietor who summons the police as part of the duty to exercise reasonable care to protect invitees does not then assume a further special duty to protect police officers from the actions of patrons on the premises."

When the proprietor commits no affirmative acts of negligence resulting in injury to the professional rescuer, no duty is owed. A proprietor is not liable for injuries sustained by professional rescuers because such injuries do not fall within the "ambit of risk" of the proprietor's original negligence.

The court in this case found that Spicuzza knowingly assumed the risk which caused the injury to his leg and that this risk was necessarily included within Spicuzza's duties as a police officer. Spicuzza had responded to a call for assistance to a rock concert attended by a large, boisterous crowd. The court noted that such calls are commonplace for police officers and that Spicuzza accepted the risks involved with such duties when he became a police officer.

The court felt that a strong policy reason justified failing to find a special duty owing to professional rescuers. The court reasoned that "if such a duty were recognized by the courts, proprietors would be reluctant to summon police to their premises when needed because of fear of liability resulting from injuries to the officers by third parties."

The court found that there was no constitutional violation in treating non-professional rescuers, to whom there is a duty to act with reasonable care, differently from professional rescuers. The court felt that the distinction was justified by the policy against discouraging proprietors from summoning professional rescuers when needed because of fear of liability resulting from injury.

The law does allow professional rescuers to recover from property owners when they are injured due to the property owners' gross (as opposed to ordinary) negligence. However, the court found that the facts of the case did not support a finding of gross negligence by Promotions.

Twenty uniformed security personnel were provided by Promotions, and the need for additional security arose only when a fight escalated into a small riot. Failure to anticipate the situation requiring additional security was not, according to the court, gross negligence. The court found that Promotions acted quickly and reasonably in contacting professionals to control the crowd in order to prevent injury to spectators. Spicuzza's injuries resulted from independent, intentional acts of third parties, not from gross negligence on the part of Promotions.

The appellate court, therefore, affirmed the trial court's dismissal of the action without a trial.

PSCLR – SEP 89 (8)

Private Security Agency Had No Duty To Protect Fan From Parking Lot Assault By Baseball Player When Assault Occurred After Security Guard's Watch Was Over:

Stephen Paul Simmons v. Baltimore Orioles, Inc., *Civ. A. No. 88-0211-A, in the United States District Court for the Western District of Virginia, Abingdon Division, decided April 20, 1989, reported at 712 F. Supp. 79.*

Paul Simmons and a friend attended a Fourth of July baseball game in Bluefield, Virginia, between the Martinsville Phillies and the Bluefield Orioles. The Orioles are a minor league team owned by the Baltimore Orioles. While at the game, Simmons moved to a seat along the third base line during the eighth inning and began heckling the Orioles players sitting in the bullpen. There is some dispute about what was said, but an Orioles player named Champ stated that Simmons was accusing the players of stealing the local women. Simmons then, according to Champ, stated that he would teach the Orioles about West Virginia manhood by blowing the players' heads off. At this point, the pitching coach asked Simmons to leave.

After the game, Champ and Simmons met in the parking lot. Simmons offered no details of what occurred, except that he was punched and kicked by Champ. Simmons also claimed that another player named Hicks then hit him in the jaw with a baseball bat, breaking his jaw in two places.

Champ claimed, however, that Simmons saw him carrying a bat and began taunting him. Champ claims to have thrown the bat down when Simmons pointed his finger as if shooting at Champ and said, "Oh, so you need a bat, huh?" Simmons then said "Let's go over to my car, and I'll blow your head off."

Another player tried to intervene, but Champ said, "Just you get out of here." According to Champ, Simmons then came towards him in a threatening manner, and Champ hit Simmons in the face. Because Simmons was not fazed by the blow, Champ then kicked Simmons in the chest. Champ claimed Simmons staggered back, and said, "I'm drunk. I didn't feel that." After which, Champ claims to have walked away.

Hicks, the other player there, then hit Simmons. Simmons claims that this blow came from a baseball bat, but Hicks stated that he used only his fist. Hicks stated that he had not been near during the heckling at the game, but that he got involved because he feared that Simmons would pull out a gun.

Simmons sued the Orioles, Champ, Hicks, and Bluestone Security Agency. The allegations against Bluestone Security Agency were that the agency was under contract to guard the field and parking lots and the agency was negligent in doing so. Simmons claimed that the agency's negligence was the cause of his injuries.

In fact, the contract was between the town and the security agency. The co-owner of the agency, Jack Asbury, was the Chief of Police for the neighboring town of Bluefield, Virginia, and he testified that in the past the agency had agreed to provide a guard for the entire game. However, Asbury also stated that doing so was too expensive for the Virginia town, so the special guard was no longer provided.

The Director of Parks and Recreation for Bluefield, Herbert Sims, stated that a few weeks before he had reduced the security guard's hours to 7:00 to 10:00 PM to save money. Sims also stated that the terms of the agreement between the town and the security agency required the agency to patrol the town park – not Bowen Field, the baseball park.

The district court focused on whether the Orioles had any duty towards Simmons which it might have breached. The law in Virginia is that the proprietor of a place of amusement is not to insure the safety of invitees. Rather, the proprietor's duty to

> exercise reasonable care for their safety and protection – such care as would be exercised by an ordinarily careful and prudent person in the same position and circumstances.

The court noted that the Virginia Supreme Court had provided some answers to the issues raised in this case in its recent case of *Wright v. Webb*. In that case the plaintiff was the victim of a violent assault in a motel parking lot. The Virginia high court ruled that, in most cases, criminal behavior is not foreseeable, and even

> knowledge of prior crimes against property does not create a duty upon a business invitor to anticipate and guard against assaults against its invitees, offenses involving a substantially different kind of criminal behavior.

The *Wright* court concluded:

> a business invitor, whose method of business does not attract or provide a climate for assaultive crimes, does not have a duty to take measures to protect an invitee against criminal assault unless he knows that criminal assaults against persons are occurring, or are about to occur, on the premises which indicate an imminent probability of harm to an invitee...

Based on these facts, the court concluded that the Bluestone Security Agency was not liable, regardless of whatever liability the other defendants might have had to Simmons. The court ruled that the security guard did not have any duty to be present at the park at any time after 10:00 PM Since the Bluestone Security Agency did not have a duty to be present, the agency was not responsible for whatever might have occurred in its absence.

Restaurants and Bars

PSCLR–AUG 83 (3)

Tavern Owner Not Liable to Patron Who Was Assaulted on Public Street Near Tavern:

Mary Katheryn Wofford v. Kennedy's 2nd Street Co., Laclede's Landing Redevelopment Corporation, *No. 45661, in the Missouri Court of Appeals, Eastern District, Division Three, decided February 15, 1983, motion for rehearing/transfer to Supreme Court denied April 15, 1983, application to transfer denied May 31, 1983, reported at 649 S.W.2d 912.*

Should a tavern owner be required to provide escorts or some other security precautions to patrons after they leave the premises? Mary Katheryn Wofford was raped, sodomized, and beaten by assailants after she left Kennedy's 2nd Street Company restaurant and tavern. She thereafter sued its owners and Laclede's Landing Redevelopment Corporation (the developer and owner of the entertainment complex where Kennedy's was located) for a total of $18,500,000.

Wofford's assault occurred after she was kidnapped by unknown criminal assailants immediately after leaving Kennedy's at about 10:30 PM on February 21, 1979. The men grabbed her as she approached her car which was parked on a street around the corner from the tavern.

Her lawsuits alleged that neither the Laclede's Landing complex nor Kennedy's had customer parking and the owners were aware that the area's street lights did not work some of the time. Moreover, she alleged that in the three-year period before her attack Laclede's Landing had been the scene of many violent crimes. Wofford contended that these facts alerted Kennedy's owners and the Laclede Corporation that certain security precautions would be necessary for the protection of its patrons and invitees. According to Wofford, her kidnapping and assault were caused by their failure to employ uniformed or plain clothes security personnel; their failure to warn her about the area's criminal history; their failure to provide female customers with escorts (as had been Kennedy's practice in the past); the absence of street lighting; and the absence of any security measures in Laclede's Landing.

The suits, which were dismissed before trial, were reviewed on appeal by the Missouri Court of Appeals. The court agreed that, as a restaurant and tavern, Kennedy's was required to provide its invitees with ordinary care to prevent injuries by third persons on the premises and on approaches to the premises. However, this duty would not extend to patrons who left the tavern premises and were subsequently attacked on a public street. "If a public street under the circumstances here were held to be an approach so the landowner's duty to prevent injuries was extended to injuries occurring on public thoroughfares, the line which would cut off the landowner's liability becomes nearly impossible to draw," according to the court.

The court expressed sympathy for the plaintiff Wofford, but held that neither of the defendants could be held liable for her attack. The court summarized its conclusions by stating:

> To extend the liability of a business to the point that the business would be responsible for crimes on the public thoroughfares in the vicinity of the business would open the flood gates to litigation until almost any crime victim in a commercial area could successfully place his or her case before a jury. It would allow a finding of negligence on the part of a commercial establishment when a crime is committed anywhere in a commercial area. Where would the liability end?

Therefore, the dismissal of her lawsuit was affirmed.

COMMENTARY:

Among other qualities, this case is an excellent example of the reason **PSCLR** *rarely reports* *on the results of trial level court decisions, and never reports on the mere filing of claims and allegations.*

The fact that an $18,500,000 suit was filed is of absolutely no legal significance to our readers (the suit could just as easily be for $18,500,000,000,000,000, depending on how many times one chooses to hit the "0" key); even the fact that the trial court ruled on the case is generally of little consequence to anyone, except the parties involved, since a trial level decision is not binding on any other court or judge. It is only when an appellate court reviews the case and renders an opinion that legal precedent is established and becomes binding on lower courts.

Therefore, since our readers rely on us to provide them with analyses of those legal cases which in law and fact affect their own security operations, we restrict our publication to the reporting of appellate court decisions which have a significant effect on the private security industry, and on the activities of supervisory personnel who have the corporate responsibility for protecting people and assets.

PSCLR – NOV-DEC 83 (12)

Owner Liable For Failing To Retain Security Services To Maintain Order In "Rough" Establishment Resulting In Patron's Death:

James Stevens, Jr. v. Patricia Jefferson, as personal representative of the estate of Earl Sidney Jefferson, Sr., deceased, *No. 61651, in the Supreme Court of Florida, decided June 2, 1983, rehearing denied September 9, 1983, reported at 436 So. 2d 33.*

The duty of a business proprietor to protect invitees from harm has been the subject of much litigation reported in this and other issues of **PSCLR**. In each case the proprietor's extent of liability typically has depended on the nature of his relationship to the injured party and the foreseeability, actual or imputed awareness, of the situation which led to the party's injury.

In this case, Earl Jefferson was shot and killed in a bar which was owned by James Stevens. Jefferson's widow brought suit alleging that Stevens either created or was aware of a dangerous situation in his bar, leading to her husband's death. The trial ended with a verdict and judgment for Jefferson in the amount of $11,400 and Stevens here appealed. (The judgment amount was not included in the appellate court's opinion but was determined by our research staff by contacting the plaintiff's counsel.)

The Supreme Court of Florida reviewed a bar owner's duty to his patrons/invitees:

> The proprietor of a place of public entertainment owes his invitee a duty to use due care to maintain his premises in a reasonably safe condition commensurate with the activities conducted thereon. ... the proprietor of a liquor saloon, although not an insurer of his patrons' safety is bound to use every reasonable effort to maintain order among his patrons, employees, or those who come upon the premises and are likely to produce disorder to the injury or inconvenience of patrons lawfully in his place of business.

Although foreseeability could be proved by showing that a proprietor was aware that a specific person of "dangerous propensities" posed a risk of harm to patrons, the court pointed out that this would not be a necessary condition. According to the evidence, Steven's bar was a "rough" place, the site of many previous fights and gunplay. The owner had, in fact, used a security service in the past for protection and to maintain order, but had stopped using the service before Jefferson was shot. At the time of the shooting, the premises was under the control of a single female employee who was unable to maintain order in the bar. On the basis of such evidence, the Florida Supreme Court held:

> ... a jury could determine that a foreseeable risk of harm to patrons existed, that the risk was either created or tolerated by Stevens, that he could have remedied the danger but failed to do so, and that because of that failure to perform his duties Jefferson was killed.

The court therefore affirmed the judgment granted to Jefferson.

COMMENTARY:

In reaching its decision, the Florida Supreme Court cited with approval holdings in similar cases in Arkansas, Colorado, Michigan, New York, and Washington.

This case could certainly be used effectively as a marketing tool: Security companies could identify "rough" establishments where previous acts of violence have occurred and point out to the owners (or their insurance carriers!) that they may be personally liable for failing to hire security services to maintain order and to protect their patrons.

PSCLR–FEB 84 (10)

Manager Of Business Must Render Reasonable Aid To Injured Or Ill Person On Business Premises:

Danny Glenn Miller v. McDonald's Corporation, et al., *No. 83 CA 0014, in the Court of Appeal of Louisiana, First Circuit, decided October 11, 1983, reported at 439 So. 2d 561.*

The plaintiff Danny Glenn Miller entered a McDonald's hamburger restaurant in Baton Rouge, Louisiana, and placed an order. He became involved in an argument with the restaurant manager when he claimed that he had been short-changed. Other customers "passed remarks" concerning the argument; after Miller did not receive additional change, he went outside and sat at one of the exterior tables on the restaurant premises. He was followed outside by three women who had been behind him at the restaurant and who had argued in favor of the manager. Once outside, they began shouting obscenities at Miller as they walked to their car. They continued to shout at him from the car and pulled up next to the table; one of the women pulled out a gun and shot Miller.

After the shooting, Miller's sister (his brother and sister were also outside the McDonald's) ran into the restaurant to call for help. She was informed that McDonald's had a policy not to allow anyone to use its telephone. She screamed and cried, but was not permitted to use the telephone. McDonald's employees kept the door locked and would not allow her to enter the area where the phone was located. She returned to her injured brother's side, and another brother ran into the restaurant, threatened to use force to get to the telephone, and was allowed with some reluctance to call for assistance.

Miller filed suit against McDonald's for the injuries he incurred from the shooting. Miller contended that the McDonald's restaurant manager failed to take action to deflate a potentially explosive situation. However, the appellate court did not find that the manager's conduct had caused the shooting.

Miller also claimed that this particular McDonald's restaurant was the location of prior criminal disturbances, and that McDonald's failure to maintain reasonable security on the premises breached a duty of care owed to Miller as one of its customers. The appellate court ruled that the owner of a business has a duty to exercise reasonable care to protect customers. When the intentional or criminal acts of third parties constitute an unreasonable risk to customers, the business owner's duty to protect its customers could be discharged by calling the police at the time the business owner knew or should have reasonably known of the third party's intention and ability to harm the customer. There was no claim in Miller's suit that the McDonald's manager knew that any of the three women had a gun or intended to use it.

Miller also argued that McDonald's should be held liable to him because of its employees' refusal to assist him and its failure to train its employees in first aid. The court noted that there was no legal duty to come to the aid of someone injured on a premises by a third person except as follows:

> The defendant is not required to take any action until he knows or has reason to know that the plaintiff is in danger or is ill or injured. He is not required to take any action beyond that which is reasonable under the circumstances. In the case of an ill or injured person, he will seldom be required to do more than give such first aid as he reasonably can, and take reasonable steps to turn the man over to a physician or to those who will look after him and see that medical assistance is obtained. He is not required to give any aid to one who is in the hands of apparently competent persons who have taken charge of him, or whose friends are present and apparently in a position to give him all necessary assistance.

The appellate court found that Miller's injuries resulted from the shooting itself and not from any act or omission by McDonald's after the shooting. Therefore, in its present posture, the plaintiff's case cannot proceed. The appellate court did, however, permit the plaintiff to amend his allegations, if possible. Miller will be able to pursue this suit only if he can allege and prove that he incurred actual damage to his person as a result of the alleged refusal to assist and/or refusal to render first aid after the McDonald's employees knew or should have known that he had been shot.

COMMENTARY:

The above-quoted statement as to the business proprietor's duties to assist an ill or injured person on his premises should certainly be included in training programs for all employees, and security directors may wish to bring this to the attention of management. It may also be helpful for this quotation to be included in an employee's handbook, where such handbooks are in use, or to be posted on the premises to insure that all employees will be familiar with the procedures to be followed under these circumstances.

PSCLR – FEB 84 (11)

Property Owner's Ability To Foresee Criminal Thought By Third Parties Can Be Shown By Proving Past Experience Of Disorderly Conduct And Assaults:

Pearl Allen v. Babrab, Inc., d/b/a the Gemini Club, and National Union Fire Insurance Company, *No. 61789, in the Supreme Court of Florida, decided August 25, 1984, rehearing denied October 26, 1983, reported at 438 So. 2d 356.*

Pearl Allen and a female friend visited the Gemini Club on October 1, 1977. As they left the club in the early morning hours, they were accosted by a male patron named Leroy Allen (no relationship). When the two women rebuffed his advances, he poured his drink on one of the woman and then hurled his empty glass at Allen, striking her in the face and permanently blinding her left eye.

Allen sued the Gemini Club's owners, and at the trial level the case went to the jury, which returned a verdict in favor of Allen. The owners of the Gemini Club then appealed that judgment to the Fourth District Court of Appeals of Florida. At that level the District Court of Appeals reversed the decision and directed that judgment be entered in favor of the defendant-owners. The basis of that reversal was that there was no evidence in the trial record showing that the assailant had previously assaulted anyone or created a disturbance at or near the Gemini Club. The District Court of Appeals found that the Florida law provided that a property owner may be liable for injuries to its patrons caused by the misconduct of third parties only if the property owner knew or should have known of the dangerous propensities of <u>that specific assailant</u>.

However, on further appeal to the Supreme Court of Florida by Pearl Allen, that court reversed the Court of Appeals and reinstated the verdict and judgment in favor of Allen. The Supreme Court ruled that the property owner will not be held liable for the <u>unforeseeable</u> criminal acts of third parties, but the property owner's specific knowledge of a particular individual's dangerous propensities is not the only method of proving foreseeability. The court went on to hold that foreseeability also can be shown by proving that a property owner knew or should have known, based on past experience, that there was a likelihood of disorderly conduct or assaults by third parties in general which could endanger the owners, customers or patrons. In this regard, the Supreme Court noted that the Gemini Club had had a history of disturbances prior to the assault on Allen. A bouncer had been employed to maintain club security; his duties included covering the parking lot and stopping patrons from taking glasses from the premises. Although the bouncer had been hired by corporate officers at the urging of the club's bartenders, he was not on duty when the two women were assaulted.

Accordingly, the Supreme Court of Florida ruled that the evidence presented at the trial was enough for the jury to conclude that the property owner Babrab knew or should have known that injuries to its customers by the misconduct of others in general was probable. In spite of this knowledge, Babrab corporate officers had failed to prevent such misconduct. Furthermore, the jury could have concluded that if the corporation had continued its prior policy of providing security guards to prevent the taking of glasses from the club and to patrol the parking lot, then the assault on and resulting injury to Allen would not have occurred. Therefore, the court sustained the verdict in favor of Allen.

COMMENTARY:

This case is an excellent example of how the law is developing in the area of property owner and proprietor liability for injuries suffered by patrons at the hands of third parties.

As a matter of fact, as recently as 1981, there were two conflicting decisions by two different District Courts of Appeal in Florida: one of them was the District Court of Appeal's decision in <u>Babrab, Inc. v. Allen</u>, which is discussed above and was a decision of the Florida <u>Fourth</u> District Court of Appeals. The other was the 1981 case of <u>Stevens v. Jefferson</u>, which was decided by the <u>Fifth</u> District Court of Appeals in Florida.

As noted above, the Fourth District had ruled that the property owner could be liable only if he knew of should have known of the dangerous propensities of the specific assailant involved in the attack; whereas the Fifth District appellate court had ruled in Stevens that the owner can be held liable if it can be shown that he could or should have foreseen the tortious misconduct by any unknown assailant. In order to resolve this conflict between two separate courts of appeal in two separate districts in the same state, the Florida Supreme Court accepted the Babrab case for review, and adopted the rationale of the Fifth District, as set forth in Stevens, thereby resulting in a reversal of the Fourth District's opinion in Babrab.

The Florida Supreme Court had previously approved the Fifth District's opinion in Stevens, as was reported in the November/December, 1983, issue of **PSCLR**, *where the full court of the Supreme Court's opinion was set forth.*

PSCLR – AUG 84 (6)

Casino May Be Liable To Patron For Robbery And Beating – Evidence Of Insufficient Security System:

Beverly W. Early and Frank C. Early v. N.L.V. Casino Corporation, d/b/a Silver Nugget Casino, *No. 14463, in the Supreme Court of Nevada, decided March 30, 1984, reported at 678 P.2d 683.*

Before entering a restaurant at the Las Vegas, Nevada, Silver Nugget Casino to have lunch with her husband and daughter, Beverly Early stopped by a restroom in the casino. She was attacked and robbed by an unknown male assailant while alone in the restroom. Because they heard a scream, Mr. Early and the daughter entered the restroom area and they discovered Mrs. Early getting up off the floor, her clothing was torn and her face appeared to have been "ripped completely open." An ambulance was summoned and Mrs. Early was taken to a hospital for treatment. Her physical injuries were accompanied by psychological trauma.

Mrs. Early and her husband sued the Silver Nugget Casino, alleging its negligence precipitated her injuries. The suit was dismissed at trial, however, and the Earlys brought this appeal.

The Supreme Court of Nevada reviewed the facts and evidence concerning the Silver Nugget's security measures. Testimony on the casino's security system was given by three of its security guards, including its chief of security. According to these casino employees, no formal training sessions or staff meetings were conducted for security guards, nor was there a written security manual for use by security employees. The guards did not have an elevated security desk and did not have closed circuit television monitors. (A security expert testified that every other casino in town conducted training programs for its guards and used elevated security desks.) There was evidence that on occasion no guards patrolled the outside of the casino for up to a four hour period. The security guards' logbook showed that 92 crimes had been committed on casino property in the two years preceding Mrs. Early's attack, although none of the crimes had occurred in the casino's restrooms.

The casino's chief of security testified that the security guards at the Silver Nugget were actually little more than "glorified porters" who carried money boxes from the counting room, read bingo machine numbers, and witnessed the changing of tape in keno machines. He further testified that although meetings for security chiefs were held each month in Las Vegas, he was unable to attend because there weren't enough guards to cover the casino without him. There was evidence that during the two to three years before Mrs. Early's attack, restroom crimes and preventive measures had been discussed at nearly every monthly meeting of the Las Vegas security chiefs.

The Supreme Court then examined whether the Silver Nugget owed to Mrs. Early a duty to protect her. The court noted that the duty owed to an invitee of the casino would be one of "ordinary care," although the casino would also be required to protect a guest from the danger of injury by a third party if such an injury was reasonably foreseeable and reasonably preventable. If a property owner reasonably anticipates "careless or criminal conduct on the part of third persons" through past experience of such incidents, or because of the "place and character" of his business, he may be required by law to take precautions to protect his guests.

Applying the previously described standard to the Silver Nugget, the Nevada Supreme Court ruled that evidence of crimes committed on the casino property and the Silver Nugget's location and type of business could indicate the requisite foreseeability:

> Indeed, a gambling casino where cash and liquor are constantly flowing may provide a fertile environment for criminal conduct such as robbery and assault. Such crimes are most likely to occur, moreover, in areas such as restrooms, which are removed from the protection offered by crowds.

Based on the evidence, the court ruled that a question of fact had indeed been raised as to whether the Silver Nugget's security measures were sufficient to reasonably protect its guests from criminal attack. The suit's dismissal was therefore reversed and the case was remanded for a new trial.

COMMENTARY:

Among other lessons to be derived from the holding in the above case, it is clear that the courts feel that it is necessary for private security personnel to stay current concerning new developments in private security techniques. The fact that the chief of security at the casino did not attend the monthly security meetings in Las Vegas because there "weren't enough guards to cover the casino without him" was cited by the court as another element which the jury could consider in determining whether adequate security measures were in effect at the casino.

The court also noted in its opinion that a witness had testified that, during the two or three years before the assault on Mrs. Early, the subject of restroom crimes and the prevention of same had been discussed at nearly every monthly meeting of the security chiefs.

It pays to keep up with recent developments in private security—and it costs those who don't.

PSCLR—MAR 85 (1)

Where Restaurant Personnel Had No Knowledge Of Police Stakeout, There Is No Duty To Warn Customer Of Danger, And No Liability For Death Of Customer Mistakenly Killed By Police:

Fernando Campos, Jr. and wife Sara Campos, individually and as natural parents and heirs at law of Fernando Campos, III, deceased v. South Texas Beverage Company d/b/a Pizza Huts of North Texas, from El Paso, *No. 08-83-00123-CV, in the Court of Appeals of Texas, El Paso, decided October 31, 1984, reported at 679 S.W.2d 739.*

Fernandos Campos, III, was mistakenly shot and killed by an El Paso police officer during a robbery of a Pizza Hut restaurant. Campos had gone to the restaurant for a meal, and was caught in gun fire which accompanied a robbery. Campos' family sued the restaurant owner.

The evidence presented at trial showed that neither the management nor the employees of the restaurant had any knowledge of the existence of a police stakeout at the restaurant. While plaintiffs attempted to show that the management of the restaurant did know that the restaurant was a stakeout location, their evidence could not be considered inasmuch as it was based on hearsay, and not on personal knowledge.

Therefore, while it is certainly true that the duty of a business owner towards a business patron is to exercise reasonable care to warn the patron of any dangers on the property, and to protect the patron from harm, the court concluded that the evidence here was sufficient to show that the restaurant did not breach any duty to Campos. Without prior knowledge of the police stakeout or of any potential danger to Campos, the restaurant management and employees thus could not breach its duty to him. The judgment of the trial court dismissing suit by the victim's parent was therefore affirmed on appeal.

PSCLR—APR 85 (2)

Business Owner/Operator Owes No Duty To Invitee To Obey Criminal's Demands So As To Avoid Further Danger:

Anthony B. Vandermost v. Alpha Beta Company, *B007194, in the Second District, Division Four of the Superior Court of Los Angeles County, California, decided February 15, 1985, reported at 164 Cal.App.3d 771.*

Anthony Vandermost won a verdict from the jury at the conclusion of his trial of his damages suit, but the trial judge set aside the jury's verdict and entered judgment in favor of the defendants notwithstanding the verdict in favor of the plaintiffs. Vandermost appealed.

One evening shortly before midnight, the plaintiff Vandermost entered the defendant's restaurant in Bellflower, California, with a group of friends. He had been drinking beer during the course of the evening, but he was not feeling the effects of alcohol at that time. Shortly after being seated, Vandermost and his friends heard a commotion near the register area. Vandermost left the booth where they were sitting, saw a man (Raymond Moreno) at the cash register and heard him tell the cashier, "Give me the money or I am going to shoot someone." Moreno then fired a shot from a handgun into the ceiling. The cashier slammed the register shut, thereby automatically locking it. She looked briefly at Vandermost and fled into the kitchen. Moreno attempted to open the cash register, but was unsuccessful.

Vandermost, seeing that the robber was angry and confused, moved toward the exit, whereupon Moreno approached him and demanded his money. When Vandermost answered, "I don't have any money," Moreno shot him. He also shot another customer who was then entering the restaurant and stated, "That will prove I'm not f___ing around." The entire incident lasted approximately two minutes.

Vandermost sued the cashier and her employer, alleging that their negligence was the cause of his gunshot wound injury. At the trial, the cashier testified that she did not recall being instructed by her employer on how to act in the event of a robbery.

It has long been the law in California and many other states, that a business proprietor has a duty to protect guests from the wrongful acts of others which threaten the safety of those guests when the proprietor reasonably expects the wrongful acts to occur and to result in injury. The defendant restaurant owner and the employee-cashier did owe Vandermost a duty. The question here lies in whether the defendants' duty to protect the plaintiff from the wrongful acts of third parties extended to requiring them to cooperate with the armed robber Moreno during the course of the robbery in order to avoid provoking him into attacking Vandermost.

The major factors in determining the existence and scope of the defendant's duty in every case are: (1) the foreseeability of harm to the plaintiff, (2) the degree of certainty that the plaintiff suffered injury, (3) the closeness of the connection between the defendant's conduct and the injury suffered by the plaintiff, (4) the blame to be attached to the defendant's conduct, (5) the policy of preventing future harm, (6) the moral blame to be attached to the defendant's conduct, (7) the public policy of preventing future harm to other guests, (8) the extent of the burden on the defendant, (9) the consequences to the community of imposing a duty to exercise care with resulting liability for breach, and (10) the availability, and prevalence of insurance for the risk involved.

Citing cases from Illinois, Kentucky and Pennsylvania, the California appellate court ruled that the proprietor of a business establishment has no duty to comply with a criminal's demands in order to avoid potential injuries to a patron. The court noted that while it was foreseeable that Vandermost might be injured by Moreno if the cashier did not comply with Moreno's demands, the defendants did not owe a duty to comply with those demands. Furthermore, in applying the test listed above, the court determined that although it was certain that the plaintiff did in fact suffer injury, the relationship between the injury and the defendant's conduct in not complying with the robber's demands was at best tenuous. The court stated that there was no moral blame attached to the defendant's conduct. But, most importantly, the court agreed with the Illinois Supreme Court which previously held as follows:

> If a duty is imposed ... to comply with such a demand the same would only inure to the benefit of the criminal without affording the desired degree of assurance that compliance with the demand will reduce the risk to the invitee. In fact, the consequence of such a holding may lead to encourage the use of hostages for such purposes, thereby generally increasing the risk to invitees upon business premises.

Accordingly, the appellate court ruled that the trial judge was correct in setting aside the jury's verdict and entering judgment for the defendants.

COMMENTARY:

It is clear that if the plaintiff's argument had prevailed, then the retailer would have little choice in deciding whether to comply with a criminal's demand and surrender the money, or to refuse the demand and be held liable in a civil action for damages brought by the patron/hostage.

Also, the knowledge of this dilemma on the part of criminals would provide criminals with additional leverage to enforce their demands. While the results of the case may appear to be harsh when viewed from the perspective of the plaintiff, it is for the protection of all future business patrons that the rule finding no duty to comply with the demands of the robber was adopted.

PSCLR – OCT 85 (7)

Restaurant May Be Liable To Customer Who Was Attacked And Injured While Assisting Restaurant's Security Guard:

John E. Welsh v. White Castle Systems, Inc., *No. 84-1126, in the Appellate Court of Illinois, First District, Second Division, decided May 21, 1985, rehearing denied June 26, 1985, reported at 479 N.E.2d 944.*

John and Cheryl Welsh brought a negligence action against White Castle Systems, Inc., to recover damages for injuries John received when he attempted to assist a security guard at a White Castle restaurant. White Castle Systems was granted a motion for summary judgment, and the Welshes appealed.

The Welshes went bowling with another couple, Paul and Josephine Angone, and afterward they went to a lounge. Upon leaving the lounge at approximately 1:00 A.M., they went to a White Castle restaurant. While the wives remained in the car, which was parked three feet from the entrance in a well-lighted parking lot, Welsh and Angone went into the restaurant to get some food. While standing in the service line, Welsh noticed a teenage boy pick up a White Castle napkin holder, place it into his pocket, and leave the restaurant with another boy. Welsh reported the theft to a uniformed security guard who was on duty in the restaurant. The guard followed the boys outside and Welsh and Angone remained in line. The two men got their food and proceeded to the car, but stopped when they heard shouts for help. Looking across the parking lot, they saw the security guard struggling with the two teenage boys. Welsh and Angone ran over to assist the guard; Welsh later testified that he never felt as though he was in any physical danger. He was six feet four inches tall and weighed 195 pounds, while the boys were each about five feet nine inches tall and weighed about 140 pounds. Further, there were no weapons visible in either the security guard's or the boys' hands. However, when Welsh reached the fracas and grabbed one of the boys, he was stabbed and slashed several times with a knife. Welsh remembered nothing of being stabbed. After grabbing the boy, he remembered waking up with his head in his wife's lap, and being taken by ambulance for emergency medical treatment and surgery. His injuries were serious and included a collapsed lung, a pierced liver, and a slashed throat. He remained in the hospital for about ten days and he missed six weeks of work. He now has scars on his side, stomach, and throat from the stab wounds and the surgery that followed.

The Welshes' lawsuit alleged that White Castle was negligent in failing to maintain its premises in a reasonably safe condition, to provide proper security and an adequate number of security guards, and to establish proper procedures to remove dangerous persons from the premises. The trial court granted the defendant's motion for summary judgment; the judge ruled that the defendant did not have a duty to protect the plaintiff from an unforeseeable criminal assault.

On review, the appellate court reversed this judgment, for it did not view Welsh merely the victim of an unforeseeable criminal attack. Nor did the court agree with the defendant's argument that it could not be held liable because the plaintiff voluntarily assumed the risk that led to his injuries. Rather, Welsh was a business invitee of White Castle and as such the defendant owed him a duty of reasonable care to provide for his safety. The court ruled that a question of fact was raised as to whether White Castle and its agent, the guard, properly discharged this duty to Welsh.

White Castle additionally argued that it could not be held liable for the acts of the security guard, because he was actually employed by the security company with which the defendant contracted for security services. The appellate court ruled that this would not prove that White Castle is not liable for the guard's conduct as a matter of law.

The Appellate Court of Illinois, finding that questions of fact have yet to be resolved in this case, therefore reversed the trial court's motion for summary judgment and remanded the issue for trial.

PSCLR – FEB 86 (2)

Restaurant May Be Liable For Customer's Injuries From Assault By Unknown Criminals While At Restaurant Drive-In Window:

Judith Antrum v. Church's Fried Chicken, Inc., *No. 288863, in the Superior Court of Connecticut, Judicial District of Hartford-New Britain at Hartford, decided March 28, 1985, reported at 499 A.2d 807.*

Judith Antrum was waiting in her car at the drive-in window line of a Church's Fried Chicken Restaurant in Hartford, Connecticut, when she was attacked by three unknown men. She sustained severe personal injuries, and brought suit against the restaurant's corporate owner, Church's Fried Chicken, Inc., alleging that the defendant owed her, an invitee, a duty to provide adequate protection. Church's moved to dismiss the suit, but the motion was denied by the court and the denial was affirmed on appeal.

Church's contended that Antrum's lawsuit should be dismissed because in Connecticut a business does not have a duty to protect its invitees from the criminal acts of third parties, unless special relationships and special circumstances exist which impose such a duty. Church's argued that there were no special relationships or circumstances in this case, and further that the plaintiff failed to allege in her complaint that there had been specific past incidents of crime at the restaurant.

The appellate court ruled that in Connecticut the business owner's duty to an invitee is to exercise reasonable care to keep the premises safe for its reasonable use by the customer. The business owner may be held liable for a customer's injuries resulting from third party crime, if it can be shown that the owner's negligence substantially contributed to the injuries. Further, specific allegations of past incidents of crime are not required in the complaint to establish a cause of action for negligence; although of course the defendant's breach of duty may be proved at trial by evidence of such prior incidents to establish foreseeability of the crime. The court went on to cite "numerous cases in other jurisdictions which have addressed the liability of a business proprietor for the injuries sustained to an invitee as the result of the criminal acts of a third party," including *Stalzer v. European American Bank* (NY), reported in the August 1982 **PSCLR** at page one; *Henley v. Pizitz Realty Co.* (AL), reported in the January 1985 issue at page nine; *Latham v. Aronov Realty Co.* (AL), reported in the August 1984 issue at page seven; and *Berdeaux v. City National Bank of Birmingham* (AL), reported in the May 1983 issue at page three. In the cases cited by the court which supported imposing liability on the business owner, the plaintiff's injury resulted from a foreseeable risk of harm, actually or constructively known to the defendant.

The defendants also argued that the action should be dismissed because the type of duty the plaintiff sought to impose on the defendants was against public policy. It would unfairly impact business in urban areas by requiring more costly security measures, and would be impossibly vague and uncertain to adhere to. In addressing this question the appellate court referred to the case of *Butler v. Acme Markets, Inc.* (NJ) which was reported in the November-December 1982 **PSCLR** at page six. In *Butler*, the court held that "it is fair that the costs of negligent failure to protect against crime be...borne by the operators and indirectly patrons of such shopping facilities." Moreover, the "duty was no more vague than is the test of reasonableness throughout our law generally."

In affirming that the defendant's motion to dismiss should be denied the appellate court ruled that the case should proceed to trial and a jury should determine whether the defendant exercised reasonable care in performing its duty to protect the plaintiff-customer from criminal attack by third parties.

PSCLR—FEB 86 (4)

Retailer Is Held Not Liable For Assault Of Patron In Parking Lot Of Shopping Mall, Where Lease Places Common Areas Under Control Of Mall Owner:

Edward St. Phillips v. Trent O'Donnell and Allan Wasicki Enterprises, Inc., d/b/a T. McGee's, *No. 2-84-0787, in the Appellate Court of Illinois, Second District, decided October 21, 1985, rehearing denied November 21, 1985, reported at 484 N.E.2d 1209.*

Edward St. Phillips brought suit against Allan Wasicki Enterprises, owner of a bar named T. McGee's, after he was assaulted upon leaving the bar by an ejected patron. He also named the ejected patron, Trent O'Donnell, as a party defendant. The trial court granted the defendant Tavern's motion for summary judgment, and the plaintiff St. Phillips appealed.

The appellate court summarized the appeal:

> The sole issue presented on appeal is whether defendant, the lessee-operator of a tavern located in a shopping center, owes a duty under the facts here to protect a patron, plaintiff, who has left the tavern premises, from the negligent or intentional harmful acts caused by a third person where defendant is placed on notice of the violent propensities of the third person and the assault upon plaintiff occurs on the common area of the shopping center used for parking for businesses of the shopping center.

St. Phillips was attacked after leaving T. McGee's by Trent O'Donnell, another patron of T. McGee's who had been ejected from the bar by the owner Wasicki and an employee Massengill because of his obnoxious behavior, including the violent assault and battery of two other patrons in the tavern. After forcibly removing him from the bar, Wasicki and Massengill watched as O'Donnell walked about fifty feet from the entrance and then they returned inside the bar. St. Phillips left a few minutes after this to move his car to a spot closer to the entrance of the bar. After doing so and as he prepared to get out of his car, St. Phillips was approached by O'Donnell who dragged him out of the car and began kicked him until he was unconscious.

The defendant's motion that the court grant a summary judgment was accompanied by a copy of T. McGee's lease with the shopping center. In it the landlord granted to defendant, its agents, employees, customers and invitees, the right to use the common areas which included the parking lots, sidewalks, and roads on the shopping center premises. The landlord reserved the right to determine the operation and maintenance of these areas, and the right to establish rules and regulations for their use. The defendant therefore maintained that it had no duty to protect the plaintiff from injury beyond its own leased premises, which did not extend to the shopping center's common areas.

The plaintiff responded by citing Illinois case law holding that a tavern operator may be held liable for the injuries of patrons due to the acts of third persons where the danger was apparent, and the tavern operator failed to protect the patron. St. Phillips contended that T. McGee's owner owed him a duty as an invitee even though the parking lot was a "shared premises" because

> the likelihood of injury was great, the magnitude of the burden to guard against assault was not great, and the consequences of placing the burden on defendant are wholly beneficial ....

On review the appellate court noted that for a plaintiff to recover in tort for a defendant's negligence, it must be shown that the defendant failed to fulfill a duty owed to a plaintiff. The existence of the duty is a question of law, to be determined by the courts, and the duty owed to invitees by a business owner is that of reasonable care. The court stated that Illinois case law has held:

> a tavern operator is in a special relationship with third persons on his premises which imposes a duty to take reasonable action to protect invitees from foreseeable dangers caused by third persons.

However, there are no Illinois appellate decisions imposing a duty on a tavern operator to safeguard patrons off the property owned or controlled by the tavern against foreseeable criminal assault by third persons.

Therefore the evidence clearly showed that St. Phillips' assault occurred off the premises controlled or leased by the defendant, and the lease specifically stated that the common areas were subject solely to the rules, regulations, maintenance, and control of the landlord. The appellate court rejected St. Phillips' argument that a duty to protect him in the shopping center's common areas should be imposed on the tenant tavern merely because of the defendant's shared right to use those common areas. Since there is no legal duty owed the plaintiff by the defendant, the summary judgment was affirmed.

PSCLR – FEB 86 (11)

Restaurant Customer May Be Guilty Of Negligence For Contributing To Own Injuries By Startling Robber Who Then Shot Customer:

John P. Lannon v. Taco Bell, Inc., *No. 82CA0102, in the Colorado Court of Appeals, Division I, decided March 21, 1985, rehearings denied May 2, 1985, certiorari granted (Taco Bell) November 4, 1985, certiorari denied (Lannon) November 4, 1985, reported at 708 P.2d 1370.*

The plaintiff John Lannon, a Denver taxicab driver, went to the one of the Taco Bell restaurants in the Denver area. As he approached the counter, he saw an armed gunman behind the counter removing cash from a safe in the floor. Lannon then backed away from the counter and bumped into a second gunman. Lannon ran out the door into the parking lot. The gunman fired at Lannon threw a window, wounding Lannon in the hand.

Lannon sued, and obtained a jury verdict in his favor; Taco Bell appealed.

On the appeal, it was determined that the trial judge erred by failing to instruct the trial jury on the issue of comparative negligence. That is, the appellate court agreed with Taco Bell that the issue of Lannon's own negligence should have been submitted to the jury which should have then been permitted to decide whether Lannon failed to exercise reasonable care for his own safety under the all the facts and circumstances surrounding the situation in which he found himself at the time he was injured. The law is in fact clear that every person

> has a duty to refrain from acts or omissions to act which may contribute to the totality of acts which cause injury to him.

Taco Bell argued, and the court noted, that Lannon had previously received training as to what actions he should take during a robbery. He had been instructed to cooperate, to make no sudden moves, and to refrain from any robbers. In spite of these instructions when he encountered the robbery in progress at the Taco Bell restaurant, he turned and ran, thereby apparently startling the gunman who wounded him. The court concluded that a jury could have found under these facts that Lannon had acted rashly in drawing the gunman's attention to himself and he may have given the gunman that he was running for the police.(The appellate court also stated, "Perhaps he should have quietly walked away instead of running.") For this reason, the appellate court reversed the judgment in favor of Lannon and remanded the case for a new trial, instructing the trial judge to permit the jury to determine whether Lannon had in fact been negligent in which case any award in his favor would have to be reduced accordingly, if not eliminated entirely.

Taco Bell had also argued on its appeal that the jury should not have been asked to decide whether Taco Bell had breached its duty to Lannon by failing to employ an armed security guard. The appellate court, however, disagreed with this contention, concluding that the trial judge had decided that the evidence as to the effectiveness of armed guards in other area restaurants, prior similar incidents at the Taco Bell restaurant and in the general area, plus the great risk of injury to customers during a robbery, all taken together established a duty on the part of Taco Bell to employ reasonable measures to alleviate a known dangerous condition. Therefore, the jury was correctly asked to decide whether Taco Bell had breached its duty to Lannon by failing to take reasonable steps to prevent foreseeable armed robberies on the premises. The test is whether a reasonably prudent restaurant owner, under the same or similar circumstances, would have taken available precautions, such as employing an armed guard.

As to the issue of causation, the court noted that the Taco Bell restaurant in question was located in a high crime area and had in fact been robbed on frequent prior occasions. Other similar restaurants in the area had also been robbed. The fact that Lannon was injured by the criminal act of a third party does not absolve Taco Bell from responsibility if the criminal misconduct of the third party is reasonably and generally foreseeable by Taco Bell, as the jury had undoubtedly decided from the evidence presented.

Accordingly, the appellate court reversed the judgment in favor of Lannon and the case will be retried.

COMMENTARY:

There was an interesting dissent filed by one of the judges, who stated that he would have reversed the judgment in favor of Lannon but, rather than ordering a retrial, would have dismissed the complaint entirely. This judge argued that the principal of law placing a duty on the defendant to protect against foreseeable harm should not be used as the basis for the legal requirement that Taco Bell hire armed guards. The dissenter pointed out that the majority decision has the far reaching implication of requiring a defendant, under threat of becoming financially liable, to equip his employees with a deadly weapon and authorizing them to use that weapon to injure or kill in order to protect customers from the consequences of criminal acts of others. The judge pointed out that while the presence of armed guards may sometimes deter an armed robber, it may also precipitate a confrontation, thereby posing an even greater danger to customers.

Furthermore, in his dissenting opinion, this judge pointed out that in all of the prior similar cases decided by this court, the specific duty imposed on the landowner "unquestionably would have alleviated or reduced the risk of injury"; whereas, he points out that in the above case, the duty imposed on the landowner by the majority opinion

> *specifically mandates the introduction of deadly weapons onto defendant's premises, and, by necessary implication, their use to kill or wound.*

PSCLR–APR 86 (8)

Restaurant Is Not Liable For Failing To Protect Customer From Unforeseeable Criminal Assault:

Edie Law v. Omelette Shop, Inc., et al., *84-664, in the Supreme Court of Alabama, decided November 22, 1985, reported at 481 So. 2d 370.*

Edie Law stopped at the Omelette Shop in Birmingham, Alabama, at two o'clock in the morning. After eating, she exited through the front door and as she stood in the front of the restaurant, a man accosted her. He pushed her to the ground, took her purse, and fled. She filed suit against the Omelette Shop, alleging that the defendant knew of the probability of third-party criminal assaults on business invitees and that it had negligently failed to take adequate steps to protect its invitees. She claimed that her personal injuries resulted from the defendant's negligence. The trial court granted the Omelette Shop's motion for summary judgment, and Law appealed. The Supreme Court of Alabama affirmed.

The appellate court reviewed Alabama case law concerning the duty of business owners to protect customers from criminal activity. The court reviewed several decisions it had recently issued on this question, including *Ortell v. Spencer Companies* (reported in the February 1986 **PSCLR**, case nine, page six), *Henley v. Pizitz Realty Co.* (reported in the January 1985 issue, case nine, page nine), *Latham v. Aronov Realty* (reported in the August 1984 issue, case eight, page seven), and *Berdeaux v. City National Bank of Birmingham* (reported in the May 1983 issue, case three, page three). These cases held that a business owner had a duty to take reasonable steps to protect customers from criminal attack if the owner had actual or constructive knowledge that customers were endangered by the risk of such attack. However, the court stated that "it is difficult to impose liability on one person for an intentional criminal act committed by a third person," and in fact had not found even one case in which the facts commanded that such a duty be imposed. For example, in *Ortell*, there was evidence of eight incidents of crime on the defendant's premises and 80 in the two-block area around the premises in the three years prior to Ortell's assault. However, the appellate court ruled:

> The occurrences of this three-year period established by the plaintiff did not alone impose a duty upon this defendant to protect this plaintiff from the crime for which she suffered at the hands of the third party criminal.

In this case there was evidence of only one prior robbery at the Omelette Shop reported to the Birmingham police in the three-year period preceding the assault on Law. Since the court held previously that the evidence in *Ortell* was insufficient to impose a duty on the defendant, the court here ruled that Law "failed to offer any proof that Omelette Shop had any knowledge of ... a probability" of criminal attack, and affirmed the dismissal.

COMMENTARY:

It is a common misconception that a business owner/operator is almost automatically liable to any business invitee who is injured on or about the owner/operator's premises. Private security managers who are frequently called upon by management to render an opinion as to a company's liability in a given situation, should be especially aware of the defenses available to the company in the event a claim or suit is prosecuted by an injured invitee.

While there is certainly a trend in the country toward holding the owner/operator liable for such third party criminal assaults, the plaintiff still must be able to prove by a preponderance of the evidence that the defendant had a duty to protect and/or warn the plaintiff, that the defendant failed to take reasonable measures to fulfill this obligation, that the plaintiff was damaged, and that the plaintiff's damage was the result of such negligence on the part of the defendant.

In this connection in order to prove the defendant's negligence, the plaintiff must also prove that the injury and damage to the plaintiff was foreseeable by the defendant as being a probable result of its failure to fulfill its duty to the plaintiff. This burden of proof is heavier in some jurisdictions, such as Alabama, than in others, due to the subjective nature of the evaluation of the circumstances. Nevertheless, the plaintiff does have the burden of proof in each and every one of these cases, and private security managers should not be too quick to recommend payment of a claim or settlement of a lawsuit without having a complete understanding of a factual situation, plus a working knowledge of the recent cases on this issue in the particular jurisdiction in question.

PSCLR – JUN-JUL 86 (3)

Restaurant Is Not Liable To Customer Who Was Mistaken For Robber And Shot By Police During Undercover Police Stake-Out At Restaurant:

John Stephen Ralls v. Noble Roman's Inc., *No. 4-985A264, in the Court of Appeals of Indiana, Fourth District, decided April 16, 1986, reported at 491 N.E.2d 205.*

John Ralls was inadvertently shot by Indianapolis police officers during a robbery of Noble Roman's restaurant. Ralls brought suit against Noble Roman's alleging that the restaurant owner had negligently failed to inform him, a customer at the restaurant, that a robbery attempt was expected that evening and that the police were conducting an undercover stake-out of the restaurant. The judge at the trial level granted Noble Roman's motion for summary judgment and Ralls appealed.

Ralls and two of his co-workers met at Noble Roman's for pizza at about 11 PM one evening. The niece of one of Ralls' companions worked at the restaurant and she told them that three men who had previously robbed Noble Roman's were in the restaurant earlier that evening, and that the police had been there. Ralls thought that she was simply trying to impress them, and he and his friends remained at the restaurant. In fact, the three robbers had been in the restaurant and the police had arranged for several officers to stake out the restaurant that evening. Because there had been several robberies of local pizza restaurants, the police planned to apprehend the suspects after they robbed Noble Roman's. The police told Noble Roman's employees to give the robbers the money and not to interfere in any way. The police decided in the interest of safety not to inform Noble Roman's customers.

The robbers did return to Noble Roman's that same evening while Ralls and his friends were still there, but had finished eating and had paid their bill. Ralls left quickly and went to his car, only to discover that he had left his car keys. Ralls walked back toward the restaurant just as the three robbers were making their getaway followed by two off-duty police deputies. The stake-out team did not know that the off-duty officers were in the restaurant and the off-duty officers did not know about the stake-out. The deputies and the robbers exchanged gunfire; Ralls fled from Noble Roman's property into a dark yard adjacent to the restaurant. He unfortunately chose to run in the direction of the getaway car and, worse, he was dressed similarly to one of the robbers. A plainclothes police officer spotted Ralls, mistook him for one of the robbers, and ordered him to stop. Thinking that the plainclothes officer was one of the robbers, Ralls did not stop and the officer shot at him.

The appellate court reviewed the trial judge's decision to grant summary judgment to the defendants. Ralls' suit alleged that Noble Roman's was guilty of negligence. In Indiana the tort of negligence is proved by showing the existence of three elements:

> (a) a duty on the part of defendant in relation to the plaintiff, (2) failure on the part of defendant to conform its conduct to the requisite standard of

> care required by the relationship,and (3) an injury to the plaintiff resulting from such failure.

The appellate court determined that Noble Roman's did owe a duty to its customer Ralls, but that the restaurant was not required to inform the customer of the stake-out, noting that to inform "the public about an undercover police operation would defeat the purpose of that operation." Moreover, Ralls had nonetheless been informed of the situation by a Noble Roman's employee. The appellate court affirmed that the defendants could not be held liable for the unforeseeable and uncontrollable chain of events leading to Ralls' shooting injuries. The restaurant did not have control over the method and manner in which the police carried out its stake-out of the restaurant or apprehension of the robbers. The shooting took place off the restaurant's property in an area not normally used by its customers.

The appellate court affirmed the summary judgment granted the defendants by the trial court.

COMMENTARY:

Generally speaking, a legal duty will not be imposed upon a person to control the conduct of a third party so as to prevent the third party from causing harm to another, unless certain conditions exist:

> *(a) there is a special relationship between the person and the third party which imposes a duty upon the person to control the third party's conduct; or (b) there is a special relationship between the person and the other who may be harmed by the third party, which relationship which gives the other the right to be protected.*

PSCLR – JUN-JUL 86 (8)

A Bar Is Not An Operation Which Necessarily Attracts Crime And Bar Owner Has No Special Duty To Protect Customers Against Third Party Criminal Acts Unless Such Acts Are Reasonably Foreseeable:

Galen Yarborough d/b/a Daniel's Club v. Michael Erway, *No. A14-85-082-CV, in the Court of Appeals of Texas, Houston (14th District), decided November 27, 1985, rehearing denied January 9, 1986, reported at 705 S.W.2d 198.*

Michael Erway suffered serious knife wounds in his neck after being assaulted by a fellow patron at Daniel's Club in Angleton, Texas. Erway recovered damages for his injuries after a trial of his lawsuit against the Club's owner, Galen Yarborough. The Texas Court of Appeals reversed the judgment granted to Erway and granted judgment to Yarborough.

Erway had apparently raised the ire of Danny Henderson after he and Henderson bumped into each other, though neither spoke to the other, while standing near the bar at Daniel's Club. Testimony by Erway and Henderson conflicted: either Henderson bumped into Erway again or angry words were exchanged, but Erway understood that Henderson "might want to fight and followed Henderson outside. Erway testified that Henderson stabbed him in the neck as soon as he walked out of the bar. Henderson testified that he pulled the knife after Erway first grabbed him by the shirt. Other patrons came outside during the altercation, but no one else saw the stabbing. After Erway was stabbed, Henderson fled and Erway was taken to the hospital. Erway spent four months in the hospital, and his medical expenses amounted to $71,152.79.

The jury at the trial of Erway's action found "all parties guilty of causal negligence," attributing 50 percent to Daniel's Club, 35 percent to Henderson, and 15 percent to Erway. Total damages awarded Erway was $71,152.79 for medical expenses incurred, $100,000.00 in damages, and $20,000.00 for future medical care. Reduced by the 15 percent attributable to Erway's own negligence, the damages awarded to Erway amounted to $162,479.87. Since Erway had already received $25,000 from the state of Texas under its Crime Victims Compensation Act, the state was granted $25,000 from Erway's award. In addition, Daniel's Club was granted the right to seek indemnification from Henderson. Henderson was convicted of criminal charges arising from the incident and was assessed ten years probation along with monthly payments of $300 for ten years.

The appellate court found that there was absolutely no evidence to support the jury's verdict. According to the evidence presented at trial, Daniel's Club is a neighborhood bar which hosted a regular group of patrons and was not the site of violent activity. The Club's owner and its employees testified that they understood their duty under the Texas Alcoholic Beverage Code to discover and to remove intoxicated patrons, and to call the police if such persons refused to leave.

Before the incident between Henderson and Erway, both men were drinking, though neither appeared to be intoxicated. There was testimony that Henderson was "talking loudly, and dancing and clowning around," but he had had only one drink in Daniel's. A Club waitress testified that he didn't do anything which made her think he posed a threat to anyone. Henderson testified that the fight began because Erway was staring at him, and that Erway threatened to kill him when they were leaving the bar. No more than two minutes separated the decision to go outside and the stabbing.

The appellate court stated,

> We have found no precedents in Texas case law directly on point concerning the duty a bar or tavern keeper owes his patrons to protect them from criminal assaults or intentional misconduct by third parties. There is no civil damage, dram shop, or similar statute in Texas providing a remedy against a bar owner for injuries suffered by person who were injured by an intoxicated patron.

Moreover, relying an decisions by the Texas appellate courts on a property owner's general duty to protect against third party criminal activity, the Texas Court of Appeals stated that criminal danger must be foreseeable in order to impose a duty on a property owner to provide security to protect invitees from crime.

> The appropriate rule in such a case is that owners whose operations do not attract or provide a climate for crime are under no duty to guard against criminal acts of third parties unless they know or have reason to know that acts are occurring or are about to occur there which pose an imminent probability of harm to an invitee. [See *Castillo v. Sears, Roebuck & Co.*, in the April 1984 **PSCLR**, case 10, page eight and *Nixon v. Mr. Property Management*, in the November-December 1984 **PSCLR**, case seven, page five.]

The appellate court acknowledged that the potential for violence may well be increased where alcohol is sold and consumed, but the evidence in this case was clear that the fight between Erway and Henderson occurred suddenly and without warning. Erway suffered injury because he voluntarily participated in the fight and Henderson intentionally assaulted him. There wasn't time or warning for Club employees to prevent the fight. Therefore, the verdict granted Erway was reversed and judgment was granted the owner of Daniel's Club.

PSCLR – AUG 86 (9)

Where Retailer Gratuitously Assumes Duty To Protect Persons Outside Retailer's Premises, Injured Person May Recover For Breach of Duty:

Roger Ember and Jane Ember v. B.F.D., Inc., *No. 2-883-A-291, in the Court of Appeals of Indiana, Second District, decided March 13, 1986, reported at 490 N.E.2d 764.*

The defendant B.F.D., Inc., is an operation which does business as O'Sullivan's Italian Pub in Fort Wayne, Indiana. The Pub is a tavern which was opened in 1978 in an older neighborhood with residential and commercial development. The Pub's parking lot is behind its building and another parking lot is located across the street which the Pub does not own. The Pub quickly became known as "the hottest bar in town" but also received complaints from nearby residents and the local neighborhood association because of problems attributable to the Pub's customers.

As a result of these complaints, the business had some difficulty in renewing its liquor license in 1979. It was required to explain to the Indiana Alcoholic Beverage Commission what measures it was taking to remedy the situations causing the complaints. In doing so, the Pub wrote to the Commission that they had hired youths to police the area for trash, bottles, cans, and other litter; to remedy parking problems, the company improved their own parking lot behind the Pub and had warnings and citations issued to illegally parked cars. Also, a police officer was stationed outside the tavern and the Pub employed two uniformed police officers, one stationed at the door and one "roving" outside to solve the excess noise problem. The Pub told the Commission that a uniformed police officer would be on duty from 10 PM to closing every Wednesday through Saturday, and on Mondays and Tuesdays as needed. They wrote, "It is the responsibility of the officer to maintain order in the parking lot as well as any adjoining streets."

The Pub also, in response to a complaint from the city, distributed a flyer to the residents in the neighborhood telling them that the Pub managers would be willing to respond and help with any disturbance in the neighborhood, and invited the residents to call the Pub to report any such problems. The flyer specifically said that the Pub managers would attempt to correct any problem, "even if it does not pertain to us." The residents were told that in addition to having the police officer on duty in the evening, the Pub managers would continue to patrol the area themselves sporadically and to check the parking lots in the area.

The Pub did indeed maintain police and civilian patrols outside its premises because of the adverse effect its customers were having on the immediate neighborhood. In addition, these patrols did occasionally render assistance with problems in the neighborhood, even though the problems occurred off the tavern premises, and even though the disturbance had nothing to do with the tavern or its customers. On St. Patrick's Day, when a large crowd was expected and the security officers were therefore on duty as early as 8:00 PM, the plaintiff, Roger Ember, made arrangements to meet his wife and another couple at the Pub to celebrate the holiday. Ember parked directly across the street from the tavern, on the sidewalk, and he was attacked as he was about to cross the street to the tavern. At the time of the attack, the off-duty police officer had not yet arrived, and the Pub manager was patrolling the rear parking lot. Ember was severely injured and required hospitalization.

Ember and his wife filed suit, but the case was dismissed by the trial judge who ruled in favor of the corporation for the reason that there was no breach of any duty owed toward Ember by the Pub. Ember and his wife appealed. They argued that the Pub had gratuitously assumed a duty to protect customers on the streets and adjacent parking lots, and had breached this duty by failing to have an officer on duty and by permitting an unruly crowd to gather outside the bar. On the other hand, the defendant denied any duty to protect customers off its premises, and further argued that even if it did assume a gratuitous duty, Ember did not rely on any such representations by the business to provide security for him.

The Indiana appellate court began its review of the case by noting that there is normally no legal duty to come to the aid of a stranger. The imposition of any such duty depends upon the existence of a "special relationship," such as the relationship of landowners to invitees, common carriers to passengers, and innkeepers to guests. In the present situation, the duty of a tavern owner who sells alcoholic beverages is to protect customers from injuries from irresponsible persons whom the tavern keeper permits to be in and around the premises. This duty generally includes providing some staff who control conduct on and about the premises.

This case can be decided based upon several well-founded principles of law:

> It is ancient learning that one who assumes to act even though gratuitously may thereby become subject to the duty of acting carefully if he acts at all.
>
> Although no assistance be exacted under the law, if any be tendered it must not be negligently rendered.
>
> The assumption of a duty creates the requisite special relationship between the parties and the corresponding duty to act in the manner of a reasonably prudent person.

The facts of the present case are sufficient for a jury to conclude that the Pub assumed a duty to patrol the area nearby its premises and to protect people in the vicinity from harmful activity. Therefore, the scope of such duty which was assumed by the Pub should be left for a jury to determine, and the trial judge should not have dismissed the case without allowing a jury to evaluate the defendant Pub's conduct and representation. The case was therefore remanded to the trial court for further proceedings.

COMMENTARY:

In reaching its conclusion in the above case, the majority of the Indiana appellate court judges noted that several other jurisdictions have had to deal with similar situations and some have adopted other approaches to consider the problem of "extra-territorial liability" of retailers:

> <u>Expansion of the premises and invitee status</u>– *when it is reasonable for customers to think that the proprietor controls the property adjacent to his own or where a proprietor knows that his customers in fact use adjacent property when coming to and from his business, then a duty to exercise reasonable care for the protection of the customers may be extended beyond the business premises proper. Arkansas and Texas courts have applied this theory in cases dealing with adjacent parking lots. A recent Minnesota case held that a tavern keeper's liability be extended to an assault in a public alley behind the tavern. Also, where children approached an ice cream truck, they were considered to be "invitees" in a California case. A West Virginia court found that minors who were swimming while their father waited to rent a boat at a marina were also "invitees."*
>
> <u>Dangerous activity</u>– *normally a business owner/operator is not responsible for injuries suffered on adjacent property because the business normally does not cause risks outside its boundaries. But if the risks which are conducted on the business property do cause risk of injury off the property, then the business owner/operator may have a duty to guard against the risk causing injuries even off the property.*
>
> <u>Nuisance</u>– *Yet other courts have found that if a business owner's activities cause unruly crowds to gather, then liability can be based upon a nuisance theory, such as a Pennsylvania case which enjoined the operation of a bar because unruly patrons were responsible for deteriorations in the surrounding*

area, and the manner of operation of the bar was determined to be responsible for the patrons' offensive conduct.

PSCLR–SEP 86 (4)

Restaurant Liable If Employee's Conduct Caused Or Contributed To Assault On Customer By Armed Robber:

Richard D. Helms, Jr., and Margaret F. Helms v. Church's Fried Chicken, Inc., *No. 8526SC1030, in the Court of Appeals of North Carolina, decided June 17, 1986, reported at 344 S.E.2d 349.*

Once again we report a case which turns on the question of foreseeability. However, here foreseeability of the crime was not an issue, but rather the foreseeable consequences of conduct by the property owner's employee when faced with a third-party criminal assault.

Richard and Margaret Helms stopped by a Church's Fried Chicken Restaurant in Charlotte, North Carolina. Richard went into the store to place an order, while Margaret remained in their truck. After he bought his food and was making his exit, he claimed that a Church's employee called to him in a loud voice, "When you leave, call the police; we are being robbed." The three robbers immediately attacked the plaintiff Helms. When his wife saw this from the truck, she ran toward the store entrance, but Helms called out to her to get back in the truck and lock the doors. However, she was also attacked, and was robbed before she made it back to their truck.

Richard and Margaret Helms suffered stab wounds and other serious injuries. They sued the Church's restaurant, alleging that the employee's negligence in announcing the robbery-in-progress caused their injuries. The trial court dismissed the suit, and plaintiffs appealed to the North Carolina Court of Appeals.

The appellate court applied North Carolina's premises liability rules to the facts. First, it noted that the Helms were invitees of the restaurant, and therefore Church's owed them a duty to exercise ordinary care to safely maintain the premises. However, a store owner is not an insurer of the safety of invitees and is not ordinarily liable for the intentional criminal acts perpetrated by third parties, unless the owner knows or has reason to know that third party criminal conduct is likely to occur, is occurring, or is about to occur. The extent of the owner's duty to protect against crime depends whether the owner could have or should have foreseen the crime. In this case, Church's argued that they were not liable for the Helms's injuries because the evidence did not show that the robbery was foreseeable.

However, the appellate court found that the foreseeability of the robbery was not the pivotal issue in this case. The crucial issue was, once the hold-up began, did the employee actually make the statement to "call the police" as claimed by plaintiffs, and if so, "whether the Church's employee should have reasonably foreseen the consequences of his or her act, which may have increased the risk of harm to the Helmses." The Helms alleged that the presence of armed robbers in the restaurant posed a danger to customers which was "obvious and apparent" to the employee. The employee's subsequent actions were unreasonable and negligent, served to increase the hazard, and thus caused injury to the Helms.

The Court of Appeals of North Carolina ruled that the Helms had a cause of action against Church's for negligence. Further, certain facts were in dispute: Did the Church's employee make the statement to Richard Helms at all? And, if so, did he or she act negligently and unreasonably in announcing to Helms that a robbery was occurring? The case will be returned to the trial court for a jury to decide whether the employee actually made the statement as alleged by the plaintiffs, and if so, whether the employee acted reasonably in the face of the emergency.

PSCLR—SEP 86 (5)

Franchisor Cannot Be Held Liable For The Wrongful Death Of A Security Guard Employed and Controlled By Franchisee, And Negligence Action Against Franchisee Is Barred By Workers' Compensation Act:

Annabelle Leonard, Administratrix of the Estate of Norris Washington, Deceased, v. All-Pro Equities, Inc., a Pennsylvania corporation; Brady Keys Food Service, Inc., a Michigan corporation; All-Pro Food Service, Inc., a Michigan corporation; and Burger King Corporation, a Florida corporation, jointly and severally, and Brady Keys, Jr., *Docket No. 81504, in the Court of Appeals of Michigan, decided February 4, 1986, released for publication May 5, 1986, reported at 386 N.W.2d 159.*

Burger King restaurant in Detroit, Michigan, employed Norris Washington as a security guard. During an attempted robbery of the restaurant, one of the assailants shot and killed Washington. His estate sued, among others, the restaurant owner-franchisee, All-Pro Food Service; the previous owner of the restaurant, All-Pro Equities; and the franchisor, Burger King Corporation. The trial judge granted summary judgment to all of the defendants, and dismissed the suit.

The plaintiff estate appealed to the Michigan Court of Appeals, alleging that the defendants owed a duty to the guard; that they intentionally, knowingly, and recklessly misrepresented the risks of his employment to him, and that they were therefore liable for his death.

The court reviewed, and affirmed the summary judgments. The plaintiffs did not prove that any surviving connection between the former owner and the current owner and current operations of the restaurant. Therefore, there was no basis for the suit against All-Pro Equities.

The exclusive remedy provision of the Michigan Workers' Compensation Act barred the claim against the current owner, All-Pro Food Services. This provision holds that an employee's recovery for injuries is limited to worker's compensation where these three questions are answered in the affirmative:

> (1) Were the 'conditions of liability under the act' present at the time of plaintiff's injury?
>
> (2) Is plaintiff seeking to recover damages for personal injuries?
>
> (3) Is plaintiff's suit based upon the employer-employee relationship?

The appellate court here found that the Act did apply, rejecting the plaintiff's contention that it was not applicable under the so-called "intentional torts" exception. This "exception" has been accepted by Michigan courts only where the plaintiff alleged a "true" intentional tort, i.e., where there is proof the employer formed a specific intent to inflict, coupled with action aimed at accomplishing injury or death to the plaintiff. In this case the plaintiff merely alleged that the employer was negligent or, at worst, grossly negligent by misrepresenting the danger of the job to get the deceased to accept the job. Accordingly, the Worker's Compensation Act was the sole avenue for the plaintiff's recovery.

The plaintiff also alleged that the defendant Burger King Corporation asserted control over the hiring, firing and supervising of the defendant restaurant's employees, and therefore owed a duty to the deceased. The court agreed that given proof of that level of control over the employees, Burger King would be liable for payment of the workers' compensation benefits, but a negligence lawsuit would be barred under the exclusive remedy provision of the Workers' Compensation Act. Since the claims against Burger King had no basis in law or fact, that summary judgment was also affirmed.

PSCLR – OCT 86 (5)

Duty Of Bar Owner To Exercise Ordinary Care To Protect Public From Injury By Third Parties Does Not Extend Beyond Business Premises:

Mary Delvaux, now Mary Karl, individually and as special administratrix of the estate of Steven Delvaux; Candy Ann Delvaux and Melissa Delvaux, by their guardian ad litem, Carl W. Kuehne v. Leon Vanden Langenberg d/b/a Van's Bar, The Shelby Mutual Insurance Company of Shelby, Ohio, Thomas Collins, Robert J. Collins, Phyllis J. Collins, and the Hanover Insurance Company, *No. 84-1404, in the Supreme Court of Wisconsin, decided June 4, 1986, reported at 387 N.W.2d 751.*

Steven Delvaux was killed in a fight which occurred outside a bar located in Green Bay, Wisconsin. He, his wife Mary, and his brother Jeff had gone to the bar, which was owned and operated by defendant Leon Vanden Langenberg. Also at the bar on that occasion was the defendant Thomas Collins who was a minor. The evidence does not show clearly whether the bar owner ever asked Collins for any proof of his age, but it does show that Collins carried a false identification card indicating that he was of adult age. Collins was there on the evening in question to shoot pool as a member of a pool team sponsored by the bar. Collins' parents were aware that he frequently went to the bar for that purpose.

Steven and Jeff Delvaux played pool with Collins several times during the evening, until an argument developed involving the three of them, based partly on Jeff Delvaux's poor pool shooting. Jeff began using abusive language and abused the bar's pool table equipment; Collins told him to stop mistreating the equipment. In response Jeff Delvaux pushed Collins. The bar owner separated them and directed that the Delvauxs sit at the opposite end of the bar from Collins. The Delvauxs and Collins, however, continued to shout abusive comments to one another.

The Delvauxs finished their drinks and left the bar, and Vanden Langenberg required Collins to remain in the bar because he did not want him to get into a fight with the Delvauxs.

Collins had consumed 10-15 glasses of beer during the four to five hours he had been in the bar. He left the bar to go home over a half hour after the Delvauxs left. Upon arrival at his house, Collins told his older brother about his encounters with the Delvauxs and the two Collins boys then returned to the bar to look for Jeff and Steven Delvaux.

While the Collins brothers were driving around looking for the Delvauxs, they saw Jeff in front of another bar, several blocks from the first bar. The evidence as to the events that followed is in conflict, with Thomas Collins testifying that his brother dropped him off, whereupon he began fighting with Jeff Delvaux. Collins said that he then saw Steven Delvaux come out of the bar, and approach him in a threatening manner. At which time, Collins said that he threw Jeff to the ground and hit Steven in the face twice and kicked him in the head.

Jeff Delvaux testified that it was impossible for Collins to have killed Steven because Jeff was on top of Collins when Collins claimed to have hit and kicked Steven. Jeff also said that while he was wrestling with Collins, he heard his brother Steven cry out for help. When Jeff looked up, he testified that he saw "two pairs of legs running away from where Steven was then lying."

Steven's death was caused by a cerebral hemorrhage resulting from the beating he received, and an action for wrongful death was brought by his widow and minor children. They claimed that Vanden Langenberg, the owner of the first bar where the defendant Collins had been playing pool with the Delvauxs was liable for the death of Steven in that the defendant bar owner had negligently sold alcoholic beverages to a minor and that he had breached his duty as a bar owner to protect the public from injury from other patrons.

The widow and children also claimed in their lawsuit that Robert and Phyllis Collins, the parents of Thomas Collins, were liable for the wrongful death of Steven Delvaux because they had negligently supervised their minor son who had intentionally struck, kicked, and killed Steven Delvaux.

Thomas Collins was convicted of homicide by reckless conduct prior to the trial for civil liability.

At the civil trial, the jury concluded that Collins had not committed a battery upon the deceased, apparently believing the testimony of Jeff Delvaux that he was on top of Collins at the time of the assault on Steven Delvaux. The jury did conclude that Delvaux, Vanden Langenberg, and the parents of Thomas Collins were all negligent as claimed by the plaintiffs. The trial court dismissed the wrongful death complaints after the jury's verdict. The plaintiffs appealed.

Before coming to the Wisconsin Supreme Court, the case was considered by the Wisconsin Court of Appeals. That court affirmed the dismissal of the lawsuits in favor of the tavern owner Vanden Langenberg, holding that the law in Wisconsin at the time of Steven's death and during the trial provided that there could be no recovery under the facts of this situation against the bar owner. The intermediate appellate court also determined that the jury's determination that there had been no negligent or intentional misconduct on the part of Thomas Collins was inconsistent with its determination that there was negligent supervision of him by his parents. Accordingly, the judgment in favor of Collins and his parents was reversed and the case was remanded for a new trial to determine the issue of liability on the part of Steven Delvaux, Collins, and Collins' parents. The case was then appealed to the Wisconsin Supreme Court.

On appeal before the Wisconsin Supreme Court, Steven's widow and children argued that the intermediate court of appeals erred in its ruling in favor of the tavern keeper by deciding that the tavern keeper's duty to protect his patrons extends only to the boundaries of the business premises. They argued further that when an owner becomes aware of danger to a patron and can protect the patron from harm caused by the danger, then the owner's duty "transcends the walls of the tavern."

At the trial, the judge instructed the jury regarding a tavern owner's duty to protect a customer from third parties as follows:

> The proprietor of a place of business (tavern, etc.) who holds it out to the public for his business purposes is not an insurer of the safety of the patrons against personal injuries inflicted by third persons, that is, other patrons on the premises. Nevertheless, he has a duty to exercise ordinary care to protect members of the public while on the premises from bodily harm caused to them by the accidental, negligent, or intentionally harmful acts of third persons if he by the exercise of ordinary care could have discovered that such acts were being done or were about to be done, and he could have protected the members of the public (injured patron) by controlling the conduct of the third person or by giving a warning adequate to enable them (injured patrons) to avoid harm.

The plaintiff widow and children argued that two earlier cases by the Wisconsin appellate court extended the owner's duty to protect his patrons beyond the business premises. The Supreme Court reviewed these two cases and the arguments of the plaintiffs and rejected their conclusions.

The plaintiffs also argued that the Wisconsin Supreme Court decision in the case of *Sorensen v. Jarvis*, decided in 1984 should be applied retroactively to the facts of the instant case. The *Sorensen* decision was a case in which the Wisconsin court designed a rule of law whereby a seller of alcoholic beverages has a duty not to sell intoxicating drinks to a minor or otherwise be liable to injured third parties, whether the injury occurs on or off the seller's business premises, when it is determined that the negligent act of selling alcoholic drinks was a "substantial factor in causing the harm to the third person." The Wisconsin Supreme Court, however, rejected the plaintiff's argument and declined to apply the *Sorensen* rationale to the present case, noting:

> We do not find that *Sorensen* in any way extends the general duty of the tavern owner to exercise ordinary care for the protection of his patrons beyond the boundaries of his business premises.

Also, the court specifically held that the *Sorensen* rule was to be given prospective effect only and not retroactive effect. The court stated in the *Sorensen* decision that a seller of alcoholic beverages would be subject to liability for negligently supplying alcoholic drinks to a minor on or after September 1, 1984, and the death of Steven Delvaux predated that effective date.

Accordingly, the decision of the Court of Appeals was affirmed by the Wisconsin Supreme Court.

COMMENTARY:

This case points up in a rather dramatic procedural fashion, the high degree of public concern over deaths and injuries caused by intoxicated persons. Since the traditional rule holding only the intoxicated persons responsible was not stemming the tide of such losses, social policy has dictated an extension of the liability to the parties serving the alcoholic beverages to the one who then commits the harmful acts. We have seen many cases across the country now where more and more often the sellers or servers of alcoholic beverages have been held liable in such instances.

Even social hosts in their homes can be held liable for injuries to third parties by persons who received alcoholic beverages in the host's home, drank to excess, and caused harm to others. Also, it is not necessary for the person who provided the alcoholic beverages to have sold the drinks to the person who then committed the harmful acts, and it is enough in some jurisdictions (such as in the social host situation just mentioned) to hold the provider of alcoholic beverages liable even where the drinks were simply given to the person who subsequently caused the harm.

This should be kept in mind by security managers and others when planning upcoming company Christmas parties, and other social events throughout the year. An employee who is provided too much to drink at a company social function and then gets into a fight or injures another with his vehicle may well be personally liable to the injured party, but the company providing the beverages to the employee may also be held liable.

And, even ignoring the legal liability consequences, the mood of the public and most employees these days seems to be that they appreciate and respect a company's decision to limit alcoholic beverages at such company functions as opposed to the time-honored, but now less respected practice of providing unlimited drinks at an open bar at such functions.

PSCLR – OCT 86 (7)

Restaurant Owner Is Held To Have No Duty To Employ Security Guards In A "High Crime Area" To Protect Its Business Invitees:

Rosie Stafford v. Church's Fried Chicken, Inc., a Texas Corporation, *No. 85-CV-70618-DT, in the United States District Court for the Eastern District of Michigan, S.D., decided March 6, 1986, reported at 629 F. Supp. 1109.*

Rosie Stafford sued Church's Fried Chicken, claiming that the restaurant had a duty to hire security guards to protect its customers at the restaurant which was located in a "high crime area." She alleged that an unknown person injured her while she was in line at the drive-through window at the defendant's restaurant located in her neighborhood. Her assailant came to her car, sprayed her with a chemical substance, robbed her, and stole her automobile.

The defendant asked the trial court to dismiss the case, arguing that a private business cannot, under the law, be liable for failing to provide security guards. The restaurant based its motion on a case decided by the Michigan Court of Appeals in 1985, *Williams v. Cunningham Drug Stores, Inc.*, (reported in the April 1986 issue of **PSCLR**, case six, page seven). The plaintiff responded that the restaurant did owe a duty to provide security guards at the particular location in question because the restaurant was located in a "high crime area," and had been the scene of other robberies before the incident in question.

The trial judge ruled that a retailer must use due care to protect its invitees. However, as to Stafford's argument that the defendant owed a duty to provide police protection, the judge ruled: "Neither the legislature nor the courts have established public policy requiring law enforcement responsibilities to be extended to commercial businesses." The judge therefore declined to hold that the restaurant owner had any duty to provide police protection for its customers. The court noted that the assault on the plaintiff occurred on the parking lot of the restaurant and concluded that if a security guard had been hired by the restaurant, then the guard could not have effectively aided the victim of an armed robbery in the restaurant parking lot.

Finally, the court concluded:

> To hold restaurant owners responsible for providing police protection against the criminal conduct of third parties outside of restaurants, especially those in high crime areas may drive businesses out of those neighborhoods.

The trial judge therefore concluded as a matter of law that it should not be left to a jury to determine what constitutes reasonable care under the circumstances because there were no circumstances which would impose liability upon a business for its failure to provide police protection.

COMMENTARY:

This decision is not an appellate court opinion. It is a decision reached by the trial judge in the federal District Court for the Eastern District of Michigan and therefore is not binding on anyone other than the parties to the case.

We will continue to monitor this case and report the appellate court's decision, if an appeal is taken.

PSCLR – NOV-DEC 86 (10)

Restaurant Manager's Failure To Warn Customer And Failure To Intervene To Prevent Injury Support Judgment And Punitive Damages Award In Customer's Negligence Action:

Rosie Kathleen Gould v. Taco Bell, *No. 57946, in the Supreme Court of Kansas, decided July 18, 1986, reported 722 P.2d 511.*

While dining late one evening at a Kansas City, Kansas, Taco Bell restaurant, Rosie Kathleen Gould was assaulted by a fellow customer. She sued Taco Bell, alleging that the restaurant was liable for her injuries because it had failed to provide adequate security. The restaurant was found partially liable for her injuries, and Gould was awarded $500 compensatory and $10,000 punitive damages. Taco Bell appealed.

Gould and Theresa Holmberg were eating at a Taco Bell restaurant in Kansas City, Kansas, at about 11:30 PM after attending a Kansas City Royals baseball game. They sat next to a booth containing six boisterous people engaged in "loud, crude and vulgar conversation, designed to be overheard and to shock." Neither Gould nor Holmberg said anything to the group, which included a Karen Brown, but a Taco Bell employee asked the group to be quiet. The group got louder, and as they all got up to leave, Karen Brown stopped and said "Those two white bitches over there think they're hot shit." Surprised at this verbal attack, Gould first asked if Brown was speaking to her and then asked Brown to "please come over here and repeat yourself." Brown's response was to dash over to Gould and punch her in the face, knocking her sideways and leaving her with bruises on her face and nose. Gould's immediate reaction was to call Brown "a nigger," which goaded Brown to beat Gould with her fists until Holmberg separated them. Gould and Holmberg tried to leave the restaurant; Brown kept verbally taunting Gould, and then began striking her again.

The assistant manager on duty at Taco Bell, Mark Wills, watched the altercation but did not try to stop Brown, "for fear Brown would strike him for interfering." Although he decided the situation didn't warrant calling the police, he did ask Brown, "Why don't you just leave? You did this two weeks before in here." Wills and another Taco Bell employee went outside to observe the fracas in the parking lot, where Brown had now attacked Holmberg by pinning her against the outside brick wall of the restaurant, and was hitting and kicking her. The Taco Bell employees ignored Holmberg's screams for help, and when she broke away from Brown and fled inside the restaurant, Wills refused to allow her to use the phone to call the police. When Holmberg threatened to jump over the counter and use the phone anyway, Wills himself called the police. While Holmberg was inside, Brown resumed her assault on Gould until Holmberg came out and announced that the police were on the way, prompting Brown and her companions to finally leave.

Gould's suit against Taco Bell alleged that Taco Bell had failed to provide adequate security to protect her as a business invitee, and that the Taco Bell employees could have prevented her injuries by preventing the assault. The jury found Gould to be 49 percent at fault, and Taco Bell to be 51 percent at fault.

On appeal, Taco Bell argued that a business owner could not be held liable for the sudden criminal attack by one customer on another customer. And, even if the business owner could be held liable, punitive damages could not be awarded.

The Kansas Supreme Court first examined the duty of care owed by a premises owner to a person on the premises; the extent of the duty depends upon the visitor's status. The restaurant patron is an invitee to whom the owner/operator owes a very high duty of care: an affirmative duty to exercise reasonable care for the invitee's safety, including the obligation to warn of any reasonably anticipated dangers. The court cited a previous case (*Kimple v. Foster*, 469 P.2d 281, 1970) which set forth the rule on a business owner's liability for an intentional assault committed by one patron upon another patron:

> A proprietor of an inn, tavern, restaurant or like business is liable for an assault upon a guest or patron by another guest or third party where the proprietor has reason to anticipate such an assault and fails to exercise reasonable care to forestall or prevent same.
>
> The duty of a proprietor of a tavern or inn to protect his patrons from injury does not arise until the impending danger becomes apparent to him, or the circumstances are such that a careful and prudent person would be put on notice of the potential danger.

The appellate court here noted that the evidence was sufficient to show that Taco Bell had reason to anticipate injury to Gould and Holmberg, but did nothing to forestall or prevent it. The manager on duty that night, Wills, knew that Karen Brown had been involved in a similar altercation two weeks before. There was evidence that "rowdyism" on the premises had led Taco Bell management to consider hiring security personnel, but they had not done so.

The appellate court also determined that the award of punitive damages in this case was not improper. Punitive damages were authorized because the defendant's employee Wills's failure to act was willful, wanton, or grossly negligent. The defendant argued that a failure to act, as opposed to an affirmative act, could not support punitive damages. The court, however, defined a "wanton act as

> something more than ordinary negligence but less than a willful act. It must indicate a realization of the imminence of danger and a reckless disregard and indifference to the consequences. [cites] 'Wantonness' refers to the mental attitude of the wrongdoer rather than a particular act of negligence. It follows that acts of omission as well as acts of commission can be wanton since reckless disregard and indifference are characterized by failure to act when action is called for to prevent injury.

The appellate court upheld the jury's finding that the evidence showed that Taco Bell knew or should have known that danger of injury to Gould was imminent, but it failed to intervene or to warn her of the danger. The manager saw Brown strike Gould inside the restaurant, but did nothing. The manager witnessed the second attack on Gould as she and Holmberg were attempting to exit, but decided not to call the police or to interfere. Finally, Wills only reluctantly called the police when Holmberg came inside and threatened to jump over the counter and use the phone herself if he did not call the police.

There was also evidence that the Taco Bell management was becoming concerned over the conduct of late night patrons at the restaurant, many of whom "originated in the neighboring bars and ... were rowdy and used loud, vulgar and obscene language, and engaged in verbal fights and occasional physical fights." Furthermore, the store manager testified that the store did not have sufficient help to handle the late night crowds, and that "Taco Bell's written policy was to call the police in case of disruptive customer behavior."

The Kansas Supreme Court therefore affirmed the judgment and punitive damages award against the plaintiff Taco Bell.

PSCLR—SEP 87 (1)

Restauranteur Not Liable For Unforeseeable Mass Murder Attack On Customers By Third Party:

Cristian Toscano Lopez, et al., v. McDonald's, et al., *No. D004619, in the Court of Appeal of California, Fourth District, Division One, decided July 9, 1987, reported at 238 Cal. Rptr. 436.*

This is the case brought by family members of the victims killed and others who were wounded in the 1984 mass murder at a McDonald's restaurant in San Ysidro, California. These plaintiffs sued the McDonald's corporation and the franchise holder, but the trial court determined that McDonald's had no legal duty to protect the plaintiffs against the massacre. Accordingly, the trial judge granted summary judgment in favor of McDonald's and the plaintiffs appealed.

The facts are best summarized by the appellate court:

> On a clear and sunny day, James Oliver Huberty entered McDonald's restaurant in San Ysidro dressed in camouflage pants and armed with a 9 mm. semi-automatic rifle, a semi-automatic 9 mm. pistol, and a .12 gauge shotgun. He immediately began indiscriminately slaughtering patrons and employees within the glass-enclosed structure. During the hour of terror before he was killed by a police sharp-shooter, Huberty showed no intent to rob the restaurant; made no demands for money; made no effort to take hostages; loaded his weapons several times, and killed 21 people in the restaurant and wounded 11 others. His single apparent purpose was to kill as many people as possible before he himself was slain.

The plaintiffs' suit for wrongful death and bodily injuries was based on claims of negligence wherein they asserted that McDonald's had failed to provide adequate security personnel or safety devices to protect its customers. The plaintiffs assert that McDonald's knew that its franchise in San Ysidro was in "high crime area" and the employees at that location had requested that a private security company contact McDonald's in an effort to provide its services to the restaurant. The security company did make a proposal to McDonald's, but the service was declined based upon economic considerations.

The defense to the plaintiffs' complaint was that such an unprecedented and totally unexpected massacre on the defendants' premises was not foreseeable and that there was no relationship between the failure to provide security personnel and devices and the resulting injuries and death to the plaintiffs. In support of its defense, McDonald's assistant manager and a San Diego police lieutenant testified as to the undisputed facts. The plaintiffs on the other hand presented evidence about crimes committed in the area over the three-and-one-half years preceding the incident, which included two robberies of the restaurant, two petty thefts, an unlawful use of vehicle, vandalism, and theft. This evidence also indicated that there were six burglaries, five batteries, a deadly weapon assault, other grand thefts, and various other crimes which occurred within a 1/10th mile radius of the restaurant.

The plaintiffs also introduced into evidence a deposition by Wilbert W. Holley of Allegiance Security Incorporated, who had been in the private security industry for about ten years and who had tried to sell security services to the restaurant only two months before the incident. He testified in his deposition that he had told Sissung in the McDonald's corporate offices that he had discussed the crime situation with employees at the McDonald's restaurant in San Ysidro and he told Sissung that security was needed at the restaurant and that a uniformed officer could be provided for a cost of $5.75 per hour but, he testified, Sissung declined, allegedly stating, "That's too much money. We don't want to spend any money. There is no problem, we don't need it anyways." Holley also stated his opinion that the presence of security at the restaurant "would have acted as a deterrent and could possibly have prevented" the massacre. There was additional evidence concerning the level of security at neighboring businesses which included roving security patrols, armed guards, closed-circuit television and alarm systems.

The court then went into great detail, examining prior legal precedents in its effort to determine whether liability should be precluded as a matter of law as opposed to being submitted to the jury for its determination based upon its consideration of the factual circumstances. After examining the general standards of care that all citizens owe one another and the specific duties owed by business owners and operators to its customers, the court concluded that the unforeseeability of the unique and horrible event at the McDonald's restaurant in San Ysidro required that the liability of the restauranteur be restricted in this case. Under all of the facts and circumstances on the summary judgment motion the court concluded that the risk of a massacre such as occurred was so remote and unexpected that McDonald's failure to provide the uniformed security officer did not facilitate or cause the event. The court also concluded that there was no causal relationship between the plaintiffs' injuries and any failure on the part of the restaurant to provide security.

The court accordingly affirmed the trial judge's grant of summary judgment in favor of McDonald's and its franchisee.

COMMENTARY:

The opinion of the appellate court in this case is rather lengthy and detailed and involves a general review of the law of negligence and the evolving law of premises liability. Readers with a special interest in the law of premises liability may wish to obtain a complete copy of the appellate court's opinion in this case.

PSCLR – OCT 87 (7)

Retailer's Duty To Protect Customers Does Not Include Duty To Provide Security Guards To Prevent Robberies Of Or Assaults On Customers:

Lucille Scurlock Belton, Leonard Belton, and Leonard Belton as next friend to Murishea Belton, a minor v. Delaware McDonald's Corporation and Percy Lewis and McDonald's Restaurant and John Doe, *No. 94904, in the Court of Appeals of Michigan, decided August 31, 1987, reported at ___ N.W.2d ___.*

This is an appeal from the dismissal of plaintiff's lawsuit prior to trial, based upon the assault and robbery of a customer at a McDonald's Restaurant. Procedurally, in determining whether the allegations of the lawsuit state a cause of action, the court must accept the plaintiff's allegations as true, although subsequent discovery may prove the facts alleged to be untrue or misstated.

In any event, the plaintiffs alleged that the plaintiff Lucille Belton, along with her son Murishea Belton, drove to the drive-up window of a McDonald's restaurant in Detroit. As sometimes happens, the plaintiffs' order could not be filled quickly and the McDonald's employee instructed Belton to pull her car up to the side of the building so that their order could be brought to them. After Belton pulled over and was waiting for her order, an unknown third party assaulted and robbed her at gun point, stealing her automobile and the personal property in the car. In the lawsuit, the plaintiffs specified that the defendants were negligent for failing to post guards on the property, to adequately light the property, and to post signs warning of the danger to customers coming on the property. The plaintiffs further alleged that the defendants were negligent for maintaining the property in such a way as to create a nuisance, for directing the plaintiffs to an unsafe area after taking their money, and for failing in general to use reasonable care to keep the property safe for its customers.

The trial judge granted the defendants' motion and dismissed the plaintiffs' lawsuit, holding in effect that even if the allegations of the complaint were subsequently supported by the evidence and proven to be true, the plaintiffs would not be legally entitled to recover. The plaintiffs appealed.

The appellate court noted that a retailer unquestionably has a duty towards its patrons to maintain the property in a reasonably safe manner and to exercise due care to protect the patrons from harm. If, however, McDonald's did not violate any duty to the plaintiffs under the facts and circumstances, then the summary dismissal of the plaintiff's complaint was proper.

In Michigan, the Michigan appellate court has previously ruled that commercial businesses have no legal duty to provide armed, uniformed security guards to stop or prevent armed robberies. This ruling was based on public policy considerations:

> Neither the legislature, nor the constitution has established public policy requiring law enforcement responsibilities to be extended to commercial businesses. We therefore decline to extend that obligation to private citizens or enterprises. Privately-hired security guards simply lack the authority or the training to effectively and safely defuse a potentially volatile situation which an armed robbery represents.

The appellate court in this case also said that it could not find any precedent suggesting that fast-food retail outlets should be required to warn their patrons merely because the establishment is located in an unsafe area. (The court explained that the assertion of such a duty would require all business in most urban areas to post such warning due to the higher crime rate in such areas.)

However, the appellate court did agree that the plaintiffs had sufficiently set out in their lawsuit a claim of lack of due care by the defendants, with regard to the plaintiffs' claim of inadequate lighting and given that they alleged that they were directed to an "unsafe area" (claims which are assumed true on the plaintiff's motion to dismiss). These allegations, if subsequently supported by admissible evidence, would be sufficient to establish a legal cause of action for the plaintiffs. If, however, pre-trial discovery fails to support plaintiffs' allegations concerning inadequate lighting, then the case could still be dismissed on summary judgment prior to trial. The court also noted that the plaintiffs would have to establish evidence to support an inference that an assault on the plaintiffs was foreseeable by the defendants.

Accordingly for the foregoing reasons, the trial court's decision to dismiss the case was reversed and remanded.

PSCLR—OCT 87 (9)

Lounge Owner Has Duty To Exercise Reasonable Care To Protect Patron From Harm By Fellow Patrons, But Is Not Guarantor Of Patron's Safety:

Carl Delgado v. Laboucherie, Inc., et al., *No. CA-6987, in the Court of Appeal of Louisiana, Fourth Circuit, decided June 3, 1987, reported at 508 So. 2d 956.*

Carl Delgado sued the owners and certain employees of LaBoucherie Lounge in New Orleans, after he was assaulted and injured by another patron at the lounge.

At approximately 2:45 AM on the evening of the assault, the plaintiff Carl Delgado was with several friends, including Lamonica Bell and Kathleen Whitlow, in the disco area of Laboucherie Lounge. Another patron, Willie Villagram, approached Delgado's table and invited Bell to dance, but she declined. Villagram began shouting obscenities at Delgado and his friends, and Bell testified that he continued harassing them for approximately 20 minutes before he left their table and returned to his place at the bar. Delgado then went to the bar himself to buy drinks for his group. According to the testimony of both Delgado and Bell, Villagram then began annoying Delgado at the bar; suddenly and without warning, Villagram struck Delgado across the face with a cocktail glass.

Delgado later testified that before he went up to the bar and was assaulted, he had attempted to alert the doorman, Mario Suazo, to the disturbance Villagram was creating. Delgado said that Suazo ignored him, but Suazo testified although he had observed what appeared to be an argument when Villagram asked one of the women in Delgado's party to dance with him, he did not intervene, because he did not think that it was necessary. He did testify that Villagram's assault on Delgado occurred "very quickly" and that "there was nothing that could have been done to prevent it. ... 'It was not a fight. It was just an attack of a guy.'" Suazo did say that Villagram seemed "weird" but he had been in the bar before and had never caused any problems. Suazo also testified that neither Delgado nor any one else at his table had asked him for help with respect to Villagram.

The bartender at the lounge also testified that the scuffle broke out very quickly, and that Villagram was not intoxicated at the time of the incident. She also said that he had been in the bar previously and had always been "well behaved."

The defendant moved to dismiss the complaint filed by Delgado for damages for his injuries. After consideration of the evidence presented on the motion, the trial judge granted the motion and dismissed the case. Delgado appealed, and the appellate court affirmed the trial judge, ruling that the proprietor's duty to protect his guests from insult, annoyance, and danger from other guests:

> does not make the proprietor the absolute insurer of the patron's safety, nor does it extend to the unforeseeable or unanticipated criminal acts of an independent third person.
>
> * * *
>
> The owner's duty to protect patrons arises when he has knowledge, or can be imputed with knowledge, of the third persons intended conduct. That duty may also encompass the requirement to call the police.

The court concluded that the assault on Delgado was indeed sudden and could not have been reasonably prevented by the doorman, the bartender, or any other employees of the lounge owner. The fact that Delgado and Villagram argued previously that evening did not sufficiently put the defendant lounge owner or his employees on notice that an attack was about to occur: "In the absence of such knowledge, there was no breach of duty."

Judgment of dismissal was affirmed.

PSCLR – MAR 88 (6)

Operator Of Tavern Has No Duty To Protect Patron From Assault One-Half Block From Premises, Even Assuming Assault Was Foreseeable:

Helen Badillo v. Theresa A. DeVivo, Felix A. DeVivo, Peter's One Two Three Lounge, Inc., d/b/a 123 Pub, an Illinois corporation, Exchange National Bank of Chicago, as trustee under trust agreement no. 25239 and Peter D'Agostino, *No. 86-2120, in the Appellate Court of Illinois, First District, First Division, decided September 21, 1987, reported at 515 N.E.2d 681.*

Helen Badillo brought suit against the owners of tavern for money damages for the injuries she allegedly incurred during an assault by a third party. She claimed that the defendants had failed to prevent the assault on her and that such failure amounted to negligence, entitling her to a recovery against them.

The trial court did not agree that she had a valid cause of action, and dismissed her complaint against the defendants; Badillo appealed. The appellate court affirmed for reasons appearing below.

According to the plaintiff's allegations, she was a customer at Peter's 123 Lounge, Inc., a defendant in the case, located at 5620 W. Diversey Street, in Chicago. While she was at the tavern, she was physically and verbally attacked by Teresa DeVivo, another customer at the tavern. Tavern employees stopped the fight between the two customers and ejected them both from the premises, and Ms. Badillo went to her car which was parked one-half block from the lounge.

As she approached her car, Badillo was once again assaulted by DeVivo who was then armed with a "police billy club." It was for these injuries that the plaintiff filed suit.

In the one count of her six-count complaint on appeal, Badillo alleged that she was entitled to compensatory and punitive damages from the defendant. Badillo argued that the tavern keeper owed a duty to her to exercise reasonable care for her safety, and that the tavern keeper failed to perform that duty in that:

(1) the defendant had not summoned the police after the plaintiff had initially been attacked by DeVivo in the lounge;

(2) the defendant had ejected both DeVivo and the plaintiff simultaneously;

(3) the defendant tavern keeper had not provided an escort or other security for the plaintiff Badillo after ejecting her from the premises.

It has been the law for sometime that a plaintiff cannot recover in a negligence action unless there is a duty which the defendant owed to the plaintiff, and the defendant breached that duty. The question as to whether a legal duty exists in a given situation is a question of law which must be determined by the trial judge.

The court noted that in those cases where a duty has been imposed on a tavern keeper to take reasonable action to protect its customers from foreseeable dangers caused by third parties, the misconduct by the third parties, which caused the plaintiff's injuries, have all occurred either on the defendant's premises, or in a parking lot operated or owned by the defendant.

In this case, however, Badillo was assaulted one-half block away from the business premises of the defendant, and the appellate court found no case which held that a business operator was required to protect its customers, even from foreseeable dangers, when such dangers are caused by third parties off the business's premises. The court found that though a business owner would have a duty to provide a safe passage for entering and exiting the property, that legal precedent would not apply to this situation where the injury occurred one-half block from the business premises.

The court concluded, therefore, that the defendants did not owe a legal duty to Badillo and affirmed the trial judge's dismissal of this count of the plaintiff's lawsuit.

COMMENTARY:

In reaching the above conclusion, it is also important to note that the issue of foreseeability is not the only issue relevant to establish the existence of a duty to protect. In deciding whether or not a duty exists in a given situation, the court will also look to the likelihood of injury posed by the foreseen danger, the burden of guarding against the danger, as well as the consequences of placing the burden on the defendant in the given situation.

As the court noted, efforts to place such liability on a business operator, and to require him or her to protect its customers from injuries occurring after the customer leaves the business premises would place an unjustifiably heavy burden on the business owner/operator, as well as on the police force. It would require the business owners to, in effect, "police the streets" to ensure that their customers got safely to their cars, or perhaps even to their homes.

Furthermore, the imposition of such a duty would require business owners and operators to determine which party was the aggressor in an altercation, and perhaps to detain that potentially dangerous third party.

For these public policy reasons, even though the danger might be foreseeable, the courts have refused to extend the liability of the business owner/operator beyond the premises, but including the passageways for ingress and egress and the parking lots (as has been seen in other cases).

There have, on the other hand, been cases where the business owners and operators have been held liable for injuries which occurred off the business premises itself: for instance, where a plaintiff was injured when he was struck by a patron who was being thrown out of the business; where the plaintiff had been struck by a bullet from a gun discharged on the premises; and where a plaintiff had been struck by a baseball which had come from the premises. Or, as in the case of Ember v. B.F.D., Inc., reported in the August 1986 **PSCLR**, *case nine, page nine, where the defendant had gratuitously and voluntarily assumed the duty for security in the off-premises area in the neighborhood around its restaurant.*

Of course, in most cases, the injuries resulted from acts which did take place on the defendant's property.

PSCLR – MAR 88 (7)

Business Owner Owes Duty To Police Officer To Refrain From Inflicting Injury On Him, To Exercise Care To Avoid Injuring Him By Negligent Acts, And To Warn Him Of Any Dangerous Situation Known To Owner Where Owner Has Reason To Believe Officer Is Unaware:

David Michael Holdsworth and Louis Ford Holdsworth v. Renegades of Louisiana, Inc., dba Cowboys Club and Restaurant, *et al., No. 19207-CA, in the Court of Appeal of California, Second Circuit, decided December 2, 1987, reported at 516 So. 2d 1299.*

A police officer, David Michael Holdsworth, was apparently called to settle a disturbance which was taking place at a party at a nightclub, and he was hurt by one of the people at the party when the two of them scuffled.

Holdsworth and his wife sued the alleged assailant and several other persons, including the owner of the club at which the party was held. All of the defendants, with the exception of the club owner, settled with the Holdsworths. The case went to a jury trial as to the claim for damages against the club itself. The jury found in favor of the club owners, and the Holdsworths appealed.

A local bail bondsman hosted the party on the second floor of the Cowboys Club in Bozier City, Louisiana. There was an open bar, and more than 100 people attended, including a number of off-duty police officers, lawyers, and court officials from the area.

A deputy city marshall by the name of Dove got into a dispute with an off-duty Bozier city police officer by the name of Bridges, and the evidence was that Dove was intoxicated. The manager alerted the Burns International security guard and an employee of Cowboys when the fight began and they immediately went to the party to break up the altercation.

The club manager notified the police, and officer Holdsworth arrived a few minutes later. When Holdsworth asked Dove for his identification, Dove refused to cooperate and the two of them got involved in a scuffle of their own, during which, Holdsworth's knee was injured.

Officer Holdsworth claimed that the club had permitted Dove to drink an excessive amount of alcoholic beverages during the party and that this constituted negligence by the club. The Holdsworths also alleged that the owners had failed to provide sufficient security, had failed to properly supervise and control the participants at the party, and had failed to keep order and control the situation.

The defendant club owners responded that Holdsworth was negligent in failing to use proper procedures in arresting Dove, causing his own injuries.

Evidence was presented by the manager of the club as to the security system employed at the Cowboys Club. The manager claimed that there was an elaborate system by which club employees monitored the situation in and outside the club to handle any disturbances on the club property.

The policy was that whenever a fight took place inside the club, the manager would immediately notify club security and would also call the police. The manager then would position him or herself at the club entrance in order to direct the police to the scene of the altercation immediately upon their arrival.

On the day of the party in question, a Monday night, the security personnel present at the club was one employee security guard plus one Burns guard. However, the bail bondsman hosting the party had also hired an additional Burns guard to monitor the second floor and to check invitations. This extra guard had departed the premises before the fight in question, and the other two guards were downstairs when it began.

The club manager said when she saw the fight between Dove and Bridges, she immediately called the two security guards from downstairs, notified the police, and went to the front door to direct the police upon arrival. When Holdsworth did arrive, she directed him to the second floor. A few minutes later, the manager went upstairs and it was then that she saw that Dove was not cooperating with Holdsworth. She then called the police for additional officers.

Evidence showed that Dove had gotten very intoxicated during the evening and that he had become increasingly belligerent. Bridges testified that, when Holdsworth arrived on the scene, Bridges answered the officer's questions as to what had happened. Bridges said that when Holdsworth asked Dove for identification and Dove refused, Holdsworth grabbed him by the collar and threw him to the floor. Bridges' testimony was corroborated by another deputy chief of police present at the party.

Holdsworth himself testified that when he arrived at the party, he found "total chaos." He said that he was told by the two security guards that there had been a fight between Bridges and someone else; that when he approached Dove for identification, Dove seemed to be "extremely intoxicated"; that Dove "drew back," apparently to strike at the officer. Holdsworth testified that he instinctively grabbed the lapel on Dove's coat in order to deflect the anticipated blow. When Dove and Holdsworth then fell to the floor, Holdsworth injured his knee.

Dove admitted that he had consumed eight to ten strong drinks during the preceding three-and-a-half hours.

After the presentation of the evidence, the jury found that Dove was 50 percent at fault, Holdsworth 50 percent at fault, and the club owners not at fault.

The plaintiff Holdsworth argued on appeal that while the club owners had a duty to call the police because of the altercation in the club, the club owners also owed Holdsworth a "special duty" to protect him against the actions of the patron he arrested.

In examining this argument, the appellate court noted that the owner or operator of a business does in fact generally have a duty to protect his invitees from injury by exercising reasonable care. It had been held that an operator of a bar owes his customers a duty to provide protection necessary to keep an "orderly and safe house." And, such a duty includes the duty to call the police when necessary. The question, then is whether the same duty is owed to the police officer who is called to the scene to put down a disturbance.

The appellate court ruled that the club owners did not owe Holdsworth any duty to keep its club free of intoxicated customers and disturbances caused by them. However, the club did have a duty to refrain from injuring the police officer, to exercise care to avoid injuring him by negligence, and to warn him of known dangers on the premises.

After reviewing the record of evidence as summarized above, the appellate court found that the club owners and operators had not breached any duty owed to officer Holdsworth.

Accordingly, the judgment of the trial court was affirmed.

COMMENTARY:

In reaching the above decision, the appellate court noted decisions in earlier cases. In one such case, police officers brought action for damages they incurred as the result of being wounded by sniper fire. The officers had been called to the hotel when the sniper began shooting.

The defendants in that case argued that a business owner/operator could not be held liable for injuries to police officers who were called in an emergency, even if the emergency itself had been caused by some negligence on the part of the business owner/operator. The court agreed with that position, holding that, "A proprietor is not liable for injuries to professional rescuers because those rescuers do not fall within the ambit of risk of the proprietor's original negligence."

The court there also noted that to rule otherwise would strongly discourage business proprietors from calling for aid from police or firemen because of a fear that they would then be liable for any injuries to the fire or police officers caused by third parties.

This situation is analogous to the "fireman's rule" which precludes liability to a property owner for injuries to a professional fire fighter during the course of a fire. The court has ruled: "We deem it unreasonable to hold that an owner owes it to fire fighters not to let his building catch on fire."

PSCLR – MAR 88 (8)

Business Proprietor's Duty To Protect Patrons Arises When Proprietor Has Knowledge Of Third Party's Intended Misconduct Which Could Endanger Patron:

Joycelyn F. Toups v. Jonthal Hawkins, Ziegfield's Inc., George Wainer Company, and XYZ Insurance Company, *No. 87-CA-297, in the Court of Appeal of Louisiana, Fifth Circuit, decided November 9, 1987, rehearing denied February 17, 1988, reported at 518 So. 2d 1077.*

The plaintiff Joycelyn F. Toups filed suit against the proprietor of a club owned by D-VIN, Inc., doing business as Ziegfield's Disco, attempting to hold the defendant liable for injuries she received when she was a patron of the club and she was shot while on the premises. Ziegfields moved for summary judgment which was granted. Toups appealed.

On appeal, the appellate court must determine whether the allegations and supporting affidavits of the respective parties establish any genuine issue of material fact which would support the plaintiff's cause of action. If they do not, the summary judgment should be affirmed; but if they do, the case should be remanded to the trial court for presentation to a jury for its determination of such issues of fact.

The appellate court first reviewed the law applicable to such cases in this jurisdiction.

First of all, it is clear that a proprietor of a business must provide a reasonably safe place for his patrons. While the proprietor is not an "insurer" of the patrons or the patrons' property, he must exercise reasonable care to protect both the person and the property of the patron. This duty imposes certain specific obligations on the proprietor:

(1) He must refrain from any conduct on his own part which would be likely to cause harm to a patron.

(2) He must keep the property and premises free from unreasonable dangers or he must warn customers of dangers which are known to exist on the property.

(3) He must also exercise reasonable care that his guests are protected from danger caused by his employees, other patrons, or third parties. This obligation encompasses additional duties:

(a) If a disturbance or probable disturbance exhibits itself, the business proprietor must try to protect his customers by calling the police, if time permits.

(b) The proprietor must warn his customers of the likely danger, when the proprietor becomes aware of the impending or possibly impending danger.

(c) Generally, a proprietor has no duty to protect others from the criminal acts of third parties; that is, his general duty to exercise reasonable care does not encompass a duty to protect his customers from <u>unanticipated</u> criminal misconduct by third parties. It is <u>only</u> when the proprietor has actual knowledge of (or knowledge that can be imputed to him) such third party's intended misconduct, that the duty to protect patrons arises.

(d) Lastly, although he has no general duty to protect against third party misconduct, once the proprietor has voluntarily assumed the obligation to protect his customers, such obligation must be performed with due care. For instance, where the proprietor hires a guard to protect the patrons, the proprietor is liable for harm caused by the guard's negligence.

The appellate court concluded that there were material issues of fact established by the pleadings and the supporting affidavits submitted by the parties in this case. The appellate court especially noted that an affidavit by Scott Toups alleged that Ziegfield's had a policy of discouraging blacks from coming onto or being in the premises and in fact encourage the ejection of black patrons for mere trivial reasons. The affidavit alleged that there was a continuing policy on the part of Ziegfield's to eject blacks if they did not drink constantly, or if they tried to talk with white customers, or if the black patrons engaged in otherwise minor activity. The affiant went on to claim that such policies created an atmosphere of hostility on the nightclub property. This claim, together with an allegation that prior disturbances had taken place on the premises, raised an issue of material fact as to whether the owner of Ziegfield's had knowledge (or whether such knowledge could be imputed) of the existence of danger to his patrons, including the plaintiff Joycelyn Toups.

Furthermore, the affidavits of both Scott and Joycelyn Toups claimed that the gunfire which injured Joycelyn Toups took place inside the Ziegfield's premises, which the court concluded raised the additional factual issue as to how the shooting occurred, where it occurred, and by whom Joycelyn Toups was shot. The appellate court also noted that Ziegfield's had not fully answered the plaintiffs' allegations that it had failed to provide "adequate security," but had merely denied the allegation without submitting any evidence as to the degree of protection provided its patrons.

Accordingly, the appellate court reversed the trial judge's decision to grant the defendant its summary judgment motion, and remanded the case for presentation to a jury.

PSCLR—APR 88 (3)

To Recover From Business Owner For Injuries Received In Assault By Third Party Customer, Plaintiff Must Show Owner Had, Or Was Imputed With, Knowledge Of Violent Propensities Of Assailants:

Jerome J. Silvio, et al. v. Pharoah's Palace, *et al., No. 86 CA 1246, in the Court of Appeal of Louisiana, First Circuit, decided November 10, 1987, reported at 517 So. 2d 185.*

This lawsuit was brought by Jerome J. Silvio, Jr., Michael A. Humphries, and John E. Fedric, for injuries they received during a fight at the Texas Dance Hall in Baton Rouge, Louisiana. They sued the owner of the Dance Hall, Pharoah's Palace, Inc., as well as one of the shareholders of the corporation and two unidentified employees. They also sued their assailants (but were unable to perfect service of the complaint on them).

After they were sued, Pharoah's filed a third party suit against Independent Guard and Security, Inc., the security guard company which provided guards for the Dance Hall at the time of the altercation.

At the close of the presentation of the plaintiffs' evidence at trial, the trial court granted Pharoah's motion to dismiss the case. This motion was granted and the judge dismissed all pending actions in the case. The plaintiffs appealed, arguing that the trial judge made a mistake when he determined that the nightclub owner had complied with its legal duty to protect the plaintiffs.

The facts are as follows: John Fedric took his wife to the Texas Dance Hall to hear a concert by a country and western singer. They had just started their first drink of the evening when Fedric apparently inadvertently took a position which blocked the view of three men and a woman seated nearby. One of the men expressed his displeasure with having his view blocked by kicking Fedric in the rear. When Fedric responded, one of the group of four called a security guard who directed Fedric to go to the bar area, which he did. Once there, he complained to the bartender. Fedric and his wife testified at trial that the bartender then told them that other patrons that night had been "run off" from the same area near the balcony by the same four customers earlier that evening.

The other plaintiffs and friends of the Fedrics, Jerome Silvio and Michael Humphries, arrived at the Dance Hall about one half hour later. Without having spoken to the Fedrics, they went to the same balcony area to get a view of the concert. Immediately upon getting to the balcony railing, the same man who had earlier kicked John Fedric jerked Humphries backward by the belt. Again, words were exchanged, but Humphries started to walk away in order to avoid a fight.

Then, the woman with the foursome made a remark and the man who had just jerked Humphries broke a beer bottle and threatened to cut Humphries. While Humphries was watching the man with the bottle, another man in the foursome hit him in the jaw, breaking it. Jerome Silvio then struck the man who had hit Humphries, and the man with the broken bottle stabbed Silvio in the throat. Fedric then tried to help his friends and was hit in the head with the beer bottle.

The security guards quickly broke up the melee, and they escorted Humphries, Silvio, and the four other customers outside the Dance Hall. Another friend took Silvio to the hospital for treatment. Two days later, Humphries and Fedric also received medical treatment for their injuries. The four customers who had engaged in the assault were never arrested or prosecuted.

The applicable law in this jurisdiction has previously been stated:

> The owner or operator of a business owes a duty to its guests to exercise reasonable care to protect them from injury.... This duty of care toward a tavern's guests does not extend to unforeseeable or unanticipated criminal acts by an independent third person. 'Only when the owner or manage-

> ment of a business has knowledge, or can be imputed with knowledge, of a third person's intended criminal conduct which is about to occur, and which is within the power of the owner or management to protect against, does such a duty towards a guest arise.'

[The court quoted from *Davenport v. Nixon*, reported in the November–December 1983 **PSCLR**, case one, page one.]

The court noted that the plaintiffs' only evidence as to the owner Pharoah's knowledge of the violent propensities of the assailants was the testimony of Mr. and Mrs. Fedric concerning the unidentified bartender's alleged statement that there had been prior incidents involving the foursome that same night. This testimony by them was, however, uncorroborated by any other testimony or evidence presented by them. The only bartender who did testify had stated that he had heard no complaints about the foursome and had had no reason to think that they were going to cause any trouble.

Two of the three security guards who were on duty at the Dance Hall that night also testified that they had had no prior knowledge of any other incidents involving the foursome. Accordingly, the plaintiffs were unable to establish by sufficient proof that Pharoah's had actual knowledge of the violent propensities of the assailants, nor that such knowledge could be imputed to them. The appellate court therefore affirmed the dismissal of the plaintiffs' claim.

PSCLR – SEP 88 (3)

Fast Food Store Not Liable For Injuries To Customer Shot In Drive-Through Lane When Acts Of Assailants Were Not Reasonably Foreseeable, Notwithstanding Expert's Opinion Testimony Of Inadequate Security:

Rose Marie Foster and Danny L. Foster, As Administrator of the Estate of Stephanie Lynn Foster v. Colonel Sanders Kentucky Fried Chicken, Inc., *No. 19560-CA, in the Court of Appeal of Louisiana, decided May 4, 1988, rehearing denied May 26, 1988, reported at 526 So. 2d 252.*

Rose Marie Foster and her husband, on behalf of their daughter Stephanie, sued the owner of a Kentucky Fried Chicken store, Holland Foods, Inc., after an incident in the store parking lot during which Ms. Foster was shot.

At 7:30 PM one evening in late December, Mrs. Foster and her 11-year-old daughter drove to the Kentucky Fried Chicken store on North Market Street in Shreveport, Louisiana. As Mrs. Foster drove into the parking lot, she noticed three young boys standing near the corner of the store. She was not alarmed by the boys' demeanor, and drove on to the rear of the store, stopping at the drive-up menu board to place her order.

Stephanie Foster warned her mother that the three boys were approaching their car, and Mrs. Foster looked back to see that the boys were surrounding the car. As she attempted to drive away, one of the boys pulled out a gun and fired a shot into the car, striking Mrs. Foster's right arm. She managed to drive to another restaurant's parking lot, where she called the police and an ambulance.

The police searched the Kentucky Fried Chicken parking lot and a large grassy field behind the lot, but did not locate the assailants. Later that night, three minors were apprehended – Anthony "Red" Thomas, Ernest Williams, and Jonathan "Johnifro"

Taylor – who were all then tried and adjudicated delinquents. Thomas testified at the trial of the Fosters' suit.

Prior to the shooting, the Shreveport Police Department had received a report of suspicious-looking persons near Midco, a machine shop next to the Kentucky Fried Chicken store. Officer Eloise Lewis, who was dispatched to the scene, checked the Midco building but found no one. She then drove to the Kentucky Fried Chicken, parked her police car in front of the store, and went inside to inform the employees that suspicious persons were reported at Midco.

Officer Lewis testified that she advised the store employees, somewhat jokingly, that there was "probably someone planning on robbing them." She did not, however, check behind the store for suspicious persons. No one from the store went outside to check for prowlers, but the assistant manager looked out the window and saw no one. The shooting occurred about 45 minutes after Officer Lewis left the store.

A Ms. Kent testified at trial that she had been at the Kentucky Fried Chicken that evening and had noticed three boys looking at cars in the parking lot. She testified that she reported this to the cashier, and that when she left, she saw the boys hiding around the garbage container. She claimed that she later called the police from her home, presumably precipitating Officer Lewis' arrival at Midco. The assistant manager did not recall a customer reporting suspicious characters that evening, and the trial judge apparently did not credit Ms. Kent's testimony.

Anthony Thomas, one of the boys involved in the shooting, lived near the field behind the Kentucky Fried Chicken store. He testified that he and his friends walked down North Market Street that night looking for someone to rob. They passed by a Burger King and Pay-Less Shoe Store, but stopped at the Kentucky Fried Chicken store, because they thought they could "get away with it" there since the area behind that building was dark and the field provided an easy escape route. They hid behind the trash bin in the rear of the building in order to stay out of sight of the store employees. After the shooting, Thomas fled through the field behind the store.

The Fosters' suit contended that the store employees were negligent in failing to prevent the assault after being warned that there were three suspicious young men on the premises. They also contended that the store breached its duty of reasonable care to protect its patrons from harm, alleging inadequate lighting behind the store and the unsafe design of the building.

At trial, the Fosters showed that the parking lot where the assailants were hiding was mostly out of sight of the store employees. A customer using the drive-through window had to drive past the lobby, around the building, and stop in the rear at the menu board to place an order. On the side of the building, the pick-up window extended out about four feet from the wall, with a window facing the rear. A small mirror was mounted close to the building so that an employee looking out the window could see whether a car was waiting at the menu board. There was a door on the rear wall of the building, but no windows.

The Fosters' security expert at trial, Mr. Ostendorff, testified that the Kentucky Fried Chicken store was unsafe because it was not fenced in, had no surveillance behind the building, had no security guard on duty, was too dark, and lacked a "safe design." His suggestions for a safer design included placing the menu board on either side of the building in fuller view of employees, placing a window in the rear wall so that employees in the kitchen could keep an eye on the rear parking lot, installing a closed circuit T.V. monitor for surveillance purposes, and hiring a security guard.

Captain Holt of the Shreveport Police Department testified on behalf of the Fosters to establish that the store was in a dangerous location, which, the plaintiffs alleged, should have alerted a reasonably prudent businessman to take precautions for customer safety. According to Captain Holt's testimony, the police beat encompassing the store registered 167 violent crimes (49 robberies) in the preceding year and 219 violent crimes (60 robberies) in the year Ms. Foster was shot. This Kentucky Fried Chicken store had been robbed about five months earlier by a gunman who entered the lobby and held up the cashier. This was the only incident at the store in the nearly three years of the store's operation.

The Fosters also introduced expert testimony by a lighting and electrical engineer, Mr. Purdle. He testified that the area around the menu board was illuminated with only 1.5 foot-candles as opposed to 12 to 16 foot-candles in the middle of the parking lot. He stated that this differential in the intensity of lighting inhibited the "see-ability" of the darker location, due to the decreased capacity to see when a person moves from a bright area to a much darker area. The nearby Burger King store's lighting, though it was lit with only two to three foot-candles, was preferable because the lighting was uniform.

The vice president of Holland Foods (owner of the store) testified that SWEPCO security lights were installed in the parking lot when the store was built and were in place and operative on the night of the shooting. The building complied with all city regulations. He stated that company policy was not to open the rear door after dark, due to the increased likelihood of robbers entering the store if the door were opened. In addition, he testified that placing the menu board on the side of the building would block traffic into North Market Street at rush hour.

The trial judge had found that the store provided a "reasonably safe place for customers to place orders." He had also found that, after Officer Lewis advised the employees that suspicious persons were reportedly in the area, neither the officer nor the employees saw any such people. Therefore, the employees did not have actual or imputed knowledge that a criminal act was about to occur. The trial judge therefore had dismissed the claims against Holland Foods.

On appeal, the court noted:

> In order to prevail in a negligence action, the plaintiff must prove that the defendant's conduct was a cause-in-fact of the plaintiff's harm, that the defendant owed a duty of care to protect against the risk involved, that the defendant breached the duty, and that actual damages resulted.

The issue contested was whether Holland Foods owed a duty of care to its patrons to protect against the risk of the assailants in the parking lot. The appellate court quoted from one of its own previous decisions:

> A business establishment . . . owes a duty to its patrons to exercise reasonable care to protect them from injury. This duty does not extend to unforeseeable or unanticipated criminal acts by independent third persons. Only when the owner, management or employees of a business have or should have knowledge of a third person's intended injurious conduct that is about to occur and which is within the power of the owner, management or employees to protect against, does the duty arise .... the duty can be discharged by summoning the police at the time the proprietor knows or should reasonably anticipate that the third person poses a probable danger.

Ballew v. Southland Corporation, reported in the May 1986 **PSCLR**, case three, page three.

The appellate court examined the trial court's finding that the employees did not know or have reason to know of the intended injurious conduct. The appellate court noted that the plaintiffs did not prove that any employee ever saw the assailants, and that Officer Lewis admitted that she was really just "joking" when she told the Kentucky Fried Chicken employees that they might be robbed. This was consistent with her decision not to check around the building. Under these circumstances, the appellate court found that the trial court was not incorrect in finding that the criminal acts of the young men were not foreseeable or anticipated by the defendant.

In addition, the trial judge had found that the duty to summon the police promptly, if one of the employees had spotted the prowlers, was fulfilled, because the police were in fact called and came to the store. The fact that the store employees did not see the assailants, who were admittedly trying to hide, was deemed irrelevant. In sum, the appellate court found that the trial court was not completely wrong in finding that the employees did not have actual or constructive knowledge of imminent tortious activity which was within their power to avert. The criminal acts of the three young men, under the circumstances, were unforeseeable and unanticipated.

As to the alleged breach of duty to maintain reasonably safe premises, the appellate court stated that the law requires a restaurant to maintain the premises "reasonably safe for the purchase and consumption of food by patrons." The appellate court noted that in a decision by the Louisiana Supreme Court, *Harris v. Pizza Hut* (reported in the January 1985 **PSCLR**, case one, page one), Pizza Hut had been held liable when its security guard first hid, and then thrust himself into the volatile situation of an armed robbery, which escalated the violence and resulted in a shooting and death. In that case, the Pizza Hut was located in an extremely high crime district.

Based on *Harris*, the court in this case held:

> [A] restaurant is not obligated to hire a security guard or to take other precautions against crime unless it decides or should decide that ordinary police protection is inadequate. The duty therefore does not extend to the protection of customers from the criminal acts of third parties unless the risk of crime on the premises was sufficiently foreseeable to require special protection through security measures.

The appellate court here noted that the Kentucky Fried Chicken store had been in business for almost three years and had had only one prior armed robbery. That robbery had occurred in the lobby and could have been averted by the "visible presence" of a security guard. The instant offense, occurring behind the store, might not have been. The court also noted that the bulk of the crime in this particular "beat" occurred in the central part of the town, and not where the store is located.

The circumstances of the case establish that the armed robbery of a customer driving through the pick-up lane was a mere possibility and not reasonably foreseeable. The court observed, "It was not enough to activate the duty of taking deterrent or preventive measures." The court did, however, warn that the evidence of this shooting, together with the evidence of two subsequent crimes, might on a future occasion be construed as grounds for knowledge that the store is in an area prone to crime and that deterrent measures should be taken.

The Louisiana Court of Appeal therefore affirmed the decision of the trial court in this case.

COMMENTARY:

In addition to the two cases referenced in the court's decision above, the Louisiana Court of Appeal also relied upon Banks v. Hyatt Corp., reported in the April 1984 **PSCLR**, *case one, page one; Davenport v. Nixon, in the November-December 1983* **PSCLR**, *case one, page one; and Hardin v. Munchies Food Store, in the October 1987* **PSCLR**, *case five, page seven.*

PSCLR – FEB 89 (3)

Restaurant Owner Has No Duty To Illuminate Rear Alleyway Of Leased Premises Used In Common With Other Tenants:

Murray Krinick, et al., v. Sharac Restaurant, Inc., etc., et al, *in the Supreme Court of New York, Appellate Division, Second Department, decided November 14, 1988, reported at 533 N.Y.S.2d 1013.*

As Murray Krinick was walking from a parking lot on Cathedral Avenue in Hempstead, New York, to Cooky's Steak Pub on Fulton Avenue, he was attacked by an unknown assailant, apparently hiding in the alleyway behind Cooky's. Cooky's shared the alleyway with the other retail establishments fronting on Fulton Avenue.

Krinick sued Sharac Restaurant, Inc., owners of Cooky's Steak Pub, alleging the restaurant had negligently failed to provide adequate lighting in the area where the attack occurred. Sharac Restaurant successfully moved to have the case dismissed, asserting that the restaurant was not obligated to illuminate areas which could conceivably conceal criminals who might attack persons using this street. Krinick appealed, and the appellate court affirmed the dismissal.

In order to prevail on a negligence claim, a plaintiff must first establish that the defendant owed a "legal duty" to him. The court held that, "Sharac was under no duty to illuminate the rear alleyway of its leased premises, which was used in common with other tenants, to protect against the type of harm incurred by the plaintiff Murray Krinick."

Since the court's decision on this issue of law left no factual issues to be decided by a jury, the court affirmed the dismissal of the case against Sharac without a trial.

COMMENTARY:

It is well known that one of the elements which must be proved by plaintiff in a negligence action is that defendant was negligent; that is, that the defendant breached the standard of care which is exercised by the reasonably prudent person in the same or similar circumstances. However, plaintiff must also prove that the defendant owed the duty to exercise such care for the benefit of the plaintiff.

Therefore, even if a defendant fails to act as a reasonably prudent person and thereby breaches the standard of care, there is still no liability unless the defendant owed the duty of care to the plaintiff. (See the case of Figueroa v. Evangelical Covenant, for another situation where a property owner was held not to owe a duty of care to the particular plaintiffs involved.)

PSCLR – SEP 89 (16)

Business Owner Has No Duty To Protect Patrons From Unforeseeable Criminal Acts Occurring On The Premises:

Christos Papadimas, Plaintiff-Appellant v. Mykonos Lounge, Demetrios Panagopoulus, Agrias Korogiannis, Athanasius A. Peristeris and Shadad Mattey, Defendants-Appellees, *Docket No.103796, in the Court of Appeals of Michigan, decided March 20, 1989, reported at 439 N.W.2d 280.*

Christos Papadimas was standing at the bar in the Mykonos Lounge, a defendant herein, when another patron of the bar, Shadad Mattey, also a defendant, assaulted him. Allegedly, Mattey struck Papadimas from behind with a handgun. Papadimas filed this case against the lounge, its owners, and Mattey. Papadimas alleged that the lounge and its owners were negligent in failing to protect him from the assault.

According to a witness who was present at the time of the attack, the assault was a "complete surprise." The witness further stated that nothing happened prior to the incident that would have led anyone to believe that such an attack might have occurred. Additionally, the owners of the lounge did not know of any violent tendencies of the assailant Mattey or of any other assaults by him in that bar or any other. Finally, there was testimony that the lounge employees intervened quickly to protect Papadimas from further injury.

Based on these facts, the court concluded that the lounge had not violated any duty owed to Papadimas. The court noted that a business owner is not an "insurer" of his customers' safety and he has no obligation to provide them police protection. Furthermore, the court recognized that criminal acts are almost inherently unforeseeable, and, as a result, a business can not reasonably be expected to prevent them. In this case, the testimony of the witnesses proved that the assault was clearly unforeseeable. Since they could not foresee the occurrence, the defendants could not have been expected to prevent it; therefore, the court held that they violated no duty that they owed to Papadimas.

Retail

PSCLR–AUG 82 (8)

Disconnection Of Burglar/Fire Alarm System By Unknown Party Is Insufficient To Hold Store Owner Liable When Fire Spreads To Other Property:

George D. Haynes, d/b/a Geo-Je's Apparel and North River Insurance Company v. Aetna Casualty & Surety Company, and Miller's Mutual of Texas, *Docket No. 14551, Court of Appeal of Louisiana, First Circuit, decided March 2, 1982, Rehearing Denied April 13, 1982, reported at 411 So. 2d 1145.*

The insurer of Lee Allen's Fashions for Men, a retail clothing store, was not held liable for damages to Geo-Je's Apparel, and adjacent women's clothing store, when an unknown third party disconnected Allen's alarm system and deliberately set a fire. Although several suits resulted from this fire, only the dismissal of Geo-Je's claim against Allen's insurer was appealed.

Geo-Je's had to prove that the insured or one for whom he is responsible was the cause of the fire to hold the insurer liable. In this case, the only evidence presented to implicate the owner of Allen's with starting the fire was gleaned from the records of the company which installed its alarm system. A fire detective reviewed these records and testified that at about 6:15 PM, ten minutes after Allen's owner and employees had left for the night, an intruder successfully disconnected the burglar alarm system. The intruder apparently understood the operation of the system, because he reset the alarm after starting the fire and leaving. The alarm company's records indicate that the alarm sound at 6:23 PM.

This evidence, based mainly on hearsay, was insufficient to identify the arsonist and impose liability on the insurer. One of Geo-Je's arguments was that Allen was strictly liable for the damages caused by the spreading fire, no matter what the cause. This position was based on a Louisiana statute which covers situations where the owner or lessee of property is carrying on some activity on his property which causes damages to his neighbor.

The Court rejected this argument, citing an older case in which a fire of unknown origin spread to a neighboring cane field:

> The [Louisiana statute] has no application to the circumstances of this case. To hold otherwise would cast liability on every landowner on whose property a fire is commenced and spread to neighboring lands even though the cause or origin of the fire was due to the act of trespassers, vandals, lightning or unknown sources not attributable to any conduct or activity on the part of the landowner.

The Court of Appeal affirmed the dismissal of the action against Allen's insurer.

PSCLR – AUG 82 (9)

Duty Of Property Owner To Warn Of Hidden Dangers:

Bud Joyce and Eleanor G. Joyce v. Charles Nash, d/b/a Thunderbird Trailer Sales and Thunderbird Trailer City, *Docket No. WD 32004, Missouri Court of Appeals, Western District, decided February 2, 1982, Motion for Rehearing and/or Transfer to Supreme Court Overruled and Denied, March 2, 1982, reported at 630 S.W.2d 219.*

The Missouri Court of Appeals reviewed a peculiar fact situation in which customers sued the operator of a motor home business for negligently locking them inside the premises and unleashing ferocious guard dogs. The jury returned compensatory damages of $3,000 for the husband and $7,000 for the wife, but denied punitive damages.

Bud and Eleanor Joyce saw a newspaper ad of Thunderbird Trailer Sales ("Thunderbird") concerning a motor home, and telephoned for information. A Thunderbird employee stated the business was open "late every night," so the Joyces and their daughter went to Thunderbird a day or so later, at about 7:30 PM, in their pickup truck. The business was surrounded by a fence topped with barbed wire. In a window of the main office, a sign stated "Yes, We're Open," but there were no signs indicating a customer parking area, or warning of guard dogs. Two cars were parked in front of the office, the office lights were on, and people could be seen in the office.

When no salesperson came out after a few minutes, the Joyces and their daughter walked over to and entered an unlocked motor home. When they left the trailer to look for the outside manual air-conditioning switch to continue their inspection, a hostile dog approached their daughter and another dog followed it. The family managed to make it to the pickup truck before five or more vicious animals surrounded them, lunging and clawing against the truck door.

The Joyces drove to the office and sounded the horn, only to discover the other cars gone and the gate chained and padlocked. Mr. Joyce tried to signal a passerby by standing on the truck seat and extending himself out the window. Two separate attempts by people alerted by Joyce to rescue the family failed, but a third person was able to get the police.

When the police arrived, the family was in the truck with the windows up and dogs attacking the truck's side. The police considered the dogs too dangerous for the Joyce family to leave the truck or for the police to enter the compound, so a fire rescue squad was called. Only by placing a ladder across the fence to the truck and by distracting the dogs were the firemen able to extricate the Joyce family. By this time, the Joyces had been trapped two and one-half hours.

When Mr. Joyce returned for his truck the next day, owner Charles Nash did not believe his story but reluctantly permitted Joyce to take his truck after an unpleasant exchange.

The Joyces maintained that Nash was guilty of false imprisonment and negligence for failing to warn or protect them, as business invitees, of the danger presented by the guard dogs, and for failing to check the compound for straggling customers before unleashing the dogs and locking the gate. Plaintiffs asked for actual and punitive damages, alleging reckless and willful conduct by Nash.

Although Nash did not deny he owed a duty to business invitees, he denied he had breached that duty. Nash contended that his nightly inspection procedures exemplified his stringent precautions. He testified he normally locked the east gate and parked his car in front of the west gate (through which the Joyces had entered) to prevent anyone else from entering. While he fed and watered the dogs, his wife closed each trailer door. Nash then took a few dogs around the premises for a final inspection of the trailer doors. While his wife waited for him at the west gate, Nash released the dogs, walked through the office building and locked its doors. After pulling out of the compound, Nash locked the west gate.

The jury disagreed with Nash's contention of reasonable care and awarded actual damages to the Joyces. Nash appealed <u>any</u> award of damages, and the Joyces appealed for not receiving punitive damages.

The Missouri Court of Appeals reversed the decision of the trial court because the instructions to the jury did not make it clear that the test was whether Nash knew or should have known the Joyces were still on the premises. As presented, the instructions "imposed liability, not for unreasonable conduct, but for consequences however reasonable the conduct: that is, without a determination of fault. The instructions submitted strict liability, not negligence." Because the inspection procedure practiced by the defendant was reasonable to the point of being painstaking, the court felt only a question of ordinary negligence was inferred by the Nashes' failure to observe the presence of the Joyce family on the Thunderbird premises. The court granted a new trial, since "[n]either the release of the dogs nor the closure of the gates was inherently negligent. That conduct becomes culpable only if the defendant knew or could have known that the Joyces were then still on the premises and—by the standard of a reasonable person—failed to take precaution commensurate with the danger to avoid risk of injury to them."

PSCLR—NOV-DEC 82 (12)

Where Customer Returns Goods to Store Owner for Repair, Store's Failure to Reasonably Protect Goods from Burglar May Be Shown by Its Failure to Provide Insurance Coverage for Such Loss:

Stefan Jewelers Inc. v. Berry et al., *No. 64519, in the Court of Appeals of Georgia, decided September 27, 1982, reported at 295 S.E.2d 373.*

A jewelry store was held liable to customers for the value of their jewelry which was stolen while being repaired at the store.

Customers Robin and Sandy Berry left five pieces of jewelry with Stefan Jewelers for repair and cleaning. When she went to pick up their jewelry, Mrs. Berry discovered that it had been included in items stolen from the store. Stefan refused to reimburse the Berrys and the couple brought a lawsuit against the store. The suit ended in judgment for the Berrys in the full amount of their loss—$7,666.

Stefan Jewelers appealed the decision, contending the trial court erred in allowing a copy of its contract with a burglar alarm company in evidence. Stefan felt this contract was irrelevant to the issue and it unfairly

> appeared to obligate Stefan Jewelers to obtain insurance against such losses because [the contract] limited the alarm company's liability in the event of theft, thereby imputing negligence in Stefan's failure to obtain such insurance.

The Court of Appeals of Georgia disagreed with Stefan's arguments, stating that such a contract is admissible and relevant in deciding whether the store "took reasonable measures to protect its store and the jewelry therein." The court continued, quoting an earlier case:

> 'A bailee [i.e., Stefan in this case] is not an insurer of bailed property, but is bound to exercise ... reasonable care, and if bailed property is damaged

or lost through negligence on the part of the bailee, the bailee properly may be held responsible.'

COMMENTARY:

It is important to understand that the customers here were not claiming that the store should have had insurance and that such insurance should have covered the loss; such an argument would be rejected because it has long been the law that the existence or non-existence of insurance is irrelevant on the issue of damages.

On the other hand, the customers argued that it was customary in the jewelry trade to carry insurance coverage for customers' goods, and the store's failure to do so was some evidence of negligence in the protection of customers' property. In other words, the customers here were simply proving that the jewelry store owner had not taken one of the usual and customary steps taken by ordinarily prudent jewelry store owners.

A store is not an insurer of its customers' property; that is, a store is not automatically liable to its customer when his property is lost. The store is liable for the loss only if the customer can show that the property was damaged or lost through the negligence on the part of the store, since the store is legally required to exercise ordinary and reasonable care for the protection of the customers' property. In trying to prove that the store did not exercise "ordinary and reasonable care," the customer can show what usual and customary steps are taken by ordinarily prudent store owners in the same industry.

PSCLR – NOV-DEC 82 (13)

Store's Failure to Reasonably Protect Customer's Goods from Burglar May Be Shown by Its Failure to Use Case-Hardened Locks as Security Measure:

James Lawrence Smith v. Byrum McRary d/b/a McRary Harley-Davidson, *No. 91PA82, Supreme Court of North Carolina, decided October 5, 1982, reported at 295 S.E.2d 444.*

This case was the result of the theft of a customer's property while it was being repaired on business premises.

Customer James Smith left his new motorcycle at McRary Harley-Davidson for routine servicing as required by the warranty. Although Smith left his motorcycle at the shop's main building, it was placed in a shed in the back after being serviced. Smith knew the main building was wired with an alarm, but specific security and storage procedures were not discussed.

Both the main building and the shed indeed had been wired with a burglar alarm and automatic lighting system after a burglary of the main building. In addition, the gate to the fence enclosing the premises and the shed door itself were both padlocked.

However, two burglaries had occurred in the main building since the security system was installed, although Smith's and another motorycycle were taken in the first burglary of the shed. McRary admitted he discovered that the alarm was not working on the day the shed was burglarized.

Furthermore, testimony by a preventive security expert indicated that padlocks are not a recommended security measure . This expert also said that only "case-hardened" padlocks could resist a bolt-cutter such as the one used to gain entry into both the fence and the shed when Smith's and the other motorcycle were taken.

Smith sued McRary to recover the value of his stolen motorcycle, alleging the burglary was due to McRary's negligent security measures. Smith also contended that his delivery of his cycle to the main building and his reliance that it would be kept there created an implied contract to that effect, even without further discussion. The jury found for the shop owner and the case was dismissed by the trial court, but the North Carolina Court of Appeals subsequently ordered a new trial.

On further discretionary review, the North Carolina Supreme Court agreed that a new trial should take place, but not on the basis of an implied contract. No authority could be found to support the idea that mere delivery of a bailed item to a particular place guaranteed its storage there. The new trial was in fact granted because of improper jury instructions at the original trial.

COMMENTARY:

This case illustrates that the courts are allowing expert witnesses to testify as to the reasonable sufficiency of preventive security measures, and that a defendant cannot absolve himself from liability by utilizing ineffective or below-standard security measures. Security equipment of reasonable quality must be used; that is, reasonably suited for the task it is intended to perform.

PSCLR – MAR 83 (8)

Customer Injured By Falling Over Floorstand In Store Aisle May Recover If Attention Was Diverted By Store Displays:

Dennis J. Schuller and Becky Schuller v. Hy-Vee Food Stores, Inc., *No. 67492, in the Supreme Court of Iowa, decided December 22, 1982, reported at 328 N.W.2d 328.*

When Dr. Dennis Schuller's wife was out of town, he went into the Hy-Vee Food Store for only the second time in his life. While surveying the merchandise down one of the aisles, Schuller walked around a sales display and into a floorstand ashtray. He fell over the ashtray and injured his right hand.

Dr. Schuller, a dentist, and his wife sued Hy-Vee Food Stores for the store's alleged negligence in blocking a pathway with a hidden hazard which directly caused his injuries. The Schullers also contended that the store had failed to put the ashtray in a safe place and had failed to warn customers of the ashtray's location.

At the trial, two store employees testified that they were not aware of anyone else having fallen over an ashtray before, even though the store usually kept ashtrays at the end of merchandise displays such as where Schuller had tripped. Judgment was entered for Hy-Vee Food Stores, and the Schullers appealed.

On review, the court noted that "...in determining the contributory negligence of a person injured while on business premises, the fact that one's attention may be distracted by merchandise on display is a factor to be weighed in the person's favor." Of course, a customer could not ignore an obvious hazard and then blame the store for a resulting injury, but the court concluded that,

> Generally... the storekeeper cannot complain if a customer looks at displayed goods instead of at the floor. The customer is not bound to the same care to avoid obstacles as a pedestrian on the street, and the storekeeper's care must be correspondingly greater.

Moreover, testimony by the Hy-Vee's assistant manager that he saw the ashtray behind the display ten minutes before Schuller tripped led the court to rule that a jury should decide whether the placement of the ashtray constituted a known hazard. In such case, the store could be held liable for failing to exercise reasonable care to protect its customers from injury. Judgment for Hy-Vee Food Stores was therefore reversed, and the case was sent back to the trial court for further proceedings.

PSCLR – SEP 83 (9)

Retailer Not Liable for Unforeseeable Injury to Customer Caused by Fleeing Shoplifter:

Mrs. Ida Mae Butler v. The K-Mart Corporation and ABC Insurance Co., *No. CA-0467, in the Court of Appeal of Louisiana, Fourth Circuit, decided May 16, 1983, reported at 432 So. 2d 968.*

As Ida Mae Butler prepared to enter a New Orleans K-Mart, she was struck in the chest by the opening door and was thrown into a bystander. The door had been flung open by a fleeing shoplifting suspect. Butler felt that the man had raced out and caused her injuries because he was being chased by K-Mart employees. She therefore sued the store's owner for its alleged negligence which created a "dangerous" situation.

However, Butler had not seen, nor was she able to prove, that the man had actually been chased out of the store by any employee of K-Mart. One of the store's security officers, Norman Caesar, testified that he had not even noticed the suspected shoplifter until after the suspect ran from the store and knocked Butler down. It was after the suspect was in the parking lot that K-Mart employees began to chase him, according to the security officer. Butler herself was unable to rebut this testimony; according to her own evidence, she saw the guard chasing the suspect only after she was injured.

The K-Mart Corporation and its insurer were granted summary judgment before a trial took place. Butler appealed, arguing that there were questions of fact yet to be determined. Louisiana's Court of Appeal disagreed and stated:

> A store owner owes a duty to its patrons to take reasonable care for their safety, but it is not the insurer of the patron's safety. ... A proprietor of a public place has a duty to protect its patrons from injuries caused by third parties when it is within its power to do so. ...

The court held that without evidence that any of the K-Mart's employees had contributed to the shoplifter's actions – bolting out of the store and throwing open the door into Butler – the store could not be held liable for failure to foresee the occurrence and its result. Therefore, the court affirmed judgment in favor of K-Mart Corporation and dismissed Butler's case.

COMMENTARY:

It is important to recognize that the key to this decision was whether or not the injury suffered by the customer was "foresseeable" by the retailer. In other words, this case does not necessarily mean that a retailer can <u>*never*</u> *be held liable for injuries to a customer caused by a fleeing shoplifter; rather, the test is whether such injuries to the customer were reasonably foreseeable by the retailer. If so, then the claim will be presented to a jury for determination.*

The appellate court here spent some time discussing whether the shoplifter was running out the door because he was being chased by the store security guards, or whether he was running without being chased. The clear implication to be drawn from the court's opinion is that if the shoplifter had been running because he was being chased, then the store could or should have been able to foresee that chasing him would cause him to run. And, a customer could therby be injured by him as he ran. If so, of course, the retailer could then be held liable for the foreseeable result of chasing the shoplifter.

In light of this decision, retailers should weigh their priorities and decide whether their policy should be to chase fleeing shoplifters and thereby incur possible liability to others who may be injured in the chase, or to forego the chase and possibly lose the suspect, in order to avoid possible liability.

PSCLR – NOV-DEC 83 (5)

Retailers/Owners Have No Duty to Protect Invitees Against Crime Absent a "Special Relationship" or Unless "Special Facts" Exist:

Mary Sue Meadows, et al. v. Friedman Railroad Salvage Warehouse, a division of Friedman Bros. Furniture Co., Inc., a corporation, and Whelan Security Co., Inc., a corporation, *No. 45684, in the Missouri Court of Appeals, Eastern District, Division Four, decided June 21, 1983, motion for rehearing and/or transfer to Supreme Court denied August 1, 1983, application to transfer denied September 20, 1983, reported at 655 S.W.2d 718.*

Mary Sue Meadows was injured when she was shot by a gun wielding assailant on Friedman Railroad Salvage Warehouse property. She brought suit against the owners of the property and against the security service company employed by Friedman. Her suit for personal injury damages and her husband's suit for the loss of his wife's consortium due to their injuries, were dismissed before trial and they appealed.

The Meadows' suits alleged that the Friedman's owners knew that Friedman's was located in a high crime area and had employed a security service in recognition of that fact. They alleged that the property owners had, however, negligently failed to provide adequate security, failed to warn customers that the area was not safe, failed to call police although they knew an assailant was on the premises and was intent on harming the plaintiff, failed to call the police when she was attacked, and failed to warn Mrs. Meadows of the violent nature of her attacker. They also sued Friedman's security firm, Whelan Security Co., on the same allegations, and also claimed that the guards' failed to help her when she was attacked. She contended that Friedman (and therefore Whelan) owed a duty to protect her as its business invitee, and that their negligence in doing so had caused her injuries.

On review, the Missouri Court of Appeals outlined the required elements of a successful negligence suit seeking damages. The plaintiff must prove:

(1) the existence of a duty on the part of the defendant to protect the plaintiff from injury,

(2) failure of the defendant to perform the duty, and

(3) injury to plaintiff resulting from such failure.

The court then pointed out that an owner of business property would not be liable for negligence for criminal acts by a third person unless such third person was either under the control of the owner (i.e., as his agent or employee) or his actions could be foreseen and prevented by the owner. But there would be no duty on the owner to protect invitees against the "intentional criminal conduct of unknown third persons," unless a special relationship existed between the property owner and the victim or other special facts were present.

Examples of special relationships requiring a higher standard of care by a property owner would include those of the innkeeper/guest, the school/student, the common carrier/passenger, and in certain circumstances, the employer/employee. Special facts requiring a higher standard of care would exist if the property owner knew that a dangerous person was on his property, or if he knew that a person on his property had acted in a way which indicated a threat of injury to other invitees. Liability would apply, however, only if there was enough time for the property owner to warn invitees before an injury occurred. The court did not agree that foreseeability of criminal activity by itself could make a property owner liable for a crime on its property. Instead, both foreseeability and a legal duty to take measures to prevent the crime would have to exist. "Whether such a duty exists depends ultimately on the question of fairness, which in turn is decided by a weighing of the relationship involved, [and] the nature of the risk ... ."

Based on the Meadows' allegations in their lawsuit, the court held that there existed no "special relationship" between the plaintiffs and the defendant, beyond that of the normal business invitee relationship. Moreover, the allegation that Friedman's was located in a high crime area was insufficient to establish "special facts." Friedman, and therefore Whelan, had no special duty to protect Meadows from the unexpected attack by an unknown criminal.

The court also rejected Meadows' argument that Whelan should have, at the very least, stopped the attacker or called the police faster, ruling that there was no evidence which showed that enough time had elapsed after Whelan was aware of the attack on Meadows to prevent her injuries. Further, the court held that the law would not require anyone, even a security employee, to risk his life and possibly that of the victim to intervene in an attack by an armed criminal.

Judgment granting dismissal of the suit was affirmed.

COMMENTARY:

This case is an important victory for retailers and security service firms in Missouri. The significance of this decision is apparent when the Missouri court's decision is contrasted with the holding set forth by the Florida court in its Stevens v. Jefferson decision reported in this issue.

Also, the above holding by the Missouri Court of Appeals should be read in conjunction with the Missouri Supreme Court's decision in Virginia D. v. Madesco Investment Corp. which was rendered in final form on April 26, 1983 (the first case reported in the September 1983 **PSCLR***). While that case did involve a hotel (the Bel Air Hilton) as defendant, it was not based on an innkeeper-guest relationship because the assaulted invitee in that case was not a hotel guest, but rather she was a diner in one of the hotel restaurants.*

Furthermore, the Missouri Supreme Court was divided in the Madesco case and the plaintiff prevailed by a vote of only four to three. Perhaps this helps to explain why the Meadows could not convince a sufficient majority of the Supreme Court to accept a further appeal of the above decision by the Missouri Court of Appeals.

In any event, the result in this case is still difficult to reconcile with the earlier holding in Madesco, and the state of the law on this issue in Missouri is, at best, confused.

For a similar case where a contract guard company was held not liable for its guard's failure to protect a client store manager and to prevent an armed robbery, see Paradiso v. Apex Investigators, reported in the May 1982 **PSCLR***, page 2.*

PSCLR—APR 84 (14)

Retail Guard Surveillance Of Fitting Rooms Does Not Constitute Invasion Of Customer's Privacy Where Signs Inform Customer Of Surveillance:

Kevin Lewis v. Dayton Hudson Corporation, d/b/a J.L. Hudson's Company, a foreign corporation, *Docket No. 64957, in the Court of Appeals of Michigan, decided August 16, 1983, released for publication October 19, 1983, reported at 339 N.W.2d 857.*

Kevin Lewis, an undercover police officer, was detained at a Michigan J.L. Hudson's Company department store after a Hudson's security guard watched him remove his gun in one of the store's dressing rooms. Lewis brought this lawsuit against the store alleging invasion of privacy, false imprisonment and intentional infliction of emotional distress. Summary judgment was granted to the store by the trial court on the invasion of privacy allegation, and the other allegations were voluntarily dismissed by the plaintiff. Lewis appealed the summary judgment to the Court of Appeals of Michigan.

The situation which brought about the suit occurred when the plaintiff was detained while shopping in the J.L. Hudson's store after a store security guard, concerned that Lewis might be a shoplifter, followed him into the dressing rooms. Each dressing room had a ceiling grill through which one could see inside the room from an overhead observation area. The guard climbed into the observation area and watched as Lewis withdrew a gun and placed it on a chair. The guard then had another employee call the police.

When police arrived, Lewis had returned to the store's main area. A police officer approached him, drew his own gun, and told him not to move. Other officers arrived and Lewis was taken out of the public area of the store. Although he attempted to explain that he was carrying a gun because he was an undercover officer, he was not believed or permitted to leave until the police examined his identification.

Considering this evidence in light of an invasion of privacy suit, the Michigan Court of Appeals defined four separate types of invasion of privacy claims:

> (1) intrusion upon plaintiff's seclusion or solitude, or into his private affairs; (2) public disclosure of embarrassing private facts about the plaintiff; (3) publicity which places the plaintiff in a false light in the public eye; and, (4) appropriation, for the defendant's advantage, of the plaintiff's name or likeness.

Of the four, only intrusion upon the plaintiff's seclusion applied to this case, and to support such a claim he would have to show that (1) Hudson's intruded (2) into a matter he had a right to keep private (3) by way of a method which would be "objectionable to a reasonable person."

The court here ruled that the conduct of Hudson's security guard did not constitute an unwarranted invasion of Lewis' privacy. Citing *In re Deborah C.* (a similar California case), the court noted that

> it was not unreasonable for store security personnel to view patrons in fitting rooms from an adjacent corridor through which customers and salesclerks frequently passed. ... [particularly where] there were large gaps above and below the fitting room doors which provided an obvious view into each fitting room.

Any reasonable expectation of privacy in a store dressing room would be further limited by "the modicum of privacy [the dressing room] appears to afford." Although in this case the surveillance was more intrusive than merely observing the inside of a dressing room from a corridor, the court pointed out that signs were posted in the Hudson's dressing rooms which informed customers that the area was under surveillance. Such signs, according to the court, gave each customer the choice to not use the dressing room if he did not want to be observed and effectively negated any expectation of privacy the customer might have had in the dressing room.

The court qualified its ruling that signs warning of surveillance negated customers' expectation of privacy by noting that such signs would not provide complete immunity from invasion of privacy claims. If the surveillance had been conducted by persons not authorized or of the opposite sex of those being observed or if the observer had used information obtained in the surveillance in an objectionable manner, the court stated that its decision might well have been different. In this case however, Lewis was warned of possible surveillance by the sign, the observer was a male security guard, and the guard merely notified the police that a suspected shoplifter had a gun in his possession. The summary judgment on the invasion of privacy claim granted to Hudson's was therefore affirmed.

PSCLR—MAY 84 (8)

Retailer's Exculpatory Agreement In Lease Cannot Protect Retailer From Its Own Negligence When Sued By Picketer Of Retailer's Lessee:

Vera Quinones v. Waldbaum's Inc. and Waldbaum's Inc. v. 21 Regent Drive Corp., *in the Supreme Court of New York, Appellate Division, First Department, decided December 20, 1983, reported at 469 N.Y.S.2d 743.*

The defendant Waldbaum's Inc., owned and operated a supermarket in the Bronx borough of New York City. 21 Regent Drive Corporation was licensed by Waldbaum's to operate a pharmacy in the supermarket. The licensing agreement provided that Regent Drive would indemnify Waldbaum's from any claims or losses to any person caused directly or indirectly by operation of the pharmacy business by Regent Drive.

Vera Quinones was a paid picket-line walker who paraded with other pickets on the walkway in front of the supermarket. The picket line was established because of a union jurisdictional dispute concerning which union local should be permitted to staff the pharmacy. She was injured when she slipped and fell as she entered the supermarket to use the bathroom. The facts showed that Waldbaum's negligence was the sole cause of her fall, and she settled her case against Waldbaum's for $7,500.

The question presented to the court was whether Regent Drive had to pay Waldbaum's $7,500 to indemnify Waldbaum's pursuant to its above-mentioned lease agreement. The trial judge ruled that the presence of the picketers in the supermarket arose "directly or indirectly" from conducting the pharmacy's business and Regent Drive therefore had to indemnify Waldbaum's for its payment to Quinones.

However, on appeal the New York Appellate Court ruled that "agreements that would exculpate one from the consequences of his own negligence are strictly construed against him ... " The court noted that another provision of the licensing agreement stated that Waldbaum's was not liable for any injury resulting from the supermarket's willful negligence. The court concluded that if the agreement was to protect the supermarket from its own negligence it should have said so and since it did not, the ambiguity should be construed against Waldbaum's supermarket. In conclusion, Regent Drive Corporation did not have to reimburse Waldbaum's.

COMMENTARY:

We have reviewed various "exculpatory" provisions and contracts in cases reported in **PSCLR**, *and we again note for our readers that it is most important that such provisions be drafted by professionals with a clear understanding of the local statutory and case law authorities concerning the efficacy of such provisions in a given jurisdiction.*

These exculpatory provisions will be given full force and effect by the courts only when they are unambiguous and when they comply with the requirements of the respective state's laws. License and lease agreements should be reviewed, if not drafted, by lawyers for the interested parties, particularly if the agreements are to include exculpatory or hold harmless or other indemnification-type provisions.

PSCLR – SEP 84 (3)

7-Eleven Convenience Store May Be Liable For Customer's Injuries Received During Armed Robbery, But Store Employee Had No Duty To Rescue The Wounded Customer:

Richard Lee Cohen v. Southland Corporation, et al., *D000734, Civ. 28677, in the Court of Appeal of California, Fourth District, Division I, decided June 14, 1984, reported at 203 Cal. Rptr. 572.*

Richard Cohen stopped at a 7-Eleven convenience store on Linda Vista Road in San Diego, California, to pick up a sandwich very early in the morning of May 14, 1978. Before Cohen got out of his car, he was accosted by an armed man who told Cohen that he would rob the store, take Cohen as a hostage, and use Cohen's car to flee. The assailant then accompanied Cohen into the store and displayed his gun to the clerk, Timothy Wolfe, who was the only person in the store at the time. Cohen then grabbed the robber from behind, struggled with the armed man, and was shot in the stomach. Wolfe fled to the back of the store, barricading himself in a back office.

Cohen filed this negligence action against Southland Corporation, the corporate owner of the store; James R. Simmons, the franchisee of the 7-Eleven store; and Wolfe, the 7-Eleven clerk, for negligence. Cohen contended that the defendants failed to protect him and other store patrons from assault or danger of criminal attack and that Wolfe failed to rescue or aid Cohen after the shooting. Summary judgment was granted to the defendants by the trial court and Cohen appealed.

The Court of Appeal of California reviewed the evidence presented by the plaintiffs and defendants, noting that Southland's assistant director of 7-Eleven store security, Edward Perkins, had testified that there had been only one criminal incident reported at the Linda Vista Road store in the six years prior to this robbery. According to his records, the single incident was an armed robbery in December, 1976, during which there were no injuries.

Perkins also outlined Southland Corporation's extensive security program offered to franchisees, which included ongoing education in robbery prevention and in dealing with robberies in progress. He also testified that 7-Eleven franchise stores typically implemented security devices such as strong boxes for excess cash, warning labels on store windows which indicated that minimal cash was kept on the premises, a "balanced lighting system" which enabled clerks to observe outside and inside, and a policy of stacking merchandise no more than four and one-half feet to allow clerks unobstructed vision inside and outside. Perkins did not however testify that the franchisee Simmons had either participated in the security training program offered by Southland or had implemented the security practices described, although he did mention that Simmons had installed a video surveillance camera and a recording device to monitor cash register activity.

Simmons also testified that there had been only one criminal incident during the period he owned the store and that it had been a robbery without injury, but that no weapons were involved. He described the premises as being equipped with lights inside and out, video surveillance, and a window sign as described by Perkins. He also pointed out that there was a police substation one-half mile from the store and that police stopped by the store about once an hour during late night hours.

Wolfe testified that the interior and the parking lot of the store were well lit on the night of the robbery and the video surveillance equipment was functioning. He testified also that when Cohen and the robber entered and began struggling with the gun, he feared for his life and fled to the back of the store. He remained there, afraid even to use the telephone, until he saw on the television monitor that police had arrived. Wolfe admitted that prior to this robbery, he had not been trained to prevent or to deal with robberies.

Cohen's testimony conflicted with the defendants; he claimed that the Linda Vista Road store was generally not well illuminated and that on the night of the robbery, the 7-Eleven signs on the store's roof and by the street were not lit. He testified that the inside lights provided the premises' only illumination, since there were no other outside lights and neighboring businesses were closed and dark. Cohen claimed that when he himself had applied for a position at the 7-Eleven on Linda Vista Road prior to this robbery, the store manager had told him that the store was robbed periodically.

Other evidence demonstrated that in 1975 each Southland store was robbed in excess of one time on the average and stores in the division which included San Diego exceeded that corporate average. A 1975 study, partially funded by Southland, indicated that robberies at 7-Eleven stores could be reduced significantly (30 percent in the controlled case study) by following the sort of security procedures outlined by Mr. Perkins.

The California appeals court reviewed the trial court's finding, based on this evidence, that the defendants did not have a duty to protect Cohen from criminal activity on the 7-Eleven premises. The court first noted that under California law a business owner/operator has a duty to protect visitors from harm by third parties when he can reasonably anticipate and prevent such danger: "Foreseeability of the harm is of primary importance in establishing the element of duty."

Southland argued that Cohen's injuries were not foreseeable because there hadn't been any prior incidents of criminal <u>injury</u> at the Linda Vista 7-Eleven in the six years preceding; the court however noted that testimony by Cohen, Simmons, and Perkins conflicted as to the prior criminal history of the store. Further, the court ruled that even if there had been only one robbery at the Linda Vista 7-Eleven, the statistical evidence concerning 7-Eleven robberies raised a question of fact as to whether future armed robberies at the Linda Vista 7-Eleven were foreseeable. The court also noted that "common experience and reason support the conclusion more robberies occur at night .... A 24-hour convenience store makes an attractive target for night time criminal activity."

The court suggested that the community would benefit from the imposition of a duty on operators of all night convenience stores. The court ruled that the issue of foreseeability, the extent of the 7-Eleven store's duty to Cohen, and the adequacy of the Linda Vista store's security measures on May 14, 1978, were questions which should be resolved by a jury. It further ruled that if Cohen proves that his injury was foreseeable, then the defendants would indeed owe him a minimal duty of care.

Though the court reversed the summary judgments and remanded for trial Cohen's actions against the defendants Southland Corporation and James Simmons, it affirmed the judgment to defendant Wolfe. The appeals court ruled that Wolfe did not have a duty to rescue Cohen after the shooting, because Wolfe did not create the hazard leading to Cohen's injuries and there wasn't a relationship between Cohen and Wolfe which gave rise to a duty to rescue. The court said, "In light of Wolfe's patently reasonable fear for his own safety we decline to depart from the general no duty rule."

COMMENTARY:

In deciding the issue of foreseeability, the court made a most interesting observation, which should be of utmost interest (and concern) to those businesses which operate at late or all-night hours:

> *In the very operation of an all-night convenience store, defendants may be said to have created 'an especial temptation and opportunity for criminal misconduct,' thus increasing the foreseeability of injury resulting from third party misconduct in the early morning hours.*

In addition, Southland argued that the imposition of a duty of care on them by the courts would require them to hire full-time armed security guards to prevent night time crime on their premises, and that would cause "economic ruin" because of their "thin profit margins." The court rejected this argument, saying that a jury could determine (depending on the facts of the particular case) that the duty could be met merely by implementing Southland's own security measures at its stores (which had resulted in a 30 percent reduction in robberies where the measures had been employed).

PSCLR – NOV-DEC 85 (3)

Oil Company Is Not Liable To Customer For Criminal Assault By Unknown Assailant At Service Station Which It Did Not Not Own Or Control:

Stephen Fillare v. Union Oil Company of California, a foreign corporation, *Docket No. 79100, in the Court of Appeals of Michigan, decided June 4, 1985, reported at 372 N.W.2d 606.*

Stephen Fillare was attacked and robbed when he approached the cashier's booth at a Union 76 service station in Detroit, Michigan. He filed suit against Union Oil Company of California, alleging that it owned and controlled the station and that it had failed to protect him, an invitee. After Union countered that it did not own or control the station, Fillare joined the station's owner as a defendant. Although the trial court ruled that the defendant Union Oil could be held liable to business invitees at the station without proof that it controlled the station, the court concluded that it could not be held liable for a sudden and unforeseeable attack at the station by an unknown third party. The trial court found no material dispute on factual matters and granted Union Oil its motion for summary judgment on the law, and Fillare appealed.

Fillare argued that Union Oil was liable to him as an invitee of the station because the oil company allowed its trademarks to be displayed at the station and its products to be sold there. The affidavit of Union's area manager indicated that Union Oil did not own, nor did it lease the service station, nor did it participate in the employment decisions of the business. It had no contractual relationship with the station's owner; indeed its only link was through an independent jobber who supplied Union Oil products to the owner of the station. The appellate court reviewed the evidence and ruled that Union Oil did not control the service station's premises or operations. "We cannot conclude that simply allowing another entity to use a trademark and indirectly supplying products to a retailer makes the defendant a business invitor for the purposes of premises liability."

The extent of Union's control over the station was critical to determining the extent of its duty, if any, to the plaintiff Fillare. Since Union had no control, it likewise had no duty to protect Fillare from the sudden and unforeseeable attack by an unknown third party.

PSCLR – FEB 86 (5)

Absent Proof Of Prior Similar Criminal Incidents, All Night Convenience Store Owed No Special Duty of Care To Customer Assaulted By Youth Gang:

Arden A. Gregorian v. National Convenience Stores, Inc., *Civ. B011574, in the Court of Appeal of California, Second District, Division 2, decided November 22, 1985, reported at 220 Cal. Rptr. 302.*

Arden Gregorian sued National Convenience Stores to recover for personal injuries he suffered in an assault by a youth gang in one of its stores. The case was dismissed at trial, and Gregorian appealed. The issue is whether the defendant store owner owed a duty to protect Gregorian against the youth gang's unexpected criminal attack.

Gregorian first encountered his assailants as he was entering the defendant's Stop 'N' Go Market in Pomona, California; two young men blocked his path and made threatening remarks. When one of the boys attempted to hit him, Gregorian blocked the blow and struck a boy in the face at which point the boys fled. Gregorian proceeded inside the store and began shopping without mentioning the incident to any of the store's employees. As he was paying for his purchases, Gregorian saw the two boys returning with six to eight other boys. Fearing that they were after him, Gregorian unsuccessfully attempted to barricade the store entrance. The boys forced their way in, knocking Gregorian to the floor, and then brutally attacked and stabbed him. A store employee was also injured in the fracas.

There were no incidents of criminal activity at the store prior to the attack on Gregorian, according to the defendant's corporate risk manager. In fact, Gregorian, who had worked as a security guard for several months at the Stop 'N' Go with the responsibility to deter shoplifting and to "observe and report life or property endangering conditions," did not report any criminal incidents there during his employment. There was no security guard on duty the night Gregorian was assaulted.

After considering the evidence cited above, the trial court determined there to be no issue of fact because defendants owed no duty to protect plaintiff against the unforeseeable criminal attack.

On review, the appellate court found the issue raised by this case to be "familiar terrain over which we have traveled on numerous occasions," stating also that in general no liability is imposed on a landowner unless there is evidence of prior similar incidents. However, the appellate court noted that the California Supreme Court had recently (1985) held in *Isaacs v. Huntington Memorial Hospital* (reported in the May 1985 **PSCLR** at page three) that "the absence of past criminal attacks did not forestall the imposition of a duty of care in every case." But the appellate court here did not see that the *Isaacs* decision changed the "fundamental rule of possessor liability" in that the property owner is not the insurer of the safety of people on his property. The property owner's duty to protect people on his property from third party crime arises only if he has reasonable cause to anticipate the risk and the probability of harm, and then fails to take action to prevent the harm.

In this case there was no evidence either of prior incidents of crime at the market nor evidence that the store was located in a high crime area. The sudden, unexpected, and vicious attack on Gregorian is not analogous to the armed robbery in an all night convenience store as occurred in a case cited by Gregorian, *Cohen v. Southland Corp.* (reported in the September 1984 **PSCLR** at page three). In *Cohen* the California Court of Appeal stated,

> In the very operation of an all night convenience store, defendants may be said to have created 'an especial temptation and opportunity for criminal misconduct,' thus increasing the foreseeability of injury resulting from third party misconduct,

In the present case the California Court of Appeal reiterated that point in summarizing its holding:

> While the proprietor of an all night; convenience store may, in this day and age, reasonably anticipate that his place of business will be the target of armed robbers, the same cannot be said for a crime resulting from gang violence. This is especially true where ... the premises has never been the focal point of gang activity.

The appellate court therefore affirmed the judgment granted to the defendant.

PSCLR – FEB 86 (9)

Convenience Store Owner Is Held Not Liable To Customer Who Was Sexually Assaulted During Robbery:

Renee Elizabeth Ortell v. Spencer Companies, Inc., *84-51, in the Supreme Court of Alabama, decided September 13, 1985, rehearing denied September 27, 1985, reported at 477 So. 2d 299.*

When Renee Ortell entered the Happy Times Convenience Store in Birmingham, Alabama, early one morning, she interrupted a robbery in progress. The young male robber was armed with a gun and took Ortell's wallet before forcing her into the restroom where he had earlier taken the female store manager. The robber returned to the restroom and sexually assaulted Ortell. She sued the owner of the store, Spencer Companies, Inc., alleging that it negligently failed to protect its customers from foreseeable criminal activity. The trial court granted the defendant's motion for summary judgment and Ortell appealed.

The Supreme Court of Alabama reviewed the duty of a store owner to protect its customers from crime and stated:

> A duty may be imposed on a storeowner to take reasonable precautions to protect invitees from criminal attack in the exceptional case where the storeowner possessed actual or constructive knowledge that criminal activity which could endanger an invitee was a probability.

However, the court also stated that "it is difficult to impose liability on one person for an intentional criminal act committed by a third person," citing previous Alabama Supreme Court decisions in *Henley v. Pizitz Realty Co.* reported in the January 1985 **PSCLR** at page nine, *Latham v. Aronov Realty Co.* reported in the August 1984 issue at page seven, and *Berdeaux v. City National Bank of Birmingham* reported in the May 1983 issue at page three.

The plaintiff attempted to establish that Spencer Companies should have foreseen and therefore should have taken steps to prevent her attack, by introducing police records of criminal incidents which were reported in the store and surrounding area for three years before Ortell's assault. Although the crimes reported in the store itself included one assault with a weapon, one theft, five robberies of the store, and one burglary, the court ruled that this alone was insufficient to impose a duty on the store owner to protect the plaintiff from the assault by the robber. Since there was no other evidence to show that Spencer Companies had actual or constructive knowledge that Ortell's assault was a probability, the appellate court affirmed the summary judgment entered on behalf of the defendants.

COMMENTARY:

The surprising thing about this case is that both the trial court and the appellate court dismissed the action on summary judgment without allowing the plaintiff to present her evidence to a jury. There was not a single dissenting opinion. In view of the decisions reached by the courts of many other states, the kindest conclusion which can be drawn with reference to the above case is that the Alabama courts are taking a very conservative approach to this developing area of premises liability law whereby landowners are increasingly being held liable for the criminal acts of third parties committed on their premises.

Certainly, in view of the evidence presented as to prior criminal activity on the premises, many courts in other states would have permitted the plaintiff to present the evidence to a jury and allowed the jury to determine whether the prior incidents should have caused the landowner to foresee the incident in question and employ reasonable measures to protect its customers from the danger.

As the Supreme Court of Alabama stated in the above case,

> *This court has not yet decided a case whose facts command the imposition of such a duty [on a storeowner].*

PSCLR – MAR 86 (9)

Convenience Store Owner Is Not Liable In Punitive Damages For Murder Of Customer Who Was Shot By Armed Robber As She Entered Store's Parking Lot Absent Proof Of Willful and Wanton Misconduct By Store Employee:

Lil' Champ Food Stores, Inc. v. Patricia Holton, as personal representative of the estate of Ruby Elizabeth Holton, *No. BA-411, in the District Court of Appeal of Florida, First District, decided September 11, 1985, rehearing denied October 9, 1985, reported at 475 So. 2d 726.*

Ruby Holton was shot and killed by a man who had robbed the Lil' Champ Food Store in Jacksonville, Florida, as she was entering the parking lot and he was fleeing the store. Her estate brought a wrongful death suit against the owners of the store, alleging that the defendants' negligence led to Holton's death. The trial ended with a judgment of punitive and compensatory damages to the plaintiff; the defendants appealed the punitive damages award and the Florida appellate court reversed.

Holton apparently entered the parking lot of the Lil' Champ just after an armed robbery had taken place. The robber had threatened the store clerk and a customer in the store with a sawed-off shotgun, but he did not shoot or injure anyone in the store. As he was leaving, however, he saw Holton and shot and killed her. The evidence was unclear as to how and why he shot Holton, though the plaintiff alleges that the robber saw the store clerk picking up a telephone inside the store and that he turned to reenter the store when he saw Holton. The plaintiff argued that if the store clerk had not acted negligently in attempting to call the police before the robber had made his getaway, then the robber would have immediately left without seeing or killing Holton.

The plaintiff also argued that Lil' Champ's security program was "grossly" inadequate and led to the shooting death of Holton. A security expert called to testify by the plaintiff stated that Lil' Champ was " 'grossly negligent' in failing to institute a formal security 'program' directed toward prevention of armed robberies at their various stores and that Lil' Champ had fallen 'below minimum dirt-floor standards' in failing to establish a comprehensive robbery prevention program." However, the expert could not make specific suggestions which would have prevented the robbery and Holton's murder. The expert admitted, and the appellate court concluded, that "no security program, no matter how comprehensive, could be 100 percent effective in preventing robberies or attempted robberies."

There was evidence that Lil' Champ's corporate officers knew that armed robberies had occurred at their stores in the past and were likely to occur again, but they had not hired a security consulting service nor had they implemented an extensive security program which included additional lighting, video monitors, silent alarm, or armed guards. However, the Lil' Champ stores were deliberately located away from high crime areas, brightly lit at all times, and were designed with large expanses of glass windows across the front to allow high visibility inside and out in order to discourage crime within the store.

Furthermore, Lil' Champ corporate officers had attended crime prevention seminars sponsored by the local police, had implemented the anti-robbery techniques from the seminars, and had distributed a memorandum to instruct employees in ways to minimize the possibility of an armed robbery or harm to customers and employees in the event of an armed robbery. Among other things, employees were instructed to cooperate fully with the robber in order that the incident and the risk would end more quickly in order to prevent harm. This instruction was proffered by the plaintiff as an example of Lil' Champ's "willful and wanton disregard for the safety of its employees and customers," claiming that it encouraged robbers to victimize Lil' Champ stores.

The appellate court however, held that the plaintiff's case against Lil' Champ was just barely sufficient to prove simple negligence and that there was clearly no proof of "willful and wanton misconduct" to support the punitive damages awarded against the defendant. The appellate court therefore reversed and set aside the award of punitive damages to the plaintiff.

COMMENTARY:

At seminars **PSCLR** *has conducted, a concern which is often expressed by participants around the country is that they or their employer may be held liable for injuries to invitees on their property no matter what security measures may have been implemented to protect the invitees. The above case is another example of the fact that no business owner/operator is required to become an insurer of the safety of invitees on its property, but have only the duty to take reasonable steps to warn and to protect invitees from reasonably anticipated or known dangers on the premises.*

In some situations the surrounding facts and circumstances would dictate the hiring of a security consulting service, additional lighting, video monitors, silent alarms, or armed guards on the premises; however, as we see from the above case, unless the particular circumstances do dictate that such measures be implemented, the business owner/operator will not be held liable when an injury to a customer takes place so long as the owner/operator has in fact implemented protective measures which are in fact reasonable given the particular circumstances of the situation.

PSCLR—MAY 86 (3)

Convenience Store Clerk Has Duty To Exercise Reasonable Care To Protect Customers From Unreasonable Risk Of Assault By Calling Police If Risk Is Apparent And Injury Is Preventable:

Ramonia Leigh Ballew v. The Southland Corporation d/b/a 7-Eleven Food Stores and Kemper Insurance Company, In Solido, *No. 17468, in the Court of Appeal of Louisiana, Second Circuit, decided January 22, 1986, reported at 462 So. 2d 890.*

Ramonia Ballew sued the Southland Corporation, after she was forcibly dragged from one of its 7-Eleven convenience stores and raped. The jury at trial held in Ballew's favor and awarded $40,000 in damages. The Southland Corporation appealed the verdict, and Ballew challenged the amount of damages awarded. The Court of Appeals of Louisiana upheld both the verdict and the damages award.

Ballew had stopped at a 7-Eleven store in Shreveport, Louisiana, at about 10 PM one evening. As she was about to get out of her car, she noticed a large man standing in front of the store. Despite the chilly February temperature, the man was only wearing a pair of yellow gym shorts. She stayed in her car until the man left but as she entered the store, the man reappeared and followed her inside. The man, later identified as Aubrey Mitchell, asked her to buy him a soft drink; she refused, but he obtained a drink and told the clerk that Ballew would pay for it. Mitchell then went outside and drank the beverage while propping one foot on Ballew's car. When Ballew took her purchases to the counter, she told the clerk that she was afraid of the man. The clerk, Judith Longoria, had seen the man come in and out of the store for 15 to 30 minutes before Ballew arrived. He had asked other customers for money and had asked to use the store's telephone. Longoria rang up Ballew's purchases and Mitchell entered the store again. Longoria told him to leave the store; he grabbed Ballew and dragged her out of the store. Longoria immediately called the police; when the police arrived minutes later, they caught Mitchell in the act of raping Ballew.

Ballew's suit against the Southland Corporation included allegations that the defendant and its employee had (a) failed to provide adequate security to protect its customers, (b) violated safety practices designed to protect customers in convenience stores, and (c) violated Shreveport city ordinances and Louisiana statutes. Southland argued that its security and safety precautions were adequate to protect customers under reasonable circumstances, and that Longoria was not negligent in failing to call the police before Ballew was dragged from the store because the attack was not foreseeable.

In order to prove the store owner liable for her injuries, Ballew had to show that Southland owed her a duty of protection, that it violated that duty, and that her injuries resulted from the breach of that duty. In addition, she had to prove that the injuries she suffered were a foreseeable result of the breach of duty, and that the type of harm she suffered was the type which the duty was designed to prevent. (The court referred to *Banks v. Hyatt Corporation*, reported in the April 1984 issue of **PSCLR**, case one page one; *Davenport v. Nixon*, reported in the November-December 1983 issue, case one, page one; and *Harris v. Pizza Hut of Louisiana*, reported in the January 1985 issue, case one, page one.)

The Louisiana appellate court therefore examined the extent of the convenience store owner's duty to protect customers from criminal assault, and whether the employee Longoria had violated that duty, and whether Ballew's assault was within the scope of Southland's duty of protection. The appellate court ruled that Southland did not have a duty to protect its customers against unforeseeable and unanticipated criminal acts. If, however, the risk of harm is known or should be known, and prevention of the harm was reasonably within their power, then such a duty would exist.

In this case, Longoria had seen Mitchell loitering around the store, wearing inappropriate clothes, and annoying the customers. Longoria was aware that the store had been the site of a theft and two armed robberies (an employee was killed in one of these incidents) during the previous year. There was also evidence that Southland instituted a security awareness program for its employees: new employees were shown a series of three films, entitled "Operation Alert," during their initial training. The films dealt with armed robberies, rape awareness, and violence avoidance; there were follow-up refresher seminars on security for all employees. Southland also had a policy that employees should call the police immediately if a suspicious person came around a store. Therefore, though Southland and Longoria could not have known of the exact nature of the risk to Ballew which was proved by Mitchell's presence, the defendants were aware that there was a risk of violent crime occurring on the store premises.

A security expert, giving evidence at the trial on behalf of the plaintiff, testified that 9 p.m. to 11 p.m. are peak hours for crimes to be committed in convenience stores, and that most convenience store crimes are committed by people who are loitering around the store without a valid reason for being there. It was his opinion that a clerk should always speak to someone loitering around the store; if the person remained around the store without good reasons for doing so, the clerk should lock up the store and call the police.

The jury here had found that the evidence showed that Longoria and Southland had a duty to protect Ballew. The jury had further determined that Longoria breached that duty by failing to identify Mitchell as a risk to customers and in failing to take the precaution of calling the police before Ballew was assaulted, particularly after Mitchell took a soft drink without paying for it.

The appellate court affirmed the jury's findings, ruling that Longoria should have recognized that Mitchell's bizarre outfit and suspicious behavior indicated he presented a potential danger to Ballew and other customers. The appellate court also held that Longoria's failure "to act by timely notifying the police and locking down the premises was clearly a 'substantial factor' in bringing about the plaintiff's harm." The court also affirmed the $40,000 damages awarded to Ballew as reasonable under the circumstances.

PSCLR – SEP 86 (11)

Retail Gun Store Owner Must Face Wrongful Death Suit For Alleged Negligent Display Of Guns And Inadequate Measures To Secure Guns Where Victim Was Killed By Stolen Gun:

Delorous Y. Kimbler v. John Vernon Stillwell, aka John Vernon Bray; G.I. Joe's, an Oregon corporation; *in the Court of Appeals of Oregon, decided April 16, 1986, reconsideration denied June 6, 1986, review by Supreme Court of Oregon allowed, June 24, 1986, reported at 717 P.2d 1223.*

John Stillwell shot and killed Michael Kimbler with a shotgun Stillwell had stolen from G.I. Joe's, a retail store offering guns and ammunition for sale to the general public in Portland, Oregon. Kimbler's mother brought a wrongful death action against G.I. Joe's, which was dismissed at the trial level. On appeal, the Oregon Court of Appeals reversed the dismissal and remanded the suit for trial.

Kimbler contended that G.I. Joe's negligence caused her son's death. Among her specific allegations, Kimbler's suit stated:

> The direct and proximate cause of said incident and the resulting death was the negligence of defendant G.I. Joe's in one or more of the following particulars:

> 1. In failing to properly guard and patrol said retail outlet.
>
> 2. In failing to outfit the retail store with security doors providing protection commensurate with the danger involved in displaying the shotguns and ammunition.
>
> 3. In failing to take reasonable security measures commensurate with the danger involved in displaying shotguns and ammunition in securing said retail outlet's external doors with proper locks and alarms.
>
> 4. In failing to secure their display of shotguns and ammunition in an unbreakable glass case or other protective barrier commensurate with the danger involved in displaying shotguns and ammunition.
>
> 5. In failing to secure their shotguns with chains, trigger guards or other protective services commensurate with the danger involved in displaying shotguns and ammunition.
>
> 6. In failing to keep shells and ammunition fitting the shotguns locked up in a separate location from the shotguns.

The defendants argued that the plaintiff seeks to impose a duty on them that Oregon law does not recognize: a duty of strict liability for firearms vendors to prevent criminal acts by third parties. Moreover, the defendants contended that they should not be held liable for Stillwell's unforeseeable criminal conduct.

The defendant's duty was a merchant's duty to take reasonable measures to safely display merchandise. The appellate court concluded that the jury should decide whether the defendant could have or should have foreseen Stillwell's criminal act, and not the court in a summary judgment.

> We cannot say, as a matter of law, that the theft of guns and the resulting injury were so highly unusual that the harm could not be foreseen.*** A gun is a dangerous instrumentality. Presumably, because it i, it calls for a greater degree of care in its display than the display of clothing or boxes of cereal; but the existence of the gun and defendant's relation to it do <u>not</u> determine the liability of defendant. Liability would arise, not from the fact that guns are displayed, but rather from proof that it was unreasonable for defendant to display guns and ammunition in the manner in which it did.

The appellate court also noted that an intervening third party criminal act is not an absolute defense to negligence claims (citing *Ollison v. Weinberg Racing Association*, reported in the February 1985 **PSCLR**, case 11, page 10).

The appellate court therefore ruled that Delorous Kimbler's suit alleged sufficient facts to sustain a wrongful death action against G.I. Joe's. The jury will have to determine whether the evidence supports the allegations, the extent of G.I. Joe's duty, whether it was negligent and violated its duty, and whether its negligence caused Kimbler's death.

PSCLR – OCT 86 (3)

Retailer Owes No Liability For Injuries Where Plaintiff Fails To Prove Similar Robberies Occurred Previously:

Loretta Fowler v. Robertson, *No. 72021, in the Court of Appeals of Georgia, decided April 8, 1985, reported at 344 S.E.2d 425.*

Eddie Fowler owned a package liquor store in Meriwether County, Georgia. Retiring after 20 years service with the local police force, Fowler's father, C.E. Fowler, spent a good deal of time helping out at his son's package store. He was not paid wages, however, and did not consider himself an employee of the store.

Eddie and C.E. Fowler had considered the possibility of a robbery attempt at the store, and kept a gun on the premises for protection. However, they had also agreed that the money would be surrendered without any resistance if and when a robbery attempt took place. The robbery which was the subject of this suit was the first to take place since the younger Fowler had begun operating a package store some six years before (for the first two years at another location).

On Christmas Eve in 1981, Mrs. Fowler brought supper to her husband and son who were working at the package store. After she had been there approximately one-half hour and was behind the counter visiting with her son and husband, two men entered the store with guns drawn and began firing without warning. The younger Fowler pushed his mother beneath the counter; he was shot in the neck and fell to the floor beside her. The elder Fowler was wounded but managed to reach his gun underneath the counter and he returned fire, killing one of the robbers. The second robber came behind the counter and continued to fire at all three Fowlers, ultimately killing the father and son, and severely wounding Mrs. Fowler.

After Mrs. Fowler recovered from her injuries, she brought suit against the administrator of her son's estate, claiming that her son had exhibited negligence in the maintenance of his business by deciding to keep a gun in the store, by inviting her to come behind the counter of the store, and by aggressivly resisting the robbers. She further contended that his negligence endangered her as an invited guest on the business premises.

There was an insurance policy protecting the premises and its invitees from the package store owner's (the younger Fowler) negligence with Continental Casualty Company. Continental intervened as a party in the lawsuit in order to assert its claims and to determine its potential liability under the insurance policy. The trial court concluded that because the robbery of the package store was unexpected and unforeseen, it could not be the basis of a negligence action against the policyholder so as to extend coverage to Mrs. Fowler. The trial court therefore granted a judgment in favor of the insurance carrier, and Mrs. Fowler filed her appeal from that ruling.

Unforeseen and unexpected acts perpetrated by third persons on the retailer's premises do not activate any liability on the part of the retailer. The only exception is if the retailer had reasonable cause to believe that the type of criminal act which resulted in injury to his guests was reasonably likely to occur. The son's estate, therefore, cannot be held liable for any injuries to his mother as a guest on the property unless there is some proof to show that he reasonably foresaw, or could have foreseen, that an armed robbery would occur and that the robbers would open gunfire on the store's occupants.

Mrs. Fowler argued that liquor store robberies are "so commonplace" that her son must have been aware of the possibility and the likelihood of an armed robbery occurring at his liquor store. She further contended that since her husband and son had discussed what to do in the event of a robbery that they must therefore have anticipated a robbery. She also argued that keeping a gun in the store proved prior knowledge that an armed robbery was likely to occur..

The appellate court, however, disagreed:

> Though these arguments are well presented and have much facial merit, we are not thereby persuaded that the exception to the rule of liability should apply rather than the general rule excepting liability for the acts of third parties.

The court noted that the facts of the situation showed that no robbery had occurred in either package store owned by Fowler in the six years he had been in such operations. He also noted that while the son had talked with his father about the possibility of a robbery, this discussion could have no greater probative value than a discussion concerning a fire, burglary, or any other catastrophic event. Also, keeping a gun in the store "is no different than having armed guards in a bank."

The court concluded that there was no evidence that the younger Fowler, as owner of the liquor store, had any reason to be aware of the probability of any armed robbery occurring in his store. Nor was there any evidence that he was aware that visiting with his mother and father in the store, either behind the counter or at any other place in the store, was in any way a disregard for safety rules. Therefore, in the absence of proof by Mrs. Fowler to show that similar robberies had occurred in the store or vicinity in the recent past, she failed to raise any issue of fact sufficient to raise a question of liability for an unexpected and unforeseen armed robbery. Accordingly, the appellate court affirmed the trial court's decision in favor of the son.

PSCLR – JUN-JUL 87 (3)

Retailer Who Displays Guns And Ammunition, Without Adequate Security, Should Foresee Theft And Subsequent Use Of Same To Assault Others, And May Be Liable For Resulting Death:

Delorous Y. Kimbler v. John Vernon Stillwell, aka John Vernon Bray and G.I. Joe's, *an Oregon Corporation, TC A8208-05113; CA A 34559, SC 232882, in the Supreme Court of Oregon, En Banc, decided March 17, 1987, reported at 734 P.2d 1344.*

Delorous Kimbler sued John Stillwell, who had shot and killed her son, Michael Kimbler. She also sued G.I. Joe's, a retail establishment from which Stillwell had stolen the firearm which he used to kill Michael Kimbler. She alleged in her action against the retail merchant that G.I. Joe's had "made it too easy" for Stillwell to steal the shotgun and ammunition from the store.

The trial judge dismissed Ms. Kimbler's suit against G.I. Joe's, and on her appeal, the Oregon Court of Appeals reversed the trial judge and ordered the case returned to the lower court for trial. (See the September 1986 **PSCLR**, case 11, page 8, for a full report of that decision). The retailer then filed this appeal to the Oregon Supreme Court.

The facts of this situation can be summarized as follows: the defendant G.I. Joe's is a corporation conducting a retail business in Portland, Oregon, where it sold shotguns and ammunition to the general public. The weapons and their corresponding ammunition were on display in close proximity to the general public. They were not behind glass, nor were they locked in cases, nor were the weapons secured by chains, trigger guards, or any other security devices, and this was readily apparent. The store itself was secured at night by locked glass doors, but without other security devices which are normally in place when firearms are displayed. Such being the case, the Oregon Supreme Court concluded that the proprietors

> knew or should have known that the lack of security measures made the shotguns and ammunition easy targets for theft, that stolen firearms often are used to commit crimes against third parties, and therefore that this manner of displaying its shotguns and ammunitions created an unreasonably high risk to third persons.

Ms. Kimbler alleged that one John Vernon Stillwell had seen the display of shotguns and ammunition at G.I. Joe's, he had entered the store, he had stolen a shotgun and ammunition, and he had used the stolen gun and ammunition to shoot and kill Michael Kimbler. She also alleged that G.I. Joe's was negligent in that it failed to guard or patrol the store, to install secured doors, locks, and alarms, to use chains, trigger guards, or other devices to secure the guns and ammunition display behind some barrier, and to store the ammunition in a different place from the shotguns.

G.I. Joe's motioned to dismiss Kimbler's case, based on the assertion that Stillwell's act was a superseding cause of Michael Kimbler's death, that the death could not have been reasonably foreseen by the retailer, and that the retailer owed no duty to Michael Kimbler to protect him from the intentional misconduct of Stillwell.

The fact that an injury or death is caused by an intentional, including criminal, act by a third party does not preclude the imposition of liability if the third party's act was a foreseeable risk which was facilitated by the defendant's negligence. And while we may under normal circumstances reasonably assume that no one will violate the law, we also recognize that such a proposition is unfortunately not the rule for all times, places, and situations. Accordingly, the plaintiff should have an opportunity to prove what actual circumstances were or should have been contemplated by the defendant given the time, place, and situation at issue. The Oregon Supreme Court therefore affirmed the Court of Appeals that the trial judge committed error when he dismissed the complaint by Ms. Kimbler against the retailer.

PSCLR – SEP 87 (6)

Retailer Responsible For Maintaining Safe Entrance To Premises For Invitees, And Is Liable For Negligence In Failing To Maintain Sidewalk Adjacent To Premises:

Ruth Vogelhut v. Waldbaum's Supermarket, *in the Supreme Court of New York, Appellate Division, Second Department, decided February 2, 1987, reported at 511 N.Y.S.2d 647.*

The plaintiff Ruth Vogelhut filed suit against Waldbaum's Supermarket alleging that she had sustained injuries as a result of the retailers negligence in failing to properly maintain the sidewalk immediately outside its place of business. She claimed that her left foot was caught in a crack in the sidewalk just outside the exit door to the store and, when she attempted to regain her footing, her other foot then slipped on some ice and she incurred a fracture to her right ankle. An eyewitness to the incident corroborated her testimony as to the location of her fall, but a police officer testified on behalf of the retailer and said that the plaintiff had not fallen on the sidewalk immediately adjacent to the retailer's store, but had rather fallen in the public roadway.

The case was tried before a judge sitting without a jury, and, after weighing the evidence and the credibility of the witnesses, he ruled in favor of Vogelhut and awarded her a verdict and judgment in the amount of $60,000, reduced to $51,000 (due to a finding of the plaintiff's 15 percent culpable conduct).

On appeal by the retailer, the appellate court found that the plaintiff had successfully established that the incident occurred on the sidewalk immediately adjacent to the defendant's place of business and under the defendant retailer's exclusive control. However, the appellate court concluded that the amount of damages awarded the plaintiff by the trial judge was excessive since the plaintiff had failed to prove that the injury she sustained was in any way permanent, and the appellate court therefore granted a new trial on the issue of damages only, conditioned upon a reduction in the amount of the plaintiff's damages to $30,000, to be further reduced by 15 percent for her comparative negligence.

COMMENTARY:

And so we see that the private security manager is responsible not only for protecting persons and property within his care from loss or injury due to fraud, embezzlement, burglary, criminal assaults, parking lot thievery, inadequate locks, negligent hiring, training, and supervision, etc., etc., etc., but also for cracks in the sidewalk adjacent to the premises! The truth is that while many corporate security directors are concentrating on preventing loss due to computer fraud and other "complex" liability theories, the basics are still in effect and still must be tended to. Such risks as cracks in the sidewalk, allowing ice to accumulate on entrances and exits where pedestrians are known to travel, allowing slippery substances to remain on floors, allowing potentially flammable rubbish to accumulate are all within the scope and purview of the knowledgeable and attentive security manager.

While it is of course necessary to remain current and up to date concerning the more complex theories of liability, it is also necessary to include in training programs, the necessity for careful observance of the elementary responsibilities to the public, and to the clients whose property and personnel are to be protected.

PSCLR – MAY 88 (7)

Even A <u>Foreseeable</u> Attack, Which <u>Could</u> Have Been Prevented By Business Owner's Reasonable Care, Will Not Warrant Recovery By Injured Customer Without Proof Of Causation:

Laura Reagin v. Charles Lawrence Terry and Troy Virgil Terry, both individually and doing business as Mt. Hope Shell Service Station; Mt. Hope Shell Service Station, a partnership; Shell Oil Company; Quality Oil Company of Statesville, Inc.; and Quality Oil Company, defendants; *Civ No. 3-84-470-G, in the United States District Court for the Middle District of North Carolina, Greensboro Division, decided September 4, 1986, reported at 675 F. Supp. 297.*

Laura Reagin stopped at the Mt. Hope Shell Service Station late one night to buy coffee. After she entered the store, she was accosted by a gun-wielding ex-convict, who was holding a 55-year-old station attendant hostage at gun point. The assailant had previously taken the attendant's pistol after "getting the drop on him from behind" with a sawed off shotgun.

The assailant forced the plaintiff out of the station at gun point to a darkened area underneath a light pole, which light did not work, where he attacked and injured her.

The plaintiff sued the station operators and Shell, among others, for negligent maintenance, supervision, and safeguarding of the gas station. The jury found that the assault was foreseeable, that the station operators had breached the duty owed the plaintiff to provide adequate security at the station, and that the breach of duty caused the plaintiff injuries. The jury further found that the station operators were agents of Shell Oil Company, and awarded the plaintiff $450,000.

The defendants moved for a judgment notwithstanding the verdict or a new trial. Shell also moved for indemnity from Mt. Hope and Quality Oil Company, their jobber. The trial court granted the motions for judgment notwithstanding the verdict, denied the motions for new trial, and ruled against Shell's motion for indemnity.

First, the court considered the demand for judgment notwithstanding the verdict. The court noted that the plaintiff was required to show that there was

> evidence before the jury ... sufficient to permit a finding that the criminal attack upon the plaintiff was not only foreseeable, thus imposing a duty of reasonable care upon the station operators, but also that the operators breached their duty of reasonable care, and that the plaintiff's injuries were causally related to their breach of duty.

First the court attempted to determine whether, based upon past experience, the store operators should have foreseen any violent acts of an assailant upon their customers in general, or upon the plaintiff in particular.

The court noted that the station stayed open 24 hours a day on a fairly empty stretch of Interstate 85 near Burlington, N.C. Another gas station and a restaurant were nearby. Both closed earlier than the Shell station. Besides those two buildings, only a single residence, woods, and fields surrounded the station.

There was evidence that the station had been robbed twice before, once with injuries to an attendant. Someone had stolen some tires from the station eight years before the attack in this case occurred. Minor break-ins had occurred at the other buildings around the station. The station attendant carried a gun and called the sheriff's office periodically to report suspicious goings on in the area.

The court ruled that these facts barely established "that it was reasonably foreseeable that criminal acts would occur on the premises which could involve a risk to any invitees present."

The court next turned to the question of whether the assailant's acts could have been prevented by the exercise of ordinary care, "considering the evidence of conditions in the surrounding area, the general character of the neighborhood, the location of the premises, the character of the business enterprise, and other criminal activity occurring on the premises."

The court noted that the sole night attendant was a 55-year-old male victim of a mild stroke who carried a pistol. He had the use of a direct line to the sheriff's department. One of the floodlights surrounding the station was out. Some displays obscured vision into and out of the store.

The court found from the record of the trial that the evidence that (1) there was a single night attendant, (2) that the light on the pole did not work, and (3) the fact that one's vision into and out of the store was blocked were all that the plaintiff presented to show a lack of care on the part of the station operators.

The court noted that the plaintiff did not notice anything strange while walking to the bathroom of the store after entering the store and the court concluded from this that the obscured field of vision did not worsen her situation. The court also held that the lack of a light on the pole did not constitute a breach of duty, since the gunman led the victim from the station at gun point and could have taken her to the woods or the fields to assault her, instead of taking her to the area near the inoperable light.

The court finally noted that the plaintiff produced no evidence that other stations customarily employed more than one attendant, or that they supplemented their single night attendant with cameras or alarms. The court thus found that the plaintiffs had failed to prove any breach of duty on the part of the station operators.

Next, the court found that the defendant's conduct was not the proximate cause, or the cause in fact of the plaintiff's injuries. The court stated,

> The evidence is uncontradicted that the assailant was an experienced criminal, just a few days out of prison, who determined while riding along the interstate with friends to 'rob something' prior to observing or knowing anything at all about the Mt. Hope Shell Service Station. Again, to determine the assailant would have been deterred or thwarted by two attendants or by additional lighting, is to speculate about mere possibilities, rather than to find reasonable probability.

The court next turned to the question of whether Shell's acts rendered the station operators agents of Shell by creating an appearance that the station operators had the authority to act as Shell's agents.

The court observed that the Shell "scallop" sign and the entire Shell advertising campaign created the appearance that Shell controlled Shell stations, including the one in question. The court also found that "the maintenance of safe premises, including restroom facilities [was] within the scope of the [station operator's] apparent authority."

The court thus denied Shell's claim that no agency relationship should be found between the station operators and Shell Oil, and that a part of that agency included the duty of the station operator, and thus the oil company, to maintain safe restrooms on their premises.

Accordingly, the verdict of the jury in favor of the plaintiff was set aside and the defendant's motion for judgment notwithstanding the verdict was granted.

PSCLR–FEB 89 (8)

Store Owner May Be Liable For Security Guard's Failure To Use Ordinary Care To Avoid Colliding With And Injuring A Customer While In Pursuit Of Shoplifter:

Pearl Brown v. Jewel Companies, Inc. and Osco Drug, Inc., *No. 4-88-0226, in the Appellate Court of Illinois, Fourth District, decided October 6, 1988, rehearing denied November 28, 1988, reported at 530 N.E.2d 57.*

Pearl Brown sued Jewel Companies, Inc., and Osco Drug, Inc., operators of a combined grocery and drug store in Decatur, Illinois, for injuries sustained during the chase of a suspected shoplifter by store security guards. Brown was in the process of leaving the store when the suspected shoplifter, in flight from the security guards, knocked her through a door and onto a sidewalk, whereupon a security guard in pursuit kicked or stepped on Brown's foot, injuring her further. The guard's effort to seize the shoplifter was unsuccessful.

Brown's complaint alleged that the defendants employed security guards to protect against shoplifting and that the guards "customarily pursued and apprehended suspected shoplifters on the store premises," some of whom had fled or attempted to flee through the exit of the store in the past. The complaint also alleged that the defendants knew that their self-service method of operation made shoplifting more likely and that they should have anticipated shoplifters would be pursued by the guards and might injure customers in their course of flight.

The complaint further alleged that the defendants were negligent and violated their duty under the Illinois Premises Liability Act in that the defendants: (1) failed to restrain the suspect; (2) failed to block the suspect's path to the exit; (3) continued to pursue the suspect when he knew or should have known that that would endanger customers; (4) failed to watch out for customers while pursuing the suspect; (5) failed to avoid plaintiff's foot; (6) attempted to apprehend the suspect at a place where her flight would jeopardize other persons on the store premises; and (7) failed to properly train its security guard in the restraint and apprehension of suspected shoplifters.

The court held that the security guard owed a duty to Brown to use care not to injure her while chasing the shoplifter. Thus, those allegations related to the guard's failure to keep a proper lookout while pursuing the suspect and his failure to avoid Brown's foot stated claims upon which Brown's case could proceed to trial.

However, the court held that the other claims in the complaint should be dismissed because the defendants owed no duty to Brown on those grounds. The court stated that placing a duty to make sure that suspected shoplifters are not able to run from security guards would place an enormous burden upon retailers, requiring a large force of guards and the use of substantial (and probably excessive) force to restrain suspected offenders of relatively minor crimes. The court also stated that the likelihood of very serious injury, resulting from a fleeing shoplifter running into a customer, is not very great.

Next, the court analyzed whether a security guard owes a duty to customers not to chase fleeing suspected shoplifters because of the danger that the shoplifter will run into a customer. The court noted that decisions from other states are not uniform on this issue (see the Commentary for this case).

The court held that, in Illinois, security guards have no duty to refrain from pursuing suspected shoplifters. Again, the likelihood of serious injury was deemed by the court not to be great and not to be significantly increased because of the guard's pursuit, because most shoplifters would flee when detected anyway. Further, a rule against pursuit would be a substantial encouragement to shoplifting and would place an unreasonable burden upon the retailer.

In sum, the court held that the duty of the guard is to use ordinary care not to collide with or injure a customer while in pursuit of a suspected shoplifter. Therefore, the court allowed the case to proceed to trial only on the allegations of negligence related to the duty to avoid colliding with or otherwise injuring the customer.

COMMENTARY:

As referred to above, the courts in various jurisdictions are not in agreement as to the duty a security guard owes customers not to chase shoplifters due to the danger that the fleeing shoplifter may run into and injure the customer. In Louisiana, it was held that such a duty does exist. In Florida, Minnesota, and Wisconsin, it has been held that such a duty does not exist.

PSCLR – MAY 89 (6)

Cash-Only Department Store Is Liable For Failure To Warn Customers Of Likelihood of Theft At Checkout Counter:

Michelle Lacelle v. Hills Department Store, *in the City Court of City of Watertown, Jefferson County, New York, decided December 13, 1988, reported at 535 N.Y.S. 1014.*

Michelle Lacelle went to Hills Department Store, a large retail establishment which, unlike other discount stores, does not accept credit cards. She had cash in her purse which was zipped and closed. While she shopped around the store with her child in her shopping cart, she kept the purse on her person.

When Lacelle went to the checkout lane, she put the purse on the cart to remove her child from the cart. While she turned her back on the purse for a few moments, her purse was taken. The purse was located, without the cash, a few minutes later.

Lacelle sued Hills Department Store to recover the amount of money lost when her purse was stolen.

To recover for the acts of criminals on a proprietor's premises, a plaintiff must show that the proprietor knew or should have known of the probability of criminal conduct likely to pose a risk of harm to third persons lawfully on the premises, and that the proprietor failed to take reasonable precautions to protect the public.

The court noted that the store advertises how a "cash only" policy keeps their prices lower for the customer. This same advertising alerts pickpockets and thieves that Hills' customers come with cash and that a purse or wallet taken from a Hills' customer will contain more than the usual amount of cash.

The Hills Department Store manager testified that the store was aware of theft problems at its checkout area. Hills had some security, but no security camera, no special security to protect its customers at the checkout area, and no warning signs concerning theft. The store had taken precautions to prevent past problems with money being taken from the cash drawers when cashiers were distracted. The store had also had a couple of purse snatchings in the year prior to Lacelle's loss.

The court framed the issue as whether a store that encourages cash-only customers has a foreseeable duty to protect customers from theft when it occurs at the one location—the cashier's desk—where the store has the greatest concern for its own cash being taken.

The court noted that because the store provides child seats in the carts to accommodate the customers' children, it was reasonable to assume that, when a shopper reaches the checkout lane, the child would be removed to empty the cart. At that moment, the Hills customer could reasonably be expected to put a cash-laden purse on the cart while attending to the child, thus becoming extremely vulnerable to having the purse snatched.

The court found that Hills should have foreseen a risk to its customers at the checkout area of the same type of theft from which it protected its cash drawers—"the sneak thief."

The court suggested that the store could post a sign at each entrance to the store and at the checkout lanes warning customers to watch their purses, that warnings could be made over the store sound system, and that the floor walkers used to protect the cash drawers could be on the lookout for situations when the customer is distracted and a thief may appear. The court felt that these simple precautions would cost the store very little and would fulfill the store's duty to warn its customers.

The court then addressed the question of the degree of Lacelle's own fault regarding the theft of her purse. The court acknowledged that Lacelle had some duty to protect her own property. The court found that Lacelle was 25 percent responsible for her loss and that the store was 75 percent responsible. The court, therefore, awarded Lacelle 75 percent of her loss of $531, leaving Lacelle with an award of $398.25, together with $5.25 in costs.

PSCLR – MAY 89 (10)

Department Store Is Liable For Negligence In Failing To Protect Customer From Injury By Large Unruly Crowd Waiting To Purchase Merchandise:

June I. Rossman v. K-Mart Corporation, *Civ. A. No. 86-0952, in the United States District Court for the Middle District of Pennsylvania, decided June 6, 1988, reported at 701 F. Supp. 1127.*

June I. Rossman was injured while waiting in a large crowd to purchase a Cabbage Patch Doll in the K-Mart Store in Lockhaven, Pennsylvania. Rossman was knocked to the ground by the other customers when the crowd rushed to obtain the dolls for which they had been waiting. Rossman sued K-Mart.

The case was tried before a jury which returned a verdict finding K-Mart 60 percent negligent and Rossman 40 percent negligent. Damages were awarded to Rossman in the amount of $167,000, reduced to $100,200 after deducting for her 40 percent contribution to the negligence leading to her injury. Both Rossman and K-Mart appealed.

On appeal, Rossman complained that the trial court erred in excluding her expert witness in retail marketing, Professor Fred Hurvitz. The appellate court, however, agreed with the trial court's decision to exclude the witness because his testimony regarding the impropriety of permitting the crowd to gather and failing to disperse it was a matter a lay jury could itself determine without the aid of an expert. The court noted that Mr. Hurvitz was neither a safety expert nor an expert on crowd control.

Rossman also contended that the trial court should have allowed the jury to consider the issue of punitive damages against K-Mart. Rossman alleged that, in the face of a large, anxious, and unruly crowd, K-Mart employees pushed a cart of Koosa Dolls out of the stockroom and into the midst of the crowd. This, Rossman contended, caused the crowd to stampede. The situation was further aggravated, according to Rossman, when two K-Mart employees began throwing boxes of dolls in the air, which added to the excitement and confusion. Two individuals in the crowd testified that they had observed K-Mart employees throwing boxes. However, the clothing worn by the person throwing the boxes (according to one individual's testimony) was not the uniform of a K-Mart employee.

The two employees responsible for wheeling the cart of Koosa Dolls from the storeroom into the crowd, Gerald Shady and Bruce Musheno, both testified that neither of them threw any boxes in the air, but instead, upon seeing the crowd rush toward them, ran in a different direction. They testified that they saw the crowd throwing boxes into the air.

Under Pennsylvania law, a plaintiff must establish that the tortious, or wrongful, conduct was outrageous in order to receive punitive damages. "Outrageous conduct" is defined as an act done with a bad motive or with reckless indifference to another's interest. A punitive damages award may be made against a corporation for an employee's tort. The conduct of the employee who inflicts the injury complained of must be clearly outrageous to justify the vicarious imposition of punitive damages upon the corporation.

The court found that the evidence in this case showed that K-Mart did not advertise the sale of the Cabbage Patch Dolls, but that word of the availability of the dolls was spread by rumor by unknown persons. Efforts made to talk to the gathering crowd were futile, and the decision to push a cart of Koosa Dolls from the storeroom was made with the intent to appease the crowd, rather than with indifference or bad motive. Accordingly, the appellate court agreed with the trial court that there was no basis to permit a claim for punitive damages to be presented to the jury.

Rossman also argued on appeal that the damages awarded by the jury were inadequate. However, the appellate court found that there was conflicting testimony as to whether Rossman's medical complaints were a result of the trauma or were more consistent with diabetic neuropathy. Thus, the appellate court upheld the jury's damages award as adequate in light of the evidence presented at trial.

K-Mart's appeal was based upon the contention that Rossman should have been barred from recovery because she "assumed the risk" of injury when she entered the large crowd waiting for Cabbage Patch Dolls.

The appellate court found that Rossman was a customer to whom K-Mart owed a duty to use reasonable care in the conduct of its business. "K-Mart had the duty to take reasonable steps to discover if third persons were likely to perform any accidental, negligent, or intentionally harmful acts and to take reasonable care to provide appropriate precautions to protect its [customers] from harm." The court found no evidence that Rossman subjectively knew that K-Mart corporation would fail to take proper precautions to control the crowd or to provide a reasonably safe method for distribution of the dolls. Nor could Rossman have anticipated that K-Mart employees would take actions that would cause the unruly crowd to stampede. The court found that due to poor judgment and lack of understanding of the risk involved, Rossman could not be said to have understood the specific harm created by K-Mart's conduct nor to have voluntarily chosen to encounter it. Rossman's conduct was at most negligent, and the matter of her negligence was considered by the jury and the damages award reduced accordingly.

Thus, the court upheld the jury's award to Rossman of $100,200, representing 60 percent of her damages.

PSCLR – SEP 89 (12)

Uniformed, Unarmed Security Guard Had No Duty To Intercede To Protect Victim Murdered During Robbery Of Employer's Store:

Huelean Milo, Personal Representative of the Estate of Willie T. Milo, deceased v. Guardian Guard Services, Inc., and The Lawson Company, jointly and severally, *Docket No. 99473, in the Court of Appeals of Michigan, decided December 12, 1988, reported at 440 N.W.2d 5.*

At 2:00 AM one morning, Willie T. Milo was murdered at The Lawson Company's store in Detroit, Michigan. Milo, the boyfriend of an employee of the store, was helping another employee close the store's front gate. A man holding a rifle came around the corner and demanded the money bags from the store. When told by the Lawson employee that there were no money bags, the gunman shot Milo.

At the time of the shooting, a security guard from Guardian Guards, Inc., who was assigned to The Lawson Company, was standing two to fifteen feet away. The guard was unarmed (as provided for in the contract between Lawson and Guardian Guards). The guard did not attempt to stop the gunman.

The plaintiff in this case argued that Lawson and Guardian Guards had a duty to protect Milo from the gunman. The court rejected this claim. Under Michigan law, no person has a duty to aid or protect another person unless there is a "special relationship" between them. The court stated that:

> ...a merchant's duty of reasonable care does not include providing armed, visible security guards to deter the criminal acts of third parties, and...the duty to provide police protection is vested in the government by constitution and statute.

As in the *Cornelius* case in this issue, where there is no duty to provide protection, there can be no liability for failing to provide protection, so the court ruled for the defendants.

Schools

PSCLR – AUG 82 (3)

School's History Of Race Riots And Campus Violence Held Sufficient Notice To Employ Security:

The School Board of Palm Beach County v. William J. Anderson, *Docket No. 80-541, District Court of Appeal of Florida, Fourth District, decided March 24, 1982, reported at 411 So. 2d 940.*

A school board was held liable for the wrongful death of a former student who was shot and killed in the school parking lot by another former student during school hours. The deceased's father brought the action to recover for the school's negligence in failing to provide adequate protection from criminal attack by third persons.

Brian Anderson (white), a former student of Palm Beach Gardens High School, went to the school at lunchtime to pick up his school records in May, 1976. While standing in the school parking lot with some friends, another former student, Gregory Ransom (black), approached them. During the resulting confrontation, Ransom took a handgun from his car trunk and began shooting at Anderson and the others. As Anderson reached into his car for a soft drink bottle, Ransom fired a fourth shot, fatally wounding Anderson. Other than one off-duty custodian, no school officials were on campus at the time of the shooting.

The jury returned a verdict for the father; the school board appealed, claiming a lack of evidence that it knew of a dangerous condition or could have known of one by the use of reasonable care. It was undisputed that Anderson, on the campus to pick up his records, was an invitee. The District Court of Appeal recognized current Florida law concerning a property owner's responsibility to protect an invitee from criminal attack from third persons, as stated in *Relyea v. State*, 385 So.2d 1378 (Fla. 4th DCA 1980):

> As a basic principle of law, a property owner has no duty to protect one on his premises from criminal attack by a third person.... Even though one's negligence may be a cause in fact of another's loss, he will not be liable if an independent, intervening and unforeseeable criminal act also causes the loss.... If, however, the criminal act is reasonably foreseeable, a duty may arise between a landowner and his invitee. But it must be borne in mind that a landowner is not an insurer of the safety of his invitee and is not required to take precautions against a sudden attack from a third person which he has no reason to anticipate....
>
> In order to impose a duty upon a landowner to protect an invitee from criminal acts of a third person, a plaintiff, invitee, must allege and prove that the landowner had actual or constructive knowledge of prior, similar criminal acts committed upon invitees. The landowner is not bound to anticipate criminal activities of third persons where, as here, [in the Relyea case] the wrongdoers were complete strangers to the landowner and to the victims, and where the incident occurred precipitously.

This situation differed substantially from *Relyea*, since the evidence reflected a history of violent crime and racial incidents on the school campus, including attempted robbery, battery, a knifing, and several race-related riots and fights. The court held that the trial judge was correct in letting the jury decide as to whether there was a duty owed to provide protection to invitees in view of the previous episodes. The jury's verdict for Anderson's father was affirmed.

[NOTE: Although the opinion in the *Anderson* case did not specify a damage amount, The editors have learned that approximately $10,000 in actual medical expenses was awarded.]

PSCLR – NOV-DEC 83 (6)

School District Cannot Be Held Liable for Assault Merely Because Assault Occurred on Its Property – Both Governmental Immunity and Negligence Law Provide Defense:

Javelle Vann, a minor, and Sevenia Vann, her mother and natural guardian v. Board of Education of the School District of Philadelphia, *in the Commonwealth Court of Pennsylvania, decided August 30, 1983, reported at 464 A.2d 684.*

While walking past the Anderson School in Philadelphia, Pennsylvania, at 11 PM on February 21, 1981, Javelle Vann, a minor, was grabbed by an assailant. He forced her to accompany him through an unsecured gate in the school's fence, took her to an unlighted area of the school ground, and beat her.

Javelle Vann's mother sued the Philadelphia Board of Education, alleging that the School District was negligent in "failing to secure the school property and in failing to maintain adequate lighting, thereby permitting the property to be used for criminal activity directed at passersby." The suit was dismissed before trial because the School District was entitled to immunity from civil liability under Pennsylvania statutory law. Mrs. Vann appealed this decision, claiming that immunity did not apply according to the law, because of a statutory exception to governmental immunity for injuries incurred on property owned by the government and arising out of "the care, custody or control of real property in the possession" of the governmental agency

The Commonwealth Court of Philadelphia reviewed the facts and circumstances of the case, and determined that immunity had not been waived because the School had not failed " ... to properly control activity permitted on the publicly owned property," as stated in the statute's exceptions to immunity. Civil liability could only be imposed on governmental entities where it could be proved that its negligence made "government-owned property unsafe for the activities for which it is regularly used, for which it is intended to be used, or for which it may be reasonably foreseen to be used." The exception would not be applicable in this case because acts of criminal violence were not a foreseeable use of the school's property.

Therefore, the appellate court concluded:

> While a school district may be held liable for negligence in maintenance and control of property when injury results to an invitee, ..., the school district here owed no duty to Javelle Vann or to the public at large to prevent the use of its property for criminal attacks such as occurred in this case. Similarly, had the attack on Javelle Vann occurred on private property, the private landowner would not have been held liable for failure to secure his property from such use, or for failure to illuminate the unlighted areas on his property. The School District as landowner, therefore, cannot be held liable here.

The trial court's decision to dismiss the complaint was affirmed.

COMMENTARY:

While the doctrine of governmental immunity from liability for negligence has been much eroded in recent years, it is still alive and kicking. But here, the court went on to say that even if governmental immunity did not provide a defense to the school district, general principles of negligence law would be sufficient under this fact situation.

PSCLR–AUG 84 (3)

School District Not Liable For Criminal Attack On Student By Non-Student On School Premises:

Samuel Rodriguez v. Inglewood Unified School District, *Civ. 68809, in the California Court of Appeal, Second District, Division 3, decided February 27, 1984, reported at 199 Cal. Rptr. 524.*

Samuel Rodriguez, a student at Inglewood High School in Los Angeles, California, was stabbed by a non-student assailant on school property. He sued the school district to recover for his injuries, contending that the school had a duty to protect its students against persons bearing weapons on school grounds because the school had knowledge of a "long history" of violent acts occurring on the school property. The suit was dismissed by the trial court and Rodriguez appealed.

The California Court of Appeal first noted that the school district was clearly not liable because of statutory immunity, but that "the question of applicability of a statutory immunity does not even arise until it is determined that a defendant otherwise owes a duty of care to the plaintiff and thus would be liable in the absence of such immunity." Thus, the court reviewed the issue of whether Inglewood School District in fact had any duty to protect Rodriguez or any other student on school property from criminal attack by persons on school property who were not students.

The court pointed out:

> As a general rule 'there is no duty so to control the conduct of a third person as to prevent him from causing the physical harm to another unless
>
> (a) a special relationship exists between the actor [School District] and the third person [non-student criminal] which imposes a duty upon the actor to control the third person's conduct, or
>
> (b) a special relationship exists between the actor [School District] and the other [Rodrigruez] which gives to the other a right to protection.

In this case, however, the court could not find such a special relationship: Inglewood did not create the risk; though it may have been aware of a general threat of violence, it was not aware of any specific danger to Rodriguez.

> Foreseeable victims are not themselves sufficient to establish a special relationship imposing upon the School District a duty to warn or protect. ... Although school authorities have a duty to reasonably supervise the conduct of students. ... we find no precedent to require the School District to be guarantors of protection from all harm. ... Even though harm may be foreseeable, the concomitant duty to prevent the harm does not automatically follow.

The court held that the likelihood of harm which resulted from the school's conduct must be balanced against the school's value to society and its ability to have taken reasonable precautions to avoid harming the plaintiff Rodriguez. Because the school is a public agency, other factors to be considered include the extent of its powers, the role imposed on it by the government, and limitations imposed by its own budget. Applying this standard, the court noted that the Inglewood School contributed to society by providing education and that it would have been difficult for the school to protect Rodriguez from danger as he alleged it should. Indeed, Rodriguez alleged that a warning from the school would have been insufficient, rather he would require protection of all students from harm by non-student trespassers. Not only would it be prohibitively expensive for the school to maintain such a standard of protection, it would be impossible to guarantee its effectiveness in completely protecting Inglewood's students.

The court concluded that Inglewood School did not owe a duty to Rodriguez to protect him from harm by third party, non-student persons on school property. Having resolved this issue, the court stated that in the final analysis the school was immune from liability due to governmental immunity because, as stated in the applicable statute, "it did not maintain a 'dangerous condition' which caused injury to Rodriguez. [and] ... as a public entity [the school district] was immune from its failure to provide a police force on the school grounds." Therefore, the judgment granted to Inglewood Unified School District was affirmed.

PSCLR – SEP 86 (8)

School Has "Special Relationship" With Students And Duty To Take Reasonable Precautions To Warn Them Of And To Protect Them From Foreseeable Criminal Activity:

Tammy Fazzolari, a minor, By and Through Ethel Fazzolari, her Guardian Ad Litem v. Portland School District No. 1J, *A8212-07615; CA 134098, in the Court of Appeals of Oregon, decided April 16, 1986, reported at 717 P.2d 1210, petition for review allowed by the Supreme Court of Oregon, July 15, 1986, reported at 722 P.2d 737.*

An unknown man assaulted and raped a student, Tammy Fazzolari, at Franklin High School in Portland, Oregon. She and her mother sued the school district to recover for her physical and emotional injuries, and the court directed a verdict to the defendant. The Fazzolaris appealed.

Though classes commenced at 8:15 AM, Tammy Fazzolari typically arrived for school at 6:50 AM, using the time to study, chat with friends, or prepare for extracurricular activities. The school janitors opened all of the school doors by 7:00 AM, and the school did not have any rule concerning what time the students were permitted on campus.

About two weeks before Fazzolari's attack, another assault and rape occurred on the campus. In this incident, the victim was a woman delivering newspapers to the school campus at about 4:30 AM. School officials changed the location and time of the newspaper delivery, but did not warn or advise students about the assault, or change their security measures in any way.

The Fazzolaris contended that the school's negligence contributed to the assault because the school failed to provide adequate security to protect students, failed to warn students that the prior rape had occurred, and failed to maintain security on the school grounds to decrease the risk of another assault. The trial court granted the defendant's motion for a directed verdict because it did not find that Fazzolari's assault was foreseeable. The trial judge ruled that the prior rape did not make Fazzolari's assault foreseeable because that victim was not a student, and the assault occurred over two hours before the school was open.

The Court of Appeals of Oregon reviewed Fazzolari's arguments that the trial court erred in granting the directed verdict and in excluding from evidence police reports which summarized criminal activity on and around the school campus. The appellate court first looked to the school's duty to students, and ruled that schools have a duty to protect students from foreseeable criminal acts "derived from its position as a possessor of land held open to the public and from its special relationship with the students who attend...." The court discussed several Oregon cases in which the state's appellate court had discussed premises liability, including *Brown v. J.C. Penney Co.* (reported in the November-December 1982 **PSCLR**, case ten, page 10; also in the January 1985 **PSCLR**, case two, page three), *Torres v. United States National Bank* (reported in the January 1984 **PSCLR**, case three, page three), and *Peterson v. San Francisco Community College District* (reported in the February 1985 **PSCLR**, case seven, page seven).

After reviewing all of the evidence, the appellate court concluded that a jury could decide that the school could have or should have foreseen Fazzolari's rape. The appellate court did not agree with the trial court's ruling that the prior rape of a non-student two hours earlier in the morning was "so different [from Fazzolari's rape] as to remove the attack on plaintiff from the realm of the reasonably foreseeable."

The Oregon Court of Appeals also ruled that a jury could find that the school's conduct was unreasonable under the circumstances. After the first rape, for instance, the school demonstrated its awareness of the danger and of the need for caution by changing the time and location of the newspaper delivery to protect the delivery person.

The trial court refused to admit a summary of police reports concerning criminal activity at the school, which compiled 18 assaults. The list included 16 "assaults – not aggravated, unspecified" and two aggravated assaults, one involving a knife and the other a gun. The appellate court affirmed that this evidence was irrelevant because at least some of the incidents were "remote in time and of a different nature from the attack on plaintiff."

One of the judges filed a dissenting opinion in which he stated that the duty imposed on private businesses to protect customers should not be imposed on public schools, because schools are established by the state to benefit students and society in general. The school's duty of care to the student, according to the dissent, is to take reasonable measures to protect the student from reasonably foreseeable harm while the student is engaged in school activities, on or off the school premises. In this case, Fazzolari's assault occurred when she voluntarily came onto the campus about an hour-and-a-half before school began for her own and her mother's convenience. The dissenting judge noted, "I know of no Oregon case that has imposed that duty when the injury occurs during a time when the school is not in session and the student is not engaged in school activities."

The dissenting judge expressed concern over the economic and administrative ramifications of the majority opinion, which he hypothesized could require schools to hire more employees or require current school personnel to work longer days to provide security before school hours. "The extent of the burden it would impose on defendant and the consequences to the community of doing so may be disproportionate to the increase in protection afforded the students." The dissent believed that calling Fazzolari's assault reasonably foreseeable because of the earlier rape was conjecture since rape can foreseeably occur anywhere, anytime.

The majority opinion, however, reversed and remanded the case, which at last report was accepted for review by the Supreme Court of Oregon.

PSCLR – MAR 87 (7)

School Has Duty To Provide Adequate Security To Students And Is Liable For Failure To Provide Supervision Resulting In Attack On Student By Fellow Students:

Broward County School Board v. Jose Ruiz, *No. 85-607, in the District Court of Appeal of Florida, Fourth District, decided July 2, 1986, rehearing denied September 30, 1986, reported at 493 So. 2d 474.*

The plaintiff Jose Ruiz won a $30,000 verdict in his lawsuit against the Broward County School Board for injuries he received when he was attacked by fellow students on the school premises. The school board appealed.

It appears from the appellate decision in the case that Ruiz was a sophomore at Hallandale High School in Fort Lauderdale, Florida, and that he stayed for a short while after school one day, in order to have his picture taken with the rest of the junior varsity football team, of which he was a member. Instead of remaining for the practice following the picture taking session, Ruiz went to the cafeteria area to telephone his father for a ride home. The junior varsity coach remained with the team at the practice.

The school cafeteria was near the center of the main building on campus, and the precise location of the attack on Ruiz was directly across from the door to the cafeteria where benches and pay phones were located. Anthony Wilson, Jimmy Wright, and Ray Sands were the three students who attacked Ruiz. The victim recalled very little about the incident, having been knocked unconscious. Another student, however, Lateasea Fredericks, was waiting for a bus along with two of her friends, and she knew the boys who attacked Ruiz. She did not know Ruiz himself. She testified that Wilson got into an argument with Ruiz, who started walking backwards away from Wilson when he was attacked from behind by the other two boys. Ruiz was severely beaten by all three of his fellow students.

It also appeared from the record in the case that two custodians who had been in the cafeteria area had been tricked into leaving the area by another group of students. The custodians found Ruiz lying on the floor when they returned some thirty minutes later; however, there was never any positive proof that there was a direct connection established between the attack and the custodians being diverted away from the area.

Linda Brown, the school principal, testified that teachers, administrators, and security personnel had the responsibility to provide security and adequate supervision for students, but that such employees were not obligated to remain on the school premises for more than fifteen to twenty minutes after the final classes ended at 2:45 PM. When students remained after school for extracurricular activities, the principal testified that the club sponsor or coach involved had supervisory authority. Also, with regard to athletic activities, the coaches had been instructed to remain on the campus premises until the student athletes had obtained rides home. Ruiz's coach testified that he had no duty to patrol the cafeteria or other common areas of the campus after practice, and his only responsibility was over the students at the practice.

The appellate court found applicable precedent in a prior case which had held that a kindegarten teacher has no duty to leave her class and to accompany a student to the bathroom (where the student had been injured when another student closed the bathroom door on her finger). The court felt that this was an analogous situation, and concluded that it could not therefore be held that the football coach had any duty to leave the student athletes at practice in order to accompany Ruiz into the cafeteria area to use the telephone. The court therefore agreed with the defendant school board's argument that the plaintiff Ruiz had failed to make out a case for any breach of supervision on the part of the coach.

On the other hand, the appellate court also concluded that there was sufficient evidence presented by the student to permit the jury to conclude that the defendant school board had in fact failed to provide adequate security to its students, including Ruiz. The school board argued that it had no such duty, based upon the earlier Florida decision of *Relyea v. State* (reported in the August 1982 issue **PSCLR**, case four, page four). That case had held that the decisions of a university regarding "whether to provide security guards, parking attendants, security gates, and the numbers thereof" are protected by sovereign immunity. The appellate court here felt that the *Relyea* decision was not controlling in the present situation.

The appellate court noted among other things that, "there are obvious differences between the supervisory and security responsibilities at a state university like the one involved in *Relyea*, and a public high school." The court felt that the authority of the school over children, as opposed to university students was a significant factor, and also noted that earlier cases had confirmed the existence of a duty on the part of the school to provide protection to minor students and supervision while students were on the school premises, particularly in an area such as the cafeteria here where students were known to be present following the end of classes.

> If anything the time period immediately after the end of school presents the greatest danger of misconduct by students who have been restrained all day in a disciplined setting.

The appellate court also rejected the school board's argument to the effect that fighting between the students when only a handful were present was not foreseeable. The decision quoted with approval a prior Supreme Court decision which had held that certain student misconduct was foreseeable and therefore could not constitute an intervening cause which would have the effect of relieving teachers and principals of liability for failing to supervise students.

After rejecting other arguments by the school board based upon certain procedural errors by the trial court, the appellate court affirmed that verdict and judgment in favor of the plaintiff and against the school board.

COMMENTARY:

The court in this case summed up the issue neatly in this quote from an earlier Florida Supreme Court case:

> *Whether a principal's or a teacher's failure to supervise a student was the proximate cause of injuries suffered by a student is the issue most consistently litigated in negligent school supervision cases. The issue is usually complicated by the simultaneous tortious conduct of fellow students, so that the question becomes one of whether the fellow students' negligence was the intervening cause of the harm. Two distinct standards have developed. The first holds that a teacher's absence leading to an injury to one student by another can be the proximate cause of the injury only if the injury could not happen while the teacher was present. These cases generally find student misbehavior is the proximate and superseding cause of the harm [citing a New York decision]. ... The second, and more reasonably supported, holds that certain student misbehavior is itself foreseeable and therefore is not an intervening cause which will relieve principals or teachers from liability for failure to supervise: '[We] should not close our eyes to the fact that ... boys of seventeen and eighteen years of age, particularly in groups where the herd instinct and competitive spirit tend naturally to relax vigilance, are not accustomed to exercise the same amount of care for their own safety as persons of more mature years.' [citing a California case] Recognizing that a principal task of supervisors is to anticipate and curb rash student behavior, our courts have often held that a failure to prevent injuries caused by intentional or reckless conduct of the victim or a fellow student may constitute negligence. Rupp v. Bryant, 417 So. 2d 658 (Fla. 1982).*

PSCLR – OCT 88 (6)

Public School District And School Officials Liable For Failing To Protect Student From Foreseeable Assault In Unsupervised School Restroom:

Jaime Leger et al. v. Stockton Unified School District et al., *No. C000367, in the Court of Appeal of California, Third District, decided July 25, 1988, reported at 249 Cal. Rptr. 688.*

Jaime Leger, a student at Franklin High School in Stockton, California, was attacked in a school restroom while changing his clothes before wrestling practice. The attacker was not a student. Leger sued the school district, the principal of Franklin High School, and his wrestling coach for negligence in failing to protect him from the attack. He contended that the bathroom was an unsupervised location, that it was unsafe for students, and that attacks by non-students were likely to occur there. The trial court dismissed the case, and Leger appealed.

The appellate court first rejected plaintiff's theory that a provision of the California Constitution entitled "Right to Safe Schools" provided him a claim for damages for violation of that right against the school district and its employees. That provision, part of the "Victim's Bill of Rights" in the California Constitution, provides: "All students and staff of public primary, elementary, junior high and senior high schools have the inalienable right to attend campuses which are safe, secure, and peaceful." The issue for the court was whether this provision of the California constitution was "self-executing" and, therefore, provided the plaintiff a right to sue for damages for its violation, without a statutory enactment by the legislature.

In other words, the appellate court had to decide whether or not the constitutional provision had any "teeth" or merely declared a general right without specifying the means by which the right could be enforced. The court concluded that the constitutional provision did not provide a right to sue for damages and that the goal of public safety in the schools was more appropriately reached through reforms in the criminal laws.

The appellate court, however, agreed with plaintiff's argument that the school district and its employees could be liable to him under traditional principles of tort law, including negligence and respondeat superior (i.e., the employer is held liable for the negligence of its employees – see *G.L. v. Kaiser Foundation Hospitals* for further explanation of this legal theory).

First, the court held that the school district and its employees owed plaintiff a duty "to use the degree of care which a person of ordinary prudence, charged with comparable duties, would exercise in the same circumstances." The special relationship between the school district and its employees and its high school students, according to the court, creates an affirmative duty on the part of the district and its employees to take reasonable steps to protect its high school students.

The appellate court next noted that the existence of a duty of care to protect against a certain harm depends in part on whether the harm was reasonably foreseeable. For the purposes of deciding whether the lower court was correct in dismissing the case prior to trial, the court was required to accept the facts stated in Leger's complaint as true.

Leger's complaint alleged that defendants knew or should have known that he was subject to an unusual risk of harm in the restroom on school grounds. The complaint alleged that the school district, the principal, and the wrestling coach knew, or should have known, that junior varsity wrestling team members were changing clothes in the unsupervised boys' restroom. Further, Leger alleged, they knew, or should have known, that the unsupervised restroom was unsafe for students and that attacks were likely to occur there.

The court deemed the facts as stated in Leger's complaint sufficient to impose liability on the school district and its employees <u>if</u> those facts could be proven at trial. The court held that Leger had no obligation to plead that prior attacks had actually occurred in the restroom.

The court further held that certain California statutes allowed liability to be imposed on the school district as a public entity under the doctrine of respondeat superior. The general rule under the California statutes is that an employee of a public entity is liable for negligence to the same extent as a private person and that the public entity is vicariously liable for any injury caused by its employee to the same extent as a private employer. Therefore, the question for the court was whether a private school and its employees could be liable in the circumstances alleged by Leger.

The court determined that a private school has a duty to supervise a particular activity if the school officials could reasonably anticipate that supervision was required. In addition, a private school is liable for the negligence of its teachers or other employees under the doctrine of respondeat superior if the negligence occurs within the scope of their employment.

The court concluded that on the facts as pleaded in this case, California law would impose liability in the private sector. Therefore, absent some government immunity, the court found that, under the applicable California statutes, the public school district and its employees could also be liable for negligence.

The court rejected the argument that imposing such liability on the school district is "tantamount to requiring them to have a police or security force." The court felt that measures short of providing police protection services, such as posting warning signs or supervising students who frequent areas of known danger more closely, might suffice to meet the duty of reasonable care to protect students.

The court thus reversed the judgment of the trial court and reinstated Leger's complaint against the school district and the principal and wrestling coach of Franklin High School.

COMMENTARY:

In discussing its findings, the court in the above case cited the California decisions in Isaacs v. Huntington Memorial Hospital, reported in the May 1985 **PSCLR**, *case three, page three; Peterson v. San Francisco Community College District, in the February 1985* **PSCLR**, *case seven, page seven; and Rodriguez v. Inglewood Unified School District, in the August 1984* **PSCLR**, *case three, page three.*

PSCLR—NOV-DEC 88 (12)

Municipality Not Liable For Rape of Public School Teacher If She Did Not Reasonably Rely Upon School's Assurances Of Precautionary Measures To Protect Her:

Carole A., (anonymous) v. City of New York, et al., *in the Supreme Court of New York, Appellate Division, Second Department, decided August 29, 1988, reported at 532 N.Y.S.2d 163.*

Carole A. ("Carole"), a psychologist for a New York City Public School, was brutally raped and sodomized one evening as she was preparing to leave work. The school principal had repeatedly assured Carole that the school custodian or his assistant would remain with her in the school annex between the hours of 3:00 PM and 4:00 PM, whenever Carole was required to remain in her office after school hours

Carole sued the city and the Department of Education for negligent breach of the "special duty" to protect her, which was voluntarily undertaken by the school. At trial, the jury awarded Carole $700,000 for personal injuries resulting from the rape. The appellate court reversed and ordered the case to be retried due to the trial court's failure to instruct the jury that plaintiff must have reasonably relied on the school's assurances that she would be protected in order to recover for her injuries.

To succeed on her negligence claim, Carole had to prove that the city-run school had undertaken a "special duty" to protect her, apart from the city's general duty to protect the public. This required her to prove: (1) assurances to her by the school that she would be protected, creating a "special duty"; (2) her reasonable reliance on those assurances; (3) breach of the special duty created by the assurances; and (4) injuries proximately caused by the breach of duty.

The appellate court held that the trial court's failure to instruct the jury concerning the requirement of reasonable reliance by Carole on the school's assurances required a reversal of the jury's verdict and retrial of the case.

The dissenting judge strongly disagreed with the appellate court's ruling. The dissenting judge stated that the testimony of the school principal, Henry Richman, established that it was mandatory for Carole to remain in her office until 4:00 PM, an hour after the normal departure time for students and teachers. Richman himself had expressed concern that intruders might enter the annex when Carole was there alone. Richman had discussed Carole's security with her at the beginning of the school year. He was unsuccessful in procuring office space for her in the main building but assured her that the custodian or his assistant would remain with her in the annex, near her office, between 3:00 PM and 4:00 PM. He further told her that the custodian and his assistant would check the annex to make sure that everyone had left the premises and to lock all entrance doors at approximately 3:15 PM. He also provided her with a special lock to secure her office door from within.

Carole repeatedly expressed her concerns that the custodian was frequently absent and often appeared intoxicated. Each time, Richman reassured Carole and spoke with the custodian and his supervisor. Richman learned that the custodian had, in fact, been intoxicated on occasion but, nevertheless, he retained the custodian in the school's employ.

On the afternoon of the attack, the custodian was either not present at all in the annex or, if present, failed to carry out his assigned duties there. As she was leaving work, Carole was struck on the back of the head by an unknown assailant, dragged into a nearby bathroom, and brutally raped and sodomized. After the attack, she dropped her career as a school psychologist. She underwent psychiatric treatment and was diagnosed as having chronic post-traumatic stress disorder. Her psychiatric expert stated that the disorder would be permanent.

The dissenting judge felt that there was ample evidence in the record to support the jury's conclusion that the school had a special duty to act on Carole's behalf and had breached that duty. Further, Carole unequivocally testified that she had relied upon the assurance given to her by the school. The dissenting judge felt that the court was penalizing Carole for relying on the assurances, as she was required to in order to perform her job. The dissenting judge pointed out that Carole had no reasonable alternative to tolerating the potentially hazardous conditions created by the requirement that she remain in the annex building after school hours. She did all that she reasonably could do to insure that the security measures devised for her protection would be implemented. The dissenting judge expressed his opinion that "to deny [Carole] the damages properly awarded to her by the jury and to subject her to the rigors of a second trial is grossly unjust."

Nonetheless, the majority of the appellate division voted to reverse the award and remand the action for determination of Carole's reliance on the special duty.

COMMENTARY:

The decision by the court to reverse the trial court's $700,000 damage award to the plaintiff in this case is based on the assumption that the trial court is required to instruct the jury regarding every element of the tort, regardless of the strength of the evidence on one or more elements. The trial court in this case was within its power to effectively direct a verdict on the issue of reasonable reliance by the plaintiff based on the apparently undisputed evidence that she was required, in order to keep her job as school psychologist, to remain in the school annex after hours.

It appears that since she was required to remain in the annex, she had no alternative but to rely on the assurances provided to her by the school principal that she would be protected by the presence of the custodian.

The trial court could have justifiably removed this issue from those presented to the jury and effectively held, as a matter of law, that the undisputed facts established that the plaintiff reasonably relied on assurances that she would be protected and that this reliance caused her injuries.

PSCLR – FEB 89 (6)

School Board And City Police Officer Hired To Provide Security At Public School May Be Liable In Negligence For Death Of Student At Special School Function:

Leonard LeRose, Jr., as Independent Administrator of the Estate of Terrance A. Hogan, deceased, and Sharon Hogan, individually and as the mother and next friend of Terrance A. Hogan, deceased, v. City of Zion/Police Department, an Illinois Municipal Corporation; Zion-Benton Township Board of Education, a local public entity, and Don Williamson, as agent/employee of the City of Zion/Police Department and/or as agent/employee of Zion-Benton Township Board of Education, *No. 85 C 7167, in the United States District Court for the Northern District of Illinois, Eastern District, decided October 4, 1988, reported at 696 F. Supp. 1222.*

The Zion-Benton Board of Education ("Board") authorized and sponsored an extracurricular activity known as "Winter Carnival," from 7:00 PM to 10:00 PM at the Zion-Benton High School in Zion, Illinois. On the afternoon of the carnival, at approximately 3:05 PM, Terrance A. Hogan and a fellow student beat up a junior high school student at the Central Junior High School in Zion. At approximately 10:10 that evening, Terrance A. Hogan received fatal injuries in a knife fight in the vicinity of the Zion-Benton High School campus. Police later determined that the knife fight was related to the earlier fight.

At the time of the incident, the Board of Education had an agreement with the City whereby the City would supply one police officer to act as a school/police counselor for the approximately 1,100 students attending the school. Officer Don Williamson acted as that counselor. The terms of the written contract provided that the counselor work with the school principal in reducing juvenile crime, including early detection. The counselor was to work regularly from 7:30 AM to 3:30 PM and deviate from that schedule as asked "[d]ue to the uniqueness of school needs."

Williamson was not present at the Carnival or in the vicinity of the knife fight and was not requested to attend the Carnival.

Hogan's mother, Sharon Hogan, and the representative of his estate sued the City of Zion/Police Department, the Zion-Benton Board of Education, and officer Don Williamson, both as an employee of the City and as an employee of the Board. The plaintiffs alleged that Williamson should have advised school officials that his presence at the Carnival was needed due to threats of disturbances, and that by failing to do so, Williamson failed to exercise reasonable care. Plaintiffs also alleged that the City and the Board were liable as Williamson's employers.

The court first analyzed whether the City and Williamson had a duty to exercise reasonable care to protect Hogan. Generally, a city's failure to provide adequate police services to the public is not actionable unless plaintiffs establish a "special duty" to provide such protection. A "special duty" exists if: (1) the municipality is uniquely aware of the particular danger or risk to which the plaintiff is exposed, (2) the complaint alleges acts or omissions by the municipality, (3) the acts or omissions are either affirmative or willful, and (4) plaintiff's injury occurs while he is under the direct and immediate control of employees or agents of the municipality.

Plaintiffs did not show that the City and its employee Williamson owed Hogan any "special duty." Plaintiffs could only point to one incident, sometime during the month prior to the Carnival, in which Williamson was made aware of threatened violence on campus. Other incidences of disturbances or fights at extracurricular events were considered by the court to be "of a completely different nature from a knife fight." Further, Hogan was not under the "direct and immediate control of the police" since no police personnel were at the Carnival.

The court distinguished this case from a case in which a village was held liable for failure to provide adequate police protection at a stadium which the village owned. In that case, the village had a duty of due care to adequately protect attendees by virtue of its ownership of the stadium because the attendees of the event were business invitees of the village and, therefore, under a "special relationship" with the village as owner of the stadium. The City of Zion/did not own the school in the above case and, thus, the business invitee theory did not apply.

The court also noted the "general reluctance of Illinois courts to find the police liable for failure to prevent crimes....[t]he pattern seems to be that the police are held liable when they create a danger, but not when they fail to prevent a danger."

The court declined to analogize the police paid to provide additional security to a private security force, which would place them outside the general rule of police non-liability. The court acknowledged that a different situation is created when a party employs "moonlighting" individuals who happen to be police officers. The court held that "so long as the police department is being paid, and the officers in question are drawing nothing beyond their regular police salaries, the general rule of police non-liability is applicable."

Therefore, the court held that Officer Williamson, as an employee of the City of Zion/Police Department, and the City of Zion/Police Department, did not owe a duty to exercise reasonable care to protect Hogan from injury.

The court next analyzed whether the Board of Education, and Williamson as an employee of the Board, could be liable. The court assumed, because neither party discussed the issue, that the Board had a special duty to protect Hogan from criminal attack.

The court also found that governmental immunity did not protect Williamson or the Board. The court, in discussing this issue, analogized Williamson's security work for the school to that of a private security guard.

The court found that there was a genuine issue as to whether Williamson had a responsibility to request or suggest additional security at the upcoming scheduled school event. The written contract did not explicitly state that other school officials were responsible for requesting Williamson's presence. A jury could conclude, the court reasoned, that it became general practice for Williamson to request additional security when and as needed.

Having found that the Board and its employee, Williamson, were not protected by government immunity, and having assumed that the Board and Williams owed Plaintiffs a duty to exercise reasonable care to provide security at the school, the court found that there were sufficient triable issues of fact to be determined by a jury.

The court, therefore, ordered the case to be tried against the Board of Education and Don Williamson as an employee of the Board, and dismissed the case against the City of Zion/Police Department and Don Williamson as an employee of the City.

Shopping Centers

PSCLR—APR 84 (10)

Shopping Mall And Retail Store Owners Not Liable For Assault On Invitee Without Evidence Of Security Lapses Or Intentional Failure To Assist Injured Invitee:

Hector Castillo, Sr., a/n/f Hector Castillo, Jr., and Ricardo Castillo v. Sears, Roebuck & Co. and Mall Del Norte, *No. 04-81-00383-CV, in the Court of Appeals of Texas, San Antonio, decided November 30, 1983, rehearing denied December 27, 1983, reported at 663 S.W.2d 60.*

While Hector and Ricardo Castillo were shopping in the Mall del Norte Sears store in Laredo, Texas, on December 21, 1979, they were approached by Rodolfo and Andres Torres who threatened them and asked them to step outside. While in the area just outside the Sears entrance, the Torreses attacked the Castillos. In the fight Hector suffered a broken jaw and Ricardo, a stab wound. Although no mall or Sears' employees witnessed the confrontation inside the store or the fight outside the store, the Castillos sued the Mall del Norte and Sears store owners. Their suit alleged that their injuries were caused by the defendant's negligence in failing to provide security forces. The defendants were granted a motion for summary judgment at trial and the Castillos appealed.

The Texas Court of Appeals reviewed the case and noted that although a store is not required to protect its invitees from criminal attacks of third parties, Texas courts have acknowledged "the intervening criminal act of another may not always relieve one of liability for his negligence." The Castillos relied on this view and contended that summary judgment was improper because a security expert's evaluation of Sears' and the mall's security practices raised factual issues which could only be resolved by a jury. The court, however, ruled that the security expert's evaluation by itself was insufficient evidence to bring the suit to trial. Without evidence that Sears or the Mall was even aware of the assault or the situation leading up to the assault, the defendants could not be held liable for failing to aid the Castillo brothers.

In affirming summary judgment to Sears and the Mall, the court reiterated:

> There is no duty upon the owners or operators of a shopping center, individually or collectively, or upon merchants and shopkeepers generally, whose mode of operation of their premises does not attract or provide a climate for crime, to guard against the criminal acts of a third party, unless they know or have reason to know that acts are occurring or are about to occur on the premises that pose imminent probability of harm to an invitee; whereupon a duty of reasonable care to protect against such act arises.

COMMENTARY:

Once again we see that the case turns upon the issue of foreseeability. The appellate court's view of the evidence here was that there was no way that the defendant retailer or the defendant shopping mall could reasonably have expected the assault by the Torres on the Castillo brothers to have occurred. Without some proof that the attempted act either was or was at least should have been foreseen by the defendants, the plaintiff cannot recover because there is then no duty by the retailer or the mall to warn the plaintiffs or to protect them.

Are criminal attacks by third persons "foreseeable" on your properties? Have such attacks taken place on any of your properties in the recent past? Have they increased in the recent past? Have you asked the local police to provide you with information concerning criminal attacks in and around your properties? Have you received an increasing number of claims alleging such assaults?

If the answer to any of these questions is "yes," are future attacks by third persons now "foreseeable" by you? If so you may well be held liable for them, unless you take steps to warn invitees of the potential dangers and to protect them from same.

PSCLR – OCT 84 (4)

Shopping Center Is Not Responsible For Injuries Suffered By Boy Who Was Thrown Thrown Through Plate Glass Window During Fight In Which He Willingly Participated:

Thomas T. Miguez, Sr., administrator of the estate of his minor child, Dudley Miguez v. Urban Developments, Inc., Sea & Sar's et al., *No. 83-CA-755, in the Court of Appeal of Louisiana, Fifth Circuit, decided May 14, 1984, reported at 451 So. 2d 614.*

Dudley Miguez, aged twelve, got involved in a fist fight with Michael Pettit at the defendant Urban Development's Chateau Village Shopping Mall. Both boys willingly participated in the fight, which ended abruptly when Dudley was thrown through a plate glass window by Michael. Dudley's foot was injured as he tried to remove himself from the bed of broken glass.

The minor plaintiff, acting by and through his father, filed suit against the shopping mall and the retailer, alleging that they had failed to employ properly-trained and sufficiently-staffed security services in the shopping mall area. At the conclusion of the presentation of the plaintiff's evidence at the trial, the trial judge granted a verdict in favor of the defendants, noting that the evidence indicated that the cause of the plaintiff's injuries was the fight the two young boys willingly engaged in. The plaintiffs filed the instant appeal.

On appeal, the court rejected the plaintiff's claim that the defendants were negligent for failing to supply security guards. Only when the owner or manager of a business or property knows (or could be imputed to have knowledge) of another person's intended criminal conduct which is about to take place, and which is within the power of the owner or manager to prevent, does a duty of care toward a customer/invitee arise. The landowner or business operator is not an "insurer" as to any and all injuries or accidents on the property, but rather he is required to take such measures as a reasonably prudent person would take in view of the probability of injury to others.

> It would indeed extend the duty-risk analysis of our law too far to require a normally safe business to employ security guards to protect against the possibility of two of its patrons coming to blows with one another.

The court did however remand the case to the trial court for the purpose of determining the retailer's counterclaim for property damage to the display window. Accordingly, the dismissal of the plaintiff's claim was affirmed.

PSCLR—APR 85 (4)

Contract Between Shopping Mall And Booth Operator Fails To Indemnify Shopping Mall For Personal Injury Losses Suffered By Slip And Fall Victim:

Northwest Mall, Inc. and N.W. Management Corp. v. Lubri-Lon International, Inc.; Lubri-Lon International, Inc. v. Madge Spencer, *No. B14-83-857-CV, in the Court of Appeals of Texas, Houston (14th District), decided October 18, 1984, reported at 681 S.W.2d 797.*

Plaintiff Madge Spencer brought this lawsuit against Northwest Mall and N.W. Management Corporation, and against Lubri-Lon International, operator of a booth in Northwest Mall. She filed the action after she slipped and fell in a puddle of oil, which had leaked from the Lubri-Lon booth located in the mall. Lubri-Lon's employee at that location, Buddy Parent, used a machine to demonstrate the use of an engine oil additive, and he disposed of the oil in a thermos jug kept in the booth. One Saturday night, he was "bone tired" and, rather than immediately disposing of the remaining oil, he left the thermos jug on a shelf underneath the counter. The next night, the mall security guard saw a youth jump out of the booth, and run away. The guard later saw that a quantity of oil was leaking from the booth, and left a note for the next guard advising him of the problem; however, he did not attempt to clean up the oil drippings.

On Monday morning, at approximately 7:15 AM, a maintenance man mopped up the oil spill. Approximately an hour and a half later, the mall manager saw a patch of oil in the same place, and radioed for a maintenance man but got no response. He then left to look for him in order to get the oil spill cleaned. While walking away from the booth, he passed Spencer, an employee of another mall store, but he said nothing to her to warn her of the oil spill. She slipped in the oil, breaking her hip.

At trial, Spencer recovered a verdict of $567,000 against Lubri-Lon International and against Northwest Mall and its management corporation; the defendants appealed. Lubri-Lon complained that the jury award should be set aside, and the mall complained that it had a contract with Lubri-Lon wherein Lubri-Lon should be required to indemnify the mall for its loss.

The appellate court first examined the evidence to determine whether Lubri-Lon's employee, Parent, had a duty to Spencer, and if so whether he violated that duty.

> The elements of negligence are the existence of a duty on the part of one party to another, a breach of that duty, and damages proximately caused by a breach of that duty ... this "duty" is a duty to act as a reasonable, prudent person under the same or similar circumstances, considering the reasonably foreseeable risk or probability of injury to persons situated as the plaintiff.

The court determined that the evidence showed that Parent violated his duty in various particulars: he testified that 60 percent of the time he locked the thermos in a bin, but that he left the jug out on the particular night in question in order to save time; this alone constituted negligence. Vandalism of the booth was foreseeable because the openness of the booth itself facilitated vandalism, and because it had occurred before. Spencer, although she was a passerby and not injured inside Lubri-Lon's booth, was nevertheless owed a duty by Lubri-Lon. The court stated, "Surely the reasonable person would recognize that a jug of oil left in an accessible location could spill and cause danger to anyone stepping in it."

Lubri-Lon also argued that Spencer had failed to prove that her injuries were "proximately caused" by any negligence on the part of Lubri-Lon's employee, Parent. In this regard, the appellate court noted,

> Proximate cause consists of two elements: (1) cause-in-fact; and (2)

> foreseeability. Cause-in-fact is satisfied by the showing that the act or omission was a substantial factor in bringing about the injury and without which no harm would have occurred. The test for foreseeability is whether the actor, as a person of ordinary intelligence, should have anticipated the danger to others created by his negligent act.

The court further noted that it is not necessary that the negligence in question be determined to be the sole producing cause of an injury. On the other hand, negligence can be a concurring cause which contributed to the result. In applying these legal principles to the facts at hand, the court found that the evidence showed that Parent was responsible for the oil and for the booth from which it leaked:the oil was on the floor because Parent failed to lock it up. The court concluded that an intervening action of a third party will not excuse the first wrongdoer if the intervening act should have been foreseen.

Also, the actions of the Northwest Mall's employees in failing to prevent the incident do not extinguish Lubri-Lon's liability; the actions of the mall's employees represented concurring causes which, combined with the original negligence on the part of Lubri-Lon's employee, caused the injury. Therefore, the original negligence by Parent is still a proximate cause of Spencer's injuries regardless of whether the mall employees' concurring negligence was foreseeable.

The court next examined the complaint that the damages awarded to Spencer was excessive, and reviewed the nature of the injuries, her lost income, and past and future medical expenses. Specifically, the court noted that an award of $107,000 for future lost earnings amounts to $600 per month over a 15-year period which is not excessive. And, considering the medical testimony that she would have to have at least three more surgeries in her lifetime to replace her artificial hip, the future medical amount for $75,000 is not excessive.

An interesting aspect of the case involves an indemnification agreement between Northwest Mall and Lubri-Lon International. Prior to the jury verdict, Northwest Mall had settled with Spencer in the amount of $350,000, and the mall claims that Lubri-Lon should pay the mall one-half of the $567,000 jury award, thereby partially indemnifying the mall for its settlement payment. The lease agreement between the mall and the booth operator included the following provision:

> The Exhibitor will indemnify the Northwest Mall Joint Venture and Northwest Mall Merchants Association, and any of the merchants leasing space in the mall and/or parking area and save them harmless from and against any and all claims, damages, actions, liability, and expense in connection with the loss of life, personal injury and/or damage to property arising from or out of the occupancy or use by Exhibitor of the premises or any part thereof, or any part of the shopping center property, or occasioned, wholly or in part, by any act or omission of Exhibitor, its agents, contractors, employees or participants.

The court noted that the agreement failed to unequivocally provide that Lubri-Lon would indemnify the mall for the mall's <u>own</u> negligence. The court stated that Texas law provides that a contract of indemnity will not protect an indemnitee against his own negligence unless the obligation of the indemnitor to do so is expressed in <u>clear and unequivocal terms</u>. This the agreement between Lubri-Lon and Northwest Mall failed to do.

The verdict in favor of the plaintiff was affirmed, and the mall's attempt to obtain indemnification from Lubri-Lon failed.

COMMENTARY:

Perhaps the most important part of the above case to be noted by our subscribers is that portion dealing with the indemnification agreement between the mall and the kiosk operator. While the the indemnity provision in the lease contract did include general language referring to the lessee's responsibility for "any and all claims," it also stated that the lessee be required to undertake the responsibility only when and if the losses were caused by the use of the property by the lessee, and the clause further provided that the lessee would indemnify the mall for those losses caused in whole or in part by the negligence of the lessee.

In order for the lessor shopping mall to have fully protected itself and obtained a valid indemnification agreement in its lease, it would then be necessary for the lease to provide specifically that the lessee would indemnify the mall for the mall's own negligence and the negligence of the mall's own employees. Furthermore, the court noted that the current trend of Texas law has been progressively stricter in applying the general rule of indemnity.

The specific language included in lease agreements which purport to provide for indemnification of the lessor by the lessee should be reviewed periodically, and revised if necessary, to insure that sufficiently specific indemnification language is included in order to properly protect the lessor from loss. Lessees must undertake to adequately insure themselves against losses caused by the negligence of the lessor if the lease agreement does contain specific language requiring the lessee to indemnify the lessor even in the event of the lessor's own negligence.

As with many other provisions in these contracts, specific language must be utilized in order to obtain the full benefit of applicable laws in each jurisdiction, and competent legal counsel should be obtained.

PSCLR – JUN-JUL 86 (6)

Without Evidence That Shopping Center Owner Had Agreed To Or Attempted To Provide Security Inside Tenant Stores, Shopping Center Is Not Liable For Death Of Tenant's Security Guard Inside Store:

In the Matter of the Estate of Guytano Pagano, deceased. Deborah Ann Pagano, personal representative of the estate of Guytano Pagano v. Norman Mesirow, trustee of Sears Shopping Center, d/b/a Mesirow and Company, and M.I.G. Management Company, jointly and severally v. The Kroger Company, a foreign corporation, *Docket No. 76852, in the Court of Appeals of Michigan, decided July 31, 1985, released for publication March 21, 1986, reported at 383 N.W.2d 103.*

While attempting to apprehend a shoplifting suspect, Guytano Pagano was shot by the suspect inside the entrance and exit area of the Kroger store located in Sears Shopping Center near Detroit. Pagano, who was employed by the Kroger Company as a security guard, died from the gunshot injuries. His widow sued the owner of the Sears Shopping Center, Norman Mesirow, and M.I.G. Management Company, alleging that the defendants were negligent in providing and supervising security personnel on the mall premises. The trial court granted the defendant's motion for summary judgment and the widow/plaintiff appealed.

The Court of Appeals of Michigan noted that Mesirow and M.I.G. were responsible for providing and supervising security personnel to patrol and secure the common areas of the shopping center, including the parking lot. Kroger undertook the sole responsibility for providing security inside its store. The shopping center security patrol of the parking areas ended nightly at 9:30 PM, and had ceased before Pagano was shot at 10:50 PM. The trial judge had in fact initially refused to grant the defendants' motion for summary judgment because he anticipated that a liability issue was raised:

> [T]he presence of security forces generally on the property provide protection for the center, and it is intended to protect all of the members of the center ... they operate for the benefit of all of the stores that are operating

> there, and if they were suddenly to take the position, we are going to close them at an arbitrary hour, knowing some of the parties were still operating their stores there, I think it is akin to the situation here, and I think they would be liable.

The trial judge changed his mind and granted the summary judgment when the defendant renewed the motion and cited a recent decision in which the Michigan Court of Appeals had ruled that the lessor of property was not liable for inadequate security in a similar case "where [the lessor] had not exercised control over the security of the leased premises."

The plaintiff/widow argued in her response that this rule was not applicable here because the defendants had undertaken to provide security services and therefore had a duty to provide reasonable care to protect invitees of the shopping center's tenants. In addition, the plaintiff asserted that the fatal assault on Pagano was caused by the defendants' negligent security provisions in the parking area around the Kroger store, which included inadequate lighting and their decision to discontinue security patrols at 9:30 PM, before Kroger's closing time.

The Michigan appellate court did not agree that the defendant shopping center owners and managers had a duty to provide security outside the Kroger store during all the hours that the store was open. In affirming the summary judgment granted to the defendants the court stated:

> Even were there a duty, we note ... that 'there would also be exceptional uncertainty with respect to the issue of causation. This is so because of the extraordinary speculation inherent in the subject of deterrence of men bent upon criminal ventures. It would be quite a guessing game to determine whether some unknown thug of unknowable character and mentality would have been deterred if the owner had furnished ... some additional police.

COMMENTARY:

The terms of the lease can be crucially important as is clearly shown by the result in the above case. Before anyone can be found to be negligent, it must first be established they owed some duty of care to the plaintiff. The crux of this case is that the owners and operators of the shopping mall owed no duty of care to anyone so long as they were inside the leased premises of one of the shopping mall tenants because of the fact that the owner and operator did not undertake to provide any security on any such leased premises. The obligation to provide security should be clearly spelled out in all shopping mall leases with its tenants.

PSCLR–APR 87 (4)

Knowledge By Shopping Mall Lessees, Operators, And Owners, Of Prior Criminal Attacks In Mall Gives Rise To Duty To Provide Adequate Security For Customers And Employees:

Helen Eichenbaum v. Rossland Real Estate, Ltd., *No. 85-2599, in the District Court of Appeal of Florida, Fourth District, decided February 25, 1987, reported at 502 So. 2d 1333.*

Helen Eichenbaum incurred injuries during an armed robbery at the Lauderhill Mall. She filed suit against the lessees, operators, and owners of the shopping mall, claiming that their knowledge of prior, similar criminal acts occurring during the year and a half preceding her injuries gave rise to a duty on their part to provide adequate security for both customers and employees, such as herself, of the shopping mall. She claimed that her injuries resulted from the defendants' breach of that duty to provide adequate security.

The defendants filed a motion to dismiss the complaint, arguing that the plaintiff's allegations were not sufficient to impose any liability on the defendants. The defendants further argued that the prior crimes set forth by the plaintiff as putting the defendants on notice were not alleged by her to have taken place in the exact location in the mall where the plaintiff's criminal attack took place. The plaintiff's complaint was dismissed by the trial court, and she appealed.

The appellate court, however, disagreed with the defendant's arguments and held that the law does not require the plaintiff to allege, or to prove, that the prior criminal misconduct occurred in the exact same place in the mall. It is sufficient for the plaintiff to allege that there were prior criminal incidents sufficiently similar to the attack on Eichenbaum "at the Lauderhill Mall."

Another issue presented on the appeal was whether the trial court properly dismissed Eichenbaum's claim for punitive damages; in this regard the appellate court agreed with the trial court. The misconduct required to sustain a claim for punitive damages goes beyond "gross negligence," and must be of a

> gross and flagrant character, evincing reckless disregard of human life, or of the safety of persons exposed to its dangerous effects, or there is that entire want of care which would raise the presumption of a conscious indifference to consequences, or which shows wantonness or recklessness or a grossly careless disregard to the safety and welfare of the public, or that reckless indifference to the rights of others which is equivalent to an intentional violation of them.

Since Eichenbaum in her lawsuit claimed only that the "defendants had a lack of concern for security," she did not sufficiently allege misconduct of such an opprobrious nature as to support a claim for punitive damages.

Accordingly, the appellate court reversed the trial judge's order dismissing the negligence claim against the mall owners, and affirmed the trial judge's order dismissing the claim for punitive damages. The appellate court further remanded the case to the trial court for trial on the negligence claim alone.

COMMENTARY:

The court cited the case of Relyea v. State as support for its holding that the plaintiff's allegations of prior criminal incidents were not limited to incidents which took place in the exact same place in the Lauderhill Mall as Eichenbaums' assault. See the August 1982 **PSCLR**, *case four, page four for the report of this case.*

PSCLR – SEP 87 (7)

Even Gross Negligence Will Not Sustain Claim For Punitive Damages Against Shopping Mall By Invitee Raped In Mall:

Bradenton Mall Associates, a Florida General Partnership; Fred A. Beshara; Fred M. Beshara; James M. Beshara; Nicholas Manos; Dorothy Manos; Bradenton Mall Holding Corporation; Dorothy Manos, as personal representative of Nicholas Manos, deceased; James M. Beshara, as personal representative of Fred A. Beshara, deceased; and the Edward J. Debartolo Corporation v. Pamela Hill, *No. 87-793, in the District Court of Appeal of Florida, Second District, decided June 12, 1987, reported at 508 So. 2d 538.*

During a shopping trip at the Bradenton Mall, Pamela Hill was accosted by a stranger who raped her in the mall. She sued the mall, alleging that it failed to provide adequate security and reasonable safety for customers in the mall, particularly claiming that the defendants had notice to take specific protective measures because of the dress and appearance of her assailant and because of previous criminal occurrences on the mall premises.

This case has not yet reached the trial stage, and came before the appellate court on a preliminary issue, dealing with the plaintiff's claim for punitive damages. In addition to claiming that the defendants are liable to her for negligence for failing to provide adequate security, she also has alleged that their failure to protect her amounted to willful, wanton, and reckless disregard for her safety, thereby entitling her to punitive damages. She therefore requested the court to permit her access to certain financial information concerning the defendants (discovery) in order to support her claim for punitive damages at the trial; however, the defendants objected to furnishing the financial information, asserting that the plaintiff's lawsuit did not state a claim upon which an award of punitive damages would be based and therefore she was not entitled to obtain the financial records she was seeking.

The trial court denied the shopping mall's motion to eliminate the plaintiff's claim for punitive damages and entered an order requiring the mall's owners to provide the financial data to the plaintiff's attorneys. The shopping mall owner then filed this interlocutory (pretrial) appeal.

The appellate court noted that there is no question and the shopping mall owners agreed, that Hill had successfully stated a cause of action for negligence. However, the appellate court noted that more than even gross negligence is required to sustain a claim for punitive damages:

> It must involve that entire want of care which would raise the presumption of a conscious indifference to consequences ... or that reckless indifference to the rights of others which is equivalent to an intentional violation of them.

The appellate court then reviewed the previous Florida appellate court decision in the case *Ten Associates v. Brunson*, 492 So. 2d 1149. The court noted that since the conduct of the defendants in that case did not amount to malice and recklessness sufficient to sustain an award of punitive damages as the *Brunson* court ruled, then Hill likewise failed to allege a case against these defendants which would justify punitive damages. Hill did not allege that the shopping mall provided no security whatever, only that the security which was provided was inadequate.

The appellate court therefore remanded the case to the trial judge with the direction that the trial judge should sustain the shopping mall owners' objections to the plaintiff's request for production of financial information. However, the appellate court also noted that the plaintiff may later petition for discovery of the financial information if factual allegations are made or evidence is produced which would sustain her claim for punitive damages.

PSCLR – MAY 88 (1)

Rule Requiring Prior History Of Similar Criminal Activity In Order To Establish Foreseeability Is Abolished Expert Witness May Establish Foreseeability Based On All Facts And Circumstances Even Absent Such History Of Similar Acts:

Bruce v. Galloway v. Banker's Trust Company, Trustee d/b/a Midlands Mall, Maenner Company, American Security Services, Inc., and First National Bank of Chicago, *No. 63/86-1879 in the Supreme Court of Iowa, decided March 16, 1988.*

Bruce Galloway, the victim of a homosexual rape in a restroom in the Midlands Mall, sued ownership and management of the mall as well as the private security company for the mall, alleging it was negligent in failing to prevent the rape. The trial court ruled that because the rape was unforeseeable, all the defendants were entitled to summary judgment.

The supreme court ruled the rape foreseeable and reversed that portion of the trial court's ruling.

The supreme court first cited an earlier case on foreseeability, *Martinko v. H-N-W Associates* (reported in the November-December 1986 **PSCLR**, case three, page three) in which the lower court had adopted the following rule:

> A possessor of land who holds it open to the public for entry for his business purposes is subject to liability to members of the public while they are upon the land for such a propose, for physical harm caused by the ... intentionally harmful acts of third persons ... and by the failure of the possessor to exercise reasonable care to (a) discover that such acts are being done or likely to be done, or (b) give warning adequate to enable the visitors to avoid the harm, or otherwise protect them against it.

In other words, a mall owner would be held liable for injuries to patrons caused by criminal acts of third parties which the mall, based on past experience, could have reasonably foreseen and prevented. In *Martinko*, the court held the mall owners not liable because the mall (in that case, a new mall) had no history of prior criminal activity.

One physical assault and no sexual assaults had been recorded at the shopping mall involved in this case. The majority of criminal acts occurring at the mall involved shoplifting, as might be expected. The supreme court, however, ruled that a history of crimes against persons was not necessary to render a rape, such as the one that occurred in this case, automatically unforeseeable, and said that a jury should determine whether the rape was foreseeable based upon the entire history of criminal activity at the mall, including the shoplifting incidents, as well as the lack of prior rapes.

The court noted that crimes, such as shoplifting, may often turn violent, and had indeed turned violent at this particular mall in the past. The court then cited a Delaware case, *Jardel Company v. Hughes* (reported in the June-July 1987 **PSCLR**, case one, page one), which stated, "[C]rimes of whatever type and whenever occurring on the premises are part of the circumstantial setting in which security needs are measured."

The court concluded that the shoplifting incidents and the one assault were sufficient evidence of past criminal activity at the mall to raise a question of fact as to whether the homosexual rape occurring in the mall bathroom could have been foreseeable.

The court accordingly returned the case to the trial court for jury trial on the foreseeability issue for all parties except First National Bank of Chicago.

The security company with which the mall owners and managers had contracted to provide security for the mall raised an additional defense, arguing that its contract required it to provide security for the mall itself and not for the mall patrons.

American Security argued that even if Bruce Galloway, the plaintiff, were treated as a "third-party beneficiary" of the security contract, he would have no case against the security company. The court examined the duties of the security company under the contract, and quoted from it as follows:

> Contractor will assist Bankers Trust's managing agent in preparing policies and procedures for the protection of Midlands Mall property assets, tenants, and customers. Specific duties include, but are not limited to, patrolling, traffic control, reporting, physical plant, arrests, and assisting customers and tenants.

Accordingly, with one judge dissenting and saying that he would continue to follow *Martinko* and hold that the past history of criminal activities at the Midlands Mall could not have caused the defendants to foresee the assault on the plaintiff, the majority of the court remanded the case to the trial judge for presentation to a jury.

As for First National Bank of Chicago, the court noted that the bank had received income from the mall, but did not exercise direct control or right of possession over the mall. The court stated that a party may lend possession rights to another party as First National Bank of Chicago had indeed done in this case, granting its rights to possession to Banker's Trust Company. The court thus ruled that First National could not be held liable to Galloway as a matter of law.

COMMENTARY:

The rule which is adopted by the Iowa Supreme Court in the above case is one which has also been proposed by a study committee for adoption in all of the states. The rule itself, which is quoted in the above summary of the case, is followed by the study committee's explanatory notes, and a portion of those notes provide the following reasons for the suggested adoption of this rule:

> *Since the possessor is not an insurer of the visitor's safety, he is ordinarily under no duty to exercise any care until he knows or has reason to know that the acts of the third person are occurring, or are about to occur. He may, however, know or have reason to know, from past experience, that there is a likelihood of conduct on the part of third persons in general which is likely to endanger the safety of the visitor, even though he has no reason to expect it on the part of any particular individual. If the place or character of his business, or his past experience, is such that he should reasonably anticipate careless or criminal conduct on the part of third persons, either generally or at some particular time, he may be under a duty to take precautions against it, and to provide a reasonably sufficient number of servants to afford a reasonable protection.*

As regular readers of **PSCLR** *are well aware, most of the prior cases in this area of law discuss the need for the plaintiff to prove a history of "prior similar acts in the geographic area in question." Indeed, in the Martinko case cited by the Iowa Supreme Court in reaching the above decision, it had been held: "In the absence of the history of similar acts in the area in question, most jurisdictions have not allowed plaintiffs to present their claims to juries."*

The shopping mall's position, therefore, was that a record of some criminal activity, but of a different type, is not sufficient to establish foreseeability. As noted above, the district court agreed with that position, citing Martinko, because the prior criminal activities occurring at the mall were not "similar" as required by Martinko.

However, the practical result of such a ruling has been in the past, and still is in those jurisdictions which still require proof of similar prior acts, that the first few victims of inadequate security in a given location are not entitled to recover. The current decision in Galloway constitutes a liberalization of the rules governing these cases, and a lessening of the protections previously provided to property owners and managers and their security companies.

It should also be noted that the fact of inadequate security, even in the absence of a prior history of similar criminal activity, was established in the mind of the court by the plaintiff's use of expert testimony. In its opinion the court noted by name the expert witness, Anthony Potter, of Atlanta, Georgia, specifically noting his "impressive credentials." The court stated:

> *He was certified by the American Society for Industrial Security and had served nineteen years as security chief, consultant, lecturer in the field of security. He had also served for fourteen years as a faculty member of the International Council of Shopping Centers Management Institute and had published over eighty articles in various professional and trade journals on the subject of security. He stated he had performed security audits of regional shopping centers throughout the United States and had testified as an expert in both federal and state courts.*

The court noted that Potter had reviewed the pleadings, depositions, and incident reports concerning the case, and had concluded from that review that the homosexual rape on the plaintiff was foreseeable.

The court also cited the case of Isaacs v. Huntington Memorial Hospital (previously reported in the May 1985 **PSCLR**, *case three, page three), reaffirming the long standing rule that questions of foreseeability are generally best left to be decided by a jury.*

PSCLR – JUN-JUL 89 (4)

An Expert Witness In Security Matters Is Not Permitted To Testify To Matters Beyond The Scope Of His Expertise:

Florence Smithson v. V.M.S. Realty, Inc., *No. 87-25, in the District Court of Appeal of Florida, Third District, decided November 8, 1988, rehearing denied February 7, 1989, reported at 536 So. 2d 260.*

Plaintiff Florence Smithson's husband, a manager at the Kendall Mall Theater, was robbed and murdered while attempting to make a deposit in the night depository of a bank located in the Kendall Mall. Smithson sued V.M.S. Realty, Inc., owner of the Kendall Mall, for the wrongful death of her husband. After a five-day trial, the jury entered a verdict in favor of V.M.S. Smithson appealed, arguing that the trial court should not have allowed certain prejudicial testimony into evidence.

The appellate court held that the trial court should not have permitted testimony by the defense's expert witness as to statements made by the perpetrators during interviews with the expert outside of the courtroom. When questioned at trial about the adequacy of V.M.S.'s security, the expert recited the perpetrators' out-of-court explanations of their scheme and motives for committing the crime. Because such out-of-court statements by witnesses not present at trial are impermissible evidence, the trial court should not have allowed the expert to testify as to such statements. An expert witness may render an opinion premised on impermissible evidence if the facts and data are of the type reasonably relied on by experts on the subject. However, the witness may not serve "merely as a conduit for the presentation of inadmissible evidence."

In this case, the witness was qualified to render an opinion on security matters and on V.M.S.'s allegedly negligent security procedures, but not on the robbers' motives for singling out Smithson as their target. The court stated that the reasoning behind the perpetrators' conduct is a matter beyond the scope of the security expert's expertise.

The appellate court also held that the trial court should not have allowed the expert witness to testify concerning whether Smithson complied with the company directive to accede to robbers' demands. Whether Smithson resisted the robbery was not an issue in the case, and the admission of such testimony unfairly prejudiced Smithson's case, even though the trial court instructed the jury that the victim "is not expected nor required to use the same judgment and prudence that is required of him in the exercise of ordinary care in calmer and more deliberate moments."

Because the trial court had allowed evidence that unfairly prejudiced Smithson's case, the appellate court reversed and remanded the case for a new trial.

COMMENTARY:

As has been reported in previous editions of **PSCLR**, *(see the COMPREHENSIVE REFERENCE INDEX, under the topic "EVIDENCE - Expert Testimony) issues concerning the use and abuse of expert witness testimony continue to be the subject of various opinions in the appellate courts.*

One of the basic rules governing expert witness testimony is that the expert is limited to issues within the scope of the witness's expertise. Accordingly, a witness with experience and expertise in the area of hotel security may or may not be qualified to testify on issues dealing with bank security.

Nevertheless, the security expert's testimony in the above case was of some value because the attorney at least succeeded in convincing the trial court judge that the witness's opinion as to the perpetrators' conduct in singling out the particular victim should be admitted to evidence.

Every professional in your organization should have a copy of

AVOIDING LIABILITY IN PREMISES SECURITY

To order Avoiding Liability in Premises ecurity for your library, your staff, or for your olleagues concerned with liability, complete and return the card at right.

For faster service, call 404/881-1141.

I NEED *THE LATEST* ON PREMISES SECURITY

Please send me ________ (quantity) copies of the casebook, AVOIDING LIABILITY IN PREMISES SECURITY, for $62.50 each (plus $2.50 Shipping & Handling)

( ) Check Enclosed
( ) Charge my: (circle one) Visa / MasterCard
Acct #: ______________________ Exp. Date: ________
Signature (required) ______________________________
( ) Bill me at the address below. Purchase Order # ______________

Name ____
Title ____
Firm ____
Address ____
City __________________________ State ________ Zip ________

PSPABF-PS-226

STRAFFORD PUBLICATIONS, INC. 1375 Peachtree Street, NE, #235 Atlanta, GA 30367

Major court decisions change your rofessional responsibilities almost daily.

To keep up with the trends, you need the eading source of key judicial decisions on all aspects of private security operations.

You need rivate Security Case Law Reporter.

art your subscription to PSCLR today. Just fill out and return the card at right.
For faster service, call 404/881-1141.

YES! I NEED *PRIVATE SECURITY CASE LAW REPORTER*

Please enter my subscription for the term marked below:

☐ One year – $267 ☐ Two years – just $474 (save $60)

( ) Check Enclosed
( ) Charge my: (circle one) Visa / MasterCard
Acct #: ______________________ Exp. Date: ________
Signature (required) ______________________________
( ) Bill me at the address below. Purchase Order # ______________

Name ____
Title ____
Firm ____
Address ____
City __________________________ State ________ Zip ________

PSSACA/PSSBCA-PS227

STRAFFORD PUBLICATIONS, INC. 1375 Peachtree Street, NE, #235 Atlanta, GA 30367

Complete your **"AVOIDING LIABILITY"** library with:

oiding Liability in Retail Security
Over 100 significant cases compiled for retail & shopping center security executives

oiding Liability in Hotel Security
Enables more effective design and management of hotel, motel and restaurant guest security programs

'ill out and return the order card at right, or call (404) 881-1141 for faster service.

COMPLETE MY "AVOIDING LIABILITY" LIBRARY

☐ Please send me _____ (quantity) copies of **Avoiding Liability in Hotel Security** at $62.50, plus $2.50 shipping & handling

☐ Please send me _____ (quantity) copies of **Avoiding Liability in Retail Security** at $62.50, plus $2.50 shipping & handling

( ) Check Enclosed
( ) Charge my: (circle one) Visa / MasterCard
Acct #: ______________________ Exp. Date: ________
Signature (required) ______________________________
( ) Bill me at the address below. Purchase Order # ______________

Name ____
Title ____
Firm ____
Address ____
City __________________________ State ________ Zip ________

PSPABB/PSPABC-PS228

STRAFFORD PUBLICATIONS, INC. 1375 Peachtree Street, NE, #235 Atlanta, GA 30367

NO POSTAGE
NECESSARY
IF MAILED
IN THE
UNITED STATES

BUSINESS REPLY MAIL

FIRST CLASS PERMIT NO. 13206 ATLANTA, GA, U.S.A.

POSTAGE WILL BE PAID BY ADDRESSEE

Strafford Publications, Inc.
1375 Peachtree St., N.E. — Suite 235
Atlanta, GA 30367-9900

MAIL THIS CARD TODAY

YOUR SATISFACTION IS GUARANTEED!

Strafford Publications, Inc.
1375 Peachtree Street, N.E., Suite 235
Atlanta, GA 30367 (404) 881-1141

NO POSTAGE
NECESSARY
IF MAILED
IN THE
UNITED STATES

BUSINESS REPLY MAIL

FIRST CLASS PERMIT NO. 13206 ATLANTA, GA, U.S.A.

POSTAGE WILL BE PAID BY ADDRESSEE

Strafford Publications, Inc.
1375 Peachtree St., N.E. — Suite 235
Atlanta, GA 30367-9900

MAIL THIS CARD TODAY

YOUR SATISFACTION IS GUARANTEED!

Strafford Publications, Inc.
1375 Peachtree Street, N.E., Suite 235
Atlanta, GA 30367 (404) 881-1141

NO POSTAGE
NECESSARY
IF MAILED
IN THE
UNITED STATES

BUSINESS REPLY MAIL

FIRST CLASS PERMIT NO. 13206 ATLANTA, GA, U.S.A.

POSTAGE WILL BE PAID BY ADDRESSEE

Strafford Publications, Inc.
1375 Peachtree St., N.E. — Suite 235
Atlanta, GA 30367-9900

MAIL THIS CARD TODAY

YOUR SATISFACTION IS GUARANTEED!

Strafford Publications, Inc.
1375 Peachtree Street, N.E., Suite 235
Atlanta, GA 30367 (404) 881-1141

Every professional in your organization should have a copy of

AVOIDING LIABILITY IN PREMISES SECURITY

To order Avoiding Liability in Premises ecurity for your library, your staff, or for your colleagues concerned with liability, complete and return the card at right.

For faster service, call 404/881-1141.

I NEED *THE LATEST* ON PREMISES SECURITY

Please send me _______ (quantity) copies of the casebook, AVOIDING LIABILITY IN PREMISES SECURITY, for $62.50 each (plus $2.50 Shipping & Handling)

( ) Check Enclosed
( ) Charge my: (circle one) Visa / MasterCard
Acct #: ________________________ Exp. Date: ________
Signature (required) ________________________________
( ) Bill me at the address below. Purchase Order # ______________

Name ____
Title ____
Firm ____
Address ____
City ____________________________ State __________ Zip ________

PSPABF-PS-226
STRAFFORD PUBLICATIONS, INC. 1375 Peachtree Street, NE, #235 Atlanta, GA 30367

Major court decisions change your professional responsibilities almost daily.

To keep up with the trends, you need the eading source of key judicial decisions on all aspects of private security operations.

You need **rivate Security Case Law Reporter.**

tart your subscription to PSCLR today. Just fill out and return the card at right.
For faster service, call 404/881-1141.

YES! I NEED *PRIVATE SECURITY CASE LAW REPORTER*

Please enter my subscription for the term marked below:

☐ One year – $267 ☐ Two years – just $474 (save $60)

( ) Check Enclosed
( ) Charge my: (circle one) Visa / MasterCard
Acct #: ________________________ Exp. Date: ________
Signature (required) ________________________________
( ) Bill me at the address below. Purchase Order # ______________

Name ____
Title ____
Firm ____
Address ____
City ____________________________ State __________ Zip ________

PSSACA/PSSBCA-PS227
STRAFFORD PUBLICATIONS, INC. 1375 Peachtree Street, NE, #235 Atlanta, GA 30367

Complete your **"AVOIDING LIABILITY"** library with:

oiding Liability in Retail Security
Over 100 significant cases compiled for retail & shopping center security executives

oiding Liability in Hotel Security
Enables more effective design and management of hotel, motel and restaurant guest security programs

ill out and return the order card at right, or call (404) 881-1141 for faster service.

COMPLETE MY "AVOIDING LIABILITY" LIBRARY

☐ Please send me_____(quantity) copies of **Avoiding Liability in Hotel Security** at $62.50, plus $2.50 shipping & handling

☐ Please send me_____(quantity) copies of **Avoiding Liability in Retail Security** at $62.50, plus $2.50 shipping & handling

( ) Check Enclosed
( ) Charge my: (circle one) Visa / MasterCard
Acct #: ________________________ Exp. Date: ________
Signature (required) ________________________________
( ) Bill me at the address below. Purchase Order # ______________

Name ____
Title ____
Firm ____
Address ____
City ____________________________ State __________ Zip ________

PSPABB/PSPABC-PS228
STRAFFORD PUBLICATIONS, INC. 1375 Peachtree Street, NE, #235 Atlanta, GA 30367

Subject Matter Index